Principles of Geotechnical Engineering

FIFTH EDITION

BRAJA M. DAS
California State University, Sacramento

916-278-6366

BROOKS/COLE

TM

THOMSON LEARNING

Australia • Canada • Mexico • Singapore • Spain • United Kingdom • United States

BROOKS/COLE

THOMSON LEARNING

Publisher: *Bill Stenquist*
Marketing Team: *Chris Kelly and Mona Weltmer*
Editorial Coordinator: *Valerie Boyajian*
Production Editor: *Mary Vezilich*
Production Service: *RPK Editorial Services*
Permissions Editor: *Sue Ewing*
Interior Design: *Carmela Pereira*
Cover Design: *Denise Davidson*
Cover Photos: *clockwise from top right, John Humble/*

The Image Bank; Vince Streano/Stone; Kim Steele/ PhotoDisc; Sami Sarkis/PhotoDisc; Mark Segal/ Stone
Interior Illustration: *Lori Heckelman*
Photo Researcher: *RPK Editorial Services*
Print Buyer: *Vena Dyer*
Typesetting: *G & S Typesetters, Inc.*
Printing and Binding: *Phoenix Color Corp.*

For more information about this or any other Brooks/Cole product, contact:
BROOKS/COLE
511 Forest Lodge Road
Pacific Grove, CA 93950 USA
www.brookscole.com
1-800-423-0563 (Thomson Learning Academic Resource Center)

Printed in the United States of America

10   9   8   7   6   5   4   3

Library of Congress Cataloging-in-Publication Data

Das, Braja M.
   Principles of geotechnical engineering / Braja M. Das. — 5th ed.
      p.   cm.
   Includes bibliographical references and index.
   ISBN 0-534-38742-X
   1. Soil mechanics.   2. Engineering geology.   I. Title.
TA710 .D264 2001
624.1'5136 — dc21   2001035126

To Janice and Valerie

Contents

4 Engineering Classification of Soil   83

5 Soil Compaction   100

6 Permeability   139

7 Seepage   178

8 *In Situ* Stresses   199

9 Stresses in a Soil Mass   224

10   Compressibility of Soil   259

11   Shear Strength of Soil   311

12 Lateral Earth Pressure: At-Rest, Rankine, and Coulomb   364

13 Lateral Earth Pressure — Curved Failure Surface   420

14   Slope Stability   445

15   Soil-Bearing Capacity for Shallow Foundations   503

Preface

Principles of Geotechnical Engineering, Fifth Edition is intended for introductory courses in soils and geotechnical engineering taken by virtually all civil engineering majors. The book also is useful for professionals and other readers wanting a general introduction to this important aspect of engineering. As in the first four editions of the book (1985, 1990, 1994, and 1998), this new edition offers an overview of soil properties and mechanics, together with coverage of field practices and basic engineering procedures. *Principles of Foundation Engineering,* Fourth Edition (1999), by the same author, goes on to apply these general concepts and procedures to earth, earth-supported, and earth-retaining structures, with an emphasis on design. *Principles of Geotechnical Engineering,* Fifth Edition provides the background information needed to support study in later design-oriented courses or in professional practice.

Changes in the Fifth Edition

The Fifth Edition, consisting of 17 chapters, includes a number of new features that were incorporated in response to suggestions made by professors, students, and professionals familiar with the earlier versions of the book. The major changes are the following:

- SI units are used as the principal units throughout the text with English units in parenthesis when required.
- Example problems and homework problems are in both SI and English units.
- Chapter 1, entitled "Geotechnical Engineering — A Historical Perspective," provides a brief history of the development of geotechnical engineering from the early 18th century to the present.
- Chapter 4 discusses only the engineering classifications of soil (AASHTO and Unified classification systems). The textural classification systems, which only provide a historical perspective, were deleted.
- The most recent ASTM specifications for laboratory compaction of soil are incorporated into Chapter 5.

- In the previous editions of the text, permeability and seepage were treated in the same chapter. In this edition, permeability is discussed in Chapter 6 and seepage is presented in Chapter 7.
- In Chapter 6 (permeability), recent methods for determination of *in situ* hydraulic conductivity of compact clay soils have been added.
- The procedure for estimating vertical stress increase due to embankment loading is included in Chapter 9.
- Chapter 10 presents the compressibility of soil. The sequence of presentation of the topics has changed. Discussions on immediate (elastic) settlement precedes consolidation settlement. Recently developed improved relationships for immediate settlement calculation which account for the rigidity of foundations, depth of embedment, linear increase in the modulus of elasticity of soil, and the effect of location of a rigid layer at a limited depth have been added. The hyperbola method for estimating the coefficient of consolidation from laboratory test results also has been included in this chapter.
- In the Fourth Edition of the text, static and dynamic lateral earth pressures were covered in Chapters 10 and 11, respectively. Some rearrangement of topics on lateral earth pressure (Chapters 12 and 13) has been included in this edition. Static and dynamic lateral earth pressures with plane failure surfaces in soil are treated in Chapter 12. Chapter 13 discusses lateral earth pressures determined with curved failure surfaces in soils including problems with braced cuts.
- Spencer's solution for slope stability analysis was replaced by Cousin's solution in this edition (Chapter 14).
- In Chapter 15 (Soil Bearing Capacity for Shallow Foundations), Meyerhof's relationships for estimating bearing capacity of sand based on settlement replaces the charts of Peck, Hanson, and Thornburn (1974). Discussions on the ultimate bearing capacity of foundations on slope were deleted since they are considered advanced topics. Theories relating to the ultimate bearing capacity of layered soil were added (Section 15.9).

Acknowledgments

I am grateful to my wife, Janice, for typing the manuscript and preparing some of the figures and tables. She has been the driving force in the completion of this edition of the text.

The previous four editions have been reviewed by 27 individuals; their helpful comments and suggestions have improved the quality of the book. For their reviews and helpful suggestions for the development of the Fifth Edition of the text, I would like to thank the following persons: Tuncer B. Edil, University of Wisconsin — Madison; Lois G. Schwarz, University of Arkansas — Fayetteville; and Raj V. Siddharthan, University of Nevada, Reno.

And finally, many thanks are due to Publisher Bill Stenquist and the production staff at Brooks/Cole for the final development and production of the book.

Braja M. Das
Sacramento, California

About the Author

Professor Braja Das received his M.S. in Civil Engineering from the University of Iowa, Iowa City, and his Ph.D. in Geotechnical Engineering from the University of Wisconsin, Madison. He is the author of several geotechnical engineering texts and reference books including *Principles of Geotechnical Engineering, Principles of Foundation Engineering,* and *Principles of Soil Dynamics,* all published by Thomson Learning. He has authored more than 200 technical papers in the area of geotechnical engineering. His primary areas of research include shallow foundations, earth anchors, and geosynthetics.

Professor Das has previously served on ASCE's Shallow Foundations Committee, Deep Foundations Committee, and Grouting Committee. He was also a member of ASCE's editorial board for the *Journal of Geotechnical Engineering.* He is the founder of the Geotechnical Engineering Division of the International Society of Offshore and Polar Engineers and has served as an associate editor of ISOPE's *International Journal of Offshore and Polar Engineering.* He now serves on the editorial board of the journal *Lowland Technology International,* which is published in Japan. Recently he has been named as the co-editor of *Geotechnical and Geological Engineering*— An International Journal published by Kluwer Academic Publishers of the Netherlands. He was the Chair of the Committee on Chemical and Mechanical Stabilization of the Transportation Research Board of the National Research Council of the United States (1995–2001).

Dr. Das has received numerous awards for teaching excellence, including the AMOCO Foundation Award, AT&T Award for Teaching Excellence from the American Society for Engineering Education, the Ralph Teetor Award from the Society of Automotive Engineers, and the Distinguished Achievement Award for Teaching Excellence from the University of Texas at El Paso.

Since 1994, Professor Das has served as Dean of the College of Engineering and Computer Science at California State University, Sacramento.

1

Geotechnical Engineering— A Historical Perspective

For engineering purposes, *soil* is defined as the uncemented aggregate of mineral grains and decayed organic matter (solid particles) with liquid and gas in the empty spaces between the solid particles. Soil is used as a construction material in various civil engineering projects, and it supports structural foundations. Thus, civil engineers must study the properties of soil, such as its origin, grain-size distribution, ability to drain water, compressibility, shear strength, and load-bearing capacity. *Soil mechanics* is the branch of science that deals with the study of the physical properties of soil and the behavior of soil masses subjected to various types of forces. *Soils engineering* is the application of the principles of soil mechanics to practical problems. *Geotechnical engineering* is the subdiscipline of civil engineering that involves natural materials found close to the surface of the earth. It includes the application of the principles of soil mechanics and rock mechanics to the design of foundations, retaining structures, and earth structures.

1.1 Geotechnical Engineering Prior to the 18th Century

The record of a person's first use of soil as a construction material is lost in antiquity. In true engineering terms, the understanding of geotechnical engineering as it is known today began early in the 18th century (Skempton, 1985). For years the art of geotechnical engineering was based on only past experiences through a succession of experimentation without any real scientific character. Based on those experimentations, many structures were built — some of which have crumbled, while others are still standing.

Recorded history tells us that ancient civilizations flourished along the banks of rivers, such as the Nile (Egypt), the Tigris and Euphrates (Mesopotamia), the Huang Ho (Yellow River, China), and the Indus (India). Dykes dating back to about 2000 B.C. were built in the basin of the Indus to protect the town of Mohenjo Dara (in what became Pakistan after 1947). During the Chan dynasty in China (1120 B.C. to 249 B.C.) many dykes were built for irrigation purposes. There is no evidence that measures were taken to stabilize the foundations or check erosion caused by floods (Kerisel,

Figure 1.1 Leaning Tower of Pisa, Italy

1985). Ancient Greek civilization used isolated pad footings and strip-and-raft foundations for building structures. Beginning around 2750 B.C., the five most important pyramids were built in Egypt in a period of less than a century (Saqqarah, Meidum, Dahshur South and North, and Cheops). This posed formidable challenges regarding foundations, stability of slopes, and construction of underground chambers. With the arrival of Buddhism in China during the Eastern Han dynasty in 68 A.D., thousands of pagodas were built. Many of these structures were constructed on silt and soft clay layers. In some cases the foundation pressure exceeded the load-bearing capacity of the soil and thereby caused extensive structural damage.

One of the most famous examples of problems related to soil-bearing capacity in the construction of structures prior to the 18th century is the Leaning Tower of Pisa in Italy. (See Figure 1.1.) Construction of the tower began in 1173 A.D. when the

Figure 1.2 Tilting of Garisenda Tower (left) in Bologna, Italy

Republic of Pisa was flourishing and continued in various stages for over 200 years. The structure weighs about 15,700 metric tons and is supported by a circular base having a diameter of 20 m ($\approx$ 66 ft). The tower has tilted in the past to the east, north, west and, finally, to the south. Recent investigations showed that a weak clay layer exists at a depth of about 11 m ($\approx$ 36 ft) below the ground surface compression, which caused the tower to tilt. It is now more than 5 m ($\approx$ 16.5 ft) out of plumb with the 54 m ($\approx$ 179 ft) height. Figure 1.2 is an example of a similar problem. The towers shown in Figure 1.2 are located in Bologna, Italy, and they were built in the 12th century. The tower on the left is usually referred to as the *Garisenda Tower*. It is 48 m ($\approx$ 157 ft) in height and has tilted severely.

After encountering several foundation-related problems during construction over centuries past, engineers and scientists began to address the properties and behavior of soils in a more methodical manner starting in the early part of the 18th century. Based on the emphasis and the nature of study in the area of geotechnical engineering, the time span extending from 1700 to 1927 can be divided into four major periods (Skempton, 1985):

1. Pre-classical (1700 to 1776 A.D.)
2. Classical soil mechanics — Phase I (1776 to 1856 A.D.)
3. Classical soil mechanics — Phase II (1856 to 1910 A.D.)
4. Modern soil mechanics (1910 to 1927 A.D.)

Brief descriptions of some significant developments during each of these four periods are discussed below.

1.2 *Preclassical Period of Soil Mechanics (1700–1776)*

This period concentrated on studies relating to natural slope and unit weights of various types of soils as well as the semiempirical earth pressure theories. In 1717 a French royal engineer, Henri *Gautier* (1660–1737), studied the natural slopes of soils when tipped in a heap for formulating the design procedures of retaining walls. The *natural slope* is what we now refer to as the *angle of repose.* According to this study, the natural slope (see Chapter 11) of *clean dry sand* and *ordinary earth* were 31° and 45°, respectively. Also, the unit weight of clean dry sand (see Chapter 3) and ordinary earth were recommended to be 18.1 kN/m^3 (115 lb/ft^3) and 13.4 kN/m^3 (85 lb/ft^3), respectively. No test results on clay were reported. In 1729, Bernard Forest de Belidor (1671–1761) published a textbook for military and civil engineers in France. In the book, he proposed a theory for lateral earth pressure on retaining walls (see Chapter 12) that was a follow-up to Gautier's (1717) original study. He also specified a soil classification system in the manner shown in the following table. (See Chapters 3 and 4.)

Classification	Unit weight	
	kN/m^3	lb/ft^3
Rock	—	—
Firm or hard sand	16.7 to	106 to
Compressible sand	18.4	117
Ordinary earth (as found in dry locations)	13.4	85
Soft earth (primarily silt)	16.0	102
Clay	18.9	120
Peat	—	—

The first laboratory model test results on a 76-mm-high ($\approx$ 3 in.) retaining wall built with sand backfill were reported in 1746 by a French engineer, Francois Gadroy (1705–1759), who observed the existence of slip planes in the soil at failure. (See Chapter 12.) Gadroy's study was later summarized by J. J. Mayniel in 1808.

1.3 *Classical Soil Mechanics—Phase I (1776–1856)*

During this period, most of the developments in the area of geotechnical engineering came from engineers and scientists in France. In the preclassical period, practically all theoretical considerations used in calculating lateral earth pressure on retaining walls were based on an arbitrarily based failure surface in soil. In his famous paper presented in 1776, French scientist Charles Augustin Coulomb (1736–1806) used the principles of calculus for maxima and minima to determine the true position of the sliding surface in soil behind a retaining wall. (See Chapter 12.) In this analysis, Coulomb used the laws of friction and cohesion for solid bodies. In 1820, special cases of Coulomb's work were studied by French engineer Jacques Frederic Francais (1775–1833) and by French applied mechanics professor Claude Louis Marie Henri Navier (1785–1836). These special cases related to inclined backfills and backfills supporting surcharge. In 1840, Jean Victor Poncelet (1788–1867), an army engineer and professor of mechanics, extended Coulomb's theory by providing a graphical method for determining the magnitude of lateral earth pressure on vertical and inclined retaining walls with arbitrarily broken polygonal ground surfaces. Poncelet was also the first to use the symbol ϕ for soil friction angle. (See Chapter 11.) He also provided the first ultimate bearing-capacity theory for shallow foundations. (See Chapter 15.) In 1846 Alexandre Collin (1808–1890), an engineer, provided the details for deep slips in clay slopes, cutting, and embankments. (See Chapter 14.) Collin theorized that in all cases the failure takes place when the mobilized cohesion exceeds the existing cohesion of the soil. He also observed that the actual failure surfaces could be approximated as arcs of cycloids.

The end of Phase I of the classical soil mechanics period is generally marked by the year (1857) of the first publication by William John Macquorn Rankine (1820–1872), a professor of civil engineering at the University of Glasgow. This study provided a notable theory on earth pressure and equilibrium of earth masses. (See Chapter 12.) Rankine's theory is a simplification of Coulomb's theory.

1.4 *Classical Soil Mechanics—Phase II (1856–1910)*

Several experimental results from laboratory tests on sand appeared in the literature in this phase. One of the earliest and most important publications is one by French engineer Henri Philibert Gaspard Darcy (1803–1858). In 1856, he published a study on the permeability of sand filters. (See Chapter 6.) Based on those tests, Darcy defined the term *coefficient of permeability* (or hydraulic conductivity) of soil, a very useful parameter in geotechnical engineering to this day.

Sir George Howard Darwin (1845–1912), a professor of astronomy, conducted laboratory tests to determine the overturning moment on a hinged wall retaining sand in loose and dense states of compaction. Another noteworthy contribution, which was published in 1885 by Joseph Valentin Boussinesq (1842–1929), was the development of the theory of stress distribution under loaded bearing areas in a homogeneous, semiinfinite, elastic, and isotropic medium. (See Chapter 9.) In 1887, Osborne Reynolds (1842–1912) demonstrated the phenomenon of dilatency in sand.

Table 1.1 Important Studies on Clays (1910–1927)

Investigator	Year	Topic
Albert Mauritz Atterberg (1846–1916), Sweden	1911	Consistency of soil, that is, liquid, plastic, and shrinkage properties (Chapter 3)
Jean Frontard (1884–1962), France	1914	Double shear tests (undrained) in clay under constant vertical load (Chapter 11)
Arthur Langtry Bell (1874–1956), England	1915	Lateral pressure and resistance of clay (Chapter 12); bearing capacity of clay (Chapter 15); and shear-box tests for measuring undrained shear strength using undisturbed specimens (Chapter 11)
Wolmar Fellenius (1876–1957), Sweden	1918, 1926	Slip-circle analysis of saturated clay slopes (Chapter 14)
Karl Terzaghi (1883–1963), Austria	1925	Theory of consolidation for clays (Chapter 10)

1.5   *Modern Soil Mechanics (1910–1927)*

In this period, results of research conducted on clays were published in which the fundamental properties and parameters of clay were established. The most notable publications are given in Table 1.1.

1.6   *Geotechnical Engineering after 1927*

The publication of *Erdbaumechanik auf Bodenphysikalisher Grundlage* by Karl Terzaghi in 1925 gave birth to a new era in the development of soil mechanics. Karl Terzaghi is known as the father of modern soil mechanics, and rightfully so. Terzaghi (Figure 1.3) was born on October 2, 1883 in Prague, which was then the capital of the Austrian province of Bohemia. In 1904 he graduated from the Technische Hochschule in Graz, Austria, with an undergraduate degree in mechanical engineering. After graduation he served one year in the Austrian army. Following his army service, Terzaghi studied one more year, concentrating on geological subjects. In January 1912, he received the degree of Doctor of Technical Sciences from his alma mater in Graz. In 1916, he accepted a teaching position at the Imperial School of Engineers in Istanbul. After the end of World War I, he accepted a lectureship at the American Robert College in Istanbul (1918–1925). There he began his research work on the behavior of soils and settlement of clays (see Chapter 10) and on the failure due to piping in sand under dams (see Chapter 8). The publication *Erdbaumechanik* is primarily the result of this research.

Figure 1.3 Karl Terzaghi (1883–1963) (photo courtesy of Ralph B. Peck)

In 1925, Terzaghi accepted a visiting lectureship at Massachusetts Institute of Technology, where he worked until 1929. During that time, he became recognized as the leader of the new branch of civil engineering called soil mechanics. In October 1929 he returned to Europe to accept a professorship at the Technical University of Vienna, which soon became the nucleus for civil engineers interested in soil mechanics. In 1939 he returned to the United States to become a professor at Harvard University.

The first conference of the International Society of Soil Mechanics and Foundation Engineering (ISSMFE) was held at Harvard University in 1936 with Karl Terzaghi presiding. It was through the inspiration and guidance of Terzaghi over the preceding quarter-century that papers were brought to that conference covering a wide range of topics, such as shear strength (Chapter 11), effective stress (Chapter 8), *in situ* testing (Chapter 17), Dutch cone penetrometer (Chapter 17), centrifuge testing, consolidation settlement (Chapter 10), elastic stress distribution (Chapter 9),

Figure 1.4 Ralph B. Peck

preloading for soil improvement, frost action, expansive clays, arching theory of earth pressure, and soil dynamics and earthquakes. For the next quarter-century, Terzaghi was the guiding spirit in the development of soil mechanics and geotechnical engineering throughout the world. To that effect, in 1985, Ralph Peck (Figure 1.4) wrote that "few people during Terzaghi's lifetime would have disagreed that he was not only the guiding spirit in soil mechanics, but that he was the clearing house for research and application throughout the world. Within the next few years he would be engaged on projects on every continent save Australia and Antarctica." Peck continued with, "Hence, even today, one can hardly improve on his contemporary assessments of the state of soil mechanics as expressed in his summary papers and presidential addresses." In 1939, Terzaghi delivered the 45th James Forrest Lecture at the Institution of Civil Engineers, London. His lecture was entitled "Soil Mechanics—A New Chapter in Engineering Science." In it, he proclaimed that most of the foundation failures that occurred were no longer "acts of God."

Following are some highlights in the development of soil mechanics and geotechnical engineering that evolved after the first conference of the ISSMFE in 1936:

- Publication of the book *Theoretical Soil Mechanics* by Karl Terzaghi in 1943 (Wiley, New York);
- Publication of the book *Soil Mechanics in Engineering Practice* by Karl Terzaghi and Ralph Peck in 1948 (Wiley, New York);
- Publication of the book *Fundamentals of Soil Mechanics* by Donald W. Taylor in 1948 (Wiley, New York);
- Start of the publication of *Geotechnique,* the international journal of soil mechanics in 1948 in England;
- Presentation of the paper on $\phi = 0$ concept for clays by A. W. Skempton in 1948 (see Chapter 11);
- Publication of A. W. Skempton's paper on A and B pore water pressure parameters in 1954 (see Chapter 11);
- Publication of the book *The Measurement of Soil Properties in the Triaxial Test* by A. W. Bishop and B. J. Henkel in 1957 (Arnold, London);
- ASCE's Research Conference on Shear Strength of Cohesive Soils held in Boulder, Colorado, in 1960.

Since the early days, the profession of geotechnical engineering has come a long way and has matured. It is now an established branch of civil engineering, and thousands of civil engineers declare geotechnical engineering to be their preferred area of speciality.

Since the first conference in 1936, except for a brief interruption during World War II, the ISSMFE conferences have been held at four-year intervals. In 1997, the ISSMFE was changed to ISSMGE (International Society of Soil Mechanics and Geotechnical Engineering) to reflect its true scope. These international conferences have been instrumental for exchange of information regarding new developments and ongoing research activities in geotechnical engineering. Table 1.2 gives the location and

Table 1.2 Details of ISSMFE (1936–1997) and ISSMGE (1997–present) Conferences

Conference	Location	Year
I	Harvard University, Boston, U.S.A.	1936
II	Rotterdam, the Netherlands	1948
III	Zurich, Switzerland	1953
IV	London, England	1957
V	Paris, France	1961
VI	Montreal, Canada	1965
VII	Mexico City, Mexico	1969
VIII	Moscow, U.S.S.R.	1973
IX	Tokyo, Japan	1977
X	Stockholm, Sweden	1981
XI	San Francisco, U.S.A.	1985
XII	Rio de Janeiro, Brazil	1989
XIII	New Delhi, India	1994
XIV	Hamburg, Germany	1997
XV	Istanbul, Turkey	2001

Table 1.3 Presidents of ISSMFE (1936–1997) and
ISSMGE (1997–present) Conferences

Year	President
1936–1957	K. Terzaghi (U. S. A.)
1957–1961	A. W. Skempton (U. K.)
1961–1965	A. Casagrande (U. S. A.)
1965–1969	L. Bjerrum (Norway)
1969–1973	R. B. Peck (U. S. A.)
1973–1977	J. Kerisel (France)
1977–1981	M. Fukuoka (Japan)
1981–1985	V. F. B. deMello (Brazil)
1985–1989	B. B. Broms (Singapore)
1989–1994	N. R. Morgenstern (Canada)
1994–1997	M. Jamiolkowski (Italy)
1997–2001	K. Ishihara (Japan)

Table 1.4 ISSMGE Technical Committees for 1997–2001 (based on Ishihara, 1999)

Committee number	Committee name
TC-1	Instrumentation for Geotechnical Monitoring
TC-2	Centrifuge Testing
TC-4	Earthquake Geotechnical Engineering
TC-5	Environmental Geotechnics
TC-6	Unsaturated Soils
TC-7	Tailing Dams
TC-8	Frost
TC-9	Geosynthetics and Earth Reinforcement
TC-10	Geophysical Site Characterization
TC-11	Landslides
TC-12	Validation of Computer Simulation
TC-14	Offshore Geotechnical Engineering
TC-15	Peat and Organic Soils
TC-16	Ground Property Characterization from In-situ Testing
TC-17	Ground Improvement
TC-18	Pile Foundations
TC-19	Preservation of Historic Sites
TC-20	Professional Practice
TC-22	Indurated Soils and Soft Rocks
TC-23	Limit State Design Geotechnical Engineering
TC-24	Soil Sampling, Evaluation and Interpretation
TC-25	Tropical and Residual Soils
TC-26	Calcareous Sediments
TC-28	Underground Construction in Soft Ground
TC-29	Stress-Strain Testing of Geomaterials in the Laboratory
TC-30	Coastal Geotechnical Engineering
TC-31	Education in Geotechnical Engineering
TC-32	Risk Assessment and Management
TC-33	Scour of Foundations
TC-34	Deformation of Earth Materials

year in which each conference of ISSMFE/ISSMGE was held, and Table 1.3 gives a list of all of the presidents of the society. In 1997, a total of 30 technical committees of ISSMGE was in place. The names of these technical committees are given in Table 1.4.

References

ATTERBERG, A. M. (1911). "Über die physikalische Bodenuntersuchung, und über die Plastizität de Tone," International Mitteilungen für Bodenkunde, *Verlag für Fachliteratur.* G.m.b.H. Berlin, Vol. 1, 10–43.

BELIDOR, B. F. (1729). *La Science des Ingenieurs dans la Conduite des Travaux de Fortification et D'Architecture Civil,* Jombert, Paris.

BELL, A. L. (1915). "The Lateral Pressure and Resistance of Clay, and Supporting Power of Clay Foundations," *Min. Proceeding of Institute of Civil Engineers,* Vol. 199, 233–272.

BISHOP, A. W. and HENKEL, B. J. (1957). *The Measurement of Soil Properties in the Triaxial Test,* Arnold, London.

BOUSSINESQ, J. V. (1885). *Application des Potentiels â L'Etude de L'Équilibre et du Mouvement des Solides Élastiques,* Gauthier-Villars, Paris.

COLLIN, A. (1846). *Recherches Expérimentales sur les Glissements Spontanés des Terrains Argileux Accompagnées de Considérations sur Quelques Principes de la Mécanique Terrestre,* Carilian-Goeury, Paris.

COULOMB, C. A. (1776). "Essai sur une Application des Règles de Maximis et Minimis à Quelques Problèmes de Statique Relatifs à L'Architecture," *Mèmoires de la Mathèmatique et de Phisique,* présentés à l'Académie Royale des Sciences, par divers savans, et lûs dans sés Assemblées, De L'Imprimerie Royale, Paris, Vol. 7, Annee 1793, 343–382.

DARCY, H. P. G. (1856). *Les Fontaines Publiques de la Ville de Dijon,* Dalmont, Paris.

DARWIN, G. H. (1883). "On the Horizontal Thrust of a Mass of Sand," *Proceedings,* Institute of Civil Engineers, London, Vol. 71, 350–378.

FELLENIUS, W. (1918). "Kaj-och Jordrasen I Göteborg," *Teknisk Tidskrift.* Vol. 48, 17–19.

FRANCAIS, J. F. (1820). "Recherches sur la Poussée de Terres sur la Forme et Dimensions des Revêtments et sur la Talus D'Excavation," *Mémorial de L'Officier du Génie,* Paris, Vol. IV, 157–206.

FRONTARD, J. (1914). "Notice sur L'Accident de la Digue de Charmes," *Anns. Ponts et Chaussées 9th Ser.,* Vol. 23, 173–292.

GADROY, F. (1746). *Mémoire sur la Poussée des Terres,* summarized by Mayniel, 1808.

GAUTIER, H. (1717). *Dissertation sur L'Epaisseur des Culées des Ponts . . . sur L'Effort et al Pesanteur des Arches . . . et sur les Profiles de Maconnerie qui Doivent Supporter des Chaussées, des Terrasses, et des Remparts.* Cailleau, Paris.

ISHIHARA, K. (1999). Personal communication.

KERISEL, J. (1985). "The History of Geotechnical Engineering up until 1700," *Proceedings,* XI International Conference on Soil Mechanics and Foundation Engineering, San Francisco, Golden Jubilee Volume, A. A. Balkema, 3–93.

MAYNIEL, J. J. (1808). *Traité Experimentale, Analytique et Pratique de la Poussé des Terres.* Colas, Paris.

NAVIER, C. L. M. (1839). *Leçons sur L'Application de la Mécanique à L'Establissement des Constructions et des Machines,* 2nd ed., Paris.

PECK, R. B. (1985). "The Last Sixty Years," *Proceedings,* XI International Conference on Soil Mechanics and Foundation Engineering, San Francisco, Golden Jubilee Volume, A. A. Balkema, 123–133.

PONCELET, J. V. (1840). *Mémoire sur la Stabilité des Revêtments et de seurs Fondations,* Bachelier, Paris.

RANKINE, W. J. M. (1857). "On the Stability of Loose Earth," *Philosophical Transactions, Royal Society,* Vol. 147, London.

REYNOLDS, O. (1887). "Experiments Showing Dilatency, a Property of Granular Material Possibly Connected to Gravitation," *Proceedings,* Royal Society, London, Vol. 11, 354–363.

SKEMPTON, A. W. (1948). "The $\phi = 0$ Analysis of Stability and Its Theoretical Basis," *Proceedings,* II International Conference on Soil Mechanics and Foundation Engineering, Rotterdam, Vol. 1, 72–78.

SKEMPTON, A. W. (1954). "The Pore Pressure Coefficients A and B," *Geotechnique,* Vol. 4, 143–147.

SKEMPTON, A. W. (1985). "A History of Soil Properties, 1717–1927," *Proceedings,* XI International Conference on Soil Mechanics and Foundation Engineering, San Francisco, Golden Jubilee Volume, A. A. Balkema, 95–121.

TAYLOR, D. W. (1948). *Fundamentals of Soil Mechanics,* John Wiley, New York.

TERZAGHI, K. (1925). *Erdbaumechanik auf Bodenphysikalisher Grundlage,* Deuticke, Vienna.

TERZAGHI, K. (1939). "Soil Mechanics — A New Chapter in Engineering Science," *Institute of Civil Engineers Journal,* London, Vol. 12, No. 7, 106–142.

TERZAGHI, K. (1943). *Theoretical Soil Mechanics,* John Wiley, New York.

TERZAGHI, K. and PECK, R. B. (1948). *Soil Mechanics in Engineering Practice,* John Wiley, New York.

2

Origin of Soil and Grain Size

In general, soils are formed by weathering of rocks. The physical properties of a soil are dictated primarily by the minerals that constitute the soil particles and, hence, the rock from which it is derived. This chapter provides an outline of the rock cycle and the origin of soil and the grain-size distribution of particles in a soil mass.

2.1 *Rock Cycle and the Origin of Soil*

The mineral grains that form the solid phase of a soil aggregate are the product of rock weathering. The size of the individual grains varies over a wide range. Many of the physical properties of soil are dictated by the size, shape, and chemical composition of the grains. To better understand these factors, one must be familiar with the basic types of rock that form the earth's crust, the rock-forming minerals, and the weathering process.

On the basis of their mode of origin, rocks can be divided into three basic types: *igneous, sedimentary,* and *metamorphic.* Figure 2.1 shows a diagram of the formation cycle of different types of rock and the processes associated with them. This is called the *rock cycle.* Brief discussions of each element of the rock cycle follow.

Igneous Rock

Igneous rocks are formed by the solidification of molten *magma* ejected from deep within the earth's mantle. After ejection by either *fissure eruption* or *volcanic eruption,* some of the molten magma cools on the surface of the earth. Sometimes magma ceases its mobility below the earth's surface and cools to form intrusive igneous rocks that are called *plutons.* Intrusive rocks formed in the past may be exposed at the surface as a result of the continuous process of erosion of the materials that once covered them.

The types of igneous rock formed by the cooling of magma depend on factors such as the composition of the magma and the rate of cooling associated with it. After conducting several laboratory tests, Bowen (1922) was able to explain the relation of the rate of magma cooling to the formation of different types of rock. This explanation — known as *Bowen's reaction principle* — describes the sequence by which new

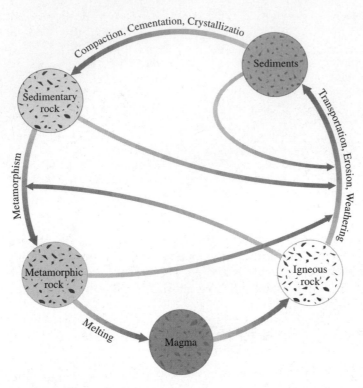

Figure 2.1 Rock cycle

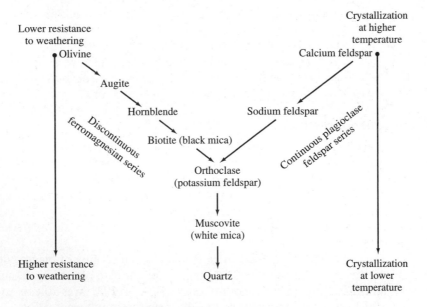

Figure 2.2 Bowen's reaction series

Table 2.1 Composition of Minerals Shown in Bowen's Reaction Series

Mineral	Composition
Olivine	$(Mg, Fe)_2SiO_4$
Augite	$Ca, Na(Mg, Fe, Al)(Al, Si_7O_6)$
Hornblende	Complex ferromagnesian silicate of Ca, Na, Mg, Ti, and Al
Biotite (black mica)	$K(Mg, Fe)_3AlSi_3O_{10}(OH)_2$
Plagioclase { calcium feldspar	$Ca(Al_2Si_2O_8)$
Plagioclase { sodium feldspar	$Na(AlSi_3O_8)$
Orthoclase (potassium feldspar)	$K(AlSi_3O_8)$
Muscovite (white mica)	$KAl_3Si_3O_{10}(OH)_2$
Quartz	SiO_2

minerals are formed as magma cools. The mineral crystals grow larger and some of them settle. The crystals that remain suspended in the liquid react with the remaining melt to form a new mineral at a lower temperature. This process continues until the entire body of melt is solidified. Bowen classified these reactions into two groups: (1) *discontinuous ferromagnesian reaction series,* in which the minerals formed are different in their chemical composition and crystalline structure, and (2) *continuous plagioclase feldspar reaction series,* in which the minerals formed have different chemical compositions with similar crystalline structures. Figure 2.2 shows Bowen's reaction series. The chemical compositions of the minerals are given in Table 2.1.

Thus, depending on the proportions of minerals available, different types of igneous rock are formed. Granite, gabbro, and basalt are some of the common types of igneous rock generally encountered in the field. Table 2.2 shows the general composition of some igneous rocks.

Table 2.2 Composition of Some Igneous Rocks

Name of rock	Mode of occurrence	Texture	Abundant minerals	Less abundant minerals
Granite	Intrusive	Coarse	Quartz, sodium feldspar, potassium feldspar	Biotite, muscovite, hornblende
Rhyolite	Extrusive	Fine		
Gabbro	Intrusive	Coarse	Plagioclase, pyroxines, olivine	Hornblende, biotite, magnetite
Basalt	Extrusive	Fine		
Diorite	Intrusive	Coarse	Plagioclase, hornblende	Biotite, pyroxenes (quartz usually absent)
Andesite	Extrusive	Fine		
Syenite	Intrusive	Coarse	Potassium feldspar	Sodium feldspar, biotite, hornblende
Trachyte	Extrusive	Fine		
Peridotite	Intrusive	Coarse	Olivine, pyroxenes	Oxides of iron

Weathering

Weathering is the process of breaking down rocks by *mechanical* and *chemical processes* into smaller pieces. Mechanical weathering may be caused by the expansion and contraction of rocks from the continuous gain and loss of heat, which results in ultimate disintegration. Frequently, water seeps into the pores and existing cracks in rocks. As the temperature drops, the water freezes and expands. The pressure exerted by ice because of volume expansion is strong enough to break down even large rocks. Other physical agents that help disintegrate rocks are glacier ice, wind, the running water of streams and rivers, and ocean waves. It is important to realize that in mechanical weathering, large rocks are broken down into smaller pieces without any change in the chemical composition. Figure 2.3 shows several examples of mechanical erosion due to ocean waves and wind at Yehliu in Taiwan. This area is located at a long and narrow sea cape at the northwest side of Keelung, about 15 kilometers between the north coast of Chin Shan and Wanli.

In chemical weathering, the original rock minerals are transformed into new minerals by chemical reaction. Water and carbon dioxide from the atmosphere form carbonic acid, which reacts with the existing rock minerals to form new minerals and soluble salts. Soluble salts present in the groundwater and organic acids formed from decayed organic matter also cause chemical weathering. An example of the chemical weathering of orthoclase to form clay minerals, silica, and soluble potassium carbonate follows:

$$H_2O + CO_2 \rightarrow H_2CO_3 \rightarrow H^+ + (HCO_3)^-$$
$$\text{Carbonic acid}$$

$$2K(AlSi_3O_8) + 2H^+ + H_2O \rightarrow 2K^+ + 4SiO_2 + Al_2Si_2O_5(OH)_4$$
$$\text{Orthoclase} \qquad\qquad\qquad\qquad \text{Silica} \qquad \text{Kaolinite}$$
$$\text{(clay mineral)}$$

Most of the potassium ions released are carried away in solution as potassium carbonate is taken up by plants.

The chemical weathering of plagioclase feldspars is similar to that of orthoclase in that it produces clay minerals, silica, and different soluble salts. Ferromagnesian minerals also form the decomposition products of clay minerals, silica, and soluble salts. Additionally, the iron and magnesium in ferromagnesian minerals result in other products such as hematite and limonite. Quartz is highly resistant to weathering and only slightly soluble in water. Figure 2.2 shows the susceptibility of rock-forming minerals to weathering. The minerals formed at higher temperatures in Bowen's reaction series are less resistant to weathering than those formed at lower temperatures.

The weathering process is not limited to igneous rocks. As shown in the rock cycle (Figure 2.1), sedimentary and metamorphic rocks also weather in a similar manner.

Thus, from the preceding brief discussion, we can see how the weathering process changes solid rock masses into smaller fragments of various sizes that can range from large boulders to very small clay particles. Uncemented aggregates of these small grains in various proportions form different types of soil. The clay minerals,

Figure 2.3 Mechanical erosion due to ocean waves and wind at Yehliu, Taiwan

Figure 2.3 (Continued)

which are a product of chemical weathering of feldspars, ferromagnesians, and micas, give the plastic property to soils. There are three important clay minerals: (1) *kaolinite,* (2) *illite,* and (3) *montmorillonite.* (We discuss these clay minerals later in this chapter.)

Transportation of Weathering Products

The products of weathering may stay in the same place or may be moved to other places by ice, water, wind, and gravity.

The soils formed by the weathered products at their place of origin are called *residual soils.* An important characteristic of residual soil is the gradation of particle size. Fine-grained soil is found at the surface, and the grain size increases with depth. At greater depths, angular rock fragments may also be found.

The transported soils may be classified into several groups, depending on their mode of transportation and deposition:

1. *Glacial soils* — formed by transportation and deposition of glaciers
2. *Alluvial soils* — transported by running water and deposited along streams
3. *Lacustrine soils* — formed by deposition in quiet lakes
4. *Marine soils* — formed by deposition in the seas
5. *Aeolian soils* — transported and deposited by wind
6. *Colluvial soils* — formed by movement of soil from its original place by gravity, such as during landslides

Sedimentary Rock

The deposits of gravel, sand, silt, and clay formed by weathering may become compacted by overburden pressure and cemented by agents like iron oxide, calcite, dolomite, and quartz. Cementing agents are generally carried in solution by groundwater. They fill the spaces between particles and form sedimentary rock. Rocks formed in this way are called *detrital sedimentary rocks.* Conglomerate, breccia, sandstone, mudstone, and shale are some examples of the detrital type.

Sedimentary rock can also be formed by chemical processes. Rocks of this type are classified as *chemical sedimentary rock.* Limestone, chalk, dolomite, gypsum, anhydrite, and others belong to this category. Limestone is formed mostly of calcium carbonate that originates from calcite deposited either by organisms or by an inorganic process. Dolomite is calcium magnesium carbonate $[CaMg(CO_3)_2]$. It is formed either by the chemical deposition of mixed carbonates or by the reaction of magnesium in water with limestone. Gypsum and anhydrite result from the precipitation of soluble $CaSO_4$ because of evaporation of ocean water. They belong to a class of rocks generally referred to as *evaporites.* Rock salt (NaCl) is another example of an evaporite that originates from the salt deposits of seawater.

Sedimentary rock may undergo weathering to form sediments or may be subjected to the process of *metamorphism* to become metamorphic rock.

Metamorphic Rock

Metamorphism is the process of changing the composition and texture of rocks, without melting, by heat and pressure. During metamorphism, new minerals are formed and mineral grains are sheared to give a foliated texture to metamorphic rocks. Granite, diorite, and gabbro become gneisses by high-grade metamorphism. Shales and mudstones are transformed into slates and phyllites by low-grade metamorphism. Schists are a type of metamorphic rock with well-foliated texture and visible flakes of platy and micaceous minerals.

Marble is formed from calcite and dolomite by recrystallization. The mineral grains in marble are larger than those present in the original rock. Quartzite is a metamorphic rock formed from quartz-rich sandstones. Silica enters into the void spaces between the quartz and sand grains and acts as a cementing agent. Quartzite is one of the hardest rocks. Under extreme heat and pressure, metamorphic rocks may melt to form magma, and the cycle is repeated.

2.2   *Soil-Particle Size*

As discussed in the preceding section, the sizes of particles that make up soil vary over a wide range. Soils are generally called *gravel, sand, silt,* or *clay,* depending on the predominant size of particles within the soil. To describe soils by their particle size, several organizations have developed particle-size classifications. Table 2.3 shows the particle-size classifications developed by the Massachusetts Institute of Technology, the U.S. Department of Agriculture, the American Association of State Highway and Transportation Officials, and the U.S. Army Corps of Engineers and U.S. Bureau of Reclamation. In this table, the MIT system is presented for illustration purposes only. This system is important in the history of the development of the size limits of particles present in soils; however, the Unified Soil Classification System is now almost universally accepted and has been adopted by the American Society for Testing and Materials (ASTM).

Gravels are pieces of rocks with occasional particles of quartz, feldspar, and other minerals. *Sand* particles are made of mostly quartz and feldspar. Other mineral

Table 2.3 Particle-Size Classifications

Name of organization	Grain size (mm)			
	Gravel	Sand	Silt	Clay
Massachusetts Institute of Technology (MIT)	>2	2 to 0.06	0.06 to 0.002	<0.002
U.S. Department of Agriculture (USDA)	>2	2 to 0.05	0.05 to 0.002	<0.002
American Association of State Highway and Transportation Officials (AASHTO)	76.2 to 2	2 to 0.075	0.075 to 0.002	<0.002
Unified Soil Classification System (U.S. Army Corps of Engineers, U.S. Bureau of Reclamation, and American Society for Testing and Materials)	76.2 to 4.75	4.75 to 0.075	Fines (i.e., silts and clays) <0.075	

Note: Sieve openings of 4.75 mm are found on a U.S. No. 4 sieve; 2-mm openings on a U.S. No. 10 sieve; 0.075-mm openings on a U.S. No. 200 sieve. See Table 2.5.

grains may also be present at times. *Silts* are the microscopic soil fractions that consist of very fine quartz grains and some flake-shaped particles that are fragments of micaceous minerals. *Clays* are mostly flake-shaped microscopic and submicroscopic particles of mica, clay minerals, and other minerals.

As shown in Table 2.3, clays are generally defined as particles smaller than 0.002 mm. However, in some cases, particles between 0.002 and 0.005 mm in size are also referred to as clay. Particles classified as clay on the basis of their size may not necessarily contain clay minerals. Clays have been defined as those particles "which develop plasticity when mixed with a limited amount of water" (Grim, 1953). (Plasticity is the puttylike property of clays that contain a certain amount of water.) Non-clay soils can contain particles of quartz, feldspar, or mica that are small enough to be within the clay classification. Hence, it is appropriate for soil particles smaller than 2 microns (2 μm), or 5 microns (5 μm) as defined under different systems, to be called clay-sized particles rather than clay. Clay particles are mostly in the colloidal size range (<1 μm), and 2 μm appears to be the upper limit.

Clay Minerals

Clay minerals are complex aluminum silicates composed of two basic units: (1) *silica tetrahedron* and (2) *alumina octahedron.* Each tetrahedron unit consists of four oxygen atoms surrounding a silicon atom (Figure 2.4a). The combination of tetrahedral silica units gives a *silica sheet* (Figure 2.4b). Three oxygen atoms at the base of each tetrahedron are shared by neighboring tetrahedra. The octahedral units consist of six hydroxyls surrounding an aluminum atom (Figure 2.4c), and the combination of the octahedral aluminum hydroxyl units gives an *octahedral sheet.* (This is also called a *gibbsite sheet*—Figure 2.4d.) Sometimes magnesium replaces the aluminum atoms in the octahedral units; in this case, the octahedral sheet is called a *brucite sheet.*

In a silica sheet, each silicon atom with a positive charge of four is linked to four oxygen atoms with a total negative charge of eight. But each oxygen atom at the base of the tetrahedron is linked to two silicon atoms. This means that the top oxygen atom of each tetrahedral unit has a negative charge of one to be counterbalanced. When the silica sheet is stacked over the octahedral sheet as shown in Figure 2.4e, these oxygen atoms replace the hydroxyls to balance their charges.

Of the three important clay minerals, *kaolinite* consists of repeating layers of elemental silica-gibbsite sheets in a 1:1 lattice as shown in Figures 2.5 and 2.6a. Each layer is about 7.2 Å thick. The layers are held together by hydrogen bonding. Kaolinite occurs as platelets, each with a lateral dimension of 1000 to 20,000 Å and a thickness of 100 to 1000 Å. The surface area of the kaolinite particles per unit mass is about 15 m²/g. The surface area per unit mass is defined as *specific surface.* Figure 2.7 shows a scanning electron micrograph of a kaolinite specimen.

Illite consists of a gibbsite sheet bonded to two silica sheets — one at the top and another at the bottom (Figures 2.8 and 2.6b). It is sometimes called *clay mica.* The illite layers are bonded by potassium ions. The negative charge to balance the potassium ions comes from the substitution of aluminum for some silicon in the tetrahedral sheets. Substitution of one element for another with no change in the crystalline

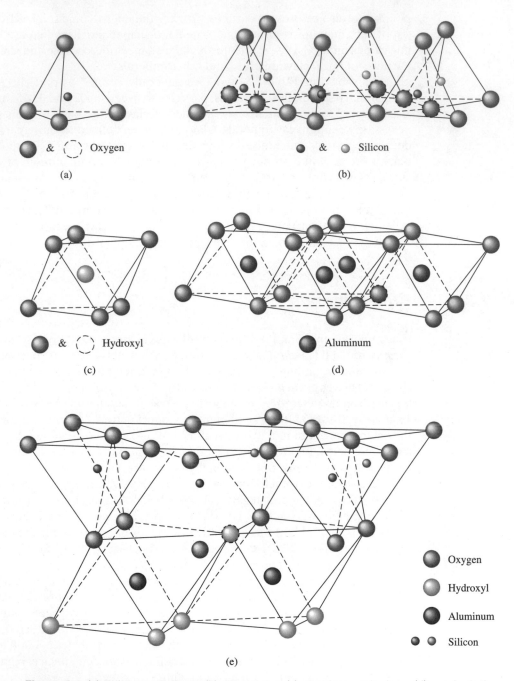

Figure 2.4 (a) Silica tetrahedron; (b) silica sheet; (c) alumina octahedron; (d) octahedral (gibbsite) sheet; (e) elemental silica-gibbsite sheet (after Grim, 1959)

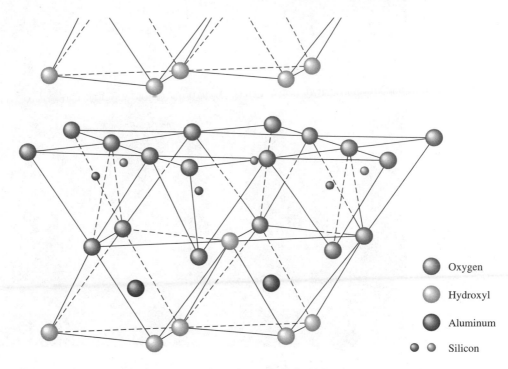

Figure 2.5 Atomic structure of kaolinite (after Grim, 1959)

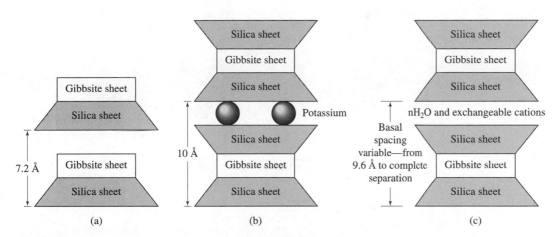

Figure 2.6 Diagram of the structures of (a) kaolinite; (b) illite; (c) montmorillonite

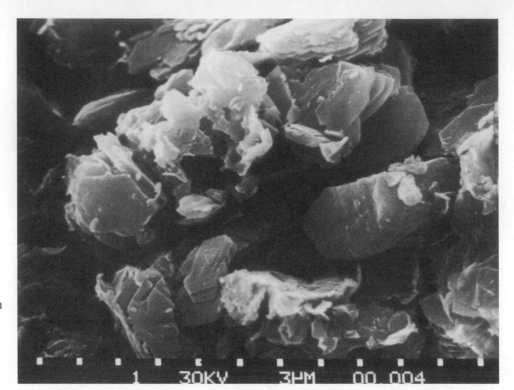

Figure 2.7
Scanning electron micrograph of a kaolinite specimen (courtesy of U.S. Geological Survey)

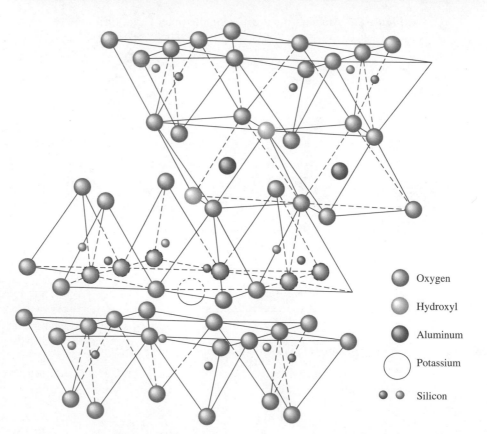

Oxygen

Hydroxyl

Aluminum

Potassium

Silicon

Figure 2.8
Atomic structure of illite

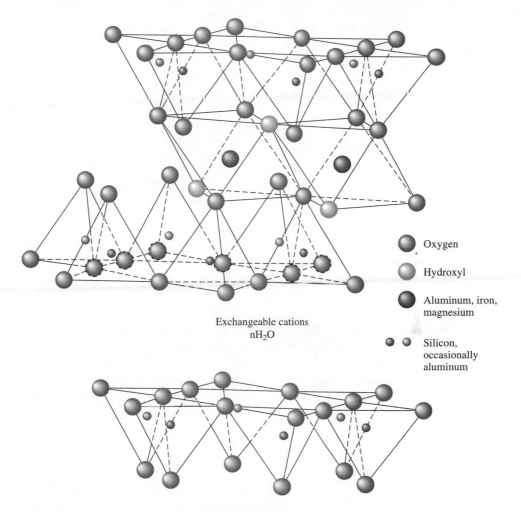

Exchangeable cations
nH$_2$O

Oxygen

Hydroxyl

Aluminum, iron,
magnesium

Silicon,
occasionally
aluminum

Figure 2.9 Atomic structure of montmorillonite (after Grim, 1959)

form is known as *isomorphous substitution.* Illite particles generally have lateral dimensions ranging from 1000 to 5000 Å and thicknesses from 50 to 500 Å. The specific
surface of the particles is about 80 m^2/g.

Montmorillonite has a structure similar to that of illite — that is, one gibbsite
sheet sandwiched between two silica sheets. (See Figures 2.9 and 2.6c). In montmorillonite there is isomorphous substitution of magnesium and iron for aluminum in
the octahedral sheets. Potassium ions are not present as in illite, and a large amount
of water is attracted into the space between the layers. Particles of montmorillonite
have lateral dimensions of 1000 to 5000 Å and thicknesses of 10 to 50 Å. The specific
surface is about 800 m^2/g.

Besides kaolinite, illite, and montmorillonite, other common clay minerals generally found are chlorite, halloysite, vermiculite, and attapulgite.

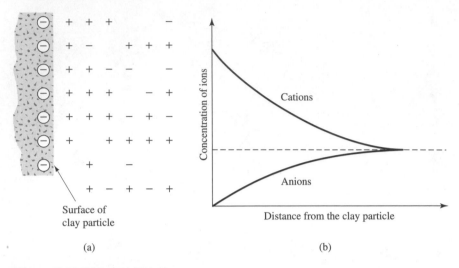

Figure 2.10 Diffuse double layer

The clay particles carry a net negative charge on their surfaces. This is the result both of isomorphous substitution and of a break in continuity of the structure at its edges. Larger negative charges are derived from larger specific surfaces. Some positively charged sites also occur at the edges of the particles. A list of the reciprocal of the average surface densities of the negative charges on the surfaces of some clay minerals follows (Yong and Warkentin, 1966):

Clay mineral	Reciprocal of average surface density of charge (Å^2/electronic charge)
Kaolinite	25
Clay mica and chlorite	50
Montmorillonite	100
Vermiculite	75

In dry clay, the negative charge is balanced by exchangeable cations like Ca^{2+}, Mg^{2+}, Na^+, and K^+ surrounding the particles being held by electrostatic attraction. When water is added to clay, these cations and a few anions float around the clay particles. This configuration is referred to as a *diffuse double layer* (Figure 2.10a). The cation concentration decreases with the distance from the surface of the particle (Figure 2.10b).

Water molecules are polar. Hydrogen atoms are not axisymmetric around an oxygen atom; instead, they occur at a bonded angle of 105° (Figure 2.11). As a result, a water molecule has a positive charge at one side and a negative charge at the other side. It is known as a *dipole*.

Dipolar water is attracted both by the negatively charged surface of the clay particles and by the cations in the double layer. The cations, in turn, are attracted to the soil particles. A third mechanism by which water is attracted to clay particles is

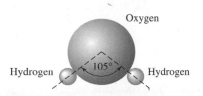

Figure 2.11 Dipolar character of water

hydrogen bonding, where hydrogen atoms in the water molecules are shared with oxygen atoms on the surface of the clay. Some partially hydrated cations in the pore water are also attracted to the surface of clay particles. These cations attract dipolar water molecules. All these possible mechanics of attraction of water to clay are shown in Figure 2.12. The force of attraction between water and clay decreases with distance from the surface of the particles. All the water held to clay particles by force of attraction is known as *double-layer water*. The innermost layer of double-layer water, which is held very strongly by clay, is known as *adsorbed water.* This water is more viscous than free water is.

Figure 2.13 shows the absorbed and double-layer water for typical montmorillonite and kaolinite particles. This orientation of water around the clay particles gives clay soils their plastic properties.

It needs to be well recognized that the presence of clay minerals in a soil aggregate has a great influence on the engineering properties of the soil as a whole. When moisture is present, the engineering behavior of a soil will change greatly as the percentage of clay mineral content increases. For all practical purposes, when the clay

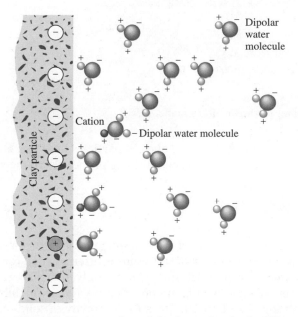

Figure 2.12 Attraction of dipolar molecules in diffuse double layer

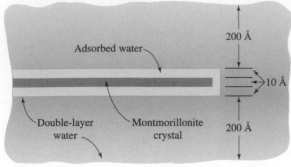

Typical montmorillonite particle, 1000 Å by 10 Å

(a)

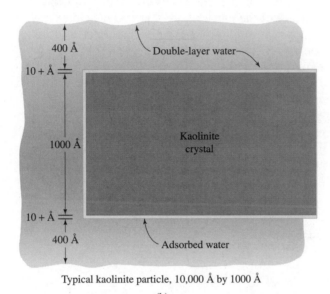

Typical kaolinite particle, 10,000 Å by 1000 Å

(b)

Figure 2.13 Clay water (redrawn after Lambe, 1958)

content is about 50% or more, the sand and silt particles float in a clay matrix, and the clay minerals primarily dictate the engineering properties of the soil.

2.4 Specific Gravity (G_s)

Specific gravity is defined as the ratio of the unit weight of a given material to the unit weight of water. The specific gravity of soil solids is often needed for various calculations in soil mechanics. It can be determined accurately in the laboratory. Table 2.4 shows the specific gravity of some common minerals found in soils. Most of the values fall within a range of 2.6 to 2.9. The specific gravity of solids of light-colored sand, which is mostly made of quartz, may be estimated to be about 2.65; for clayey and silty soils, it may vary from 2.6 to 2.9.

Table 2.4 Specific Gravity of Common Minerals

Mineral	Specific gravity, G_s
Quartz	2.65
Kaolinite	2.6
Illite	2.8
Montmorillonite	2.65–2.80
Halloysite	2.0–2.55
Potassium feldspar	2.57
Sodium and calcium feldspar	2.62–2.76
Chlorite	2.6–2.9
Biotite	2.8–3.2
Muscovite	2.76–3.1
Hornblende	3.0–3.47
Limonite	3.6–4.0
Olivine	3.27–3.7

2.5 *Mechanical Analysis of Soil*

Mechanical analysis is the determination of the size range of particles present in a soil, expressed as a percentage of the total dry weight. Two methods are generally used to find the particle-size distribution of soil: (1) *sieve analysis* — for particle sizes larger than 0.075 mm in diameter, and (2) *hydrometer analysis* — for particle sizes smaller than 0.075 mm in diameter. The basic principles of sieve analysis and hydrometer analysis are briefly described in the following two sections.

Sieve Analysis

Sieve analysis consists of shaking the soil sample through a set of sieves that have progressively smaller openings. U.S. standard sieve numbers and the sizes of openings are given in Table 2.5.

The sieves used for soil analysis are generally 203 mm (8 in.) in diameter. To conduct a sieve analysis, one must first oven-dry the soil and then break all lumps into small particles. The soil is then shaken through a stack of sieves with openings of decreasing size from top to bottom (a pan is placed below the stack). Figure 2.14 shows a set of sieves in a shaker used for conducting the test in the laboratory. The smallest-size sieve that should be used for this type of test is the U.S. No. 200 sieve. After the soil is shaken, the mass of soil retained on each sieve is determined. When cohesive soils are analyzed, breaking the lumps into individual particles may be difficult. In this case, the soil may be mixed with water to make a slurry and then washed through the sieves. Portions retained on each sieve are collected separately and oven-dried before the mass retained on each sieve is measured.

1. Determine the mass of soil retained on each sieve (i.e., $M_1, M_2, \cdots M_n$) and in the pan (i.e., M_p).
2. Determine the total mass of the soil: $M_1 + M_2 + \cdots + M_i + \cdots + M_n + M_p = \Sigma M$.

Table 2.5 U.S. Standard Sieve Sizes

Sieve no.	Opening (mm)
4	4.75
5	4.00
6	3.35
7	2.80
8	2.36
10	2.00
12	1.70
14	1.40
16	1.18
18	1.00
20	0.850
25	0.710
30	0.600
35	0.500
40	0.425
50	0.355
60	0.250
70	0.212
80	0.180
100	0.150
120	0.125
140	0.106
170	0.090
200	0.075
270	0.053

3. Determine the cumulative mass of soil retained above each sieve. For the ith sieve, it is $M_1 + M_2 + \cdots + M_i$.
4. The mass of soil passing the ith sieve is $\Sigma M - (M_1 + M_2 + \cdots + M_i)$.
5. The percent of soil passing the ith sieve (or *percent finer*) is

$$F = \frac{\Sigma M - (M_1 + M_2 + \cdots + M_i)}{\Sigma M} \times 100$$

Once the percent finer for each sieve is calculated (step 5), the calculations are plotted on semilogarithmic graph paper (Figure 2.15) with percent finer as the ordinate (arithmetic scale) and sieve opening size as the abscissa (logarithmic scale). This plot is referred to as the *particle-size distribution curve*.

Hydrometer Analysis

Hydrometer analysis is based on the principle of sedimentation of soil grains in water. When a soil specimen is dispersed in water, the particles settle at different velocities, depending on their shape, size, and weight, and the viscosity of the water.

Figure 2.14 A set of sieves for a test in the laboratory

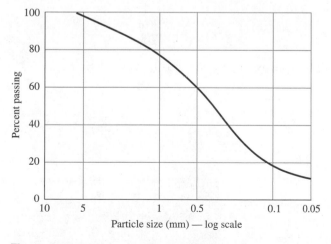

Figure 2.15 Particle-size distribution curve

For simplicity, it is assumed that all the soil particles are spheres and that the velocity of soil particles can be expressed by *Stokes' law,* according to which

$$v = \frac{\rho_s - \rho_w}{18\eta} D^2 \tag{2.1}$$

where v = velocity
  ρ_s = density of soil particles
  ρ_w = density of water
  η = viscosity of water
  D = diameter of soil particles

Thus, from Eq. (2.1),

$$D = \sqrt{\frac{18\eta v}{\rho_s - \rho_w}} = \sqrt{\frac{18\eta}{\rho_s - \rho_w}} \sqrt{\frac{L}{t}} \tag{2.2}$$

where $v = \dfrac{\text{Distance}}{\text{Time}} = \dfrac{L}{t}$.
  Note that

$$\rho_s = G_s \rho_w \tag{2.3}$$

Thus, combining Eqs. (2.2) and (2.3) gives

$$D = \sqrt{\frac{18\eta}{(G_s - 1)\rho_w}} \sqrt{\frac{L}{t}} \tag{2.4}$$

If the units of η are $(\text{g} \cdot \text{sec})/\text{cm}^2$, ρ_w is in g/cm^3, L is in cm, t is in min, and D is in mm, then

$$\frac{D(\text{mm})}{10} = \sqrt{\frac{18\eta\,[(\text{g}\cdot\text{sec})/\text{cm}^2]}{(G_s - 1)\rho_w\,(\text{g/cm}^3)}} \sqrt{\frac{L\,(\text{cm})}{t\,(\text{min}) \times 60}}$$

or

$$D = \sqrt{\frac{30\eta}{(G_s - 1)\rho_w}} \sqrt{\frac{L}{t}}$$

Assume ρ_w to be approximately equal to $1\ \text{g/cm}^3$, so that

$$D\,(\text{mm}) = K\sqrt{\frac{L\,(\text{cm})}{t\,(\text{min})}} \tag{2.5}$$

where

$$K = \sqrt{\frac{30\eta}{(G_s - 1)}} \tag{2.6}$$

Table 2.6 Values of K from Eq. (2.6)[a]

Temperature (°C)	G_s							
	2.45	2.50	2.55	2.60	2.65	2.70	2.75	2.80
16	0.01510	0.01505	0.01481	0.01457	0.01435	0.01414	0.01394	0.01374
17	0.01511	0.01486	0.01462	0.01439	0.01417	0.01396	0.01376	0.01356
18	0.01492	0.01467	0.01443	0.01421	0.01399	0.01378	0.01359	0.01339
19	0.01474	0.01449	0.01425	0.01403	0.01382	0.01361	0.01342	0.01323
20	0.01456	0.01431	0.01408	0.01386	0.01365	0.01344	0.01325	0.01307
21	0.01438	0.01414	0.01391	0.01369	0.01348	0.01328	0.01309	0.01291
22	0.01421	0.01397	0.01374	0.01353	0.01332	0.01312	0.01294	0.01276
23	0.01404	0.01381	0.01358	0.01337	0.01317	0.01297	0.01279	0.01261
24	0.01388	0.01365	0.01342	0.01321	0.01301	0.01282	0.01264	0.01246
25	0.01372	0.01349	0.01327	0.01306	0.01286	0.01267	0.01249	0.01232
26	0.01357	0.01334	0.01312	0.01291	0.01272	0.01253	0.01235	0.01218
27	0.01342	0.01319	0.01297	0.01277	0.01258	0.01239	0.01221	0.01204
28	0.01327	0.01304	0.01283	0.01264	0.01244	0.01225	0.01208	0.01191
29	0.01312	0.01290	0.01269	0.01249	0.01230	0.01212	0.01195	0.01178
30	0.01298	0.01276	0.01256	0.01236	0.01217	0.01199	0.01182	0.01169

[a] After ASTM (1999)

Note that the value of K is a function of G_s and η, which are dependent on the temperature of the test. Table 2.6 gives the variation of K with the test temperature and the specific gravity of soil solids.

In the laboratory, the hydrometer test is conducted in a sedimentation cylinder usually with 50 g of oven-dried sample. Sometimes 100-g samples can also be used. The sedimentation cylinder is 457 mm (18 in.) high and 63.5 mm (2.5 in.) in diameter. It is marked for a volume of 1000 ml. Sodium hexametaphosphate is generally used as the *dispersing agent*. The volume of the dispersed soil suspension is increased to 1000 ml by adding distilled water. Figure 2.16 shows an ASTM 152H type of hydrometer.

When a hydrometer is placed in the soil suspension at a time t, measured from the start of sedimentation it measures the specific gravity in the vicinity of its bulb at a depth L (Figure 2.17). The specific gravity is a function of the amount of soil particles present per unit volume of suspension at that depth. Also, at a time t, the soil particles in suspension at a depth L will have a diameter smaller than D as calculated in Eq. (2.5). The larger particles would have settled beyond the zone of measurement. Hydrometers are designed to give the amount of soil, in grams, that is still in suspension. They are calibrated for soils that have a specific gravity, G_s, of 2.65; for soils of other specific gravity, a correction must be made.

By knowing the amount of soil in suspension, L, and t, we can calculate the percentage of soil by weight finer than a given diameter. Note that L is the depth measured from the surface of the water to the center of gravity of the hydrometer bulb at which the density of the suspension is measured. The value of L will change with time t. Hydrometer analysis is effective for separating soil fractions down to a size of

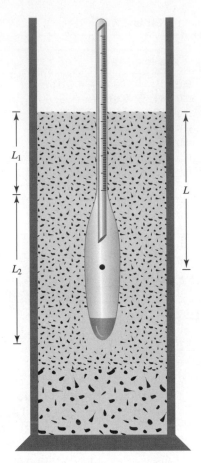

Figure 2.16
ASTM 152H hydrometer
(courtesy of Soiltest, Inc.,
Lake Bluff, Illinois)

Figure 2.17 Definition of L in hydrometer test

about 0.5 μm. The value of L (cm) for the ASTM 152H hydrometer can be given by the expression (see Figure 2.17)

$$L = L_1 + \frac{1}{2}\left(L_2 - \frac{V_B}{A}\right) \tag{2.7}$$

where L_1 = distance along the stem of the hydrometer from the top of the
bulb to the mark for a hydrometer reading (cm)
  L_2 = length of the hydrometer bulb = 14 cm
  V_B = volume of the hydrometer bulb = 67 cm^3
  A = cross-sectional area of the sedimentation cylinder = 27.8 cm^2

The value of L_1 is 10.5 cm for a reading of $R = 0$ and 2.3 cm for a reading of $R = 50$. Hence, for any reading R,

$$L_1 = 10.5 - \frac{(10.5 - 2.3)}{50}R = 10.5 - 0.164R \text{ (cm)}$$

Table 2.7 Variation of L with Hydrometer Reading —
ASTM 152H Hydrometer

Hydrometer reading, R	L (cm)	Hydrometer reading, R	L (cm)
0	16.3	31	11.2
1	16.1	32	11.1
2	16.0	33	10.9
3	15.8	34	10.7
4	15.6	35	10.6
5	15.5	36	10.4
6	15.3	37	10.2
7	15.2	38	10.1
8	15.0	39	9.9
9	14.8	40	9.7
10	14.7	41	9.6
11	14.5	42	9.4
12	14.3	43	9.2
13	14.2	44	9.1
14	14.0	45	8.9
15	13.8	46	8.8
16	13.7	47	8.6
17	13.5	48	8.4
18	13.3	49	8.3
19	13.2	50	8.1
20	13.0	51	7.9
21	12.9	52	7.8
22	12.7	53	7.6
23	12.5	54	7.4
24	12.4	55	7.3
25	12.2	56	7.1
26	12.0	57	7.0
27	11.9	58	6.8
28	11.7	59	6.6
29	11.5	60	6.5
30	11.4		

Thus, from Eq. (2.7),

$$L = 10.5 - 0.164R + \frac{1}{2}\left(14 - \frac{67}{27.8}\right) = 16.29 - 0.164R \tag{2.8}$$

where R = hydrometer reading corrected for the meniscus.

On the basis of Eq. (2.8), the variations of L with the hydrometer readings R are given in Table 2.7.

In many instances, the results of sieve analysis and hydrometer analysis for finer fractions for a given soil are combined on one graph, such as the one shown in Figure 2.18. When these results are combined, a discontinuity generally occurs in the range where they overlap. This discontinuity occurs because soil particles are generally irregular in shape. Sieve analysis gives the intermediate dimensions of a

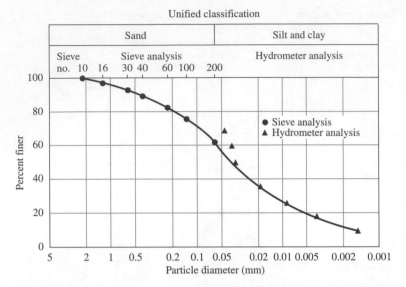

Figure 2.18 Particle-size distribution curve — sieve analysis and hydrometer analysis

particle; hydrometer analysis gives the diameter of an equivalent sphere that would settle at the same rate as the soil particle.

2.6 *Particle-Size Distribution Curve*

A particle-size distribution curve can be used to determine the following four parameters for a given soil (Figure 2.19):

1. *Effective size* (D_{10}): This parameter is the diameter in the particle-size distribution curve corresponding to 10% finer. The effective size of a granular soil is a good measure to estimate the hydraulic conductivity and drainage through soil.

2. *Uniformity coefficient* (C_u): This parameter is defined as

$$C_u = \frac{D_{60}}{D_{10}} \qquad (2.9)$$

where D_{60} = diameter corresponding to 60% finer.

3. *Coefficient of gradation* (C_z): This parameter is defined as

$$C_z = \frac{D_{30}^2}{D_{60} \times D_{10}} \qquad (2.10)$$

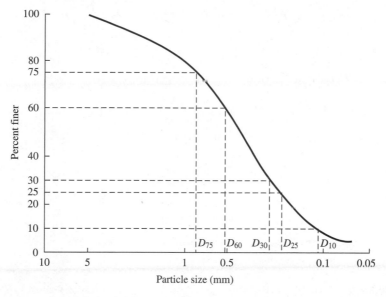

Figure 2.19 Definition of D_{75}, D_{60}, D_{30}, D_{25}, and D_{10}

4. *Sorting coefficient* (S_0): This parameter is another measure of uniformity and is generally encountered in geologic works and expressed as

$$S_0 = \sqrt{\frac{D_{75}}{D_{25}}} \qquad (2.11)$$

The sorting coefficient is not frequently used as a parameter by geotechnical engineers.

The percentages of gravel, sand, silt, and clay-size particles present in a soil can be obtained from the particle-size distribution curve. As an example, we will use the particle-size distribution curve shown in Figure 2.18 to determine the gravel, sand, silt, and clay-size particles as follows (according to the Unified Soil Classification System — see Table 2.3):

Size (mm)		% finer	
76.2	100		100 − 100 = 0% gravel
4.75	100		100 − 62 = 38% sand
0.075	62		62 − 0 = 62% silt and clay
—	0		

The particle-size distribution curve shows not only the range of particle sizes present in a soil, but also the type of distribution of various-size particles. Such types of distributions are demonstrated in Figure 2.20. Curve I represents a type of soil in

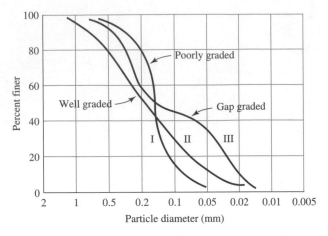

Figure 2.20 Different types of particle-size distribution curves

which most of the soil grains are the same size. This is called *poorly graded* soil. Curve II represents a soil in which the particle sizes are distributed over a wide range, termed *well graded.* A well-graded soil has a uniformity coefficient greater than about 4 for gravels and 6 for sands, and a coefficient of gradation between 1 and 3 (for gravels and sands). A soil might have a combination of two or more uniformly graded fractions. Curve III represents such a soil. This type of soil is termed *gap graded.*

Example 2.1

Following are the results of a sieve analysis. Make the necessary calculations and draw a particle-size distribution curve.

U.S. sieve size	Mass of soil retained on each sieve (g)
4	0
10	40
20	60
40	89
60	140
80	122
100	210
200	56
Pan	12

Solution
The following table can now be prepared.

U.S. sieve (1)	Opening (mm) (2)	Mass retained on each sieve (g) (3)	Cumulative mass retained above each sieve (g) (4)	Percent finer[a] (5)
4	4.75	0	0	**100**
10	2.00	40	0 + 40 = 40	**94.5**
20	0.850	60	40 + 60 = 100	**86.3**
40	0.425	89	100 + 89 = 189	**74.1**
60	0.250	140	189 + 140 = 329	**54.9**
80	0.180	122	329 + 122 = 451	**38.1**
100	0.150	210	451 + 210 = 661	**9.3**
200	0.075	56	661 + 56 = 717	**1.7**
Pan	—	12	717 + 12 = 729 = ΣM	**0**

$$^a \frac{\Sigma M - \text{col. } 4}{\Sigma M} \times 100 = \frac{729 - \text{col. } 4}{729} \times 100$$

The particle-size distribution curve is shown in Figure 2.21.   ■

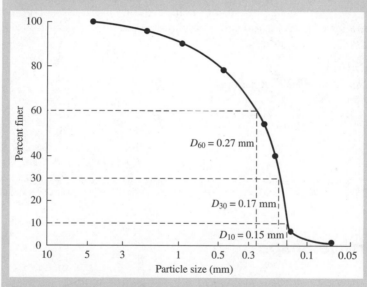

Figure 2.21 Particle-size distribution curve

Example 2.2

For the particle-size distribution curve shown in Figure 2.21, determine

a. D_{10}, D_{30}, and D_{60}
b. Uniformity coefficient, C_u
c. Coefficient of gradation, C_z

Solution

a. From Figure 2.21,

$$D_{10} = \textbf{0.15 mm}$$

$$D_{30} = \textbf{0.17 mm}$$

$$D_{60} = \textbf{0.27 mm}$$

b. $C_u = \dfrac{D_{60}}{D_{10}} = \dfrac{0.27}{0.15} = \textbf{1.8}$

$$C_z = \dfrac{D_{30}^2}{D_{60} \times D_{10}} = \dfrac{(0.17)^2}{(0.27)(0.15)} = \textbf{0.71}$$

■

Example 2.3

For the particle-size distribution curve shown in Figure 2.21, determine the percentages of gravel, sand, silt, and clay-size particles present. Use the Unified Soil Classification System.

Solution

From Figure 2.21, we can prepare the following table.

Size (mm)		% finer	
76.2	100	100 − 100 = **0% gravel**	
4.75	100	100 − 1.7 = **98.3% sand**	
0.075	1.7	1.7 − 0 = **1.7% silt and clay**	
—	0		

■

2.7 *Particle Shape*

The shape of particles present in a soil mass is equally as important as the particle-size distribution because it has significant influence on the physical properties of a given soil. However, not much attention is paid to particle shape because it is more difficult to measure. The particle shape can generally be divided into three major categories:

1. Bulky
2. Flaky
3. Needle shaped

Bulky particles are mostly formed by mechanical weathering of rock and minerals. Geologists use such terms as *angular, subangular, rounded,* and *subrounded*

Figure 2.22 Electron micrograph of some fine subangular and subrounded quartz sand

to describe the shapes of bulky particles. Figure 2.22 shows a scanning electron micrograph of some subangular and subrounded quartz sand. The *angularity, A,* is defined as

$$A = \frac{\text{Average radius of corners and edges}}{\text{Radius of the maximum inscribed sphere}} \tag{2.12}$$

The *sphericity* of bulky particles is defined as

$$S = \frac{D_e}{L_p} \tag{2.13}$$

where D_e = equivalent diameter of the particle = $\sqrt[3]{\dfrac{6V}{\pi}}$

V = volume of particle
L_p = length of particle

Flaky particles have very low sphericity — usually 0.01 or less. These particles are predominantly clay minerals.

Needle-shaped particles are much less common than the other two particle types. Examples of soils containing needle-shaped particles are some coral deposits and attapulgite clays.

| 2.8 | Summary |

In this chapter, we discussed the rock cycle, the origin of soil by weathering, the particle-size distribution in a soil mass, the shape of particles, and clay minerals. Some important points include the following:

1. Rocks can be classified into three basic categories: (a) igneous, (b) sedimentary, and (c) metamorphic.
2. Soils are formed by chemical and mechanical weathering of rocks.
3. Based on the size of the soil particles, soil can be classified as gravel, sand, silt, or clay.
4. Clays are mostly flake-shaped microscopic and submicroscopic particles of mica, clay minerals, and other minerals.
5. Clay minerals are complex aluminum silicates that develop plasticity when mixed with a limited amount of water.
6. Mechanical analysis is a process for determining the size range of particles present in a soil mass. Sieve analysis and hydrometer analysis are two tests used in the mechanical analysis of soil.

Problems

2.1 For a soil with $D_{60} = 0.42$ mm, $D_{30} = 0.21$ mm, and $D_{10} = 0.16$ mm, calculate the uniformity coefficient and the coefficient of gradation.

2.2 Repeat Problem 2.1 with the following values: $D_{10} = 0.27$ mm, $D_{30} = 0.41$ mm, and $D_{60} = 0.81$ mm.

2.3 Following are the results of a sieve analysis:

U.S. sieve no.	Mass of soil retained on each sieve (g)
4	0
10	18.5
20	53.2
40	90.5
60	81.8
100	92.2
200	58.5
Pan	26.5

a. Determine the percent finer than each sieve size and plot a grain-size distribution curve.
b. Determine D_{10}, D_{30}, and D_{60} from the grain-size distribution curve.
c. Calculate the uniformity coefficient C_u.
d. Calculate the coefficient of gradation, C_z.

2.4 Repeat Problem 2.3 with the following results of a sieve analysis.

U.S. sieve no.	Mass of soil retained on each sieve (g)
4	0
10	41.2
20	55.1
40	80.0
60	91.6
100	60.5
200	35.6
Pan	21.5

2.5 Repeat Problem 2.3 with the following results for a sieve analysis.

U.S. sieve no.	Mass of soil retained on each sieve (g)
4	0
6	0
10	20.1
20	19.5
40	210.5
60	85.6
100	22.7
200	15.5
Pan	23.5

2.6 The particle-size characteristics of a soil are given in this table. Draw the particle-size distribution curve.

Size (mm)	Percent finer
0.425	100
0.033	90
0.018	80
0.01	70
0.0062	60
0.0035	50
0.0018	40
0.001	35

Determine the percentages of gravel, sand, silt, and clay:
a. According to the USDA system.
b. According to the AASHTO system.

2.7 Repeat Problem 2.6 with the following data:

Size (mm)	Percent finer
0.425	100
0.1	92
0.052	84
0.02	62
0.01	46
0.004	32
0.001	22

2.8   Repeat Problem 2.6 with the following values:

Size (mm)	Percent finer
0.425	100
0.1	79
0.04	57
0.02	48
0.01	40
0.002	35
0.001	33

2.9   Repeat Problem 2.6 with the following data:

Size (mm)	Percent finer
0.425	100
0.07	90
0.046	80
0.034	70
0.026	60
0.019	50
0.014	40
0.009	30
0.0054	20
0.0019	10

2.10   A hydrometer test has the following results: $G_s = 2.7$, temperature of water = 24°C, and $L = 9.2$ cm at 60 minutes after the start of sedimentation. (See Figure 2.17.) What is the diameter D of the smallest-size particles that have settled beyond the zone of measurement at that time (that is, $t = 60$ min)?

2.11   Repeat Problem 2.10 with the following values: $G_s = 2.75$, temperature of water = 23°C, $t = 100$ min, and $L = 12.8$ cm.

References

AMERICAN SOCIETY FOR TESTING AND MATERIALS (1999). *ASTM Book of Standards,* Sec. 4, Vol. 04.08, West Conshohocken, Pa.

BOWEN, N. L. (1922). "The Reaction Principles in Petrogenesis," *Journal of Geology,* Vol. 30, 177–198.

GRIM, R. E. (1953). *Clay Mineralogy,* McGraw-Hill, New York.

GRIM, R. E. (1959). "Physico-Chemical Properties of Soils: Clay Minerals," *Journal of the Soil Mechanics and Foundations Division,* ASCE, Vol. 85, No. SM2, 1–17.

LAMBE, T. W. (1958). "The Structure of Compacted Clay," *Journal of the Soil Mechanics and Foundations Division,* ASCE, Vol. 84, No. SM2, 1655–1 to 1655–35.

YONG, R. N., and WARKENTIN, B. P. (1966). *Introduction of Soil Behavior,* Macmillan, New York.

3

Weight–Volume Relationships, Plasticity, and Structure of Soil

Chapter 2 presented the geologic processes by which soils are formed, the description of limits on the sizes of soil particles, and the mechanical analysis of soils. In natural occurrence, soils are three-phase systems consisting of soil solids, water, and air. This chapter discusses the weight–volume relationships of soil aggregates, along with their structures and plasticity.

3.1 Weight–Volume Relationships

Figure 3.1a shows an element of soil of volume V and weight W as it would exist in a natural state. To develop the weight–volume relationships, we must separate the three phases (that is, solid, water, and air) as shown in Figure 3.1b. Thus, the total volume of a given soil sample can be expressed as

$$V = V_s + V_v = V_s + V_w + V_a \tag{3.1}$$

where V_s = volume of soil solids
V_v = volume of voids
V_w = volume of water in the voids
V_a = volume of air in the voids

Assuming that the weight of the air is negligible, we can give the total weight of the sample as

$$W = W_s + W_w \tag{3.2}$$

where W_s = weight of soil solids
W_w = weight of water

The *volume relationships* commonly used for the three phases in a soil element are *void ratio, porosity,* and *degree of saturation. Void ratio* (e) is defined as the ratio of the volume of voids to the volume of solids. Thus,

$$e = \frac{V_v}{V_s} \tag{3.3}$$

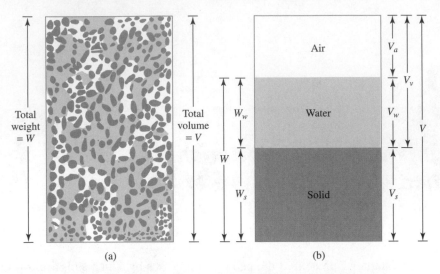

Figure 3.1 (a) Soil element in natural state; (b) three phases of the soil element

Porosity (n) is defined as the ratio of the volume of voids to the total volume, or

$$n = \frac{V_v}{V} \tag{3.4}$$

The *degree of saturation* (S) is defined as the ratio of the volume of water to the volume of voids, or

$$S = \frac{V_w}{V_v} \tag{3.5}$$

It is commonly expressed as a percentage.

The relationship between void ratio and porosity can be derived from Eqs. (3.1), (3.3), and (3.4) as follows:

$$e = \frac{V_v}{V_s} = \frac{V_v}{V - V_v} = \frac{\left(\dfrac{V_v}{V}\right)}{1 - \left(\dfrac{V_v}{V}\right)} = \frac{n}{1 - n} \tag{3.6}$$

Also, from Eq. (3.6),

$$n = \frac{e}{1 + e} \tag{3.7}$$

The common terms used for *weight relationships* are *moisture content* and *unit weight*. *Moisture content* (w) is also referred to as *water content* and is defined as the ratio of the weight of water to the weight of solids in a given volume of soil:

$$w = \frac{W_w}{W_s} \qquad (3.8)$$

Unit weight (γ) is the weight of soil per unit volume. Thus,

$$\gamma = \frac{W}{V} \qquad (3.9)$$

The unit weight can also be expressed in terms of the weight of soil solids, the moisture content, and the total volume. From Eqs. (3.2), (3.8), and (3.9),

$$\gamma = \frac{W}{V} = \frac{W_s + W_w}{V} = \frac{W_s\left[1 + \left(\dfrac{W_w}{W_s}\right)\right]}{V} = \frac{W_s(1 + w)}{V} \qquad (3.10)$$

Soils engineers sometimes refer to the unit weight defined by Eq. (3.9) as the *moist unit weight.*

Often, to solve earthwork problems, one must know the weight per unit volume of soil, excluding water. This weight is referred to as the *dry unit weight, γ_d.* Thus,

$$\gamma_d = \frac{W_s}{V} \qquad (3.11)$$

From Eqs. (3.10) and (3.11), the relationship of unit weight, dry unit weight, and moisture content can be given as

$$\gamma_d = \frac{\gamma}{1 + w} \qquad (3.12)$$

Unit weight is expressed in English units (a gravitational system of measurement) as pounds per cubic foot (lb/ft^3). In SI (Système International), the unit used is kilo Newtons per cubic meter (kN/m^3). Because the Newton is a derived unit, working with mass densities (ρ) of soil may sometimes be convenient. The SI unit of mass density is kilograms per cubic meter (kg/m^3). We can write the density equations [similar to Eqs. (3.9) and (3.11)] as

$$\rho = \frac{M}{V} \qquad (3.13)$$

and

$$\rho_d = \frac{M_s}{V} \qquad (3.14)$$

where ρ = density of soil (kg/m^3)
ρ_d = dry density of soil (kg/m^3)
M = total mass of the soil sample (kg)
M_s = mass of soil solids in the sample (kg)

The unit of total volume, V, is m^3.

The unit weight in kN/m³ can be obtained from densities in kg/m³ as

$$\gamma \,(kN/m^3) = \frac{g\rho(kg/m^3)}{1000}$$

and

$$\gamma_d \,(kN/m^3) = \frac{g\rho_d(kg/m^3)}{1000}$$

where g = acceleration due to gravity = 9.81 m/sec².

Note that unit weight of water (γ_w) is equal to 9.81 kN/m³ or 62.4 lb/ft³ or 1000 kgf/m³.

3.2   *Relationships among Unit Weight, Void Ratio, Moisture Content, and Specific Gravity*

To obtain a relationship among unit weight (or density), void ratio, and moisture content, let us consider a volume of soil in which the volume of the soil solids is one, as shown in Figure 3.2. If the volume of the soil solids is one, then the volume of voids is numerically equal to the void ratio, e [from Eq. (3.3)]. The weights of soil solids and water can be given as

$$W_s = G_s\gamma_w$$

$$W_w = wW_s = wG_s\gamma_w$$

where G_s = specific gravity of soil solids
  w = moisture content
  γ_w = unit weight of water

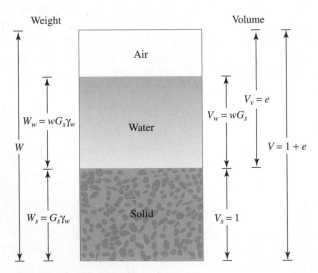

Figure 3.2 Three separate phases of a soil element with volume of soil solids equal to one

Now, using the definitions of unit weight and dry unit weight [Eqs. (3.9) and (3.11)], we can write

$$\gamma = \frac{W}{V} = \frac{W_s + W_w}{V} = \frac{G_s\gamma_w + wG_s\gamma_w}{1 + e} = \frac{(1 + w)G_s\gamma_w}{1 + e} \qquad (3.15)$$

and

$$\gamma_d = \frac{W_s}{V} = \frac{G_s\gamma_w}{1 + e} \qquad (3.16)$$

or

$$e = \frac{G_s\gamma_w}{\gamma_d} - 1 \qquad (3.17)$$

Because the weight of water for the soil element under consideration is $wG_s\gamma_w$, the volume occupied by water is

$$V_w = \frac{W_w}{\gamma_w} = \frac{wG_s\gamma_w}{\gamma_w} = wG_s$$

Hence, from the definition of degree of saturation [Eq. (3.5)],

$$S = \frac{V_w}{V_v} = \frac{wG_s}{e}$$

or

$$Se = wG_s \qquad (3.18)$$

This equation is useful for solving problems involving three-phase relationships.

If the soil sample is *saturated*— that is, the void spaces are completely filled with water (Figure 3.3)— the relationship for saturated unit weight (γ_{sat}) can be derived in a similar manner:

$$\gamma_{sat} = \frac{W}{V} = \frac{W_s + W_w}{V} = \frac{G_s\gamma_w + e\gamma_w}{1 + e} = \frac{(G_s + e)\gamma_w}{1 + e} \qquad (3.19)$$

Also, from Eq. (3.18) with $S = 1$,

$$e = wG_s \qquad (3.20)$$

As mentioned before, due to the convenience of working with densities in the SI system, the following equations, similar to unit-weight relationships given in Eqs. [3.15], [3.16], and [3.19], will be useful:

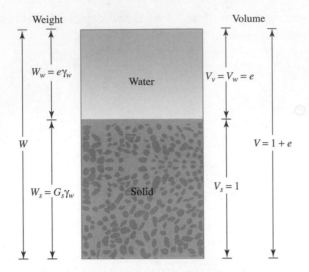

Figure 3.3 Saturated soil element with volume of soil solids equal to one

$$\text{Density} = \rho = \frac{(1 + w)G_s\rho_w}{1 + e} \qquad (3.21)$$

$$\text{Dry density} = \rho_d = \frac{G_s\rho_w}{1 + e} \qquad (3.22)$$

$$\text{Saturated density} = \rho_{\text{sat}} = \frac{(G_s + e)\rho_w}{1 + e} \qquad (3.23)$$

where ρ_w = density of water = 1000 kg/m³.

Equation (3.21) may be derived by referring to the soil element shown in Figure 3.4, in which the volume of soil solids is equal to 1 and the volume of voids is

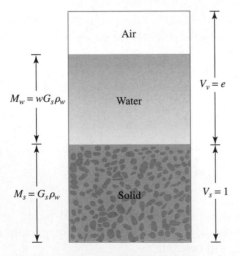

Figure 3.4 Three separate phases of a soil element showing mass-volume relationship

equal to e. Hence, the mass of soil solids, M_s, is equal to $G_s\rho_w$. The moisture content has been defined in Eq. (3.8) as

$$w = \frac{W_w}{W_s} = \frac{(\text{mass of water}) \cdot g}{(\text{mass of solid}) \cdot g}$$

$$= \frac{M_w}{M_s}$$

where M_w = mass of water.

Since the mass of soil in the element is equal to $G_s\rho_w$, the mass of water

$$M_w = wM_s = wG_s\rho_w$$

From Eq. (3.13), density

$$\rho = \frac{M}{V} = \frac{M_s + M_w}{V_s + V_v} = \frac{G_s\rho_w + wG_s\rho_w}{1 + e}$$

$$= \frac{(1 + w)G_s\rho_w}{1 + e}$$

Equations (3.22) and (3.23) can be derived similarly.

3.3   Relationships among Unit Weight, Porosity, and Moisture Content

The relationship among *unit weight, porosity,* and *moisture content* can be developed in a manner similar to that presented in the preceding section. Consider a soil that has a total volume equal to one, as shown in Figure 3.5. From Eq. (3.4),

$$n = \frac{V_v}{V}$$

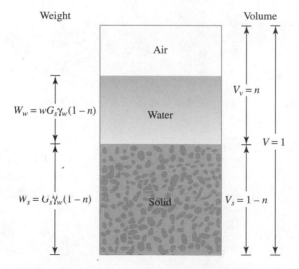

Figure 3.5 Soil element with total volume equal to one

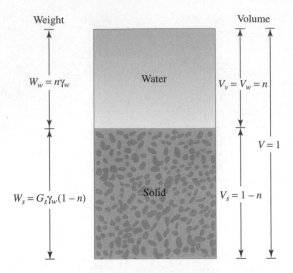

Figure 3.6 Saturated soil element with total volume equal to one

If V is equal to 1, then V_v is equal to n, so $V_s = 1 - n$. The weight of soil solids (W_s) and the weight of water (W_w) can then be expressed as follows:

$$W_s = G_s\gamma_w(1 - n) \tag{3.24}$$

$$W_w = wW_s = wG_s\gamma_w(1 - n) \tag{3.25}$$

So, the dry unit weight equals

$$\gamma_d = \frac{W_s}{V} = \frac{G_s\gamma_w(1 - n)}{1} = G_s\gamma_w(1 - n) \tag{3.26}$$

The moist unit weight equals

$$\gamma = \frac{W_s + W_w}{V} = G_s\gamma_w(1 - n)(1 + w) \tag{3.27}$$

Figure 3.6 shows a soil sample that is saturated and has $V = 1$. According to this figure,

$$\gamma_{\text{sat}} = \frac{W_s + W_w}{V} = \frac{(1 - n)G_s\gamma_w + n\gamma_w}{1} = [(1 - n)G_s + n]\gamma_w \tag{3.28}$$

The moisture content of a saturated soil sample can be expressed as

$$w = \frac{W_w}{W_s} = \frac{n\gamma_w}{(1 - n)\gamma_wG_s} = \frac{n}{(1 - n)G_s} \tag{3.29}$$

3.4 *Various Unit-Weight Relationships*

In Sections 3.2 and 3.3, we derived the fundamental relationships for the moist unit weight, dry unit weight, and saturated unit weight of soil. Several other forms of relationships that can be obtained for γ, γ_d, and γ_{sat} are given in Table 3.1. Some typical values of void ratio, moisture content in a saturated condition, and dry unit weight for soils in a natural state are given in Table 3.2.

Table 3.1 Various Forms of Relationships for γ, γ_d, and γ_{sat}

Moist unit weight (γ)		Dry unit weight (γ_d)		Saturated unit weight (γ_{sat})	
Given	**Relationship**	**Given**	**Relationship**	**Given**	**Relationship**
w, G_s, e	$\dfrac{(1 + w)G_s\gamma_w}{1 + e}$	γ, w	$\dfrac{\gamma}{1 + w}$	G_s, e	$\dfrac{(G_s + e)\gamma_w}{1 + e}$
S, G_s, e	$\dfrac{(G_s + Se)\gamma_w}{1 + e}$	G_s, e	$\dfrac{G_s\gamma_w}{1 + e}$	G_s, n	$[(1 - n)G_s + n]\gamma_w$
w, G_s, S	$\dfrac{(1 + w)G_s\gamma_w}{1 + \dfrac{wG_s}{S}}$	G_s, n	$G_s\gamma_w(1 - n)$	G_s, w_{sat}	$\left(\dfrac{1 + w_{sat}}{1 + w_{sat}G_s}\right)G_s\gamma_w$
		G_s, w, S	$\dfrac{G_s\gamma_w}{1 + \left(\dfrac{wG_s}{S}\right)}$	e, w_{sat}	$\left(\dfrac{e}{w_{sat}}\right)\left(\dfrac{1 + w_{sat}}{1 + e}\right)\gamma_w$
w, G_s, n	$G_s\gamma_w(1 - n)(1 + w)$				
S, G_s, n	$G_s\gamma_w(1 - n) + nS\gamma_w$	e, w, S	$\dfrac{eS\gamma_w}{(1 + e)w}$	n, w_{sat}	$n\left(\dfrac{1 + w_{sat}}{w_{sat}}\right)\gamma_w$
		γ_{sat}, e	$\gamma_{sat} - \dfrac{e\gamma_w}{1 + e}$	γ_d, e	$\gamma_d + \left(\dfrac{e}{1 + e}\right)\gamma_w$
				γ_d, n	$\gamma_d + n\gamma_w$
		γ_{sat}, n	$\gamma_{sat} - n\gamma_w$		
		γ_{sat}, G_s	$\dfrac{(\gamma_{sat} - \gamma_w)G_s}{(G_s - 1)}$	γ_d, S	$\left(1 - \dfrac{1}{G_s}\right)\gamma_d + \gamma_w$
				γ_d, w_{sat}	$\gamma_d(1 + w_{sat})$

Table 3.2 Void Ratio, Moisture Content, and Dry Unit Weight for Some Typical Soils in a Natural State

Type of soil	Void ratio, e	Natural moisture content in a saturated state (%)	Dry unit weight, γ_d	
			lb/ft^3	kN/m^3
Loose uniform sand	0.8	30	92	14.5
Dense uniform sand	0.45	16	115	18
Loose angular-grained silty sand	0.65	25	102	16
Dense angular-grained silty sand	0.4	15	121	19
Stiff clay	0.6	21	108	17
Soft clay	0.9–1.4	30–50	73–93	11.5–14.5
Loess	0.9	25	86	13.5
Soft organic clay	2.5–3.2	90–120	38–51	6–8
Glacial till	0.3	10	134	21

Example 3.1

For a soil, show that

$$\gamma_{sat} = \left(\frac{e}{w}\right)\left(\frac{1+w}{1+e}\right)\gamma_w$$

Solution
From Eqs. (3.19) and (3.20),

$$\gamma_{sat} = \frac{(G_s + e)\gamma_w}{1+e} \tag{a}$$

and

$$e = wG_s$$

or

$$G_s = \frac{e}{w} \tag{b}$$

Combining Eqs. (a) and (b) gives

$$\gamma_{sat} = \frac{\left(\dfrac{e}{w} + e\right)\gamma_w}{1+e} = \left(\frac{e}{w}\right)\left(\frac{1+w}{1+e}\right)\gamma_w \qquad \blacksquare$$

Example 3.2

The mass of a moist soil sample having a volume of 0.0057 m³ is 10.5 kg. The moisture content (w) and the specific gravity of soil solids (G_s) were determined to be 13% and 2.68, respectively. Determine

   a. Moist density, ρ (kg/m³)
   b. Dry density, ρ_d (kg/m³)
   c. Void ratio, e
   d. Porosity, n
   e. Degree of saturation, S (%)

Solution
   a. From Eq. (3.13),

$$\rho = \frac{M}{V} = \frac{10.5}{0.0057} = \textbf{1842 kg/m}^3$$

   b. From Eqs. (3.21) and (3.22),

$$\rho_d = \frac{\rho}{1+w} = \frac{1842}{1 + \dfrac{13}{100}} = \textbf{1630 kg/m}^3$$

c. From Eq. (3.22),

$$e = \frac{G_s \gamma_w}{\rho_d} - 1 = \frac{(2.68)(1000)}{1630} - 1 = \mathbf{0.64}$$

d. From Eq. (3.7),

$$n = \frac{e}{1 + e} = \frac{0.64}{1 + 0.64} = \mathbf{0.39}$$

e. From Eq. (3.18),

$$S(\%) = \frac{wG_s}{e} \times 100 = \frac{(0.13)(2.68)}{0.64} \times 100 = \mathbf{54.4\%}$$ ∎

Example 3.3

The saturated unit weight, γ_{sat}, of a soil is 19.5 kN/m³, and the specific gravity of soil solids is 2.65.

a. Derive an expression for γ_d in terms of γ_{sat}, γ_w, and G_s.
b. Using the expression derived in part (a), determine the dry unit weight of the soil.

Solution

a. From Eq. (3.19),

$$\gamma_{sat} = \frac{G_s \gamma_w + e\gamma_w}{1 + e}$$

$$\gamma_{sat} - \gamma_w = \frac{G_s \gamma_w + e\gamma_w}{1 + e} - \gamma_w = \frac{G_s \gamma_w + e\gamma_w - \gamma_w - e\gamma_w}{1 + e} = \frac{\gamma_w(G_s - 1)}{1 + e}$$

$$\gamma_{sat} - \gamma_w = \frac{\gamma_w(G_s - 1)G_s}{(1 + e)G_s} = \frac{\gamma_d(G_s - 1)}{G_s}$$

or

$$\gamma_d = \frac{(\gamma_{sat} - \gamma_w)G_s}{G_s - 1}$$

b. Given that $\gamma_{sat} = 19.5$ kN/m³ and $G_s = 2.65$,

$$\gamma_d = \frac{(\gamma_{sat} - \gamma_w)G_s}{G_s - 1} = \frac{(19.5 - 9.81)(2.65)}{2.65 - 1} = 15.56 \text{ kN/m}^3$$ ∎

Example 3.4

In its natural state, a moist soil has a volume of 0.33 ft^3 and weighs 39.93 lb. The oven-dried weight of the soil is 34.54 lb. If $G_s = 2.67$, calculate

   a. Moisture content (%)
   b. Moist unit weight (lb/ft^3)
   c. Dry unit weight (lb/ft^3)
   d. Void ratio
   e. Porosity
   f. Degree of saturation (%)

Solution

   a. From Eq. (3.8),

$$w = \frac{W_w}{W_s} = \frac{39.93 - 34.54}{34.54}(100) = \textbf{15.6\%}$$

   b. From Eq. (3.9),

$$\gamma = \frac{W}{V} = \frac{39.93}{0.33} = \textbf{121 lb/ft}^3$$

   c. From Eq. (3.11),

$$\gamma_d = \frac{W_s}{V} = \frac{34.54}{0.33} = \textbf{104.7 lb/ft}^3$$

   d. The volume of solids is

$$V_s = \frac{W_s}{G_s\gamma_w} = \frac{34.54}{(2.67)(62.4)} = 0.207 \text{ ft}^3$$

Thus,

$$V_v = V - V_s = 0.33 - 0.207 = 0.123 \text{ ft}^3$$

The volume of water is

$$V_w = \frac{W_w}{\gamma_w} = \frac{39.93 - 34.54}{62.4} = 0.086 \text{ ft}^3$$

Now, refer to Figure 3.7. From Eq. (3.3),

$$e = \frac{V_v}{V_s} = \frac{0.123}{0.207} = \textbf{0.59}$$

   e. From Eq. (3.4),

$$n = \frac{V_v}{V} = \frac{0.123}{0.33} = \textbf{0.37}$$

   f. From Eq. (3.5),

$$S = \frac{V_w}{V_v} = \frac{0.086}{0.123} = 0.699 = \textbf{69.9\%}$$ ■

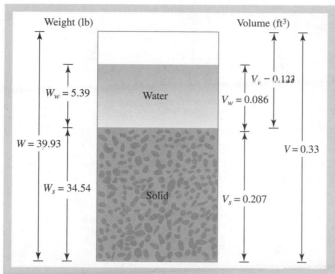

Figure 3.7 Diagram for Example 3.4

Example 3.5

For a saturated soil, given $w = 40\%$ and $G_s = 2.71$, determine the saturated and dry unit weights in lb/ft^3 and kN/m^3.

Solution

For saturated soil, from Eq. (3.20),

$$e = wG_s = (0.4)(2.71) = 1.084$$

From Eq. (3.19),

$$\gamma_{sat} = \frac{(G_s + e)\gamma_w}{1 + e} = \frac{(2.71 + 1.084)62.4}{1 + 1.084} = \textbf{113.6 lb/ft}^3$$

Also,

$$\gamma_{sat} = (113.6)\left(\frac{9.81}{62.4}\right) = \textbf{17.86 kN/m}^3$$

From Eq. (3.16),

$$\gamma_d = \frac{G_s\gamma_w}{1 + e} = \frac{(2.71)(62.4)}{1 + 1.084} = \textbf{81.1 lb/ft}^3$$

Also,

$$\gamma_d = (81.1)\left(\frac{9.81}{62.4}\right) = \textbf{12.75 kN/m}^3$$

■

Example 3.6

The mass of a moist soil sample collected from the field is 465 grams, and its oven dry mass is 405.76 grams. The specific gravity of the soil solids was determined in the laboratory to be 2.68. If the void ratio of the soil in the natural state is 0.83, find the following:

a. The moist density of the soil in the field (kg/m^3)
b. The dry density of the soil in the field (kg/m^3)
c. The mass of water, in kilograms, to be added per cubic meter of soil in the field for saturation

Solution
Part a

$$w = \frac{M_w}{M_s} = \frac{465 - 405.76}{405.76} = \frac{59.24}{405.76} = 14.6\%$$

From Eq. (3.21),

$$\rho = \frac{G_s\rho_w + wG_s\rho_w}{1 + e} = \frac{G_s\rho_w(1 + w)}{1 + e} = \frac{(2.68)(1000)(1.146)}{1.83}$$

$$= \mathbf{1678.3 \ kg/m^3}$$

Part b
From Eq. (3.22),

$$\rho_d = \frac{G_s\rho_w}{1 + e} = \frac{(2.68)(1000)}{1.83} = \mathbf{1468.48 \ kg/m^3}$$

Part c
Mass of water to be added $= \rho_{sat} - \rho$
From Eq. (3.23),

$$\rho_{sat} = \frac{G_s\rho_w + e\rho_w}{1 + e} = \frac{\rho_w(G_s + e)}{1 + e} = \frac{(1000)(2.68 + 0.83)}{1.83} = 1918 \ kg/m^3$$

So the mass of water to be added $= 1918 - 1678.3 = \mathbf{239.7 \ kg/m^3}$.   ■

3.5   *Relative Density*

The term *relative density* is commonly used to indicate the *in situ* denseness or looseness of granular soil. It is defined as

$$D_r = \frac{e_{max} - e}{e_{max} - e_{min}} \tag{3.30}$$

where D_r = relative density, usually given as a percentage

e = *in situ* void ratio of the soil

e_{max} = void ratio of the soil in the loosest state

e_{min} = void ratio of the soil in the densest state

The values of D_r may vary from a minimum of 0% for very loose soil to a maximum of 100% for very dense soils. Soils engineers qualitatively describe the granular soil deposits according to their relative densities, as shown in Table 3.3. In-place soils seldom have relative densities less than 20 to 30%. Compacting a granular soil to a relative density greater than about 85% is difficult.

Table 3.3 Qualitative Description of Granular Soil Deposits

Relative density (%)	Description of soil deposit
0–15	Very loose
15–50	Loose
50–70	Medium
70–85	Dense
85–100	Very dense

The relationships for relative density can also be defined in terms of porosity, or

$$e_{max} = \frac{n_{max}}{1 - n_{max}} \tag{3.31}$$

$$e_{min} = \frac{n_{min}}{1 - n_{min}} \tag{3.32}$$

$$e = \frac{n}{1 - n} \tag{3.33}$$

where n_{max} and n_{min} = porosity of the soil in the loosest and densest conditions, respectively. Substituting Eqs. (3.31), (3.32), and (3.33) into Eq. (3.30), we obtain

$$D_r = \frac{(1 - n_{min})(n_{max} - n)}{(n_{max} - n_{min})(1 - n)} \tag{3.34}$$

By using the definition of dry unit weight given in Eq. (3.16), we can express relative density in terms of maximum and minimum possible dry unit weights. Thus,

$$D_r = \frac{\left[\dfrac{1}{\gamma_{d(min)}}\right] - \left[\dfrac{1}{\gamma_d}\right]}{\left[\dfrac{1}{\gamma_{d(min)}}\right] - \left[\dfrac{1}{\gamma_{d(max)}}\right]} = \left[\frac{\gamma_d - \gamma_{d(min)}}{\gamma_{d(max)} - \gamma_{d(min)}}\right]\left[\frac{\gamma_{d(max)}}{\gamma_d}\right] \tag{3.35}$$

where $\gamma_{d(min)}$ = dry unit weight in the loosest condition (at a void ratio of e_{max})

γ_d = *in situ* dry unit weight (at a void ratio of e)

$\gamma_{d(max)}$ = dry unit weight in the densest condition (at a void ratio of e_{min})

ASTM Test Designation D-2049 (1999) provides a procedure for determining the minimum and maximum dry unit weights of granular soils so that they can be used in Eq. (3.35) to measure the relative density of compaction in the field. For sands, this procedure involves using a mold with a volume of 2830 cm^3 (0.1 ft^3). For a determination of the *minimum dry unit weight,* sand is loosely poured into the mold from a funnel with a 12.7 mm ($\frac{1}{2}$ in.) diameter spout. The average height of the fall of sand into the mold is maintained at about 25.4 mm (1 in.). The value of $\gamma_{d(min)}$ can then be calculated by using the following equation

$$\gamma_{d(min)} = \frac{W_s}{V_m} \tag{3.36}$$

where W_s = weight of sand required to fill the mold
      V_m = volume of the mold

The *maximum dry unit weight* is determined by vibrating sand in the mold for 8 min. A surcharge of 14 kN/m^2 (2 lb/in^2) is added to the top of the sand in the mold. The mold is placed on a table that vibrates at a frequency of 3600 cycles/min and that has an amplitude of vibration of 0.635 mm (0.025 in.). The value of $\gamma_{d(max)}$ can be determined at the end of the vibrating period with knowledge of the weight and volume of the sand. Several factors control the magnitude of $\gamma_{d(max)}$: the magnitude of acceleration, the surcharge load, and the geometry of acceleration. Hence, one can obtain a larger-value $\gamma_{d(max)}$ than that obtained by using the ASTM standard method described earlier.

Example 3.7

For a given sandy soil, $e_{max} = 0.82$ and $e_{min} = 0.42$. Let $G_s = 2.66$. In the field, the soil is compacted to a moist density of 1720 kg/m^3 at a moisture content of 9%. Determine the relative density of compaction.

Solution
From Eq. (3.21),

$$\rho = \frac{(1 + w)G_s\rho_w}{1 + e}$$

or

$$e = \frac{G_s\rho_w(1 + w)}{\rho} - 1 = \frac{(2.66)(1000)(1 + 0.09)}{1720} - 1 = 0.686$$

From Eq. (3.30),

$$D_r = \frac{e_{max} - e}{e_{max} - e_{min}} = \frac{0.82 - 0.686}{0.82 - 0.42} = \mathbf{0.335} = \mathbf{33.5\%} \qquad \blacksquare$$

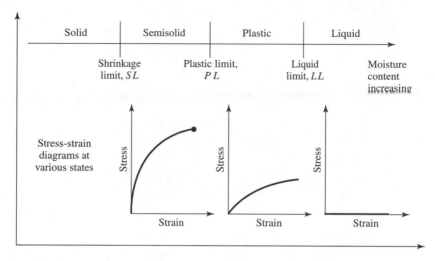

Figure 3.8 Atterberg Limits

3.6   *Consistency of Soil — Atterberg Limits*

When clay minerals are present in fine-grained soil, the soil can be remolded in the presence of some moisture without crumbling. This cohesive nature is caused by the adsorbed water surrounding the clay particles. In the early 1900s, a Swedish scientist named Atterberg developed a method to describe the consistency of fine-grained soils with varying moisture contents. At a very low moisture content, soil behaves more like a solid. When the moisture content is very high, the soil and water may flow like a liquid. Hence, on an arbitrary basis, depending on the moisture content, the behavior of soil can be divided into four basic states —*solid, semisolid, plastic,* and *liquid*— as shown in Figure 3.8.

     The moisture content, in percent, at which the transition from solid to semi-solid state takes place is defined as the *shrinkage limit.* The moisture content at the point of transition from semisolid to plastic state is the *plastic limit,* and from plastic to liquid state is the *liquid limit.* These parameters are also known as *Atterberg limits.* In the following sections, we describe the procedures for laboratory determination of Atterberg limits.

3.7   *Liquid Limit (LL)*

A schematic diagram (side view) of a liquid limit device is shown in Figure 3.9a. This device consists of a brass cup and a hard rubber base. The brass cup can be dropped onto the base by a cam operated by a crank. To perform the liquid limit test, one must place a soil paste in the cup. A groove is then cut at the center of the soil pat with the standard grooving tool (Figure 3.9b). By the use of the crank-operated cam, the cup is lifted and dropped from a height of 10 mm (0.394 in.). The moisture content, in

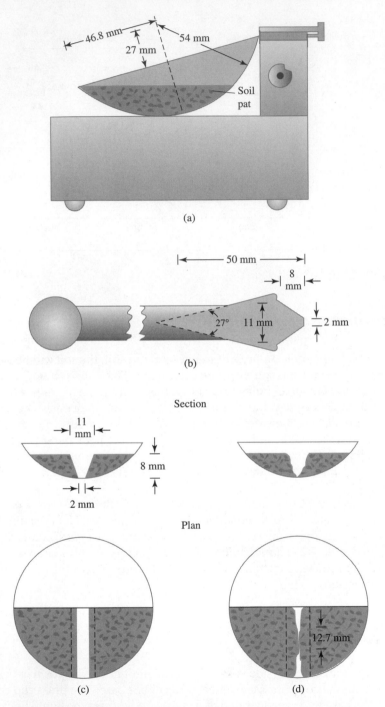

Figure 3.9 Liquid limit test: (a) liquid limit device; (b) grooving tool; (c) soil pat before test; (d) soil pat after test

Figure 3.10 Liquid limit test device and grooving tools (courtesy of Soiltest, Inc., Lake
          Bluff, Illinois)

percent, required to close a distance of 12.7 mm (0.5 in.) along the bottom of the
groove (see Figures 3.9c and 3.9d) after 25 blows is defined as the *liquid limit.*
     It is difficult to adjust the moisture content in the soil to meet the required 12.7
mm (0.5 in.) closure of the groove in the soil pat at 25 blows. Hence, at least three
tests for the same soil are conducted at varying moisture contents, with the number of
blows, *N,* required to achieve closure varying between 15 and 35. Figure 3.10 shows
a photograph of a liquid limit test device and grooving tools. The moisture content
of the soil, in percent, and the corresponding number of blows are plotted on semilogarithmic graph paper (Figure 3.11). The relationship between moisture content
and log *N* is approximated as a straight line. This line is referred to as the *flow curve.*
The moisture content corresponding to $N = 25$, determined from the flow curve,
gives the liquid limit of the soil. The slope of the flow line is defined as the *flow index*
and may be written as

$$I_F = \frac{w_1 - w_2}{\log\left(\dfrac{N_2}{N_1}\right)} \tag{3.37}$$

where I_F = flow index
     w_1 = moisture content of soil, in percent, corresponding to N_1 blows
     w_2 = moisture content corresponding to N_2 blows

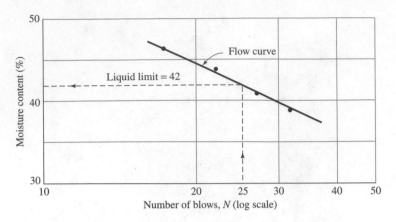

Figure 3.11 Flow curve for liquid limit determination of a clayey silt

Note that w_2 and w_1 are exchanged to yield a positive value even though the slope of the flow line is negative. Thus, the equation of the flow line can be written in a general form as

$$w = -I_F \log N + C \tag{3.38}$$

where C = a constant.

From the analysis of hundreds of liquid limit tests, the U.S. Army Corps of Engineers (1949) at the Waterways Experiment Station in Vicksburg, Mississippi, proposed an empirical equation of the form

$$LL = w_N \left(\frac{N}{25} \right)^{\tan \beta} \tag{3.39}$$

where N = number of blows in the liquid limit device for a 12.7 mm (0.5 in.) groove closure
w_N = corresponding moisture content
$\tan \beta$ = 0.121 (but note that $\tan \beta$ is not equal to 0.121 for all soils)

Equation (3.39) generally yields good results for the number of blows between 20 and 30. For routine laboratory tests, it may be used to determine the liquid limit when only one test is run for a soil. This procedure is generally referred to as the *one-point method* and was also adopted by ASTM under designation D-4318. The reason that the one-point method yields fairly good results is that a small range of moisture content is involved when $N = 20$ to $N = 30$.

Another method of determining liquid limit that is popular in Europe and Asia is the *fall cone method* (British Standard — BS1377). In this test the liquid limit is defined as the moisture content at which a standard cone of apex angle 30° and weight of 0.78 N (80 gf) will penetrate a distance $d = 20$ mm in 5 seconds when allowed to drop from a position of point contact with the soil surface (Figure 3.12a). Due to the difficulty in achieving the liquid limit from a single test, four or more tests can be conducted at various moisture contents to determine the fall cone penetration, d. A semilogarithmic graph can then be plotted with moisture content (w) versus cone

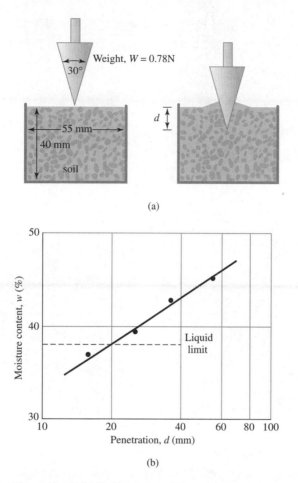

Figure 3.12 (a) Fall cone test (b) plot of moisture content vs. cone penetration for determination of liquid limit

penetration d. The plot results in a straight line. The moisture content corresponding
to $d = 20$ mm is the liquid limit (Figure 3.12b). From Figure 3.12(b), the *flow index*
can be defined as

$$I_{FC} = \frac{w_2(\%) - w_1(\%)}{\log d_2 - \log d_1} \tag{3.40}$$

where w_1, w_2 = moisture contents at cone penetrations of d_1 and d_2, respectively.

3.8   *Plastic Limit (PL)*

The *plastic limit* is defined as the moisture content in percent, at which the soil
crumbles, when rolled into threads of 3.2 mm ($\frac{1}{8}$ in.) in diameter. The plastic limit is

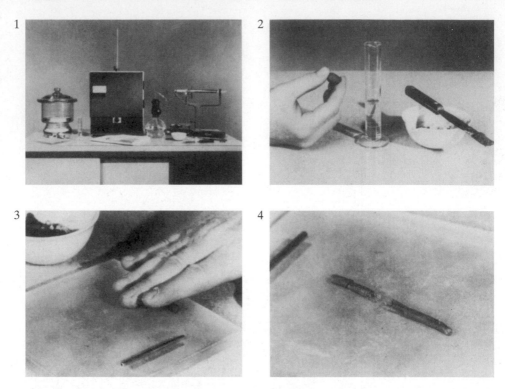

Figure 3.13 Plastic limit test: (1) equipment; (2) beginning of test; (3) thread being rolled; (4) crumbled soil (courtesy of Soiltest, Inc., Lake Bluff, Illinois)

the lower limit of the plastic stage of soil. The plastic limit test is simple and is performed by repeated rollings of an ellipsoidal-size soil mass by hand on a ground glass plate (Figure 3.13). The procedure for the plastic limit test is given by ASTM in Test Designation D-4318.

As in the case of liquid limit determination, the fall cone method can be used to obtain the plastic limit. This can be achieved by using a cone of similar geometry but with a mass of 2.35 N (240 gf). Three to four tests at varying moisture contents of soil are conducted, and the corresponding cone penetrations (d) are determined. The moisture content corresponding to a cone penetration of $d = 20$ mm is the plastic limit. Figure 3.14 shows the liquid and plastic limit determination of Cambridge Gault clay reported by Worth and Wood (1978).

The *plasticity index (PI)* is the difference between the liquid limit and the plastic limit of a soil, or

$$PI = LL - PL \tag{3.41}$$

Table 3.4 gives the ranges of liquid limit, plastic limit, and activity (Section 3.11) of some clay minerals (Mitchell, 1976; Skempton, 1953).

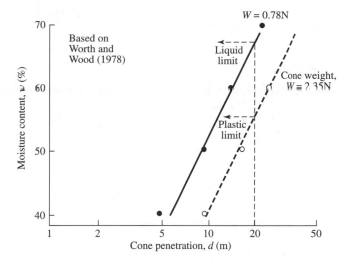

Figure 3.14
Liquid and plastic limits for Cambridge Gault clay determined by fall cone test

Table 3.4 Typical Values of Liquid Limit, Plastic Limit, and Activity of Some Clay Minerals

Mineral	Liquid limit, LL	Plastic limit, PL	Activity, A
Kaolinite	35–100	20–40	0.3–0.5
Illite	60–120	35–60	0.5–1.2
Montmorillonite	100–900	50–100	1.5–7.0
Halloysite (hydrated)	50–70	40–60	0.1–0.2
Halloysite (dehydrated)	40–55	30–45	0.4–0.6
Attapulgite	150–250	100–125	0.4–1.3
Allophane	200–250	120–150	0.4–1.3

Burmister (1949) classified the plasticity index in a qualitative manner as follows:

PI	Description
0	Nonplastic
1–5	Slightly plastic
5–10	Low plasticity
10–20	Medium plasticity
20–40	High plasticity
>40	Very high plasticity

The plasticity index is important in classifying fine-grained soils. It is fundamental to the Casagrande plasticity chart (presented in Section 3.12), which is currently the basis for the Unified Soil Classification System. (See Chapter 4.)

Sridharan et al. (1999) showed that the plasticity index can be correlated to the flow index as obtained from the liquid limit tests (Section 3.7). According to their study,

$$PI\,(\%) = 4.12 I_F\,(\%) \qquad (3.42)$$

and

$$PI\,(\%) = 0.74 I_{FC}\,(\%) \qquad (3.43)$$

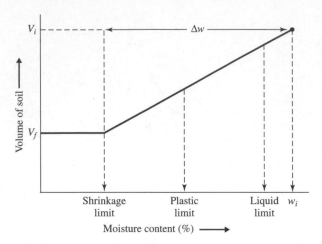

Figure 3.15 Definition of shrinkage limit

3.9   *Shrinkage Limit (SL)*

Soil shrinks as moisture is gradually lost from it. With continuing loss of moisture, a stage of equilibrium is reached at which more loss of moisture will result in no further volume change (Figure 3.15). The moisture content, in percent, at which the volume of the soil mass ceases to change is defined as the *shrinkage limit.*

Shrinkage limit tests (ASTM Test Designation D-427) are performed in the laboratory with a porcelain dish about 44 mm (1.75 in.) in diameter and about 12.7 mm ($\frac{1}{2}$ in.) high. The inside of the dish is coated with petroleum jelly and is then filled completely with wet soil. Excess soil standing above the edge of the dish is struck off with a straightedge. The mass of the wet soil inside the dish is recorded. The soil pat in the dish is then oven-dried. The volume of the oven-dried soil pat is determined by the displacement of mercury. Because handling mercury may be hazardous, ASTM D-4943 describes a method of dipping the oven-dried soil pat in a melted pot of wax. The wax-coated soil pat is then cooled. Its volume is determined by submerging it in water.

By reference to Figure 3.15, the shrinkage limit can be determined as

$$SL = w_i(\%) - \Delta w(\%) \tag{3.44}$$

where w_i = initial moisture content when the soil is placed in the shrinkage limit dish

Δw = change in moisture content (that is, between the initial moisture content and the moisture content at the shrinkage limit)

However,

$$w_i(\%) = \frac{M_1 - M_2}{M_2} \times 100 \tag{3.45}$$

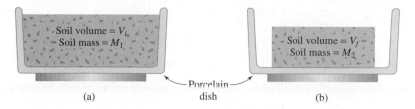

Figure 3.16 Shrinkage limit test: (a) soil pat before drying; (b) soil pat after drying

where M_1 = mass of the wet soil pat in the dish at the beginning of the test (g)
M_2 = mass of the dry soil pat (g) (see Figure 3.16)

Also,

$$\Delta w\,(\%) = \frac{(V_i - V_f)\rho_w}{M_2} \times 100 \qquad (3.46)$$

where V_i = initial volume of the wet soil pat (that is, inside volume of the dish, cm^3)
V_f = volume of the oven-dried soil pat (cm^3)
ρ_w = density of water (g/cm^3)

Finally, combining Eqs. (3.44), (3.45), and (3.46) gives

$$SL = \left(\frac{M_1 - M_2}{M_2}\right)(100) - \left(\frac{V_i - V_f}{M_2}\right)(\rho_w)(100) \qquad (3.47)$$

Another parameter that can be determined from a shrinkage limit test is the *shrinkage ratio,* which is the ratio of the volume change of soil as a percentage of the dry volume to the corresponding change in moisture content, or

$$SR = \frac{\left(\dfrac{\Delta V}{V_f}\right)}{\left(\dfrac{\Delta M}{M_2}\right)} = \frac{\left(\dfrac{\Delta V}{V_f}\right)}{\left(\dfrac{\Delta V \rho_w}{M_2}\right)} = \frac{M_2}{V_f \rho_w} \qquad (3.48)$$

where ΔV = change in volume
ΔM = corresponding change in the mass of moisture

It can also be shown that

$$G_s = \frac{1}{\dfrac{1}{SR} - \left(\dfrac{SL}{100}\right)} \qquad (3.49)$$

where G_s = specific gravity of soil solids.

Example 3.8

Following are the results of a shrinkage limit test:

- Initial volume of soil in a saturated state = 24.6 cm³
- Final volume of soil in a dry state = 15.9 cm³
- Initial mass in a saturated state = 44 g
- Final mass in a dry state = 30.1 g

Determine the shrinkage limit of the soil.

Solution
From Eq. (3.47),

$$SL = \left(\frac{M_1 - M_2}{M_2}\right)(100) - \left(\frac{V_i - V_f}{M_2}\right)(\rho_w)(100)$$

$$M_1 = 44\text{g} \qquad V_i = 24.6 \text{ cm}^3 \qquad \rho_w = 1 \text{ g/cm}^3$$

$$M_2 = 30.1\text{g} \qquad V_f = 15.9 \text{ cm}^3$$

$$SL = \left(\frac{44 - 30.1}{30.1}\right)(100) - \left(\frac{24.6 - 15.9}{30.1}\right)(1)(100)$$

$$= 46.18 - 28.9 = \mathbf{17.28}\%$$

■

3.10   *Liquidity Index and Consistency Index*

The relative consistency of a cohesive soil in the natural state can be defined by a ratio called the *liquidity index*, which is given by

$$LI = \frac{w - PL}{LL - PL} \tag{3.50}$$

where w = *in situ* moisture content of soil.

The *in situ* moisture content for a sensitive clay may be greater than the liquid limit. In this case (Figure 3.17),

$$LI > 1$$

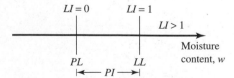

Figure 3.17 Liquidity index

These soils, when remolded, can be transformed into a viscous form to flow like a liquid.

Soil deposits that are heavily overconsolidated may have a natural moisture content less than the plastic limit. In this case (Figure 3.17),

$$LI < 0$$

Activity

Because the plasticity of soil is caused by the adsorbed water that surrounds the clay particles, we can expect that the type of clay minerals and their proportional amounts in a soil will affect the liquid and plastic limits. Skempton (1953) observed that the plasticity index of a soil increases linearly with the percentage of clay-size fraction (% finer than 2 μm by weight) present (Figure 3.18). The correlations of PI with the clay-size fractions for different clays plot separate lines. This difference is due to the diverse plasticity characteristics of the various types of clay minerals. On the basis of these results, Skempton defined a quantity called *activity,* which is the slope of the line correlating PI and % finer than 2 μm. This activity may be expressed as

$$A = \frac{PI}{(\,\%\,\text{of clay-size fraction, by weight})} \tag{3.51}$$

where A = activity. Activity is used as an index for identifying the swelling potential of clay soils. Typical values of activities for various clay minerals are given in Table 3.4.

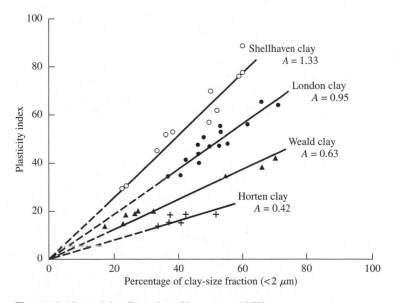

Figure 3.18 Activity (based on Skempton, 1953)

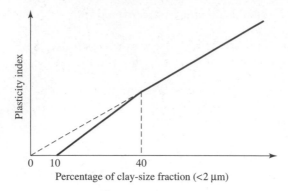

Figure 3.19 Simplified relationship between plasticity index and percentage of clay-size fraction by weight (after Seed, Woodward, and Lundgren, 1964b)

Seed, Woodward, and Lundgren (1964a) studied the plastic property of several artificially prepared mixtures of sand and clay. They concluded that although the relationship of the plasticity index to the percentage of clay-size fraction is linear, as observed by Skempton, it may not always pass through the origin. Thus, the activity can be redefined as in Eq. (3.52), viz.,

$$A = \frac{PI}{\% \text{ of clay-size fraction} - C'} \qquad (3.52)$$

where C' is a constant for a given soil. For the experimental results of Seed et al. (1964a), $C' = 9$.

Further works of Seed, Woodward, and Lundgren (1964b) showed that the relationship of the plasticity index to the percentage of clay-size fraction present in a soil can be represented by two straight lines. This finding is shown qualitatively in Figure 3.19. For clay-size fractions greater than 40%, the straight line passes through the origin when it is projected back.

3.12   *Plasticity Chart*

Liquid and plastic limits are determined by relatively simple laboratory tests that provide information about the nature of cohesive soils. Engineers have used the tests extensively for the correlation of several physical soil parameters as well as for soil identification. Casagrande (1932) studied the relationship of the plasticity index to the liquid limit of a wide variety of natural soils. On the basis of the test results, he proposed a plasticity chart as shown in Figure 3.20. The important feature of this chart is the empirical A-line that is given by the equation $PI = 0.73(LL - 20)$. An A-line separates the inorganic clays from the inorganic silts. Inorganic clay values lie above the A-line, and values for inorganic silts lie below the A-line. Organic silts plot in the same region (below the A-line and with LL ranging from 30 to 50) as the inorganic silts of medium compressibility. Organic clays plot in the same region as

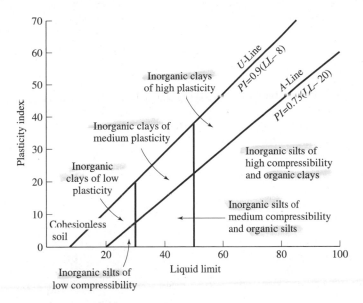

Figure 3.20 Plasticity chart

inorganic silts of high compressibility (below the *A*-line and *LL* greater than 50). The information provided in the plasticity chart is of great value and is the basis for the classification of fine-grained soils in the Unified Soil Classification System. (See Chapter 4.)

Note that a line called the *U*-line lies above the *A*-line. The *U*-line is approximately the upper limit of the relationship of the plasticity index to the liquid limit for any currently known soil. The equation for the *U*-line can be given as

$$PI = 0.9(LL - 8) \qquad (3.53)$$

3.13   *Soil Structure*

Soil structure is defined as the geometric arrangement of soil particles with respect to one another. Among the many factors that affect the structure are the shape, size, and mineralogical composition of soil particles, and the nature and composition of soil water. In general, soils can be placed into two groups: cohesionless and cohesive. The structures found in soils in each group are described next.

Structures in Cohesionless Soil

The structures generally encountered in cohesionless soils can be divided into two major categories: *single grained* and *honeycombed*. In single-grained structures, soil particles are in stable positions, with each particle in contact with the surrounding ones. The shape and size distribution of the soil particles and their relative positions influence the denseness of packing (Figure 3.21); thus, a wide range of void ratios is possible. To get an idea of the variation of void ratios caused by the relative positions

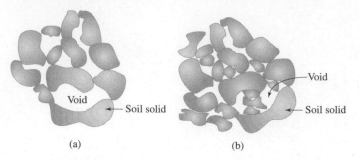

Figure 3.21 Single-grained structure: (a) loose; (b) dense

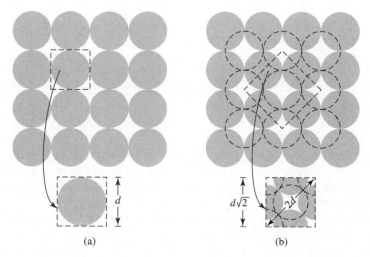

Figure 3.22 Mode of packing of equal spheres (plan views): (a) very loose packing ($e = 0.91$); (b) very dense packing ($e = 0.35$)

of the particles, let us consider the mode of packing of equal spheres shown in Figure 3.22.

Figure 3.22a shows the case of a very loose state of packing. If we isolate a cube with each side measuring d, which is equal to the diameter of each sphere as shown in the figure, the void ratio can be calculated as

$$e = \frac{V_v}{V_s} = \frac{V - V_s}{V_s}$$

where V = volume of the cube = d^3

V_s = volume of sphere (i.e., solid) inside the cube

Noting that $V = d^3$ and $V_s = \pi d^3/6$ yields

$$e = \frac{d^3 - \left(\dfrac{\pi d^3}{6}\right)}{\left(\dfrac{\pi d^3}{6}\right)} = 0.91$$

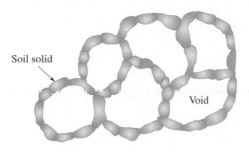

Figure 3.23 Honeycombed structure

Similarly, Figure 3.22b shows the case of a very dense state of packing. Figure 3.22b also shows an isolated cube, for which each side measures $d\sqrt{2}$. It can be shown that, for this case, $e = 0.35$.

Real soil differs from the equal-spheres model in that soil particles are neither equal in size nor spherical. The smaller-size particles may occupy the void spaces between the larger particles, thus the void ratio of soils is decreased compared with that for equal spheres. However, the irregularity in the particle shapes generally yields an increase in the void ratio of soils. As a result of these two factors, the void ratios encountered in real soils have approximately the same range as those obtained in equal spheres.

In the honeycombed structure (Figure 3.23). relatively fine sand and silt form small arches with chains of particles. Soils that exhibit a honeycombed structure have large void ratios, and they can carry an ordinary static load. However, under a heavy load or when subjected to shock loading, the structure breaks down, which results in a large amount of settlement.

Structures in Cohesive Soils

To understand the basic structures in cohesive soils, we need to know the types of forces that act between clay particles suspended in water. In Chapter 2, we discussed the negative charge on the surface of the clay particles and the diffuse double layer surrounding each particle. When two clay particles in suspension come close to each other, the tendency for interpenetration of the diffuse double layers results in repulsion between the particles. At the same time, an attractive force exists between the clay particles that is caused by van der Waals forces and is independent of the characteristics of water. Both repulsive and attractive forces increase with decreasing distance between the particles, but at different rates. When the spacing between the particles is very small, the force of attraction is greater than the force of repulsion. These are the forces treated by colloidal theories.

The fact that local concentrations of positive charges occur at the edges of clay particles was discussed in Chapter 2. If the clay particles are very close to each other, the positively charged edges can be attracted to the negatively charged faces of the particles.

Let us consider the behavior of clay in the form of a dilute suspension. When the clay is initially dispersed in water, the particles repel one another. This repulsion

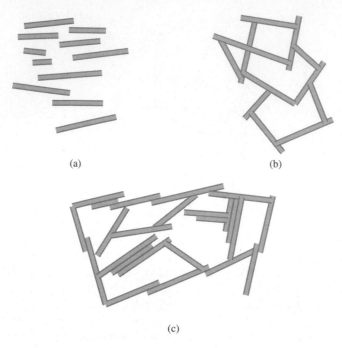

(a)   (b)

(c)

Figure 3.24 Sediment structures: (a) dispersion; (b) nonsalt flocculation; (c) salt flocculation (adapted from Lambe, 1958)

occurs because with larger interparticle spacing, the forces of repulsion between the particles are greater than the forces of attraction (van der Waals forces). The force of gravity on each particle is negligible. Thus, the individual particles may settle very slowly or remain in suspension, undergoing *Brownian motion* (a random zigzag motion of colloidal particles in suspension). The sediment formed by the settling of the individual particles has a dispersed structure, and all particles are oriented more or less parallel to one another (Figure 3.24a).

If the clay particles initially dispersed in water come close to one another during random motion in suspension, they might aggregate into visible flocs with edge-to-face contact. In this instance, the particles are held together by electrostatic attraction of positively charged edges to negatively charged faces. This aggregation is known as *flocculation.* When the flocs become large, they settle under the force of gravity. The sediment formed in this manner has a flocculent structure (Figure 3.24b).

When salt is added to a clay-water suspension that has been initially dispersed, the ions tend to depress the double layer around the particles. This depression reduces the interparticle repulsion. The clay particles are attracted to one another to form flocs and settle. The flocculent structure of the sediments formed is shown in Figure 3.24c. In flocculent sediment structures of the salt type, the particle orientation approaches a large degree of parallelism, which is due to van der Waals forces.

Clays that have flocculent structures are lightweight and possess high void ratios. Clay deposits formed in the sea are highly flocculent. Most of the sediment de-

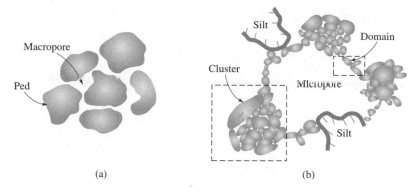

Figure 3.25 Soil structure: (a) arrangement of peds and macropore spaces; (b) arrangement of domains and clusters with silt-size particles

posits formed from freshwater possess an intermediate structure between dispersed and flocculent.

A deposit of pure clay minerals is rare in nature. When a soil has 50% or more particles with sizes of 0.002 mm or less, it is generally termed *clay*. Studies with scanning electron microscopes (Collins and McGown, 1974; Pusch, 1978; Yong and Sheeran, 1973) have shown that individual clay particles tend to be aggregated or flocculated in submicroscopic units. These units are referred to as *domains*. The domains then group together, and these groups are called *clusters*. Clusters can be seen under a light microscope. This grouping to form clusters is caused primarily by interparticle forces. The clusters, in turn, group to form *peds*. Peds can be seen without a microscope. Groups of peds are macrostructural features along with joints and fissures. Figure 3.25a shows the arrangement of the peds and macropore spaces. The arrangement of domains and clusters with silt-size particles is shown in Figure 3.25b.

From the preceding discussion, we can see that the structure of cohesive soils is highly complex. Macrostructures have an important influence on the behavior of soils from an engineering viewpoint. The microstructure is more important from a fundamental viewpoint. Table 3.5 summarizes the macrostructures of clay soils.

Table 3.5 Structure of Clay Soils

Item	Remarks
Dispersed structures	Formed by settlement of individual clay particles. More or less parallel orientation (see Figure 3.24a).
Flocculent structures	Formed by settlement of flocs of clay particles (see Figures 3.24b and 3.24c).
Domains	Aggregated or flocculated submicroscopic units of clay particles.
Clusters	Domains group to form clusters. Can be seen under light microscope.
Peds	Clusters group to form peds. Can be seen without microscope.

3.14   Summary

This chapter discussed three major components in the study of soil mechanics. They are (a) weight–volume relationships (Sections 3.1 to 3.5), (b) plasticity of soil and related topics (Sections 3.6 to 3.12), and (c) structure of soil (Section 3.13).

Weight–volume relationships include relationships among parameters such as void ratio, porosity, degree of saturation, moisture content, and unit weight. The parameters are fundamental to the study of geotechnical engineering.

Liquid limit, plastic limit, and shrinkage limit tests of fine-grained soil are indicators of the nature of its plasticity. The difference between the liquid limit and plastic limit is called the plasticity index. Liquid limit and plasticity index are required parameters for classification of fine-grained soils.

The structure of cohesionless soils can be single grained or honeycombed. Honeycombed structures are encountered in relatively fine sands and silts. The macrostructure of clay soils can be broadly divided into categories such as dispersed structures, flocculent structures, domains, clusters, and peds.

Problems

3.1   For a given soil, show that

a. $\gamma_{sat} = \gamma_d + n\gamma_w$

b. $\gamma_{sat} = n\left(\dfrac{1 + w_{sat}}{w_{sat}}\right)\gamma_w$

where w_{sat} = moisture content at saturated state

c. $\gamma_d = \dfrac{eS\gamma_w}{(1 + e)w}$

3.2   For a given soil, show that

$$e = \frac{\gamma_{sat} - \gamma_d}{\gamma_d - \gamma_{sat} + \gamma_w}$$

3.3   For a given soil, show that

$$G_s = \frac{\gamma_{sat}}{\gamma_w - w_{sat}(\gamma_{sat} - \gamma_w)}$$

3.4   For a given soil, show that

$$w_{sat} = \frac{n\gamma_w}{\gamma_{sat} - n\gamma_w}$$

3.5   For a moist soil, given that
- Volume of moist soil = 0.25 ft³
- Weight of moist soil = 30 lb
- Weight of dry soil = 26.1 lb
- G_s = 2.63

calculate
a. Moisture content
b. Moist unit weight
c. Dry unit weight
d. Void ratio

      e. Porosity

      f. Degree of saturation

3.6 For a moist soil, given that Volume = 5660 cm^3; Mass = 10.4 kg; Moisture content = 10%; G_s = 2.7, calculate the following:

      a. Moist density (kg/m^3)

      b. Dry density (kg/m^3)

      c. Void ratio

      d. Porosity

      e. Degree of saturation (%)

      f. Volume occupied by water (m^3)

3.7 The saturated unit weight of a soil is 126 lb/ft^3. The moisture content of the soil is 18.2%. Determine the following:

      a. Dry unit weight

      b. Void ratio

      c. Specific gravity of soil solids

3.8 The unit weight of a soil is 14.9 kN/m^3. The moisture content of this soil is 17% when the degree of saturation is 60%. Determine

      a. Saturated unit weight

      b. Void ratio

      c. Specific gravity of soil solids

3.9 For a soil, the following are given: G_s = 2.67, moist unit weight γ = 17.6 kN/m^3, and moisture content w = 10.8%. Determine

      a. Dry unit weight

      b. Void ratio

      c. Porosity

      d. Degree of saturation

3.10 Refer to Problem 3.9. Determine the mass of water, in kilograms, to be added per cubic meter of soil for

      a. 80% degree of saturation

      b. 100% degree of saturation

3.11 The moist unit weight of a soil is 105 lb/ft^3. Given that w = 15% and G_s = 2.7, determine

      a. Dry unit weight

      b. Porosity

      c. Degree of saturation

      d. Weight of water, in lb/ft^3, to be added to reach full saturation

3.12 The dry density of a soil is 1760 kg/m^3. Given that G_s = 2.66, what would be the moisture content of the soil when saturated?

3.13 The porosity of a soil is 0.35. If G_s = 2.69, calculate

      a. Saturated unit weight (kN/m^3)

      b. Moisture content when moist unit weight = 17.6 kN/m^3

3.14 A saturated soil has w = 28% and G_s = 2.66. Determine its saturated and dry unit weights in lb/ft^3.

3.15 A soil has e = 0.75, w = 21.5%, and G_s = 2.71. Determine

      a. Moist unit weight (lb/ft^3)

      b. Dry unit weight (lb/ft^3)

      c. Degree of saturation (%)

3.16   Repeat Problem 3.15 with the following: $e = 0.6$, $w = 6\%$, and $G_s = 2.65$.

3.17   The moist densities and degrees of saturation of a soil are given in the following table:

$\rho(kg/m^3)$	$S(\%)$
1690	50
1808	75

Determine

a. G_s

b. e

3.18   Refer to Problem 3.17. Determine the weight of water, in kg, that will be in 70.8×10^{-3} m^3 of the soil when it is saturated.

3.19   For a given sand, the maximum and minimum void ratios are 0.78 and 0.43, respectively. Given that $G_s = 2.67$, determine the dry unit weight of the soil when the relative density is 65% (in lb/ft^3).

3.20   For a given sandy soil, $e_{max} = 0.75$, $e_{min} = 0.46$, and $G_s = 2.68$. What will be the moist unit weight of compaction (kN/m^3) in the field if $D_r = 78\%$ and $w = 9\%$?

3.21   For a given sandy soil, the maximum and minimum dry unit weights are 108 lb/ft^3 and 92 lb/ft^3, respectively. Given that $G_s = 2.65$, determine the moist unit weight of this soil when the relative density is 60% and the moisture content is 8%.

3.22   A loose, uncompacted sand fill 2 m in depth has a relative density of 40%. Laboratory tests indicated that the minimum and maximum void ratios of the sand are 0.46 and 0.90, respectively. The specific gravity of solids of the sand is 2.65.

a. What is the dry unit weight of the sand?

b. If the sand is compacted to a relative density of 75%, what is the decrease in thickness of the 2 m fill?

3.23   A soil at a constant moisture content shows the following properties when compacted:

Degree of saturation (%)	Dry unit weight (lb/ft^3)
40	92.1
70	113.7

Determine the moisture content of the soil.

3.24   Following are the results from the liquid and plastic limit tests for a soil:
Liquid limit test:

Number of blows, N	Moisture content (%)
15	42
20	40.8
28	39.1

Plastic limit = 17.2%

    a. Draw the flow curve and obtain the liquid limit.

    b. What is the plasticity index of the soil?

3.25 Refer to Problem 3.24 Determine the liquidity index of the soil when the *in situ* moisture content is 30%.

3.26 Repeat Problem 3.24 with the following values:

Number of blows, N	Moisture content (%)
13	33
18	27
29	22

    Plastic limit = 15.5%

3.27 Determine the liquidity index of the soil referred to in Problem 3.26 when the *in situ* moisture content is 14%.

3.28 A saturated soil used to determine the shrinkage limit has initial volume, $V_i = 20.2$ cm^3, final volume, $V_f = 14.3$ cm^3, mass of wet soil, $M_1 = 34$ g, and mass of dry soil, $M_2 = 24$ g. Determine the shrinkage limit.

References

AMERICAN SOCIETY FOR TESTING AND MATERIALS (1999). *Annual Book of ASTM Standards,* Sec. 4, Vol. 04.08, West Conshohocken, Pa.

BS:1377 (1990). *British Standard Methods of Tests for Soil for Engineering Purposes,* Part 2, BSI, London.

BURMISTER, D. M. (1949). "Principles and Techniques of Soil Identification," *Proceedings,* Annual Highway Research Board Meeting, National Research Council, Washington, D.C., Vol. 29, 402–433.

CASAGRANDE, A. (1932). "Research of Atterberg Limits of Soils," *Public Roads,* Vol. 13, No. 8, 121–136.

COLLINS, K., and McGOWN, A. (1974). "The Form and Function of Microfabric Features in a Variety of Natural Soils," *Geotechnique,* Vol. 24, No. 2, 223–254.

LAMBE, T. W. (1958). "The Structure of Compacted Clay," *Journal of the Soil Mechanics and Foundations Division,* ASCE, Vol. 85, No. SM2, 1654-1 to 1654-35.

MITCHELL, J. K. (1976). *Fundamentals of Soil Behavior,* Wiley, New York.

PUSCH, R. (1978). "General Report on Physico-Chemical Processes Which Affect Soil Structure and Vice Versa," *Proceedings,* International Symposium on Soil Structure, Gothenburg, Sweden, Appendix, 33.

SEED, H. B., WOODWARD, R. J., and LUNDGREN, R. (1964a). "Clay Mineralogical Aspects of Atterberg Limits," *Journal of the Soil Mechanics and Foundations Division,* ASCE, Vol. 90, No. SM4, 107–131.

SEED, H. B., WOODWARD, R. J., and LUNDGREN, R. (1964b). "Fundamental Aspects of the Atterberg Limits," *Journal of the Soil Mechanics and Foundations Division,* ASCE, Vol. 90, No. SM6, 75–105.

SKEMPTON, A. W. (1953). "The Colloidal Activity of Clays," *Proceedings,* 3rd International Conference on Soil Mechanics and Foundation Engineering, London, Vol. 1, 57–61.

SRIDHARAN, A., NAGARAJ, H. B., and PRAKASH, K. (1999). "Determination of the Plasticity Index from Flow Index," *Geotechnical Testing Journal,* ASTM, Vol. 22, No. 2, 175–181.

U.S. ARMY CORPS OF ENGINEERS (1949). *Technical Memo 3-286,* U.S. Waterways Experiment Station, Vicksburg, Miss.

WORTH, C. P., and WOOD, D. M. (1978). "The Correlation of Index Properties with Some Basic Engineering Properties of Soils," *Canadian Geotechnical Journal,* Vol. 15, No. 2, 137–145.

YONG, R. N., and SHEERAN, D. E. (1973). "Fabric Unit Interaction and Soil Behavior," *Proceedings,* International Symposium on Soil Structure, Gothenburg, Sweden, 176–183.

4

Engineering Classification of Soil

Different soils with similar properties may be classified into groups and sub-groups according to their engineering behavior. Classification systems provide a common language to concisely express the general characteristics of soils, which are infinitely varied, without detailed descriptions. Currently, two elaborate classification systems are commonly used by soils engineers. Both systems take into consideration the particle-size distribution and Atterberg limits. They are the American Association of State Highway and Transportation Officials (AASHTO) classification system and the Unified Soil Classification System. The AASHTO classification system is used mostly by state and county highway departments. Geotechnical engineers generally prefer the Unified system.

4.1 AASHTO Classification System

The AASHTO system of soil classification was developed in 1929 as the Public Road Administration Classification System. It has undergone several revisions, with the present version proposed by the Committee on Classification of Materials for Sub-grades and Granular Type Roads of the Highway Research Board in 1945 (ASTM designation D-3282; AASHTO method M145).

The AASHTO classification in present use is given in Table 4.1. According to this system, soil is classified into seven major groups: A-1 through A-7. Soils classified under groups A-1, A-2, and A-3 are granular materials of which 35% or less of the particles pass through the No. 200 sieve. Soils of which more than 35% pass through the No. 200 sieve are classified under groups A-4, A-5, A-6, and A-7. These soils are mostly silt and clay-type materials. The classification system is based on the following critiera:

1. *Grain size*
   a. *Gravel:* fraction passing the 75 mm (3-in.) sieve and retained on the No. 10 (2-mm) U.S. sieve
   b. *Sand:* fraction passing the No. 10 (2-mm) U.S. sieve and retained on the No. 200 (0.075-mm) U.S. sieve
   c. *Silt and clay:* fraction passing the No. 200 U.S. sieve

Table 4.1 Classification of Highway Subgrade Materials

General classification	Granular materials (35% or less of total sample passing No. 200)						
	A-1				A-2		
Group classification	A-1-a	A-1-b	A-3	A-2-4	A-2-5	A-2-6	A-2-7
Sieve analysis (percentage passing)							
No. 10	50 max.						
No. 40	30 max.	50 max.	51 min.				
No. 200	15 max.	25 max.	10 max.	35 max.	35 max.	35 max.	35 max.
Characteristics of fraction passing No. 40							
Liquid limit				40 max.	41 min.	40 max.	41 min.
Plasticity index	6 max.		NP	10 max.	10 max.	11 min.	11 min.
Usual types of significant constituent materials	Stone fragments, gravel, and sand		Fine sand	Silty or clayey gravel and sand			
General subgrade rating	Excellent to good						

General classification	Silt-clay materials (more than 35% of total sample passing No. 200)			
Group classification	A-4	A-5	A-7 A-7-5[a] A-6	A-7-6[b]
Sieve analysis (percentage passing)				
No. 10				
No. 40				
No. 200	36 min.	36 min.	36 min.	36 min.
Characteristics of fraction passing No. 40				
Liquid limit	40 max.	41 min.	40 max.	41 min.
Plasticity index	10 max.	10 max.	11 min.	11 min.
Usual types of significant constituent materials	Silty soils		Clayey soils	
General subgrade rating	Fair to poor			

[a]For A-7-5, $PI \leq LL - 30$
[b]For A-7-6, $PI > LL - 30$

2. *Plasticity:* The term *silty* is applied when the fine fractions of the soil have a plasticity index of 10 or less. The term *clayey* is applied when the fine fractions have a plasticity index of 11 or more.
3. If cobbles and *boulders* (size larger than 75 mm) are encountered, they are excluded from the portion of the soil sample from which classification is made. However, the percentage of such material is recorded.

To classify a soil according to Table 4.1, one must apply the test data from left to right. By process of elimination, the first group from the left into which the test data fit is the correct classification. Figure 4.1 shows a plot of the range of the liq-

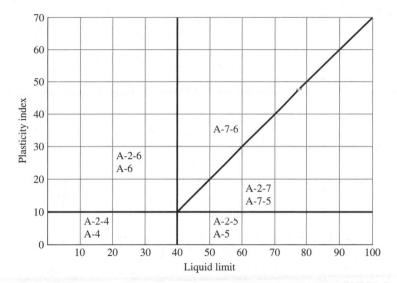

Figure 4.1 Range of liquid limit and plasticity index for soils in groups A-2, A-4, A-5, A-6, and A-7

uid limit and the plasticity index for soils that fall into groups A-2, A-4, A-5, A-6, and A-7.

   To evaluate the quality of a soil as a highway subgrade material, one must also incorporate a number called the *group index* (*GI*) with the groups and subgroups of the soil. This index is written in parentheses after the group or subgroup designation. The group index is given by the equation

$$GI = (F_{200} - 35)[0.2 + 0.005(LL - 40)] + 0.01(F_{200} - 15)(PI - 10) \quad (4.1)$$

where F_{200} = percentage passing through the No. 200 sieve
   LL = liquid limit
   PI = plasticity index

The first term of Eq. (4.1)—that is, $(F_{200} - 35)[0.2 + 0.005(LL - 40)]$—is the partial group index determined from the liquid limit. The second term—that is, $0.01(F_{200} - 15)(PI - 10)$—is the partial group index determined from the plasticity index. Following are some rules for determining the group index:

1. If Eq. (4.1) yields a negative value for *GI,* it is taken as 0.
2. The group index calculated from Eq. (4.1) is rounded off to the nearest whole number (for example, $GI = 3.4$ is rounded off to 3; $GI = 3.5$ is rounded off to 4).
3. There is no upper limit for the group index.
4. The group index of soils belonging to groups A-1-a, A-1-b, A-2-4, A-2-5, and A-3 is always 0.

5. When calculating the group index for soils that belong to groups A-2-6 and A-2-7, use the partial group index for *PI,* or

$$GI = 0.01(F_{200} - 15)(PI - 10) \qquad (4.2)$$

In general, the quality of performance of a soil as a subgrade material is inversely proportional to the group index.

Example 4.1

The results of the particle-size analysis of a soil are as follows:

Percent passing through the No. 10 sieve = 100
Percent passing through the No. 40 sieve = 80
Percent passing through the No. 200 sieve = 58

The liquid limit and plasticity index of the minus No. 40 fraction of the soil are 30 and 10, respectively. Classify the soil by the AASHTO system.

Solution
Using Table 4.1, since 58% of the soil is passing through the No. 200 sieve, it falls under silt-clay classifications — that is, it falls under group A-4, A-5, A-6, or A-7. Proceeding from left to right, it falls under group A-4.
  From Eq. (4.1),

$$GI = (F_{200} - 35)[0.2 + 0.005(LL - 40)] + 0.01(F_{200} - 15)(PI - 10)$$

$$= (58 - 35)[0.2 + 0.005(30 - 40)] + (0.01)(58 - 15)(10 - 10)$$

$$= 3.45 \simeq 3$$

So, the soil will be classified as **A-4(3)**.   ∎

Example 4.2

Ninety-five percent of a soil passes through the No. 200 sieve and has a liquid limit of 60 and plasticity index of 40. Classify the soil by the AASHTO system.

Solution
According to Table 4.1, this soil falls under group A-7. (Proceed in a manner similar to Example 4.1.) Since

$$40 > 60 - 30$$

$$\uparrow \quad \uparrow$$

$$PI \quad LL$$

this is an A-7-6 soil. Hence,

$$GI = (F_{200} - 35)[0.2 + 0.005(LL - 40)] + 0.01(F_{200} - 15)(PI - 10)$$
$$= (95 - 35)[0.2 + 0.005(60 - 40)] + (0.01)(95 - 15)(40 - 10)$$
$$- 42$$

So, the classification is **A-7-6(42)**. ∎

Example 4.3

For a soil, given

Sieve No.	Percent passing
4	90
10	76
200	34

Liquid limit = 37
Plasticity index = 12
Classify the soil by the AASHTO system.

Solution
The percentage passing through the No. 200 sieve is less than 35, so the soil is a granular material. From Table 4.1, we see that it is type A-2-6. From Eq. (4.2),

$$GI = 0.01(F_{200} - 15)(PI - 10)$$

For this soil, $F_{200} = 34$ and $PI = 12$, so

$$GI = 0.01(34 - 15)(12 - 10) = 0.38 \approx 0$$

Thus, the soil is type **A-2-6(0)**. ∎

4.2 Unified Soil Classification System

The original form of the Unified Soil Classification System was proposed by Casagrande in 1942 during World War II for use in airfield construction undertaken by the Army Corps of Engineers. In cooperation with the U.S. Bureau of Reclamation, the Corps revised this system in 1952. At present, it is widely used by engineers (ASTM designation D-2487). In order to use the classification system, the following points must be kept in mind:

1. The classification is based on material passing a 75 mm (3 in.) sieve.
2. Coarse fraction = percent retained above No. 200 sieve = $100 - F_{200} = R_{200}$.

Table 4.2 Unified Classification System (Based on Materials Passing 75 mm (3 in.) Sieve (Based on ASTM-2487)

Major division		Group symbol	Criteria
$F_{200} < 50$	Gravels $\dfrac{R_4}{R_{200}} > 0.5$	GW	$F_{200} < 5$; $C_u \geq 4$; $1 \leq C_z \leq 3$
		GP	$F_{200} < 5$; Not meeting the GW criteria of C_u and C_z
		GM	$F_{200} > 12$; $PI < 4$ or plots *below* A-line (Fig. 4.2)
		GC	$F_{200} > 12$; $PI > 7$ and plots *on or above* A-line (Fig. 4.2)
		GM-GC	$F_{200} > 12$; PI plots in the hatched area (Fig. 4.2)
		GW-GM	$5 \leq F_{200} \leq 12$; satisfies C_u and C_z criteria of GW and meets the PI criteria for GM
		GW-GC	$5 \leq F_{200} \leq 12$; satisfies C_u and C_z criteria of GW and meets the PI criteria for GC
		GP-GM	$5 \leq F_{200} \leq 12$; does not satisfy C_u and C_z criteria of GW and meets the PI criteria for GM
		GP-GC	$5 \leq F_{200} \leq 12$; does not satisfy C_u and C_z criteria of GW and meets the PI criteria for GC
	Sands $\dfrac{R_4}{R_{200}} \leq 0.5$	SW	$F_{200} < 5$; $C_u \geq 6$; $1 \leq C_z \leq 3$
		SP	$F_{200} < 5$; Not meeting the SW criteria of C_u and C_z
		SM	$F_{200} > 12$; $PI < 4$ or plots *below* A-line (Fig. 4.2)
		SC	$F_{200} > 12$; $PI > 7$ and plots *on or above* A-line (Fig. 4.2)
		SM-SC	$F_{200} > 12$; PI plots in the hatched area (Fig. 4.2)
		SW-SM	$5 \leq F_{200} \leq 12$; satisfies C_u and C_z criteria of SW and meets the PI criteria for SM
		SW-SC	$5 \leq F_{200} \leq 12$; satisfies C_u and C_z criteria of SW and meets the PI criteria for SC
		SP-SM	$5 \leq F_{200} \leq 12$; does not satisfy C_u and C_z criteria of SW and meets the PI criteria for SM
		SP-SC	$5 \leq F_{200} \leq 12$; does not satisfy C_u and C_z criteria of SW and meets the PI criteria for SC
$F_{200} \geq 50$	Silts and Clays $LL < 50$	ML	$PI < 4$ or plots *below* A-line (Fig. 4.2)
		CL	$PI > 7$ and plots *on or above* A-line (Fig. 4.2)
		CL-ML	PI plots in the hatched area (Fig. 4.2)
		OL	$\dfrac{LL_{\text{(oven dried)}}}{LL_{\text{(not dried)}}} < 0.75$; PI plots in the OL area in Fig. 4.2
	Silts and Clays $LL \geq 50$	MH	PI plots *below* A-line (Fig. 4.2)
		CH	PI plots *on or above* A-line (Fig. 4.2)
		OH	$\dfrac{LL_{\text{(oven dried)}}}{LL_{\text{(not dried)}}} < 0.75$; PI plots in the OH area in Fig. 4.2
	Highly organic matter	Pt	Peat

Note: C_u = uniformity coefficient = $\dfrac{D_{60}}{D_{10}}$; C_z = coefficient of gradation = $\dfrac{D_{30}^2}{D_{60} \times D_{10}}$

LL = liquid limit on minus 40 sieve fraction

PI = plasticity index on minus 40 sieve fraction

3. Fine fraction = percent passing No. 200 sieve = F_{200}.
4. Gravel fraction = percent retained above No. 4 sieve = R_4.

According to the Unified Soil Classification System, the soils are divided into two major categories:

1. *Coarse-grained soils* that are gravelly and sandy in nature with less than 50% passing through the No. 200 sieve (that is, $F_{200} < 50$). The group symbols start with prefixes of either G or S. G stands for gravel or gravelly soil, and S for sand or sandy soil.
2. *Fine-grained soils* with 50% or more passing through the No. 200 sieve (that is, $F_{200} \geq 50$). The group symbols start with prefixes of M, which stands for inorganic silt, C for inorganic clay, and O for organic silts and clays. The symbol Pt is used for peat, muck, and other highly organic soils.

Other symbols used for the classification are:

- W — well graded
- P — poorly graded
- L — low plasticity (liquid limit less than 50)
- H — high plasticity (liquid limit more than 50)

Table 4.2 gives the details of the soil classification system to determine the *group symbols*.

More recently, ASTM designation D-2487 created an elaborate system to assign *group names* to soils. These names are summarized in Figures 4.3, 4.4, and 4.5. In using these figures, it is important to remember that, in a given soil, percentage of gravel = R_4 and percentage of sand = $R_{200} - R_4$.

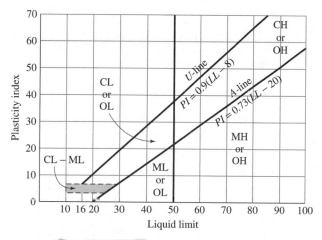

Figure 4.2 Plasticity chart

Group Symbol   **Group Name**

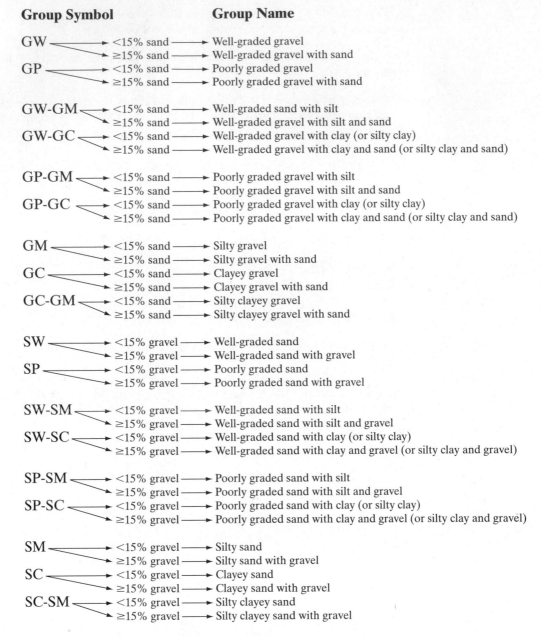

Figure 4.3 Flowchart group names for gravelly and sandy soil. *Source:* From "Annual Book of ASTM Standards, 04.08." Copyright © 1999 American Society for Testing and Materials. Reprinted with permission.

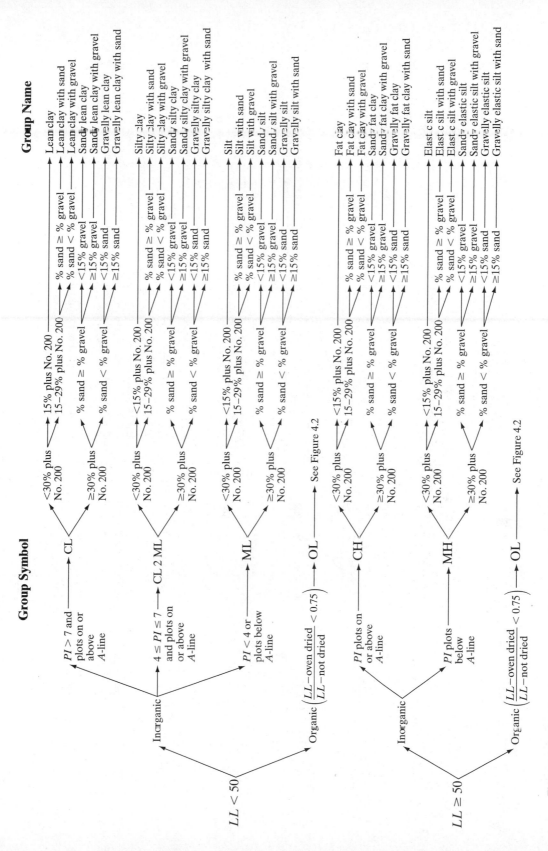

Figure 4.4 Flowchart group names for inorganic silty and clayey soils. *Source*: From "Annual Book of ASTM Standards, 04.08." Copyright © 1999 American Society for Testing and Materials. Reprinted with permission.

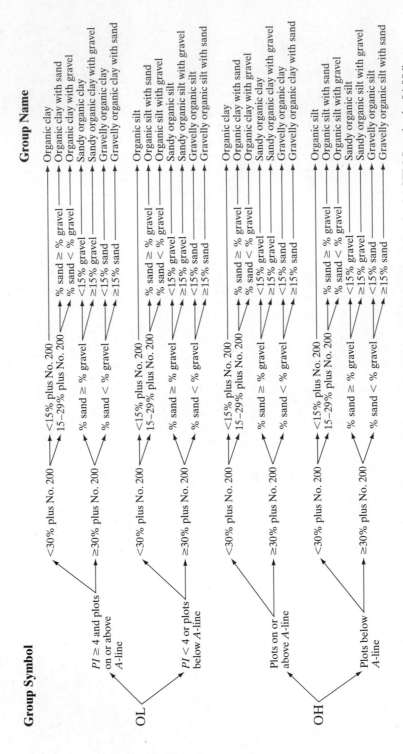

Figure 4.5 Flowchart group names for organic silty and clayey soils. *Source:* From "Annual Book of ASTM Standards, 04.08."
Copyright © 1999 American Society for Testing and Materials. Reprinted with permission.

Example 4.4

Figure 4.6 gives the grain-size distribution of two soils. The liquid and plastic limits of minus No. 40 sieve fraction of the soil are as follows:

	Soil A	Soil B
Liquid limit	30	26
Plastic limit	22	20

Determine the group symbols and group names according to the Unified Soil Classification System.

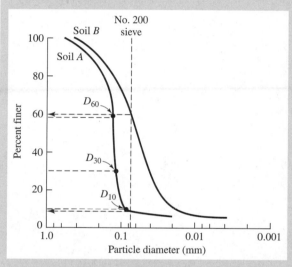

Figure 4.6 Particle-size distribution of two soils

Solution
Soil A
The grain-size distribution curve indicates that $F_{200} = 8$. So this is a coarse-grained soil, and

$$R_{200} = 100 - F_{200} = 100 - 8 = 92$$

From Figure 4.6, $R_4 \approx 0$; therefore,

$$\frac{R_4}{R_{200}} = \frac{0}{92} < 0.5$$

Hence, it is a sandy soil (Table 4.2). From Figure 4.6, $D_{10} = 0.085$ mm, $D_{30} = 0.12$ m, and $D_{60} = 0.135$ mm. Thus,

$$C_u = \frac{D_{60}}{D_{10}} = \frac{0.135}{0.085} = 1.59 < 6$$

$$C_z = \frac{D_{30}^2}{D_{60} \times D_{10}} = \frac{(0.12)^2}{(0.135)(0.085)} = 1.25 > 1$$

With $LL = 30$ and $PI = 30 - 22 = 8$ (which is greater than 7), it plots above the A-line in Figure 4.2. Hence, the group symbol is **SP-SC**.

In order to determine the group name, we refer to Figure 4.3.

$$\text{Percentage of gravel} = R_4 = 0 \, (\text{which is} < 15\%)$$

So, the group name is **poorly graded sand with clay**.

Soil B

From the grain-size distribution curve, $F_{200} = 61$. Hence, this is a fine-grained soil. Given: $LL = 26$ and $PI = 26 - 20 = 6$. In Figure 4.2, the PI plots in the hatched area. So, from Table 4.2, the group symbol is **CL-ML**.

For group name, assuming that the soil is inorganic, we go to Figure 4.4 and obtain

Plus No. 200 sieve = $R_{200} = 100 - F_{200} = 100 - 61 = 39$ (which is greater than 30)

Percentage of gravel = $R_4 = 0$; percentage of sand = $R_{200} - R_4 = 39$

Thus, because the percentage of sand is greater than the percentage of gravel, the soil is **sandy silty clay**. ∎

Example 4.5

The grain-size analysis for a soil is given next:

Sieve no.	% passing
4	94
10	63
20	21
40	10
60	7
100	5
200	3

Given that the soil is nonplastic, classify the soil by using the Unified Soil Classification System.

Solution

$$F_{200} = 3$$

$$R_{200} = 100 - 3 = 97$$

$$R_4 = 100 - F_4 = 100 - 94 = 6$$

$$\frac{R_4}{R_{200}} = \frac{6}{97} < 0.5$$

Thus, this soil is sandy. The grain-size distribution is shown in Figure 4.7. From this figure, we obtain

$$D_{60} = 1.41 \text{ mm} \qquad D_{30} = 0.96 \text{ mm} \qquad D_{10} = 0.41 \text{ mm}$$

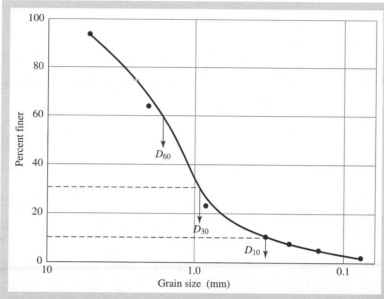

Figure 4.7 Grain-size distribution curve

Thus,

$$C_u = \frac{D_{60}}{D_{10}} = \frac{1.41}{0.41} = 3.44$$

$$C_z = \frac{D_{30}^2}{D_{60} \times D_{10}} = \frac{0.96^2}{1.41 \times 0.41} = 1.59$$

From Table 4.2, we see that the group symbol is **SP**. For this soil, $R_4 = 6$. Referring to Figure 4.3, we find that the group name is **poorly graded sand**. ∎

4.3

Summary and Comparison between the AASHTO and Unified Systems

Both soil classification systems, AASHTO and Unified, are based on the texture and plasticity of soil. Also, both systems divide the soils into two major categories, coarse grained and fine grained, as separated by the No. 200 sieve. According to the AASHTO system, a soil is considered fine grained when more than 35% passes through the No. 200 sieve. According to the Unified system, a soil is considered fine grained when more than 50% passes through the No. 200 sieve. A coarse-grained soil that has about 35% fine grains will behave like a fine-grained material. This is because enough fine grains exist to fill the voids between the coarse grains and hold them apart. In this respect, the AASHTO system appears to be more appropriate. In the AASHTO system, the No. 10 sieve is used to separate gravel from sand; in the Unified system, the No. 4 sieve is used. From the viewpoint of soil-separate size limits, the No. 10 sieve is the more accepted upper limit for sand. This limit is used in concrete and highway base-course technology.

In the Unified system, the gravelly and sandy soils are clearly separated; in the AASHTO system, they are not. The A-2 group, in particular, contains a large variety of soils. Symbols like GW, SM, CH, and others that are used in the Unified system are more descriptive of the soil properties than the A symbols used in the AASHTO system.

The classification of organic soils as OL, OH, and Pt is provided in the Unified system. Under the AASHTO system, there is no place for organic soils. Peats usually have a high moisture content, low specific gravity of soil solids, and low unit weight. Figure 4.8 shows the scanning electron micrographs of four peat samples collected in Wisconsin. Some of the properties of the peats are given in Table 4.3.

Liu (1967) compared the AASHTO and Unified systems. The results of his study are presented in Tables 4.4 and 4.5.

Figure 4.8 Scanning electron micrographs for four peat samples (after Dhowian and Edil, 1980)

Table 4.3 Properties of the Peats Shown in Figure 4.8

Source of peat	Moisture content (%)	Unit weight kN/m³	Unit weight lb/ft³	Specific gravity, G_s	Ash content (%)
Middleton	510	9.1	57.9	1.41	12.0
Waupaca County	460	9.6	61.1	1.68	15.0
Portage	600	9.6	61.1	1.72	19.5
Fond du Lac County	240	10.2	64.9	1.94	39.8

Table 4.4 Comparison of the AASHTO System with the Unified System*

Soil group in AASHTO system	Comparable soil groups in Unified system Most probable	Possible	Possible but improbable
A-1-a	GW, GP	SW, SP	GM, SM
A-1-b	SW, SP, GM, SM	GP	—
A-3	SP	—	SW, GP
A-2-4	GM, SM	GC, SC	GW, GP, SW, SP
A-2-5	GM, SM	—	GW, GP, SW, SP
A-2-6	GC, SC	GM, SM	GW, GP, SW, SP
A-2-7	GM, GC, SM, SC	—	GW, GP, SW, SP
A-4	ML, OL	CL, SM, SC	GM, GC
A-5	OH, MH, ML, OL	—	SM, GM
A-6	CL	ML, OL, SC	GC, GM, SM
A-7-5	OH, MH	ML, OL, CH	GM, SM, GC, SC
A-7-6	CH, CL	ML, OL, SC	OH, MH, GC, GM, SM

*After Liu (1967)

Table 4.5 Comparison of the Unified System with the AASHTO System*

Soil group in Unified system	Comparable soil groups in AASHTO system Most probable	Possible	Possible but improbable
GW	A-1-a	—	A-2-4, A-2-5, A-2-6, A-2-7
GP	A-1-a	A-1-b	A-3, A-2-4, A-2-5, A-2-6, A-2-7
GM	A-1-b, A-2-4, A-2-5, A-2-7	A-2-6	A-4, A-5, A-6, A-7-5, A-7-6, A-1-a
GC	A-2-6, A-2-7	A-2-4	A-4, A-6, A-7-6, A-7-5
SW	A-1-b	A-1-a	A-3, A-2-4, A-2-5, A-2-6, A-2-7
SP	A-3, A-1-b	A-1-a	A-2-4, A-2-5, A-2-6, A-2-7
SM	A-1-b, A-2-4, A-2-5, A-2-7	A-2-6, A-4	A-5, A-6, A-7-5, A-7-6, A-1-a
SC	A-2-6, A-2-7	A-2-4, A-6, A-4, A-7-6	A-7-5
ML	A-4, A-5	A-6, A-7-5, A-7-6	—
CL	A-6, A-7-6	A-4	—
OL	A-4, A-5	A-6, A-7-5, A-7-6	—
MH	A-7-5, A-5	—	A-7-6
CH	A-7-6	A-7-5	—
OH	A-7-5, A-5	—	A-7-6
Pt	—	—	—

*After Liu (1967)

Problems

4.1 The sieve analysis of 10 soils and the liquid and plastic limits of the fraction passing through the No. 40 sieve are given in the following table:

Soil	Sieve analysis, % finer			Liquid limit	Plastic limit
	No. 10	No. 40	No. 200		
1	95	79	53	36	21
2	100	95	78	65	26
3	100	80	62	35	20
4	90	55	45	28	20
5	90	71	60	40	26
6	95	65	32	25	16
7	100	55	8	—	NP
8	96	82	65	40	24
9	85	60	23	20	15
10	100	92	86	70	38

Classify the soils by the AASHTO classification system and give the group indices for each.

4.2 Classify the following soils using the AASHTO classification system, and give the group indices:

Soil	Sieve analysis, % finer			Liquid limit	Plastic limit
	No. 10	No. 40	No. 200		
A	62	30	8	—	NP
B	90	67	35	32	8
C	90	76	34	37	12
D	100	78	8	—	NP
E	85	68	45	38	9

4.3 Classify the following soils using the AASHTO classification system, and give the group indices also:

Sieve size	Percent passing				
	A	B	C	D	E
No. 4	94	98	100	100	100
No. 10	63	86	100	100	100
No. 20	21	50	98	100	100
No. 40	10	28	93	99	94
No. 60	7	18	88	95	82
No. 100	5	14	83	90	66
No. 200	3	10	77	86	45
0.01 mm	—	—	65	42	26
0.002 mm	—	—	60	47	21
Liquid limit	—	—	63	55	36
Plasticity index	NP	NP	25	28	22

4.4 Classify soils 1–6 given in Problem 4.1 using the Unified classification system. Give the group symbols and the group names.

4.5 Classify the soils given in Problem 4.3 using the Unified classification system. Give the group symbols and the group names.

4.6 Classify the following soils using the Unified classification system:

Soil	Sieve analysis, % finer		Liquid limit	Plasticity index
	No. 4	No. 200		
A	80	52	30	8
B	79	45	26	4
C	91	80	60	32
D	95	75	41	12
E	82	41	24	2

Give the group symbols and the group names.

4.7 For an inorganic soil, the following grain-size analysis is given:

U.S. sieve no.	Percent passing
4	100
10	90
20	64
40	38
80	18
200	13

For this soil, $LL = 23$ and $PL = 19$. Classify the soil according to
a. the AASHTO soil classification system;
b. the Unified soil classification system. Give group names and group symbols.

References

AMERICAN ASSOCIATION OF STATE HIGHWAY AND TRANSPORTATION OFFICIALS (1982). *AASHTO Materials, Part I, Specifications,* Washington, D.C.

AMERICAN SOCIETY FOR TESTING AND MATERIALS (1999). *Annual Book of ASTM Standards,* Sec. 4, Vol. 04.08, West Conshohoken, Pa.

CASAGRANDE, A. (1948). "Classification and Identification of Soils," *Transactions,* ASCE, Vol. 113, 901–930.

DHOWIAN, A. W., and EDIL, T. B. (1980). "Consolidation Behavior of Peats," *Geotechnical Testing Journal,* ASTM, Vol. 3, No. 3, 105–114.

LIU, T. K. (1967). "A Review of Engineering Soil Classification Systems," *Highway Research Record No. 156,* National Academy of Sciences, Washington, D.C., 1–22.

5

Soil Compaction

In the construction of highway embankments, earth dams, and many other engineering structures, loose soils must be compacted to increase their unit weights. Compaction increases the strength characteristics of soils, which increase the bearing capacity of foundations constructed over them. Compaction also decreases the amount of undesirable settlement of structures and increases the stability of slopes of embankments. Smooth-wheel rollers, sheepsfoot rollers, rubber-tired rollers, and vibratory rollers are generally used in the field for soil compaction. Vibratory rollers are used mostly for the densification of granular soils. Vibroflot devices are also used for compacting granular soil deposits to a considerable depth. Compaction of soil in this manner is known as *vibroflotation*. This chapter discusses in some detail the principles of soil compaction in the laboratory and in the field.

5.1 Compaction—General Principles

Compaction, in general, is the densification of soil by removal of air, which requires mechanical energy. The degree of compaction of a soil is measured in terms of its dry unit weight. When water is added to the soil during compaction, it acts as a softening agent on the soil particles. The soil particles slip over each other and move into a densely packed position. The dry unit weight after compaction first increases as the moisture content increases. (See Figure 5.1.) Note that at a moisture content $w = 0$, the moist unit weight (γ) is equal to the dry unit weight (γ_d), or

$$\gamma = \gamma_{d(w=0)} = \gamma_1$$

When the moisture content is gradually increased and the same compactive effort is used for compaction, the weight of the soil solids in a unit volume gradually increases. For example, at $w = w_1$,

$$\gamma = \gamma_2$$

However, the dry unit weight at this moisture content is given by

$$\gamma_{d(w=w_1)} = \gamma_{d(w=0)} + \Delta\gamma_d$$

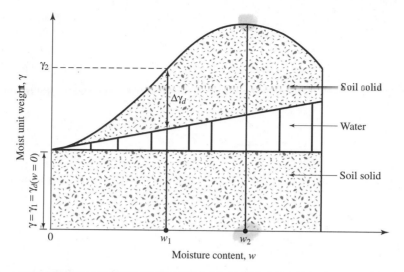

Figure 5.1 Principles of compaction

Beyond a certain moisture content $w = w_2$ (Figure 5.1), any increase in the moisture content tends to reduce the dry unit weight. This phenomenon occurs because the water takes up the spaces that would have been occupied by the solid particles. The moisture content at which the maximum dry unit weight is attained is generally referred to as the *optimum moisture content*.

The laboratory test generally used to obtain the maximum dry unit weight of compaction and the optimum moisture content is called the *Proctor compaction test* (Proctor, 1933). The procedure for conducting this type of test is described in the following section.

5.2   Standard Proctor Test

In the Proctor test, the soil is compacted in a mold that has a volume of 944 cm³ ($\frac{1}{30}$ ft³). The diameter of the mold is 101.6 mm (4 in.). During the laboratory test, the mold is attached to a baseplate at the bottom and to an extension at the top (Figure 5.2a). The soil is mixed with varying amounts of water and then compacted in three equal layers by a hammer (Figure 5.2b) that delivers 25 blows to each layer. The hammer has a mass of 2.5 kg (5.5 lb) and has a drop of 30.5 mm (12 in.). Figure 5.2c is a photograph of the laboratory equipment required for conducting a standard Proctor test.

For each test, the moist unit weight of compaction, γ, can be calculated as

$$\gamma = \frac{W}{V_{(m)}} \tag{5.1}$$

where W = weight of the compacted soil in the mold
$V_{(m)}$ = volume of the mold [944 cm³ ($\frac{1}{30}$ ft³)]

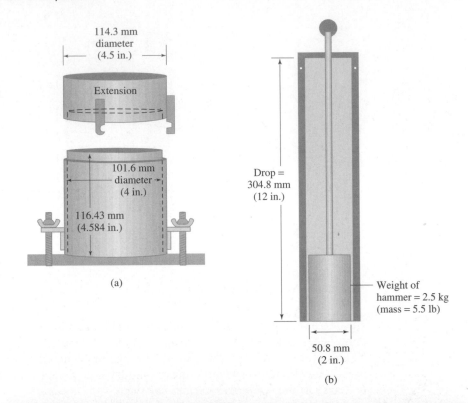

(a)

(b)

(c)

Figure 5.2 Standard Proctor test equipment: (a) mold; (b) hammer (c) photograph of laboratory equipment used for test

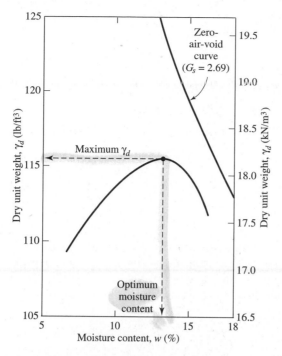

Figure 5.3 Standard Proctor compaction test results for a silty clay

For each test, the moisture content of the compacted soil is determined in the laboratory. With the known moisture content, the dry unit weight can be calculated as

$$\gamma_d = \frac{\gamma}{1 + \dfrac{w\,(\%)}{100}} \qquad (5.2)$$

where $w\,(\%)$ = percentage of moisture content.

The values of γ_d determined from Eq. (5.2) can be plotted against the corresponding moisture contents to obtain the maximum dry unit weight and the optimum moisture content for the soil. Figure 5.3 shows such a plot for a silty-clay soil.

The procedure for the standard Proctor test is elaborated in ASTM Test Designation D-698 (ASTM, 1999) and AASHTO Test Designation T-99 (AASHTO, 1982).

For a given *moisture content w* and *degree of saturation S,* the dry unit weight of compaction can be calculated as follows: From Chapter 3 [Eq. (3.16)], for any soil,

$$\gamma_d = \frac{G_s \gamma_w}{1 + e}$$

where G_s = specific gravity of soil solids

γ_w = unit weight of water

e = void ratio

and, from Eq. (3.18),

$$Se = G_s w$$

or

$$e = \frac{G_s w}{S}$$

Thus,

$$\gamma_d = \frac{G_s \gamma_w}{1 + \dfrac{G_s w}{S}} \tag{5.3}$$

For a given moisture content, the theoretical maximum dry unit weight is obtained when no air is in the void spaces — that is, when the degree of saturation equals 100%. Hence, the maximum dry unit weight at a given moisture content with zero air voids can be obtained by substituting $S = 1$ into Eq. (5.3), or

$$\gamma_{zav} = \frac{G_s \gamma_w}{1 + wG_s} = \frac{\gamma_w}{w + \dfrac{1}{G_s}} \tag{5.4}$$

where γ_{zav} = zero-air-void unit weight.

To obtain the variation of γ_{zav} with moisture content, use the following procedure:

1. Determine the specific gravity of soil solids.
2. Know the unit weight of water (γ_w).
3. Assume several values of w, such as 5%, 10%, 15%, and so on.
4. Use Eq. (5.4) to calculate γ_{zav} for various values of w.

Figure 5.3 also shows the variation of γ_{zav} with moisture content and its relative location with respect to the compaction curve. Under no circumstances should any part of the compaction curve lie to the right of the zero-air-void curve.

5.3   Factors Affecting Compaction

The preceding section showed that moisture content has a strong influence on the degree of compaction achieved by a given soil. Besides moisture content, other important factors that affect compaction are soil type and compaction effort (energy per unit volume). The importance of each of these two factors is described in more detail in the following two sections.

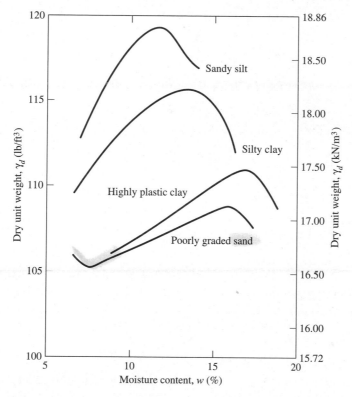

Figure 5.4 Typical compaction curves for four soils (ASTM D-698)

Effect of Soil Type

The soil type — that is, grain-size distribution, shape of the soil grains, specific gravity of soil solids, and amount and type of clay minerals present — has a great influence on the maximum dry unit weight and optimum moisture content. Figure 5.4 shows typical compaction curves obtained from four soils. The laboratory tests were conducted in accordance with ASTM Test Designation D-698.

Note also that the bell-shaped compaction curve shown in Figure 5.3 is typical of most clayey soils. Figure 5.4 shows that for sands, the dry unit weight has a general tendency first to decrease as moisture content increases, and then to increase to a maximum value with further increase of moisture. The initial decrease of dry unit weight with increase of moisture content can be attributed to the capillary tension effect. At lower moisture contents, the capillary tension in the pore water inhibits the tendency of the soil particles to move around and be densely compacted.

Lee and Suedkamp (1972) studied compaction curves for 35 soil samples. They observed that four types of compaction curves can be found. These curves are shown in Figure 5.5. Type A compaction curves are those that have a single peak. This type of curve is generally found for soils that have a liquid limit between 30 and 70. Curve type B is a one-and-one-half-peak curve, and curve type C is a double-peak curve.

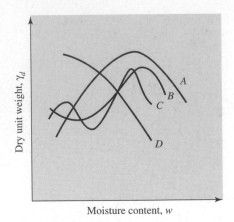

Figure 5.5 Types of compaction curve

Compaction curves of types B and C can be found for soils that have a liquid limit less than about 30. Compaction curves of type D do not have a definite peak. They are termed *odd shaped.* Soils with a liquid limit greater than about 70 may exhibit compaction curves of type C or D. Such soils are uncommon.

Effect of Compaction Effort

The compaction energy per unit volume used for the standard Proctor test described in Section 5.2 can be given as

$$E = \frac{\left(\begin{array}{c}\text{Number} \\ \text{of blows} \\ \text{per layer}\end{array}\right) \times \left(\begin{array}{c}\text{Number} \\ \text{of} \\ \text{layers}\end{array}\right) \times \left(\begin{array}{c}\text{Weight} \\ \text{of} \\ \text{hammer}\end{array}\right) \times \left(\begin{array}{c}\text{Height of} \\ \text{drop of} \\ \text{hammer}\end{array}\right)}{\text{Volume of mold}} \qquad (5.5)$$

or, in SI units,

$$E = \frac{(25)(3)\left(\dfrac{2.5 \times 9.81}{1000}\,\text{kN}\right)(0.305\ \text{m})}{944 \times 10^{-6}\ \text{m}^3} = 594\ \text{kN-m/m}^3 \approx 600\ \text{kN-m/m}^3$$

In English units,

$$E = \frac{(25)(3)(5.5)(1)}{\left(\dfrac{1}{30}\right)} = 12{,}375\ \text{ft-lb/ft}^3 \approx 12{,}400\ \text{ft-lb/ft}^3$$

If the compaction effort per unit volume of soil is changed, the moisture-unit weight curve also changes. This fact can be demonstrated with the aid of Figure 5.6, which shows four compaction curves for a sandy clay. The standard Proctor mold and hammer were used to obtain these compaction curves. The number of layers of soil used for compaction was three for all cases. However, the number of hammer blows per each layer varied from 20 to 50, which varied the energy per unit volume.

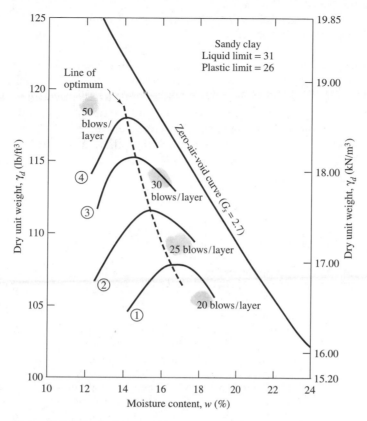

Figure 5.6 Effect of compaction energy on the compaction of a sandy clay

From the preceding observation and Figure 5.6, we can see that

1. As the compaction effort is increased, the maximum dry unit weight of compaction is also increased.
2. As the compaction effort is increased, the optimum moisture content is decreased to some extent.

The preceding statements are true for all soils. Note, however, that the degree of compaction is not directly proportional to the compaction effort.

5.4 *Modified Proctor Test*

With the development of heavy rollers and their use in field compaction, the standard Proctor test was modified to better represent field conditions. This revised version is sometimes referred to as the *modified Proctor test* (ASTM Test Designation D-1557 and AASHTO Test Designation T-180). For conducting the modified Proctor test, the same mold is used with a volume of 944 cm³ (1/30 ft³) as in the case of the standard Proctor test. However, the soil is compacted in five layers by a hammer that has a mass of 4.54 kg (10 lb). The drop of the hammer is 457 mm (18 in.). The number of hammer blows for each layer is kept at 25 as in the case of the standard Proctor test.

The compaction energy for this type of compaction test can be calculated as 2700 kN-m/m^3 (56,000 ft-lb/lb^3).

Because it increases the compactive effort, the modified Proctor test results in an increase in the maximum dry unit weight of the soil. The increase in the maximum dry unit weight is accompanied by a decrease in the optimum moisture content.

In the preceding discussions, the specifications given for Proctor tests adopted by ASTM and AASHTO regarding the volume of the mold and the number of blows are generally those adopted for fine-grained soils that pass through the U.S. No. 4 sieve. However, under each test designation, there are three suggested methods that reflect the mold size, the number of blows per layer, and the maximum particle size in a soil aggregate used for testing. A summary of the test methods is given in Table 5.1.

Table 5.1 Summary of Standard and Modified Proctor Compaction Test Specifications (ASTM D-698 and D-1557)

	Description	Method A	Method B	Method C
Physical Data for the Tests	Material	Passing No. 4 sieve	Passing 9.5 mm ($\frac{3}{8}$ in.) sieve	Passing 19 mm ($\frac{3}{4}$ in.) sieve
	Use	Used if 20% or less by weight of material is retained on No. 4 (4.75 mm) sieve	Used if more than 20% by weight of material is retained on No. 4 (4.75 mm) sieve and 20% or less by weight of material is retained on 9.5 mm ($\frac{3}{8}$ in.) sieve	Used if more than 20% by weight of material is retained on 9.5 mm ($\frac{3}{8}$ in.) sieve and less than 30% by weight of material is retained on 19 mm ($\frac{3}{4}$ in.) sieve
	Mold volume	944 cm^3 ($\frac{1}{30}$ ft^3)	944 cm^3 ($\frac{1}{30}$ ft^3)	944 cm^3 ($\frac{1}{30}$ ft^3)
	Mold diameter	101.6 mm (4 in.)	101.6 mm (4 in.)	101.6 mm (4 in.)
	Mold height	116.4 mm (4.584 in.)	116.4 mm (4.584 in.)	116.4 mm (4.584 in.)
Standard Proctor Test	Weight of hammer	24.4 N (5.5 lb)	24.4 N (5.5 lb)	24.4 N (5.5 lb)
	Height of drop	305 mm (12 in.)	305 mm (12 in.)	305 mm (12 in.)
	Number of soil layers	3	3	3
	Number of blows/layer	25	25	56
Modified Proctor Test	Weight of hammer	44.5 N (10 lb)	44.5 N (10 lb)	44.5 N (10 lb)
	Height of drop	457 mm (18 in.)	457 mm (18 in.)	457 mm (18 in.)
	Number of soil layers	5	5	5
	Number of blows/layer	25	25	56

Example 5.1

For a compacted soil, $G_s = 2.72$, $w = 18\%$, and $\gamma_d = 0.9\gamma_{zav}$. Determine the dry unit weight of the compacted soil.

Solution
From Eq. (5.4),

$$\gamma_{zav} = \frac{\gamma_w}{w + \dfrac{1}{G_s}} = \frac{9.81}{\dfrac{18}{100} + \dfrac{1}{2.72}} = 17.9 \text{ kN/m}^3$$

Hence, for the compacted soil,

$$\gamma_d = 0.9\gamma_{zav} = (0.9)(17.9) = 16.1 \text{ kN/m}^3 \qquad \blacksquare$$

Example 5.2

The laboratory test results of a standard Proctor test are given in the following table:

Volume of mold (ft^3)	Weight of moist soil in mold (lb)	Moisture content, w (%)
$\frac{1}{30}$	3.63	10
$\frac{1}{30}$	3.86	12
$\frac{1}{30}$	4.02	14
$\frac{1}{30}$	3.98	16
$\frac{1}{30}$	3.88	18
$\frac{1}{30}$	3.73	20

Determine the maximum dry unit weight of compaction and the optimum moisture content.

Solution
The following table can be prepared:

Volume of mold, V (ft^3)	Weight of soil, W (lb)	Moist unit weight, γ $(\text{lb/ft}^3)^a$	Moisture content, w (%)	Dry unit weight, γ_d $(\text{lb/ft}^3)^b$
$\frac{1}{30}$	3.63	108.9	10	99.0
$\frac{1}{30}$	3.86	115.8	12	103.4
$\frac{1}{30}$	4.02	120.6	14	105.8
$\frac{1}{30}$	3.98	119.4	16	102.9
$\frac{1}{30}$	3.88	116.4	18	98.6
$\frac{1}{30}$	3.73	111.9	20	93.3

$^a\gamma = W/V$
$^b\gamma_d = \gamma/\{1 + [w\,(\%)/100]\}$

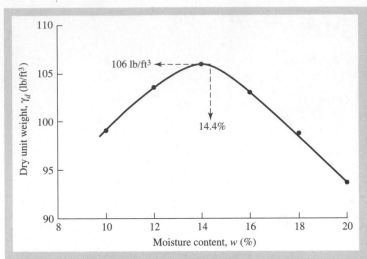

Figure 5.7 Moisture content, w (%)

The plot of γ_d versus w is shown in Figure 5.7. From the plot, we see that the maximum dry unit weight ($\gamma_{d(max)}$) = **106 lb/ft³** and that the optimum moisture content is **14.4%**. ∎

5.5   *Structure of Compacted Clay Soil*

Lambe (1958) studied the effect of compaction on the structure of clay soils, and the results of his study are illustrated in Figure 5.8. If clay is compacted with a moisture content on the dry side of the optimum, as represented by point *A,* it will possess a flocculent structure. This type of structure results because, at low moisture content, the diffuse double layers of ions surrounding the clay particles cannot be fully developed; hence, the interparticle repulsion is reduced. This reduced repulsion results in a more random particle orientation and a lower dry unit weight. When the moisture content of compaction is increased, as shown by point *B,* the diffuse double layers around the particles expand, which increases the repulsion between the clay particles and gives a lower degree of flocculation and a higher dry unit weight. A continued increase in moisture content from *B* to *C* expands the double layers more. This expansion results in a continued increase of repulsion between the particles and thus a still greater degree of particle orientation and a more or less dispersed structure. However, the dry unit weight decreases because the added water dilutes the concentration of soil solids per unit volume.

At a given moisture content, higher compactive effort yields a more parallel orientation to the clay particles, which gives a more dispersed structure. The particles are closer and the soil has a higher unit weight of compaction. This phenomenon can be seen by comparing point *A* with point *E* in Figure 5.8

Figure 5.9 shows the variation in the degree of particle orientation with molding water content for compacted Boston blue clay. Works of Seed and Chan (1959) have shown similar results for compacted kaolin clay.

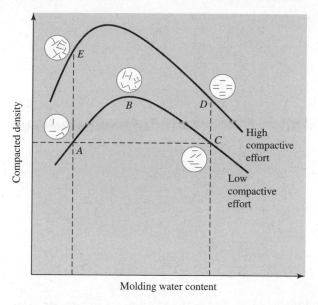

Figure 5.8 Effect of compaction on structure of clay soils (redrawn after Lambe, 1958)

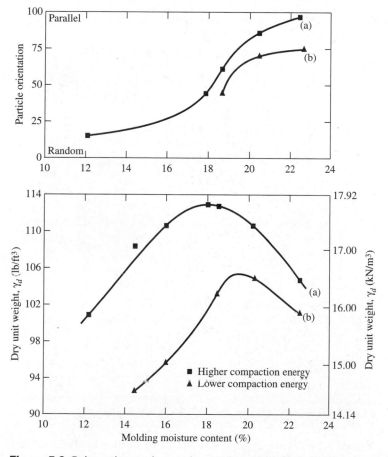

Figure 5.9 Orientation against moisture content for Boston blue clay (after Lambe, 1958)

Figure 5.10 Smooth-wheel roller (courtesy of David A. Carroll, Austin, Texas)

Figure 5.11 Pneumatic rubber-tired roller (courtesy of David A. Carroll, Austin, Texas)

5.6   *Field Compaction*

Compaction Equipment

Most of the compaction in the field is done with rollers. The four most common types of rollers are

1. Smooth-wheel rollers (or smooth-drum rollers)
2. Pneumatic rubber-tired rollers
3. Sheepsfoot rollers
4. Vibratory rollers

Smooth-wheel rollers (Figure 5.10) are suitable for proof rolling subgrades and for finishing operation of fills with sandy and clayey soils. These rollers provide 100% coverage under the wheels, with ground contact pressures as high as 310 to 380 kN/m^2 (45 to 55 lb/in^2). They are not suitable for producing high unit weights of compaction when used on thicker layers.

Pneumatic rubber-tired rollers (Figure 5.11) are better in many respects than the smooth-wheel rollers. The former are heavily loaded with several rows of tires. These tires are closely spaced — four to six in a row. The contact pressure under the tires can range from 600 to 700 kN/m^2 (85 to 100 lb/in^2), and they produce about 70 to 80% coverage. Pneumatic rollers can be used for sandy and clayey soil compaction. Compaction is achieved by a combination of pressure and kneading action.

Sheepsfoot rollers (Figure 5.12) are drums with a large number of projections. The area of each projection may range from 25 to 85 cm^2 ($\approx$ 4 to 13 in^2). These rollers

Figure 5.12 Sheepsfoot roller (courtesy of David A. Carroll, Austin, Texas)

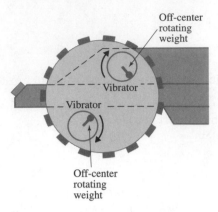

Figure 5.13 Principles of vibratory rollers

are most effective in compacting clayey soils. The contact pressure under the projections can range from 1400 to 7000 kN/m² (200 to 1000 lb/in²). During compaction in the field, the initial passes compact the lower portion of a lift. Compaction at the top and middle of a lift is done at a later stage.

Vibratory rollers are extremely efficient in compacting granular soils. Vibrators can be attached to smooth-wheel, pneumatic rubber-tired, or sheepsfoot rollers to provide vibratory effects to the soil. Figure 5.13 demonstrates the principles of vibratory rollers. The vibration is produced by rotating off-center weights.

Handheld vibrating plates can be used for effective compaction of granular soils over a limited area. Vibrating plates are also gang-mounted on machines. These plates can be used in less restricted areas.

Factors Affecting Field Compaction

In addition to soil type and moisture content, other factors must be considered to achieve the desired unit weight of compaction in the field. These factors include the thickness of lift, the intensity of pressure applied by the compacting equipment, and the area over which the pressure is applied. These factors are important because the pressure applied at the surface decreases with depth, which results in a decrease in the degree of soil compaction. During compaction, the dry unit weight of soil is also affected by the number of roller passes. Figure 5.14 shows the growth curves for a silty clay soil. The dry unit weight of a soil at a given moisture content increases to a certain point with the number of roller passes. Beyond this point, it remains approximately constant. In most cases, about 10 to 15 roller passes yield the maximum dry unit weight economically attainable.

Figure 5.15a shows the variation in the unit weight of compaction with depth for a poorly graded dune sand for which compaction was achieved by a vibratory drum roller. Vibration was produced by mounting an eccentric weight on a single rotating shaft within the drum cylinder. The weight of the roller used for this compaction was 55.6 kN (12.5 kip), and the drum diameter was 1.19 m (47 in). The lifts were kept at 2.44 m (8 ft). Note that, at any given depth, the dry unit weight of compaction increases with the number of roller passes. However, the rate of increase in unit

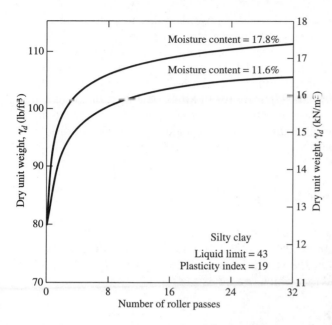

Figure 5.14
Growth curves for a silty clay — relationship between dry unit weight and number of passes of 84.5 kN (19 kip) three-wheel roller when the soil is compacted in 229 mm (9 in) loose layers at different moisture contents (redrawn after Johnson and Sallberg, 1960)

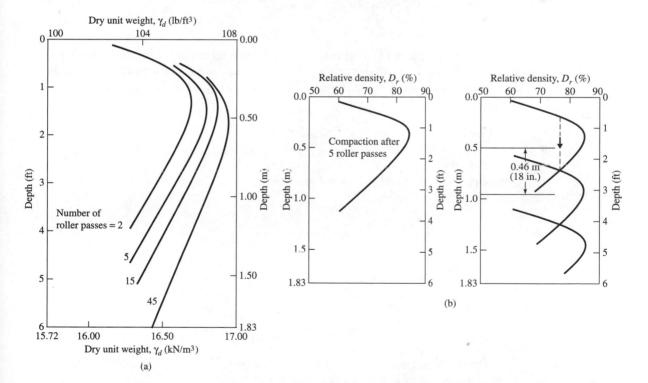

Figure 5.15 (a) Vibratory compaction of a sand — variation of dry unit weight with number of roller passes; thickness of lift = 2.45 m (8 ft); (b) estimation of compaction lift thickness for minimum required relative density of 75% with five roller passes (after D'Appolonia, Whitman, and D'Appolonia, 1969)

weight gradually decreases after about 15 passes. Another fact to note from Figure 5.15a is the variation of dry unit weight with depth for any given number of roller passes. The dry unit weight and hence the relative density, D_r, reach maximum values at a depth of about 0.5 m (1.5 ft) and gradually decrease at lesser depths. This decrease occurs because of the lack of confining pressure toward the surface. Once the relationship between depth and relative density (or dry unit weight) for a given soil with a given number of roller passes is determined, estimating the approximate thickness of each lift is easy. This procedure is shown in Figure 5.15b (D'Appolonia, Whitman, and D'Appolonia, 1969).

5.7   *Specifications for Field Compaction*

In most specifications for earthwork, the contractor is instructed to achieve a compacted field dry unit weight of 90 to 95% of the maximum dry unit weight determined in the laboratory by either the standard or modified Proctor test. This is a specification for relative compaction, which can be expressed as

$$R(\%) = \frac{\gamma_{d(\text{field})}}{\gamma_{d(\text{max}-\text{lab})}} \times 100 \tag{5.6}$$

For the compaction of granular soils, specifications are sometimes written in terms of the required relative density D_r or the required relative compaction. Relative density should not be confused with relative compaction. From Chapter 3, we can write

$$D_r = \left[\frac{\gamma_{d(\text{field})} - \gamma_{d(\text{min})}}{\gamma_{d(\text{max})} - \gamma_{d(\text{min})}} \right] \left[\frac{\gamma_{d(\text{max})}}{\gamma_{d(\text{field})}} \right] \tag{5.7}$$

Comparing Eqs. (5.6) and (5.7), we see that

$$R = \frac{R_0}{1 - D_r(1 - R_0)} \tag{5.8}$$

where

$$R_0 = \frac{\gamma_{d(\text{min})}}{\gamma_{d(\text{max})}} \tag{5.9}$$

On the basis of observation of 47 soil samples, Lee and Singh (1971) devised a correlation between R and D_r for granular soils:

$$R = 80 + 0.2D_r \tag{5.10}$$

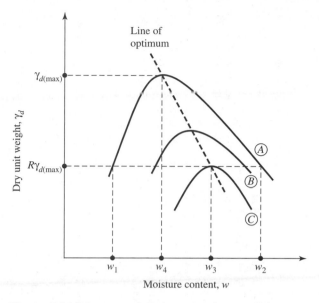

Figure 5.16 Most economical compaction condition

The specification for field compaction based on relative compaction or on relative density is an end-product specification. The contractor is expected to achieve a minimum dry unit weight regardless of the field procedure adopted. The most economical compaction condition can be explained with the aid of Figure 5.16. The compaction curves *A, B,* and *C* are for the same soil with varying compactive effort. Let curve *A* represent the conditions of maximum compactive effort that can be obtained from the existing equipment. Let the contractor be required to achieve a minimum dry unit weight of $\gamma_{d(field)} = R\gamma_{d(max)}$. To achieve this, the contractor must ensure that the moisture content w falls between w_1 and w_2. As can be seen from compaction curve *C*, the required $\gamma_{d(field)}$ can be achieved with a lower compactive effort at a moisture content $w = w_3$. However, for most practical conditions, a compacted field unit weight of $\gamma_{d(field)} = R\gamma_{d(max)}$ cannot be achieved by the minimum compactive effort. Hence, equipment with slightly more than the minimum compactive effort should be used. The compaction curve *B* represents this condition. Now we can see from Figure 5.16 that the most economical moisture content is between w_3 and w_4. Note that $w = w_4$ is the optimum moisture content for curve *A*, which is for the maximum compactive effort.

The concept described in the preceding paragraph, along with Figure 5.16, is historically attributed to Seed (1964), who was a giant in modern geotechnical engineering. This concept is elaborated on in more detail in Holtz and Kovacs (1981).

Table 5.2 gives some of the requirements to achieve 95-to-100% relative compaction (based on standard Proctor maximum dry unit weight) by various field compaction equipment (U.S. Department of Navy, 1971).

Table 5.2 Requirements to Achieve R = 95 to 100% (based on standard Proctor maximum dry unit-weight)*

Equipment type	Applicability	Compacted lift thickness	Passes or coverages	Dimensions and weight of equipment			Possible variations in equipment
				Soil type	*Foot contact area*	*Foot contact pressures*	
Sheepsfoot rollers	For fine-grained soils or dirty coarse-grained soils with more than 20% passing the No. 200 sieve. Not suitable for clean, coarse-grained soils. Particularly appropriate for compaction of impervious zone for earth dam or linings where bonding of lifts is important.	150 mm (6 in.)	4 to 6 passes for fine-grained soil	Fine-grained soil, $PI > 30$	30 to 80 cm² (5 to 12 in²)	1700 to 3400 kN/m² (250 to 500 psi)	For earth dam, highway, and airfield work, a drum of 1.5 m (60-in.) diameter, loaded to 45 to 90 kN per linear meter (1.5 to 3 tons per linear ft) of drum is generally utilized. For smaller projects, a 1 m (40-in.) diameter drum, loaded to 22 to 50 kN per linear meter (0.75 to 1.75 tons per linear ft) of drum, is used. Foot contact pressure should be regulated to avoid shearing the soil on the third or fourth pass.
				Fine-grained soil, $PI < 30$	45 to 90 cm² (7 to 14 in²)	1400 to 2800 kN/m² (200 to 400 psi)	
			6 to 8 passes for coarse-grained soil	Coarse-grained soil	65 to 90 cm² (10 to 14 in²)	1000 to 1800 kN/m² (150 to 250 psi)	
				Efficient compaction of soils on the wet side of the optimum requires less contact pressure than that required by the same soils at lower moisture contents.			
Rubber tired rollers	For clean, coarse-grained soils with 4 to 8% passing the No. 200 sieve.	250 mm (10 in)	3 to 5 coverages	Tire inflation pressures of 400 to 550 kN/m² (60 to 80 psi) for clean granular material or base course and subgrade compaction. Wheel load, 80 to 110 kN (18,000 to 25,000 lb.)			A wide variety of rubber-tired compaction equipment is available. For cohesive soils, light-wheel loads, such as provided by wobble-wheel equipment, may be substituted for heavy-wheel loads if lift thickness is decreased. For cohesionless soils, large-size tires are desirable to avoid shear and rutting.
	For fine-grained soils or well-graded, dirty, coarse-grained soils with more than 8% passing the No. 200 sieve.	150 to 200 mm (6 to 8 in)	4 to 6 coverages	Tire inflation pressures in excess of 450 kN/m² (65 psi) for fine-grained soils of high plasticity. For uniform clean sands or silty fine sands, use larger-size tires with pressures of 280 to 350 kN/m² (40 to 50 psi.)			

Equipment	Applicability	Lift thickness	Coverages	Requirements	Possible variations in equipment
Smooth-wheel rollers	Appropriate for subgrade or base course compaction of well-graded sand-gravel mixtures.	200 to 300 mm (8 to 12 in.)	4 coverages	Tandem-type rollers for base-course or sub-grade compaction, 90 to 135 kN (10 to 15 ton) weight, 53 to 88 kN per linear meter (300 to 500 lb per linear in.) of width of rear roller.	Three-wheel rollers are obtainable in a wide range of sizes Two-wheel tandem rollers are available in the 9 to 80 kN (1- to 20-ton) weight range. Three-axle tandem rollers are generally used in the 90 to 180 kN (10- to 20-ton) weight range. Very heavy rollers are used for proof rolling of subgrade or base course.
	May be used for fine-grained soils other than in earth dams. Not suitable for clean, well-graded sands or silty, uniform sands.	150 to 200 mm (6 to 8 in.)	6 coverages	Three-wheel roller for compaction of fine-grained soil; weights from 45 to 55 kN (5 to 6 tons) for materials of low plasticity to 90 kN (10 tons) for materials of high plasticity.	
Vibrating baseplate compactors	For coarse-grained soils with less than about 12% passing the No. 200 sieve. Best suited for materials with 4 to 8% passing the No. 200 sieve, placed thoroughly wet.	200 to 250 mm (8 to 10 in.)	3 coverages	Single pads or plates should weigh no less than 0.9 kN (200 lb) May be used in tandem where working space is available. For clean, coarse-grained soil, vibration frequency should be no less than 1600 cycles per minute.	Vibrating pads or plates are available, hand-propelled or self-propelled, single or in gangs with width of coverage from 0.45 to 4.5 m ($1\frac{1}{2}$ to 15 ft). Various types of vibrating-drum equipment should be considered for compaction in large areas.
Crawler tractor	Best suited for coarse-grained soils with less than 4 to 8% passing the No. 200 sieve, placed thoroughly wet.	250 to 300 mm (10 to 12 in.)	3 to 4 coverages	No smaller than D8 tractor with blade, 153 kN (34,500 lb) weight, for high compaction.	Tractor weights up to 265 kN (60,000 lb.)
Power tamper or rammer	For difficult access, trench backfill. Suitable for all inorganic soils.	100 to 150 mm (4 to 6 in.) for silt or clay; 150 mm (6 in.) for coarse-grained soils	2 coverages	130 N (30 lb) minimum weight. Considerable range is tolerable, depending on materials and conditions.	Weighs up to 1.1 kN (250 lb); foot diameter, 100 to 250 mm (4 to 10 in.)

* After U.S. Navy (1971). Published by U.S. Government Printing Office

Determination of Field Unit Weight of Compaction

When the compaction work is progressing in the field, knowing whether the specified unit weight has been achieved is useful. The standard procedures for determining the field unit weight of compaction include

1. Sand cone method
2. Rubber balloon method
3. Nuclear method

Following is a brief description of each of these methods.

Sand Cone Method (ASTM Designation D-1556)

The sand cone device consists of a glass or plastic jar with a metal cone attached at its top (Figure 5.17). The jar is filled with uniform dry Ottawa sand. The combined weight of the jar, the cone, and the sand filling the jar is determined (W_1). In the field, a small hole is excavated in the area where the soil has been compacted. If the weight of the moist soil excavated from the hole (W_2) is determined and the moisture content of the excavated soil is known, the dry weight of the soil can be obtained as

$$W_3 = \frac{W_2}{1 + \frac{w\,(\%)}{100}} \tag{5.11}$$

where w = moisture content.

Figure 5.17 Glass jar filled with Ottawa sand with sand cone attached

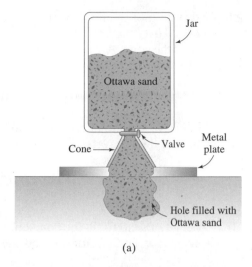

(a)

(b)

Figure 5.18 Field unit weight determined by sand cone method: (a) schematic diagram; (b) a test in progress in the field

After excavation of the hole, the cone with the sand-filled jar attached to it is inverted and placed over the hole (Figure 5.18). Sand is allowed to flow out of the jar to fill the hole and the cone. After that, the combined weight of the jar, the cone, and the remaining sand in the jar is determined (W_4), so

$$W_5 = W_1 - W_4 \tag{5.12}$$

where W_5 = weight of sand to fill the hole and cone.

The volume of the excavated hole can then be determined as

$$V = \frac{W_5 - W_c}{\gamma_{d(\text{sand})}} \tag{5.13}$$

where W_c = weight of sand to fill the cone only
$\gamma_{d(\text{sand})}$ = dry unit weight of Ottawa sand used

The values of W_c and $\gamma_{d(\text{sand})}$ are determined from the calibration done in the laboratory. The dry unit weight of compaction made in the field can then be determined as follows:

$$\gamma_d = \frac{\text{Dry weight of the soil excavated from the hole}}{\text{Volume of the hole}} = \frac{W_3}{V} \tag{5.14}$$

Rubber Balloon Method (ASTM Designation D-2167)

The procedure for the rubber balloon method is similar to that for the sand cone method; a test hole is made and the moist weight of soil removed from the hole and its moisture content are determined. However, the volume of the hole is determined by introducing into it a rubber balloon filled with water from a calibrated vessel, from which the volume can be read directly. The dry unit weight of the compacted soil can be determined by using Eq. (5.14). Figure 5.19 shows a calibrated vessel that would be used with a rubber balloon.

Nuclear Method

Nuclear density meters are often used for determining the compacted dry unit weight of soil. The density meters operate either in drilled holes or from the ground surface. The instrument measures the weight of wet soil per unit volume and the weight of water present in a unit volume of soil. The dry unit weight of compacted soil can be determined by subtracting the weight of water from the moist unit weight of soil. Figure 5.20 shows a photograph of a nuclear density meter.

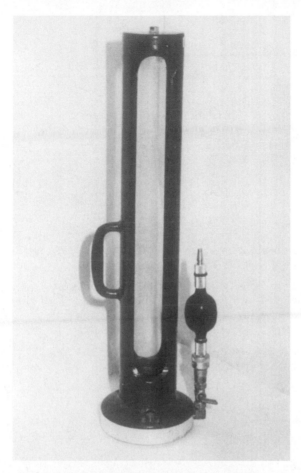

Figure 5.19
Calibrated vessel used with rubber balloon (not shown) (courtesy of John Hester, Carterville, Illinois)

Figure 5.20
Nuclear density meter (courtesy of David A. Carroll, Austin, Texas)

Example 5.3

Laboratory compaction test results for a clayey silt are given in the following table:

Moisture content (%)	Dry unit weight (kN/m³)
6	14.80
8	17.45
9	18.52
11	18.9
12	18.5
14	16.9

Following are the results of a field unit weight determination test performed on the same soil by means of the sand-cone method:

- Calibrated dry density of Ottawa sand = 1570 kg/m³
- Calibrated mass of Ottawa sand to fill the cone = 0.545 kg
- Mass of jar + cone + sand (before use) = 7.59 kg
- Mass of jar + cone + sand (after use) = 4.78 kg
- Mass of moist soil from hole = 3.007 kg
- Moisture content of moist soil = 10.2%

Determine

a. Dry unit weight of compaction in the field
b. Relative compaction in the field

Solution

a. In the field,

Mass of sand used to fill the hole and cone = 7.59 kg − 4.78 kg = 2.81 kg

Mass of sand used to fill the hole = 2.81 kg − 0.545 kg = 2.265 kg

$$\text{Volume of the hole } (V) = \frac{2.265 \text{ kg}}{\text{Dry density of Ottawa sand}}$$

$$= \frac{2.265 \text{ kg}}{1570 \text{ kg/m}^3} = 0.0014426 \text{ m}^3$$

$$\text{Moist density of compacted soil} = \frac{\text{Mass of moist soil}}{\text{Volume of hole}}$$

$$= \frac{3.007}{0.0014426} = 2084.4 \text{ kg/m}^3$$

$$\text{Moist unit weight of compacted soil} = \frac{(2084.4)(9.81)}{1000} = 20.45 \text{ kN/m}^3$$

Hence,

$$\gamma_d = \frac{\gamma}{1 + \dfrac{w\,(\%)}{100}} = \frac{20.45}{1 + \dfrac{10.2}{100}} = \mathbf{18.56 \text{ kN/m}^3}$$

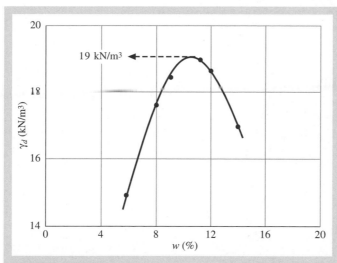

Figure 5.21 Plot of laboratory compaction test results

b. The results of the laboratory compaction test are plotted in Figure 5.21. From the plot, we see that $\gamma_{d(\text{max})} = 19$ kN/m^3. Thus, from Eq. (5.6),

$$R = \frac{\gamma_{d(\text{field})}}{\gamma_{d(\text{max})}} = \frac{18.56}{19.0} = \mathbf{97.7\%}$$

∎

5.9 *Compaction of Organic Soil and Waste Materials*

The presence of organic materials in a soil reduces its strength. In many cases, soils with a high organic content are generally discarded as fill material; however, in certain economic circumstances, slightly organic soils are used for compaction. In fact, organic soils are desirable in many circumstances (e.g., for agriculture, decertification, mitigation, and urban planning). More recently, the high costs of waste disposal have sparked an interest in the possible use of waste materials (e.g., bottom ash obtained from coal burning, copper slag, paper mill sludge, shredded waste tires mixed with inorganic soil, and so forth) in various landfill operations. Such use of waste materials is one of the major thrusts of present-day environmental geotechnology. Following is a discussion of the compaction characteristics of some of these materials.

Organic Soil

Franklin, Orozco, and Semrau (1973) conducted several laboratory tests to observe the effect of organic content on the compaction characteristics of soil. In the test program, various natural soils and soil mixtures were tested. Figure 5.22 shows the effect of organic content on the maximum dry unit weight. When the organic content exceeds 8 to 10%, the maximum dry unit weight of compaction decreases rapidly. Conversely, the optimum moisture content for a given compactive effort increases with an increase in organic content. This trend is shown in Figure 5.23. Likewise, the maximum unconfined compression strength (see Chapter 10) obtained from a compacted soil (with a given compactive effort) decreases with increasing organic content of a soil. From these facts, we can see that soils with organic contents higher than about 10% are undesirable for compaction work.

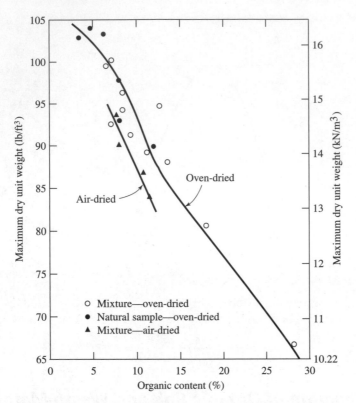

Figure 5.22 Variation of maximum dry unit weight with organic content (after Franklin, Orozco, and Semrau, 1973)

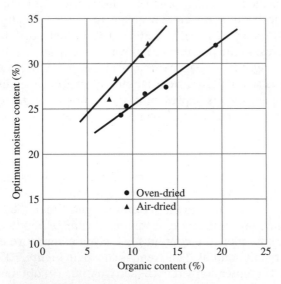

Figure 5.23 Variation of optimum moisture content with organic content (after Franklin, Orozco, and Semrau, 1973)

Soil and Organic Material Mixtures

Lancaster et al. (1996) conducted several modified Proctor tests to determine the effect of organic content on the maximum dry unit weight and optimum moisture content of soil and organic material mixtures. The soils tested consisted of a poorly graded sandy soil (SP-SM) mixed with either shredded redwood bark, shredded rice hulls, or municipal sewage sludge. Figures 5.24 and 5.25 show the variations of

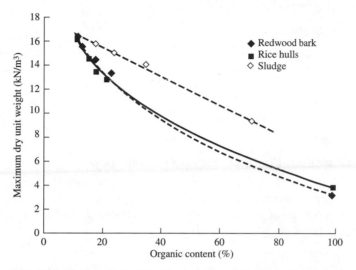

Figure 5.24 Variation of maximum dry unit weight of compaction with organic content — soil and organic material mixtures. *Source:* After "The Effect of Organic Content on Soil Compaction," by J. Lancaster, R. Waco, J. Towle, and R. Chaney, 1996. In *Proceedings, Third International Symposium on Environmental Geotechnology,* p. 159. Used with permission of the author.

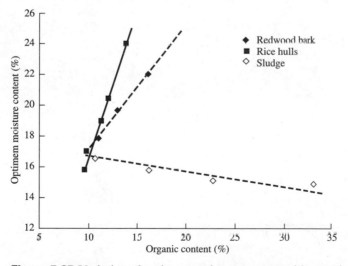

Figure 5.25 Variation of optimum moisture content with organic content — soil and organic material mixtures. *Source:* After "The Effect of Organic Content on Soil Compaction," by J. Lancaster, R. Waco, J. Towle, and R. Chaney, 1996. In *Proceedings, Third International Symposium on Environmental Geotechnology,* p. 159. Used with permission of the author.

maximum dry unit weight of compaction and optimum moisture content, respectively, with organic content. As in Figure 5.22, the maximum dry unit weight decreased with organic content in all cases (see Figure 5.24). Conversely, the optimum moisture content increased with organic content for soil mixed with shredded redwood or rice hulls (see Figure 5.25), similar to the pattern shown in Figure 5.23. However, for soil and municipal sewage sludge mixtures, the optimum moisture content remained practically constant (see Figure 5.25).

Paper Mill Sludge

Paper mill sludge, despite a high water content and low solid contents, can be compacted and used for landfill. The states of Wisconsin and Massachusetts have both used paper mill sludge to cap landfills. Moo-Young and Zimmie (1996) provided the standard Proctor compaction characteristics for several paper mill sludges, and these are shown in Figure 5.26. The physical properties of these sludges are shown in Table 5.3.

Bottom Ash from Coal Burning and Copper Slag

Laboratory standard Proctor test results for bottom ash from coal-burning power plants and for copper slag are also available in the literature. These waste products have been shown to be environmentally safe for use as landfill. A summary of some of these test results is given in Table 5.4.

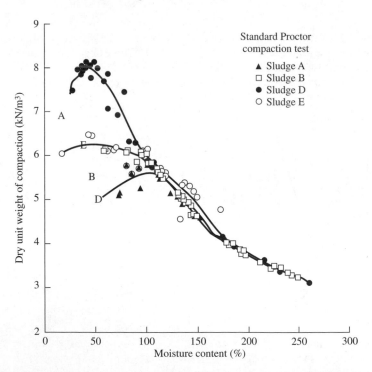

Figure 5.26 Variation of dry unit weight of compaction with moisture content for paper mill sludge. *Source:* From "Geotechnical Properties of Paper Mill Sludges for Use in Landfill Covers," by H. K. Moo-Young, T. F. Zimmie, 1996, *Journal of Geotechnical Engineering, 122* (9), p. 768–775. Copyright © 1996 American Society of Civil Engineers. Used by permission.

Table 5.3 Physical Properties of Sludges Shown in Figure 5.26

Sludge	Moisture content (%)	Organic content (%)	Specific gravity of solids, G_s	Plasticity index
A	150–250	45–50	1.88–1.96	191
B	200–250	56	1.83–1.85	15
D	150–200	44	1.93–1.95	117.5
E	150–200	35–44	1.96–2.08	—

Table 5.4 Standard Proctor Test Results of Bottom Ash and Copper Slag

Type	Location	Maximum dry unit weight kN/m^3	Maximum dry unit weight lb/ft^3	Optimum moisture content (%)	Source
Bottom ash — bituminous coal (West Virginia)	Fort Martin	13.4	85	24.8	Seals, Moulton, and Ruth (1972)
	Kammer	16.0	102	13.8	
	Kanawha River	11.4	72.6	26.2	
	Mitchell	18.3	116.6	14.6	
	Muskingham	14.3	91.1	22.0	
	Willow Island	14.5	92.4	21.2	
Bottom ash — lignite coal	Big Stone Power Plant, South Dakota	16.4	104.4	20.5	Das, Selim, and Pfeifle (1978)
Copper slag	American Smelter and Refinery Company, El Paso, Texas	19.8	126	18.8	Das, Tarquin, and Jones (1983)

5.10   Special Compaction Techniques

Several special types of compaction techniques have been developed for deep compaction of in-place soils, and these techniques are used in the field for large-scale compaction works. Among these, the popular methods are vibroflotation, dynamic compaction, and blasting. Details of these methods are provided in the following sections.

Vibroflotation

Vibroflotation is a technique for *in situ* densification of thick layers of loose granular soil deposits. It was developed in Germany in the 1930s. The first vibroflotation device was used in the United States about 10 years later. The process involves the use of a *Vibroflot* 5.27 (also called the *vibrating unit*), which is about 2.1 m ($\approx$7 ft) long. (as shown in Figure 5.27.) This vibrating unit has an eccentric weight inside it and can develop a centrifugal force, which enables the vibrating unit to vibrate horizontally. There are openings at the bottom and top of the vibrating unit for water jets. The vibrating unit is attached to a follow-up pipe. Figure 5.27 shows the entire assembly of equipment necessary for conducting the field compaction.

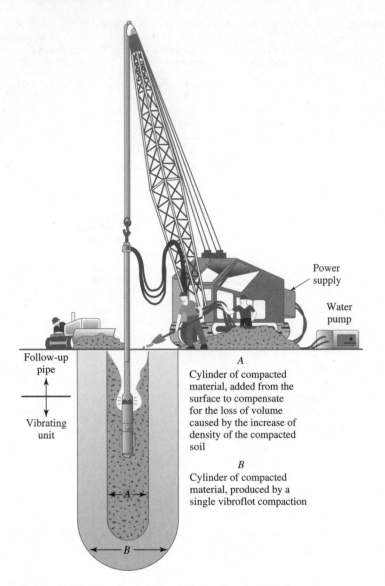

Figure 5.27 Vibroflotation unit (after Brown, 1977)

The entire vibroflotation compaction process in the field can be divided into four stages (Figure 5.28):

Stage 1: The jet at the bottom of the Vibroflot is turned on and lowered into the ground.

Stage 2: The water jet creates a quick condition in the soil and it allows the vibrating unit to sink into the ground.

Stage 3: Granular material is poured from the top of the hole. The water from the lower jet is transferred to the jet at the top of the vibrating unit. This water carries the granular material down the hole.

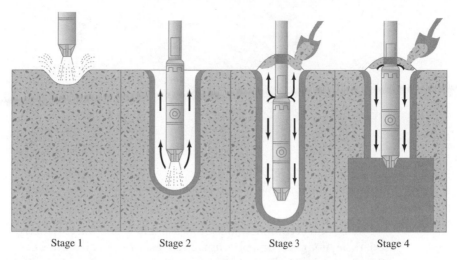

Stage 1   Stage 2   Stage 3   Stage 4

Figure 5.28 Compaction by vibroflotation process (after Brown, 1977)

Table 5.5 Types of Vibroflot Units*

Motor type	75 kW electric and hydraulic	23 kW electric
a. Vibrating tip		
Length	2.1 m (7.0 ft)	1.86 m (6.11 ft)
Diameter	406 mm (16 in.)	381 mm (15 in)
Weight	17.8 kN (4000 lb)	17.8 kN (4000 lb)
Maximum movement when full	12.5 mm (0.49 in)	7.6 mm (0.3 in.)
Centrifugal force	160 kN (18 ton)	89 kN (10 ton)
b. Eccentric		
Weight	1.2 kN (260 lb)	0.76 kN (170 lb)
Offset	38 mm (1.5 in)	32 mm (1.25 in)
Length	610 mm (24 in)	390 mm (15.25 in.)
Speed	1800 rpm	1800 rpm
c. Pump		
Operating flow rate	0–1.6 m³/min (0–400 gal/min)	0–0.6 m³/min (0–150 gal/min)
Pressure	700–1050 kN/m² (100–150 lb/in²)	700–1050 kN/m² (100 150 lb/in²)
d. Lower follow-up pipe and extensions		
Diameter	305 mm (12 in.)	305 mm (12 in.)
Weight	3.65 kN/m (250 lb/ft)	3.65 kN/m (250 lb/ft)

*After Brown (1977)

Stage 4: The vibrating unit is gradually raised in about 0.3 m ($\approx$1 ft) lifts and held vibrating for about 30 seconds at each lift. This process compacts the soil to the desired unit weight.

The details of various types of Vibroflot units used in the United States are given in Table 5.5. Note that 23 kW (30-hp) electric units have been used since the latter part of the 1940s. The 75 kW (100-hp) units were introduced in the early 1970s.

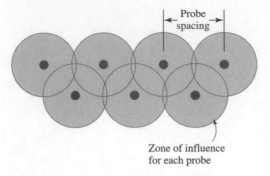

Figure 5.29 Probe spacing for vibroflotation

The zone of compaction around a single probe varies with the type of Vibroflot used. the cylindrical zone of compaction has a radius of about 2 m (≈6 ft) for a 23 kW (30-hp) unit. This radius can extend to about 3 m (≈10 ft) for a 75 kW (100-hp) unit.

Compaction by vibroflotation is done in various probe spacings, depending on the zone of compaction. This spacing is shown in Figure 5.29. The capacity for successful densification of *in situ* soil depends on several factors, the most important of which is the grain-size distribution of the soil and the type of backfill used to fill the holes during the withdrawal period of the Vibroflot. The range of the grain-size distribution of *in situ* soil marked Zone 1 in Figure 5.30 is most suitable for compaction by vibroflotation. Soils that contain excessive amounts of fine sand and silt-size particles are difficult to compact, and considerable effort is needed to reach the proper relative density of compaction. Zone 2 in Figure 5.30 is the approximate lower limit of grain-size distribution for which compaction by vibroflotation is effective. Soil deposits whose grain-size distributions fall in Zone 3 contain appreciable amounts of gravel. For these soils, the rate of probe penetration may be slow and may prove uneconomical in the long run.

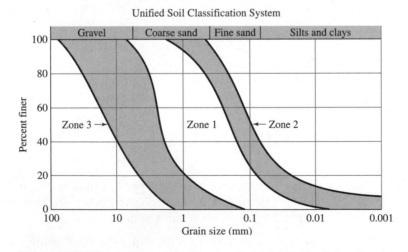

Figure 5.30 Effective range of grain-size distribution of soil for vibroflotation

The grain-size distribution of the backfill material is an important factor that controls the rate of densification. Brown (1977) has defined a quantity called the *suitability number* for rating backfill as

$$S_N = 1.7\sqrt{\frac{3}{(D_{50})^2} + \frac{1}{(D_{20})^2} + \frac{1}{(D_{10})^2}} \qquad (5.15)$$

where D_{50}, D_{20}, and D_{10} are the diameters (in mm) through which, respectively, 50, 20, and 10% of the material passes.

The smaller the value of S_N, the more desirable the backfill material. Following is a backfill rating system proposed by Brown:

Range of S_N	Rating as backfill
0–10	Excellent
10–20	Good
20–30	Fair
30–50	Poor
>50	Unsuitable

Dynamic Compaction

Dynamic compaction is a technique that has gained popularity in the United States for the densification of granular soil deposits. This process consists primarily of dropping a heavy weight repeatedly on the ground at regular intervals. The weight of the hammer used varies over a range of 80 to 360 kN (18 to 80 kip), and the height of the hammer drop varies between 7.5 and 30.5 m (2.5 and 100 ft). The stress waves generated by the hammer drops aid in the densification. The degree of compaction achieved at a given site depends on the following three factors:

1. Weight of hammer
2. Height of hammer drop
3. Spacing of locations at which the hammer is dropped

Leonards, Cutter, and Holtz (1980) suggested that the significant depth of influence for compaction can be approximated by using the equation

$$D \simeq \left(\tfrac{1}{2}\right)\sqrt{W_H h} \qquad (5.16)$$

where D = significant depth of densification (m)
W_H = dropping weight (metric ton)
h = height of drop (m)

In English units, the preceding equation takes the form

$$D = 0.61\sqrt{W_H h} \qquad (5.17)$$

where the units of D and h are ft, and the unit of W_H is kip.

Blasting

Blasting is a technique that has been used successfully in many projects (Mitchell, 1970) for the densification of granular soils. The general soil grain sizes suitable for compaction by blasting are the same as those for compaction by vibroflotation. The process involves the detonation of explosive charges such as 60% dynamite at a certain depth below the ground surface in saturated soil. The lateral spacing of the charges varies from about 3 to 10 m (10 to 30 ft). Three to five successful detonations are usually necessary to achieve the desired compaction. Compaction up to a relative density of about 80% and up to a depth of about 20 m (60 ft) over a large area can easily be achieved by using this process. Usually, the explosive charges are placed at a depth of about two-thirds of the thickness of the soil layer desired to be compacted.

Example 5.4

Following are the details for the backfill material used in a vibroflotation project:

- $D_{10} = 0.36$ mm
- $D_{20} = 0.52$ mm
- $D_{50} = 1.42$ mm

Determine the suitability number S_N. What would be its rating as a backfill material?

Solution
From Eq. (5.15),

$$S_N = 1.7\sqrt{\frac{3}{(D_{50})^2} + \frac{1}{(D_{20})^2} + \frac{1}{(D_{10})^2}}$$

$$= 1.7\sqrt{\frac{3}{(1.42)^2} + \frac{1}{(0.52)^2} + \frac{1}{(0.36)^2}}$$

$$= 6.1$$

Rating: **Excellent**   ∎

Example 5.5

For a dynamic compaction test we are given the following: weight of hammer = 15 metric tons and height of drop = 12 m. Determine the significant depth D of influence for compaction, in meters.

Solution
From Eq. (5.16),

$$D = \left(\tfrac{1}{2}\right)\sqrt{W_H h} = \left(\tfrac{1}{2}\right)\sqrt{(15)(12)} = \mathbf{6.71\,m}$$ ∎

5.11   *Summary and General Comments*

Laboratory standard and modified Proctor compaction tests described in this chapter are essentially for *impact* or *dynamic* compaction of soil; however, in the laboratory, *static compaction* and *kneading compaction* can also be used. It is important to realize that the compaction of clayey soils achieved by rollers in the field is essentially the kneading type. The relationships of dry unit weight (γ_d) and moisture content (w) obtained by dynamic and kneading compaction are not the same. Proctor compaction test results obtained in the laboratory are used primarily to determine whether the roller compaction in the field is sufficient. The structures of compacted cohesive soil at a similar dry unit weight obtained by dynamic and kneading compaction may be different. This difference, in turn, affects physical properties such as hydraulic conductivity, compressibility, and strength.

For most fill operations, the final selection of the borrow site depends on such factors as the soil type and the cost of excavation and hauling.

Fill materials for compaction are generally brought to the site by trucks and wagons. The fill material may be *end-dumped, side-dumped,* or *bottom-dumped* at the site in piles. If the material is too wet, it may be cut and turned to aerate and dry before being spread in lifts for compaction. If it is too dry, the desired amount of water is added by sprinkling irrigation.

Problems

5.1   Given $G_s = 2.72$, calculate the zero-air-void unit weight for a soil in lb/ft^3 at $w = 5\%, 8\%, 10\%, 12\%$, and 15%.

5.2   Repeat Problem 5.1 with $G_s = 2.62$. Plot a graph of γ_{max} (kN/m^3) against w.

5.3   Calculate the variation of dry unit weight (kN/m^3) of a soil ($G_s = 2.65$) at $w = 10\%$ and 20% for degree of saturation (S) = $80\%, 90\%$, and 100%.

5.4   The results of a standard Proctor test are given below. Determine the maximum dry unit weight of compaction and the optimum moisture content.

Volume of Proctor mold (ft^3)	Weight of wet soil in the mold (lb)	Moisture content (%)
1/30	3.26	8.4
1/30	4.15	10.2
1/30	4.67	12.3
1/30	4.02	14.6
1/30	3.63	16.8

5.5   For the soil described in Problem 5.4, if $G_s = 2.72$, determine the void ratio and the degree of saturation at optimum moisture content.

5.6   The results of a standard Proctor test are given in the following table. Determine the maximum dry unit weight of compaction and the optimum moisture content. Also, determine the moisture content required to achieve 95% of $\gamma_{d(\max)}$.

Volume of Proctor mold (cm³)	Mass of wet soil in the mold (kg)	Moisture content (%)
943.3	1.68	9.9
943.3	1.71	10.6
943.3	1.77	12.1
943.3	1.83	13.8
943.3	1.86	15.1
943.3	1.88	17.4
943.3	1.87	19.4
943.3	1.85	21.2

5.7   A field unit weight determination test for the soil described in Problem 5.6 yielded the following data: moisture content = 10.2% and moist unit weight = 16.7 kN/m³. Determine the relative compaction.

5.8   The *in situ* moisture content of a soil is 18% and the moist unit weight is 105 lb/ft³. The specific gravity of soil solids is 2.75. This soil is to be excavated and transported to a construction site for use in a compacted fill. If the specifications call for the soil to be compacted to a minimum dry unit weight of 103.5 lb/ft³ at the same moisture content of 18%, how many cubic yards of soil from the excavation site are needed to produce 10,000 yd³ of compacted fill? How many 20-ton truckloads are needed to transport the excavated soil?

5.9   A proposed embankment fill requires 5000 m³ of compacted soil. The void ratio of the compacted fill is specified as 0.7. Four borrow pits are available as described in the following table, which lists the respective void ratios of the soil and the cost per cubic meter for moving the soil to the proposed construction site. Make the necessary calculations to select the pit from which the soil should be bought to minimize the cost. Assume G_s to be the same at all pits.

Borrow pit	Void ratio	Cost ($/m³)
A	0.85	9
B	1.2	6
C	0.95	7
D	0.75	10

5.10   The maximum and minimum dry unit weights of a sand were determined in the laboratory to be 104 lb/ft³ and 93 lb/ft³, respectively. What would be the relative compaction in the field if the relative density is 78%?

5.11   The maximum and minimum dry unit weights of a sand were determined in the laboratory to be 16.5 kN/m³ and 14.6 kN/m³, respectively. In the field, if the relative density of compaction of the same sand is 70%, what are its relative compaction (%) and dry unit weight (kN/m³)?

5.12 The relative compaction of a sand in the field is 94%. The maximum and minimum dry unit weights of the sand are 103 lb/ft^3 and 95 lb/ft^3, respectively. For the field condition, determine

a. Dry unit weight

b. Relative density of compaction

c. Moist unit weight at a moisture content of 10%

5.13 Laboratory compaction test results on a clayey silt are given in the following table:

Moisture content (%)	Dry unit weight (kN/m³)
6	14.80
8	17.45
9	18.52
11	18.9
12	18.6
14	16.9

Following are the results of a field unit weight determination test on the same soil with the sand cone method:

- Calibrated dry density of Ottawa sand = 1667 kg/m^3
- Calibrated mass of Ottawa sand to fill the cone = 0.117 kg
- Mass of jar + cone + sand (before use) = 5.99 kg
- Mass of jar + cone + sand (after use) = 2.81 kg
- Mass of moist soil from hole = 3.331 kg
- Moisture content of moist soil = 11.6%

Determine

a. Dry unit weight of compaction in the field

b. Relative compaction in the field

5.14 The backfill material for a vibroflotation project has the following grain sizes:

- D_{10} = 0.11 mm
- D_{20} = 0.19 mm
- D_{50} = 1.3 mm

Determine the suitability number, S_N, for each

5.15 Repeat Problem 5.14 using the following values:

D_{10} = 0.09 mm

D_{20} = 0.25 mm

D_{50} = 0.61 mm

References

AMERICAN ASSOCIATION OF STATE HIGHWAY AND TRANSPORTATION OFFICIALS (1982). *AASHTO Materials, Part II,* Washington, D.C.

AMERICAN SOCIETY FOR TESTING AND MATERIALS (1999). *ASTM Standards,* Vol 04.08, West Conshohocken, Pa.

BROWN, E. (1977). "Vibroflotation Compaction of Cohesionless Soils," *Journal of the Geotechnical Engineering Division,* ASCE, Vol. 103, No. GT12, 1437–1451.

D'APPOLONIA, D. J., WHITMAN, R. V., and D'APPOLONIA, E. D. (1969). "Sand Compaction with Vibratory Rollers," *Journal of the Soil Mechanics and Foundations Division,* ASCE, Vol. 95, No. SM1, 263–284.

DAS, B. M., SELIM, A. A., and PFEIFLE, T. W. (1978). "Effective Use of Bottom Ash as a Geotechnical Material," *Proceedings,* 5th Annual UMR-DNR Conference and Exposition on Energy, University of Missouri, Rolla, 342–348.

DAS, B. M., TARQUIN, A. J., and JONES, A. D. (1983). "Geotechnical Properties of a Copper Slag," *Transportation Research Record No. 941,* National Research Council, Washington, D.C., 1–4.

FRANKLIN, A. F., OROZCO, L. F., and SEMRAU, R. (1973). "Compaction of Slightly Organic Soils," *Journal of the Soil Mechanics and Foundations Division,* ASCE, Vol. 99, No. SM7, 541–557.

HOLTZ, R. D., and KOVACS, W. D. (1981). *An Introduction to Geotechnical Engineering,* Prentice-Hall, Englewood Cliffs, N.J.

JOHNSON, A. W., and SALLBERG, J. R. (1960). "Factors That Influence Field Compaction of Soil," Highway Research Board, *Bulletin No. 272.*

LAMBE, T. W. (1958). "The Structure of Compacted Clay," *Journal of the Soil Mechanics and Foundations Division,* ASCE, Vol. 84, No. SM2, 1654–1 to 1654–34.

LANCASTER, J., WACO, R., TOWLE, J., and CHANEY, R. (1996). "The Effect of Organic Content on Soil Compaction," *Proceedings,* 3rd International Symposium on Environmental Geotechnology, San Diego, 152–161.

LEE, K. W., and SINGH, A. (1971). "Relative Density and Relative Compaction," *Journal of the Soil Mechanics and Foundations Division,* ASCE, Vol. 97, No. SM7, 1049–1052.

LEE, P. Y., and SUEDKAMP, R. J. (1972). "Characteristics of Irregularly Shaped Compaction Curves of Soils," *Highway Research Record No. 381,* National Academy of Sciences, Washington, D.C., 1–9.

LEONARDS, G. A., CUTTER, W. A., and HOLTZ, R. D. (1980). "Dynamic Compaction of Granular Soils," *Journal of the Geotechnical Engineering Division,* ASCE, Vol. 106, No. GT1, 35–44.

MITCHELL, J. K. (1970). "In-Place Treatment of Foundation Soils," *Journal of the Soil Mechanics and Foundations Division,* ASCE, Vol. 96, No. SM1, 73–110.

MOO-YOUNG, H. K., and ZIMMIE, T. F. (1996). "Geotechnical Properties of Paper Mill Sludges for Use in Landfill Covers," *Journal of Geotechnical Engineering,* ASCE, Vol. 122, No. 9, 768–775.

PROCTOR, R. R. (1933). "Design and Construction of Rolled Earth Dams," *Engineering News Record,* Vol. 3, 245–248, 286–289, 348–351, 372–376.

SEALS, R. K. MOULTON, L. K., and RUTH, E. (1972). "Bottom Ash: An Engineering Material," *Journal of the Soil Mechanics and Foundations Division,* ASCE, Vol. 98, No. SM4, 311–325.

SEED, H. B. (1964). Lecture Notes, CE 271, Seepage and Earth Dam Design, University of California, Berkeley.

U.S. DEPARTMENT OF NAVY (1971). "Design Manual — Soil Mechanics, Foundations, and Structures," *NAVFAC DM-7,* U.S. Government Printing Office, Washington, D.C.

6

Permeability

Soils are permeable due to the existence of interconnected voids through which water can flow from points of high energy to points of low energy. The study of the flow of water through permeable soil media is important in soil mechanics. It is necessary for estimating the quantity of underground seepage under various hydraulic conditions, for investigating problems involving the pumping of water for underground construction, and for making stability analyses of earth dams and earth-retaining structures that are subject to seepage forces.

6.1 Bernoulli's Equation

From fluid mechanics, we know that, according to Bernoulli's equation, the total head at a point in water under motion can be given by the sum of the pressure, velocity, and elevation heads, or

$$h = \underbrace{\frac{u}{\gamma_w}}_{\substack{\uparrow \\ \text{Pressure} \\ \text{head}}} + \underbrace{\frac{v^2}{2g}}_{\substack{\uparrow \\ \text{Velocity} \\ \text{head}}} + \underbrace{Z}_{\substack{\uparrow \\ \text{Elevation} \\ \text{head}}} \qquad (6.1)$$

where h = total head
u = pressure
v = velocity
g = acceleration due to gravity
γ_w = unit weight of water

Note that the elevation head, Z, is the vertical distance of a given point above or below a datum plane. The pressure head is the water pressure, u, at that point divided by the unit weight of water, γ_w.

If Bernoulli's equation is applied to the flow of water through a porous soil

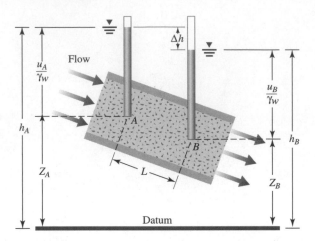

Figure 6.1 Pressure, elevation, and total heads for flow of water through soil

medium, the term containing the velocity head can be neglected because the seepage velocity is small, and the total head at any point can be adequately represented by

$$h = \frac{u}{\gamma_w} + Z \tag{6.2}$$

Figure 6.1 shows the relationship among pressure, elevation, and total heads for the flow of water through soil. Open standpipes called *piezometers* are installed at points A and B. The levels to which water rises in the piezometer tubes situated at points A and B are known as the *piezometric levels* of points A and B, respectively. The pressure head at a point is the height of the vertical column of water in the piezometer installed at that point.

The loss of head between two points, A and B, can be given by

$$\Delta h = h_A - h_B = \left(\frac{u_A}{\gamma_w} + Z_A\right) - \left(\frac{u_B}{\gamma_w} + Z_B\right) \tag{6.3}$$

The head loss, Δh, can be expressed in a nondimensional form as

$$i = \frac{\Delta h}{L} \tag{6.4}$$

where i = hydraulic gradient
L = distance between points A and B—that is, the length of flow over which the loss of head occurred

In general, the variation of the velocity v with the hydraulic gradient i is as shown in Figure 6.2. This figure is divided into three zones:

1. Laminar flow zone (Zone I)
2. Transition zone (Zone II)
3. Turbulent flow zone (Zone III)

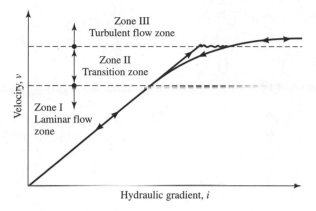

Figure 6.2 Nature of variation of v with hydraulic gradient, i

When the hydraulic gradient is gradually increased, the flow remains laminar in Zones I and II, and the velocity, v, bears a linear relationship to the hydraulic gradient. At a higher hydraulic gradient, the flow becomes turbulent (Zone III). When the hydraulic gradient is decreased, laminar flow conditions exist only in Zone I.

In most soils, the flow of water through the void spaces can be considered laminar; thus,

$$v \propto i \tag{6.5}$$

In fractured rock, stones, gravels, and very coarse sands, turbulent flow conditions may exist, and Eq. (6.5) may not be valid.

6.2   Darcy's Law

In 1856, Darcy published a simple equation for the discharge velocity of water through saturated soils, which may be expressed as

$$v = ki \tag{6.6}$$

where v = *discharge velocity*, which is the quantity of water flowing in unit time through a unit gross cross-sectional area of soil at right angles to the direction of flow

k = hydraulic conductivity (otherwise known as the coefficient of permeability)

This equation was based primarily on Darcy's observations about the flow of water through clean sands. Note that Eq. (6.6) is similar to Eq. (6.5); both are valid for laminar flow conditions and applicable for a wide range of soils.

In Eq. (6.6), v is the discharge velocity of water based on the gross cross-sectional area of the soil. However, the actual velocity of water (that is, the seepage velocity)

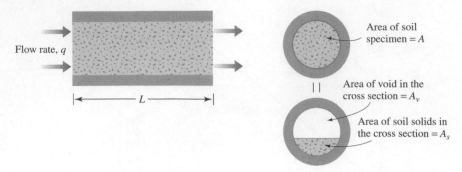

Figure 6.3 Derivation of Eq. (6.10)

through the void spaces is greater than v. A relationship between the discharge velocity and the seepage velocity can be derived by referring to Figure 6.3, which shows a soil of length L with a gross cross-sectional area A. If the quantity of water flowing through the soil in unit time is q, then

$$q = vA = A_v v_s \tag{6.7}$$

where v_s = seepage velocity
A_v = area of void in the cross section of the specimen

However,

$$A = A_v + A_s \tag{6.8}$$

where A_s = area of soil solids in the cross section of the specimen.
Combining Eqs. (6.7) and (6.8) gives

$$q = v(A_v + A_s) = A_v v_s$$

or

$$v_s = \frac{v(A_v + A_s)}{A_v} = \frac{v(A_v + A_s)L}{A_v L} = \frac{v(V_v + V_s)}{V_v} \tag{6.9}$$

where V_v = volume of voids in the specimen
V_s = volume of soil solids in the specimen

Equation (6.9) can be rewritten as

$$v_s = v \left[\frac{1 + \left(\dfrac{V_v}{V_s} \right)}{\dfrac{V_v}{V_s}} \right] = v \left(\frac{1 + e}{e} \right) = \frac{v}{n} \tag{6.10}$$

where e = void ratio
n = porosity

Darcy's law as defined by Eq. (6.6) implies that the discharge velocity v bears a linear relationship to the hydraulic gradient i and passes through the origin as shown in Figure 6.4. Hansbo (1960), however, reported the test results for four undisturbed

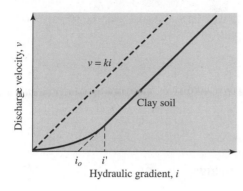

Figure 6.4
Variation of discharge velocity
with hydraulic gradient in clay

natural clays. On the basis of his results, a hydraulic gradient i' (see Figure 6.4) appears to exist, at which

$$v = k(i - i_0) \qquad \text{(for } i \geq i') \tag{6.11}$$

and

$$v = ki^m \qquad \text{(for } i < i') \tag{6.12}$$

The preceding equation implies that for very low hydraulic gradients, the relationship between v and i is nonlinear. The value of m in Eq. (6.12) for four Swedish clays was about 1.5. However, several other studies refute the preceding findings. Mitchell (1976) discussed these studies in detail. Taking all points into consideration, he concluded that Darcy's law is valid.

6.3   *Hydraulic Conductivity*

Hydraulic conductivity is generally expressed in cm/sec or m/sec in SI units and in ft/min or ft/day in English units.

The hydraulic conductivity of soils depends on several factors: fluid viscosity, pore-size distribution, grain size distribution, void ratio, roughness of mineral particles, and degree of soil saturation. In clayey soils, structure plays an important role in hydraulic conductivity. Other major factors that affect the permeability of clays are the ionic concentration and the thickness of layers of water held to the clay particles.

The value of hydraulic conductivity (k) varies widely for different soils. Some typical values for saturated soils are given in Table 6.1. The hydraulic conductivity of unsaturated soils is lower and increases rapidly with the degree of saturation.

Table 6.1 Typical Values of Hydraulic Conductivity of Saturated Soils

Soil type	k	
	cm/sec	ft/min
Clean gravel	100−1.0	200−2.0
Coarse sand	1.0−0.01	2.0−0.02
Fine sand	0.01−0.001	0.02−0.002
Silty clay	0.001−0.00001	0.002−0.00002
Clay	<0.000001	<0.000002

The hydraulic conductivity of a soil is also related to the properties of the fluid flowing through it by the equation

$$k = \frac{\gamma_w}{\eta}\overline{K}$$ (6.13)

where γ_w = unit weight of water
 η = viscosity of water
 $\overline{K}$ = absolute permeability

The *absolute permeability* $\overline{K}$ is expressed in units of L^2 (that is, cm², ft², and so forth).

Equation (6.13) showed that hydraulic conductivity is a function of the unit weight and the viscosity of water, which is in turn a function of the temperature at which the test is conducted. So, from Eq. (6.13),

$$\frac{k_{T_1}}{k_{T_2}} = \left(\frac{\eta_{T_2}}{\eta_{T_1}}\right)\left(\frac{\gamma_{w(T_1)}}{\gamma_{w(T_2)}}\right)$$ (6.14)

where k_{T_1}, k_{T_2} = hydraulic conductivity at temperatures T_1 and T_2, respectively
 η_{T_1}, η_{T_2} = viscosity of water at temperatures T_1 and T_2, respectively
 $\gamma_{w(T_1)}, \gamma_{w(T_2)}$ = unit weight of water at temperatures T_1 and T_2, respectively

It is conventional to express the value of k at a temperature of 20°C. Within the range of test temperatures, we can assume that $\gamma_{w(T_1)} \simeq \gamma_{w(T_2)}$. So, from Eq. (6.14)

$$k_{20°C} = \left(\frac{\eta_{T°C}}{\eta_{20°C}}\right)k_{T°C}$$ (6.15)

The variation of $\eta_{T°C}/\eta_{20°C}$ with the test temperature T varying from 15 to 30°C is given in Table 6.2.

Table 6.2 Variation of $\eta_{T°C}/\eta_{20°C}$

Temperature, T (°C)	$\eta_{T°C}/\eta_{20°C}$	Temperature, T (°C)	$\eta_{T°C}/\eta_{20°C}$
15	1.135	23	0.931
16	1.106	24	0.910
17	1.077	25	0.889
18	1.051	26	0.869
19	1.025	27	0.850
20	1.000	28	0.832
21	0.976	29	0.814
22	0.953	30	0.797

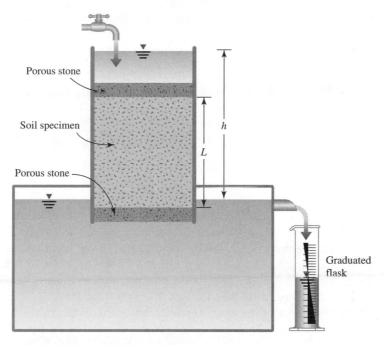

Figure 6.5 Constant-head permeability test

Laboratory Determination of Hydraulic Conductivity

Two standard laboratory tests are used to determine the hydraulic conductivity of soil — the constant-head test and the falling-head test. A brief description of each follows.

Constant-Head Test

A typical arrangement of the constant-head permeability test is shown in Figure 6.5. In this type of laboratory setup, the water supply at the inlet is adjusted in such a way that the difference of head between the inlet and the outlet remains constant during the test period. After a constant flow rate is established, water is collected in a graduated flask for a known duration.

The total volume of water collected may be expressed as

$$Q = Avt = A(ki)t \tag{6.16}$$

where Q = volume of water collected
A = area of cross section of the soil specimen
t = duration of water collection

And because

$$i = \frac{h}{L} \tag{6.17}$$

where L = length of the specimen, Eq. (6.17) can be substituted into Eq. (6.16) to yield

$$Q = A\left(k\frac{h}{L}\right)t \tag{6.18}$$

or

$$k = \frac{QL}{Aht} \tag{6.19}$$

Falling-Head Test

A typical arrangement of the falling-head permeability test is shown in Figure 6.6. Water from a standpipe flows through the soil. The initial head difference h_1 at time $t = 0$ is recorded, and water is allowed to flow through the soil specimen such that the final head difference at time $t = t_2$ is h_2.

The rate of flow of the water through the specimen at any time t can be given by

$$q = k\frac{h}{L}A = -a\frac{dh}{dt} \tag{6.20}$$

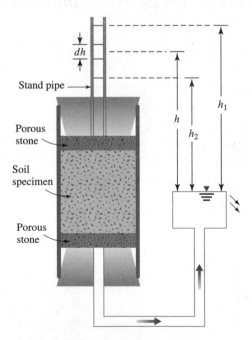

Figure 6.6 Falling-head permeability test

where q = flow rate
 a = cross-sectional area of the standpipe
 A = cross-sectional area of the soil specimen

Rearrangement of Eq. (6.20) gives

$$dt = \frac{aL}{Ak}\left(-\frac{dh}{h}\right) \tag{6.21}$$

Integration of the left side of Eq. (6.21) with limits of time from 0 to t and the right side with limits of head difference from h_1 to h_2 gives

$$t = \frac{aL}{Ak}\log_e \frac{h_1}{h_2}$$

or

$$k = 2.303 \frac{aL}{At}\log_{10}\frac{h_1}{h_2} \tag{6.22}$$

Example 6.1

Find the flow rate in m³/sec/m length (at right angles to the cross section shown) through the permeable soil layer shown in Figure 6.7 given H = 8m, H_1 = 3m, h = 4m, L = 50m, α = 8°, and k = 0.08 cm/sec.

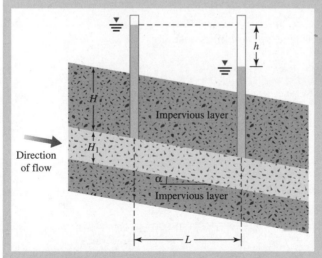

Figure 6.7 Flow-through permeable layer

Solution

$$\text{Hydraulic gradient } (i) = \frac{h}{\dfrac{L}{\cos \alpha}}$$

From Eqs. (6.17) and (6.18),

$$q = kiA = k\left(\frac{h \cos \alpha}{L}\right)(H_1 \cos \alpha \times 1)$$

$$= (0.08 \times 10^{-2} \text{ m/sec})\left(\frac{4 \cos 8°}{50}\right)(3 \cos 8° \times 1)$$

$$= 0.19 \times 10^{-3} \text{ m}^3/\text{sec/m} \qquad\qquad \blacksquare$$

Example 6.2

The results of a constant-head permeability test for a fine sand sample having a diameter of 150 mm and a length of 300 mm are as follows:

Constant head difference = 500 mm
Time of collection of water = 5 min
Volume of water collected = 350 cm^3
Temperature of water = 24°C

Find the hydraulic conductivity for the soil at 20°C.

Solution
For constant-head permeability test,

$$k = \frac{QL}{Aht}$$

Given that $Q = 350$ cm^3, $L = 300$ mm, $A = (\pi/4)(150)^2 = 17671.46$ mm^2, $h = 500$ mm, and $t = 5 \times 60 = 300$ sec, we have

$$\overset{\text{change to mm}^3}{\underset{\downarrow}{}}$$

$$k = \frac{(350 \times 10^3) \times 300}{17671.46 \times 500 \times 300} = 3.96 \times 10^{-2} \text{ mm/sec}$$

$$= 3.96 \times 10^{-3} \text{ cm/sec}$$

$$k_{20} = k_{24}\frac{\eta_{24}}{\eta_{20}}$$

From Table 6.2,

$$\frac{\eta_{24}}{\eta_{20}} = 0.91$$

So, $k_{20} = (3.96 \times 10^{-3}) \times 0.91 = 3.6 \times 10^{-3}$ cm/sec.   ∎

Example 6.3

For a variable-head permeability test, the following are given: length of specimen = 15 in., area of specimen = 3 in.^{2}, and k = 0.0688 in./min. What should be the area of the standpipe for the head to drop from 25 to 12 in. in 8 min.?

Solution
From Eq. (6.22),

$$k = 2.303 \frac{aL}{At} \log_{10} \frac{h_1}{h_2}$$

$$0.0688 = 2.303 \left(\frac{a \times 15}{3 \times 8} \right) \log_{10} \left(\frac{25}{12} \right)$$

$$a = \textbf{0.15 in.}^2$$ ∎

Example 6.4

The hydraulic conductivity of a clayey soil is 3×10^{-7} cm/sec. The viscosity of water at 25°C is 0.0911×10^{-4} g · sec/cm^2. Calculate the absolute permeability $\overline{K}$ of the soil.

Solution
From Eq. (6.13),

$$k = \frac{\gamma_w}{\eta} \overline{K} = 3 \times 10^{-7} \text{ cm/sec}$$

so

$$3 \times 10^{-7} = \left(\frac{1 \text{ g/cm}^3}{0.0911 \times 10^{-4}} \right) \overline{K}$$

$$\overline{K} = \textbf{0.2733} \times \textbf{10}^{-11} \textbf{ cm}^2$$ ∎

6.5   *Empirical Relations for Hydraulic Conductivity*

Several empirical equations for estimating hydraulic conductivity have been proposed in the past. Some of these are briefly discussed in this section.

For fairly uniform sand (that is, sand with a small uniformity coefficient), Hazen (1930) proposed an empirical relationship for hydraulic conductivity in the form

$$k \text{ (cm/sec)} = cD_{10}^2 \tag{6.23}$$

where c = a constant that varies from 1.0 to 1.5
  D_{10} = the effective size, in mm

Equation (6.23) is based primarily on Hazen's observations of loose, clean, filter sands. A small quantity of silts and clays, when present in a sandy soil, may change the hydraulic conductivity substantially.

Casagrande proposed a simple relationship for hydraulic conductivity for fine-to-medium clean sand in the form

$$k = 1.4e^2k_{0.85} \tag{6.24}$$

where k = hydraulic conductivity at a void ratio e
  $k_{0.85}$ = the corresponding value at a void ratio of 0.85

Another form of equation that gives fairly good results in estimating the hydraulic conductivity of sandy soils is based on the Kozeny-Carman equation. The derivation of this equation is not presented here. Interested readers are referred to any advanced soil mechanics book (for example, Das, 1997). An application of the Kozeny-Carman equation yields

$$k \propto \frac{e^3}{1 + e} \tag{6.25}$$

where k = hydraulic conductivity at a void ratio of e. This equation can be rewritten as

$$k = C_1 \frac{e^3}{1 + e} \tag{6.26}$$

where C_1 = a constant.

Mention was made at the end of Section 6.1 that turbulent flow conditions may exist in very coarse sands and gravels, and that Darcy's law may not be valid for these materials. However, under a low hydraulic gradient, laminar flow conditions usually exist. Kenney, Lau, and Ofoegbu (1984) conducted laboratory tests on granular soils in which the particle sizes in various specimens ranged from 0.074 to 25.4 mm. The uniformity coefficients, C_u, of these specimens ranged from 1.04 to 12. All permeability tests were conducted at a relative density of 80% or more. These tests showed that for laminar flow conditions,

$$\overline{K} \text{ (mm}^2\text{)} = (0.05 \text{ to } 1)D_5^2 \tag{6.27}$$

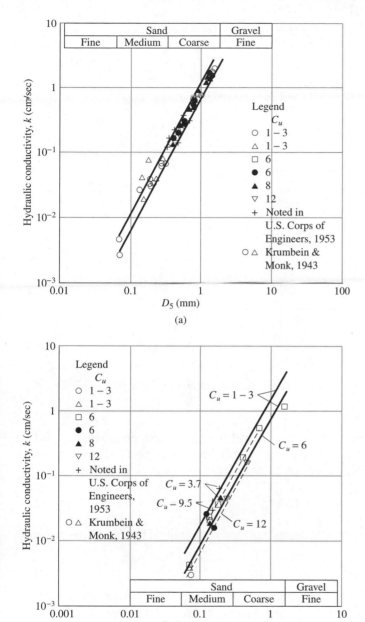

(a)

(b)

Figure 6.8

Results of permeability tests on which
Eq. (6.27) is based: (a) results for $C_u = 1$–3;
(b) results for $C_u > 3$ (after Kenney, Lau, and
Ofoegbu, 1984)

where D_5 = diameter (mm) through which 5% of soil passes. Figures 6.8a and 6.8b show the results on which Eq. (6.27) is based.

On the basis of laboratory experiments, the U.S. Department of Navy (1971) provided an empirical correlation between k (ft/min) and D_{10} (mm) for granular soils with the uniformity coefficient varying between 2 and 12 and $D_{10}/D_5 < 1.4$. This correlation is shown in Figure 6.9.

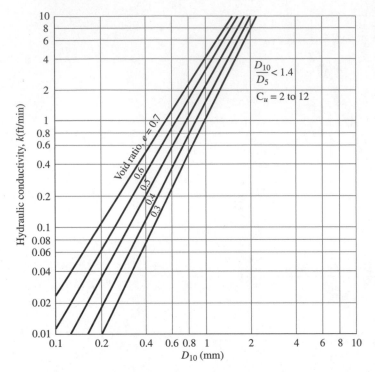

Figure 6.9 Permeability of granular soils (after U.S. Department of Navy, 1971)

According to their experimental observations, Samarasinghe, Huang, and Drnevich (1982) suggested that the hydraulic conductivity of normally consolidated clays (see Chapter 10 for definition) can be given by

$$k = C_3 \left(\frac{e^n}{1 + e} \right) \tag{6.28}$$

where C_3 and n are constants to be determined experimentally. This equation can be rewritten as

$$\log[k(1 + e)] = \log C_3 + n \log e \tag{6.29}$$

Hence, for any given clayey soil, if the variation of k with the void ratio is known, a log-log graph can be plotted with $k(1 + e)$ against e to determine the values of C_3 and n.

Some other empirical relationships for estimating the hydraulic conductivity in sand and clayey soils are given in Table 6.3. One should keep in mind, however, that any empirical relationship of this type is for estimation only, because the magnitude of k is a highly variable parameter and depends on several factors.

Tavenas et al. (1983) also gave a correlation between the void ratio and the hydraulic conductivity of clayey soil. This correlation is shown in Figure 6.10. An important point to note, however, is that in Figure 6.10, *PI*, the plasticity index, and *CF*, the clay-size fraction in the soil, are in *fraction* (decimal) form.

Table 6.3 Empirical Relationships for Estimating Hydraulic Conductivity

Type of Soil	Source	Relationship[a]	Comments
Sand	Amer and Awad (1974)	$k = C_a D_{10}^{2.32} C_u^{0.6} \dfrac{e^3}{1 + e}$	
	Shahabi, Das, Tarquin (1984)	$k = 1.2 C_2^{0.735} D_{10}^{0.89} \dfrac{e^3}{1 + e}$	Medium to fine sand
Clay	Mesri and Olson (1971)	$\log k = A' \log e + B'$	
	Taylor (1948)	$\log k = \log k_0 - \dfrac{e_0 - e}{C_k}$ $C_k \approx 0.5 e_0$	For $e < 2.5$,

[a] D_{10} — effective size
C_u = uniformity coefficient
C_2 = a constant
k_0 = *in situ* hydraulic conductivity at void ratio e_0
k = hydraulic conductivity at void ratio e
C_k = permeability change index

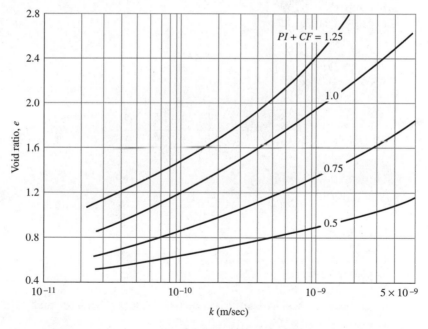

Figure 6.10 Variation of void ratio with hydraulic conductivity of clayey soils (based on Tavenas et al., 1983)

Example 6.5

The hydraulic conductivity of a sand at a void ratio of 0.8 is 0.047 cm/sec. Estimate the hydraulic conductivity of this sand at a void ratio of 0.5. Use Eq. (6.24).

Solution
From Eq. (6.24), $k = 1.4\, e^2 k_{0.85}$. Thus,

$$\frac{k_{0.8}}{k_{0.5}} = \frac{(0.8)^2}{(0.5)^2}$$

So

$$k_{0.5} = k_{0.8}\left(\frac{0.5}{0.8}\right)^2 = 0.047\left(\frac{0.5}{0.8}\right)^2$$

$$= 0.018 \text{ cm/sec} \qquad \blacksquare$$

Example 6.6

Redo Example Problem 6.5 using Eq. (6.26).

Solution
From Eq. (6.26),

$$k = C_1\frac{e^3}{1 + e}$$

So

$$\frac{k_{0.8}}{k_{0.5}} = \frac{\left[\dfrac{0.8^3}{1 + 0.8}\right]}{\left[\dfrac{0.5^3}{1 + 0.5}\right]} = \frac{0.284}{0.083} = 3.42$$

Hence,

$$k_{0.5} = \frac{k_{0.8}}{3.42} = \frac{0.047}{3.42} \approx 0.014 \text{ cm/sec} \qquad \blacksquare$$

Example 6.7

The void ratio and hydraulic conductivity relation for a normally consolidated clay are given below.

Void ratio	k (cm/sec)
1.2	0.6×10^{-7}
1.52	1.519×10^{-7}

Estimate the value of k for the same clay with a void ratio of 1.4.

Solution
From Eq. (6.28)

$$\frac{k_1}{k_2} = \frac{\left[\dfrac{e_1^n}{1 + e_1}\right]}{\left[\dfrac{e_2^n}{1 + e_2}\right]}$$

Substitution of $e_1 = 1.2$, $k_1 = 0.6 \times 10^{-7}$ cm/sec, $e_2 = 1.52$, $k_2 = 1.519 \times 10^{-7}$ cm/sec in the preceding equation gives

$$\frac{0.6}{1.519} = \left(\frac{1.2}{1.52}\right)^n \left(\frac{2.52}{2.2}\right)$$

or

$$n = 4.5$$

Again, from Eq. (6.28),

$$k_1 = C_3 \left(\frac{e_1^n}{1 + e_1}\right)$$

$$0.6 \times 10^{-7} = C_3 \left(\frac{1.2^{4.5}}{1 + 1.2}\right)$$

or

$$C_3 = 0.581 \times 10^{-7} \text{ cm/sec}$$

So

$$k = (0.581 \times 10^{-7}) \left(\frac{e^{4.5}}{1 + e}\right) \text{ cm/sec}$$

Now, substituting $e = 1.4$ in the preceding equation yields

$$k = (0.581 \times 10^{-7}) \left(\frac{1.4^{4.5}}{1 + 1.4}\right) = 1.1 \times 10^{-7} \text{ cm/scc} \quad \blacksquare$$

6.6   *Directional Variation of Permeability*

Most soils are not isotropic with respect to permeability. In a given soil deposit, the magnitude of k changes with respect to the direction of flow. Figure 6.11 shows a soil layer through which water flows in a direction inclined at an angle α with the vertical. Let the hydraulic conductivity in the vertical ($\alpha = 0$) and horizontal ($\alpha = 90°$) directions be k_V and k_H, respectively. The magnitudes of k_V and k_H in a given soil depend on several factors, including the method of deposition in the field. Basak (1972)

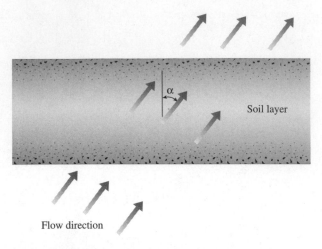

Figure 6.11 Directional variation of permeability

and Al-Tabbaa and Wood (1987) conducted laboratory tests on kaolin to determine the variation of the anisotropy ratio, $r_k = k_H/k_V$, with the void ratio. The specimens for these tests were subjected to *unidimensional* (oedometric) consolidation. The variations of r_k and void ratio (e) from the tests of Basak, and from those of Al-Tabbaa and Wood, are shown in Figure 6.12. Similar test results on a sand ($C_u = 3.5$, $e_{max} = 0.824$, and $e_{min} = 0.348$) were provided by Chapuis, Gill, and Baass (1989). The specimens for these tests were prepared by unidimensional static and dynamic compaction. The variations of r_k with e for these tests are also shown in Figure 6.12.

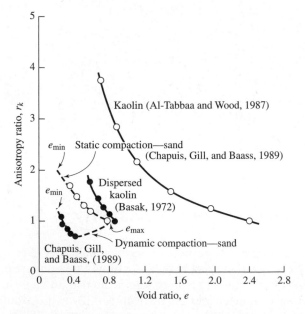

Figure 6.12 Variation of anisotropy ratio with void ratio for various soils

According to Figure 6.12, the following three general conclusions can be drawn:

1. For static compaction conditions, the magnitude of r_k decreases with the increase in void ratio.
2. For sand, the anisotropy ratio is equal to one at $e = e_{max}$.
3. For sand, with $e < e_{max}$, the magnitude of r_k is greater than one when the specimens are formed by unidimensional static compaction. However, for dynamically compacted sand, r_k may be less than one for $e < e_{max}$.

6.7 *Equivalent Hydraulic Conductivity in Stratified Soil*

In a stratified soil deposit where the hydraulic conductivity for flow in a given direction changes from layer to layer, an equivalent hydraulic conductivity can be computed to simplify calculations. The following derivations relate to the equivalent hydraulic conductivities for flow in vertical and horizontal directions through multilayered soils with horizontal stratification.

Figure 6.13 shows n layers of soil with flow in the *horizontal direction*. Let us consider a cross section of unit length passing through the n layer and perpendicular to the direction of flow. The total flow through the cross section in unit time can be written as

$$q = v \cdot 1 \cdot H$$
$$= v_1 \cdot 1 \cdot H_1 + v_2 \cdot 1 \cdot H_2 + v_3 \cdot 1 \cdot H_3 + \cdots + v_n \cdot 1 \cdot H_n \qquad (6.30)$$

where v = average discharge velocity

$v_1, v_2, v_3, \ldots, v_n$ = discharge velocities of flow in layers denoted by the subscripts

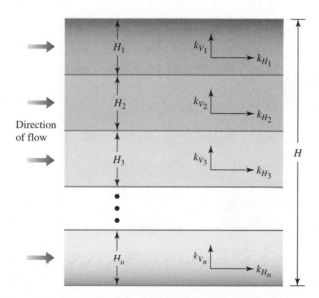

Figure 6.13 Equivalent hydraulic conductivity determination — horizontal flow in stratified soil

If k_{H_1}, k_{H_2}, k_{H_3}, . . ., k_{H_n} are the hydraulic conductivities of the individual layers in the horizontal direction and $k_{H(eq)}$ is the equivalent hydraulic conductivity in the horizontal direction, then, from Darcy's law,

$$v = k_{H(eq)}i_{eq}; \quad v_1 = k_{H_1}i_1; \quad v_2 = k_{H_2}i_2; \quad v_3 = k_{H_3}i_3; \quad \cdots \quad v_n = k_{H_n}i_n$$

Substituting the preceding relations for velocities into Eq. (6.30) and noting that $i_{eq} = i_1 = i_2 = i_3 = \cdots = i_n$ results in

$$k_{H(eq)} = \frac{1}{H}(k_{H_1}H_1 + k_{H_2}H_2 + k_{H_3}H_3 + \cdots + k_{H_n}H_n) \qquad (6.31)$$

Figure 6.14 shows n layers of soil with flow in the vertical direction. In this case, the velocity of flow through all the layers is the same. However, the total head loss, h, is equal to the sum of the head losses in all layers. Thus,

$$v = v_1 = v_2 = v_3 = \cdots = v_n \qquad (6.32)$$

and

$$h = h_1 + h_2 + h_3 + \cdots + h_n \qquad (6.33)$$

Using Darcy's law, we can rewrite Eq. (6.32) as

$$k_{V(eq)}\left(\frac{h}{H}\right) = k_{V_1}i_1 = k_{V_2}i_2 = k_{V_3}i_3 = \cdots = k_{V_n}i_n \qquad (6.34)$$

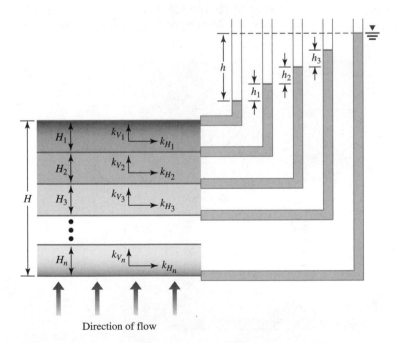

Direction of flow

Figure 6.14 Equivalent hydraulic conductivity determination — vertical flow in stratified soil

where k_{V_1}, k_{V_2}, k_{V_3}, . . . , k_{V_n} are the hydraulic conductivities of the individual layers in the vertical direction and $k_{V(eq)}$ is the equivalent hydraulic conductivity.

Again, from Eq. (6.33),

$$h = H_1 i_1 + H_2 i_2 + H_3 i_3 + \cdots + H_n i_n \tag{6.35}$$

Solving Eqs. (6.34) and (6.35) gives

$$k_{V(eq)} = \frac{H}{\left(\dfrac{H_1}{k_{V_1}}\right) + \left(\dfrac{H_2}{k_{V_2}}\right) + \left(\dfrac{H_3}{k_{V_3}}\right) + \cdots + \left(\dfrac{H_n}{k_{V_n}}\right)} \tag{6.36}$$

An excellent example of naturally deposited layered soil is *varved soil,* which is a rhythmically layered sediment of coarse and fine minerals. Varved soils result from annual seasonal fluctuation of sediment conditions in glacial lakes. Figure 6.15 shows the variation of moisture content and grain-size distribution in New Liskeard, Canada, varved soil. Each varvey is about 41 to 51 mm (1.6 to 2.0 in.) thick and consists of two homogeneous layers of soil — one coarse and one fine — with a transition layer between.

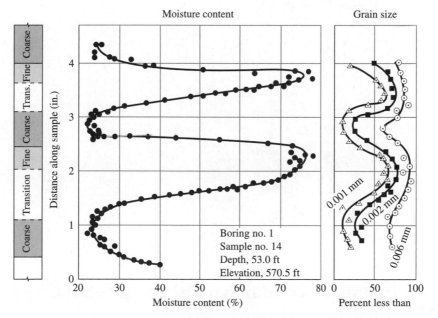

Figure 6.15 Variation of moisture content and grain-size distribution in New Liskeard varved soil. *Source:* After "Laboratory Investigation of Permeability Ratio of New Liskeard Varved Clay," by H. T. Chan and T. C. Kenney, 1973, *Canadian Geotechnical Journal, 10*(3), p. 453–472. Copyright © 1973 National Research Council of Canada. Used by permission.

Example 6.8

A layered soil is shown in Figure 6.13. Given that

- $H_1 = 1$ m $k_1 = 10^{-4}$ cm/sec
- $H_2 = 1.5$ m $k_2 = 3.2 \times 10^{-2}$ cm/sec
- $H_3 = 2$ m $k_3 = 4.1 \times 10^{-5}$ cm/sec

estimate the ratio of equivalent permeability,

$$\frac{k_{H(eq)}}{k_{V(eq)}}$$

Solution
From Eq. (6.31),

$$k_{H(eq)} = \frac{1}{H}(k_{H_1}H_1 + k_{H_2}H_2 + k_{H_3}H_3)$$

$$= \frac{1}{(1 + 1.5 + 2)}[(10^{-4})(1) + (3.2 \times 10^{-2})(1.5) + (4.1 \times 10^{-5})(2)]$$

$$= 107.07 \times 10^{-4} \text{ cm/sec}$$

Again, from Eq. (6.36),

$$k_{V(eq)} = \frac{H}{\left(\dfrac{H_1}{k_{V_1}}\right) + \left(\dfrac{H_2}{k_{V_2}}\right) + \left(\dfrac{H_3}{k_{V_3}}\right)}$$

$$= \frac{1 + 1.5 + 2}{\left(\dfrac{1}{10^{-4}}\right) + \left(\dfrac{1.5}{3.2 \times 10^{-2}}\right) + \left(\dfrac{2}{4.1 \times 10^{-5}}\right)}$$

$$= 0.765 \times 10^{-4} \text{ cm/sec}$$

Hence,

$$\frac{k_{H(eq)}}{k_{V(eq)}} = \frac{107.07 \times 10^{-4}}{0.765 \times 10^{-4}} \approx \mathbf{140}$$

∎

6.8   *Hydraulic Conductivity of Compacted Clayey Soils*

It was shown in Chapter 5 (Section 5.5) that when a clay is compacted at a lower moisture content it possesses a flocculent structure. Approximately at optimum moisture content of compaction, the clay particles have a lower degree of flocculation. A further increase in the moisture content at compaction provides a greater degree of par-

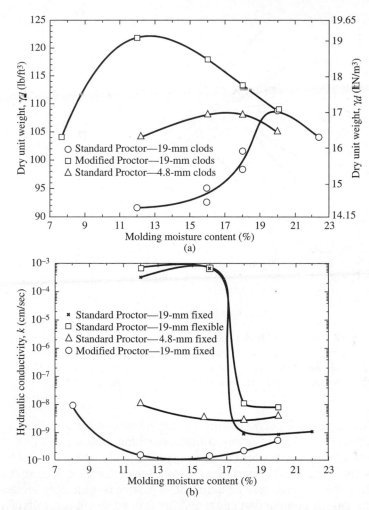

Figure 6.16 Tests on a clay soil: (a) Standard and modified Proctor compaction curves; (b) variation of k with molding moisture content. *Source:* After "Influence of Clods on Hydraulic Conductivity of Compacted Clay," by C. H. Benson and D. E. Daniel, 1990, *Journal of Geotechnical Engineering, 116*(8), p. 1231–1248. Copyright © 1990 American Society of Civil Engineers. Used by permission.

ticle orientation; however, the dry unit weight decreases because the added water dilutes the concentration of soil solids per unit volume.

Figure 6.16 shows the results of laboratory compaction tests on a clay soil as well as the variation of hydraulic conductivity on the compacted clay specimens. The compaction tests and thus the specimens for hydraulic conductivity tests were prepared from clay clods that were 19 mm and 4.8 mm. From the laboratory test results shown, the following observations can be made:

1. For similar compaction effort and molding moisture content, the magnitude of k decreases with the decrease in clod size.

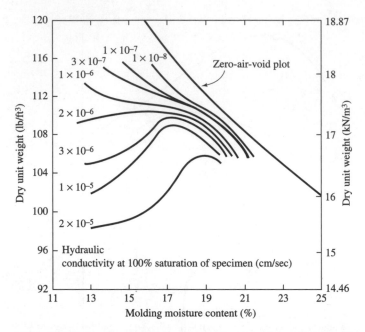

Figure 6.17 Contours of hydraulic conductivity for a silty clay. *Source:* After "Permeability of Compacted Clay," by J. K. Mitchell, D. R. Hooper, and R. B. Campenella, 1965, *Journal of the Soil Mechanics and Foundations Division, 91* (SM4), p. 41–65. Copyright © 1965 American Society of Civil Engineers. Used by permission.

2. For a given compaction effort, the hydraulic conductivity decreases with the increase in molding moisture content, reaching a minimum value at about the optimum moisture content (that is, approximately where the soil has a higher unit weight with the clay particles having a lower degree of flocculation). Beyond the optimum moisture content, the hydraulic conductivity increases slightly.
3. For similar compaction effort and dry unit weight, a soil will have a lower hydraulic conductivity when it is compacted on the wet side of the optimum moisture content. This fact is further illustrated in Figure 6.17, which shows a summary of hydraulic conductivity test results on a silty clay (Mitchell, Hooper and Campanella, 1965).

6.9 Considerations for Hydraulic Conductivity of Clayey Soils in Field Compaction

In some compaction work in clayey soils, the compaction must be done in a manner so that a certain specified upper level of hydraulic conductivity of the soil is achieved. Examples of such works are compaction of the core of an earth dam and installation of clay liners in solid-waste disposal sites.

To prevent groundwater pollution from leachates generated from solid-waste disposal sites, the U.S. Environmental Protection Agency (EPA) requires that clay liners have a hydraulic conductivity of 10^{-7} cm/sec or less. To achieve this value, the contractor must ensure that the soil meets the following criteria (Environmental Protection Agency, 1989):

1. The soil should have at least 20% fines (fine silt and clay-size particles).
2. The plasticity index *(PI)* should be greater than 10. Soils that have a *PI* greater than about 30 are difficult to work with in the field.
3. The soil should not include more than 10% gravel-size particles.
4. The soil should not contain any particles or chunks of rock that are larger than 25 to 50 mm (1 to 2 in.).

In many instances, the soil found at the construction site may be somewhat nonplastic. Such soil may be blended with imported clay minerals like sodium bentonite to achieve the desired range of hydraulic conductivity. In addition, during field compaction, a heavy sheepsfoot roller can introduce larger shear strains during compaction that create a more dispersed structure in the soil. This type of compacted soil will have an even lower hydraulic conductivity. Small lifts should be used during compaction so that the feet of the compactor can penetrate the full depth of the lift.

The size of the clay clods has a strong influence on the hydraulic conductivity of a compacted clay. Hence, during compaction, the clods must be broken down mechanically to as small as possible. A very heavy roller used for compaction helps to break them down.

Bonding between successive lifts is also an important factor; otherwise, permeant can move through a vertical crack in the compacted clay and then travel along the interface between two lifts until it finds another crack, as is schematically shown in Figure 6.18. Bonding can substantially reduce the overall hydraulic conductivity of a compacted clay. An example of poor bonding was seen in a trial pad construction in Houston in 1986. The trial pad was 0.91 m (3 ft) thick and built in six, 15.2 mm (6 in.) lifts. The results of the hydraulic conductivity tests for the compact soil from the trial pad are given in Table 6.4. Note that although the laboratory-determined values of k for various lifts are on the order of 10^{-7} to 10^{-9} cm/sec, the actual overall value of k increased to the order of 10^{-4}. For this reason, scarification and control of

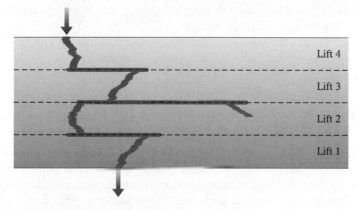

Figure 6.18 Pattern of flow through a compacted clay with improper bonding between lifts (after Environmental Protection Agency, 1989)

Table 6.4 Hydraulic Conductivity from Houston Liner Tests*

Location	Sample	Laboratory k (cm/sec)
Lower lift	76 mm ($\approx$3 in.) tube	4×10^{-9}
Upper lift	76 mm ($\approx$3 in.) tube	1×10^{-9}
Lift interface	76 mm ($\approx$3 in.) tube	1×10^{-7}
Lower lift	Block	8×10^{-5}
Upper lift	Block	1×10^{-8}
Actual overall $k = 1 \times 10^{-4}$ cm/sec		

*After Environmental Protection Agency (1989)

the moisture content after compaction of each lift are extremely important in achieving the desired hydraulic conductivity.

6.10 Moisture Content—Unit Weight Criteria for Clay Liner Construction

As mentioned in Section 6.9, for construction of clay liners for solid-waste disposal sites, the compacted clay is required to have a hydraulic conductivity of 10^{-7} cm/sec or less. Daniel and Benson (1990) developed a procedure to establish the moisture content — unit weight criteria for clayey soils to meet the hydraulic conductivity requirement. Following is a step-by-step procedure to develop the criteria:

1. Conduct *modified, standard,* and *reduced* Proctor tests to establish the dry unit weight versus molding moisture content relationships (Figure 6.19a). Modified and standard Proctor tests were discussed in Chapter 5. The *reduced* Proctor test is similar to the standard Proctor test, except the hammer is dropped only 15 times per lift instead of the usual 25 times. Modified, standard, and reduced Proctor efforts represent, respectively, the upper, medium, and minimum levels of compaction energy for a typical clayey soil liner.
2. Conduct permeability tests on the compacted soil specimens (from step 1), and plot the results, as shown in Figure 6.19b. In this figure, also plot the maximum allowable value of k (that is, k_{all}).
3. Replot the dry unit weight–moisture content points (Figure 6.19c) with different symbols to represent the compacted specimens with $k > k_{all}$ and $k \leq k_{all}$.
4. Plot the acceptable zone for which k is less than or equal to k_{all} (Figure 6.19c).

6.11 Permeability Test in the Field by Pumping from Wells

In the field, the average hydraulic conductivity of a soil deposit in the direction of flow can be determined by performing pumping tests from wells. Figure 6.20a shows a case where the top permeable layer, whose hydraulic conductivity has to be determined, is unconfined and underlain by an impermeable layer. During the test, water is pumped out at a constant rate from a test well that has a perforated casing. Several observation wells at various radial distances are made around the test well.

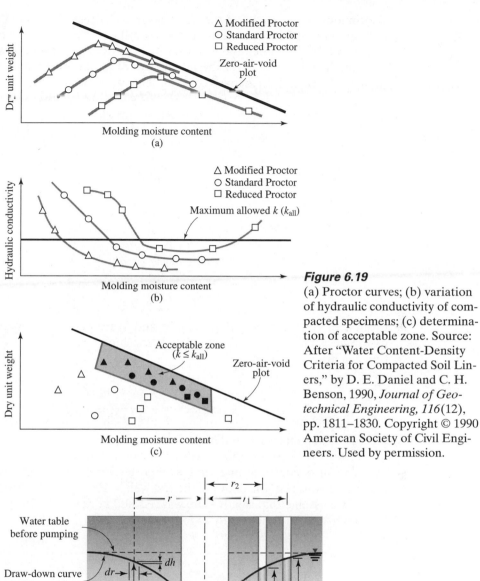

Figure 6.19
(a) Proctor curves; (b) variation of hydraulic conductivity of compacted specimens; (c) determination of acceptable zone. Source: After "Water Content-Density Criteria for Compacted Soil Liners," by D. E. Daniel and C. H. Benson, 1990, *Journal of Geotechnical Engineering, 116*(12), pp. 1811–1830. Copyright © 1990 American Society of Civil Engineers. Used by permission.

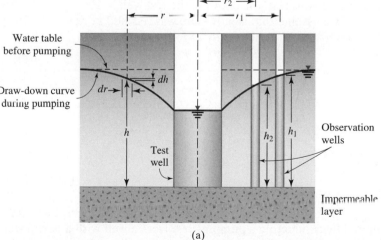

Figure 6.20 (a) Pumping test from a well in an unconfined permeable layer underlain by an impermeable stratum. Figure 6.20 continued on page 167.

Continuous observations of the water level in the test well and in the observation wells are made after the start of pumping, until a steady state is reached. The steady state is established when the water level in the test and observation wells becomes constant. The expression for the rate of flow of groundwater into the well, which is equal to the rate of discharge from pumping, can be written as

$$q = k\left(\frac{dh}{dr}\right)2\pi rh \qquad (6.37)$$

or

$$\int_{r_2}^{r_1} \frac{dr}{r} = \left(\frac{2\pi k}{q}\right)\int_{h_2}^{h_1} h\, dh$$

Thus,

$$k = \frac{2.303q \log_{10}\left(\dfrac{r_1}{r_2}\right)}{\pi(h_1^2 - h_2^2)} \qquad (6.38)$$

From field measurements, if q, r_1, r_2, h_1, and h_2 are known, the hydraulic conductivity can be calculated from the simple relationship presented in Eq. (6.38). This equation can also be written as

$$k \text{ (cm/sec)} = \frac{2.303q \log_{10}\left(\dfrac{r_1}{r_2}\right)}{14.7\pi(h_1^2 - h_2^2)} \qquad (6.39)$$

where q is in gpm and h_1 and h_2 are in ft.

Figure 6.20b shows the drawdown versus distance plots for a field pumping test in a deposit of coarse to fine sand, as reported by Ahmad, Lacroix, and Steinback (1975). For this test, the depth of the test well = 100 ft and $q = 1515$ gpm. From the plot, if we assume that steady state was reached at time $t = 6064$ min, we can calculate the hydraulic conductivity as follows:

r^2 (ft^2)	r (ft)	Drawdown (ft)	h (ft)
1,000	31.6	5	$100 - 5 = 95$
10,000	100	3.5	$100 - 3.5 = 96.5$

From Eq. (6.39),

$$k = \frac{(2.303)(1515) \log\left(\dfrac{100}{31.6}\right)}{(14.7)(\pi)(96.5^2 - 95^2)} = 0.132 \text{ cm/sec}$$

The average hydraulic conductivity for a confined aquifer can also be determined by conducting a pumping test from a well with a perforated casing that penetrates the full depth of the aquifer and by observing the piezometric level in a number of observation wells at various radial distances (Figure 6.21). Pumping is continued at a uniform rate q until a steady state is reached.

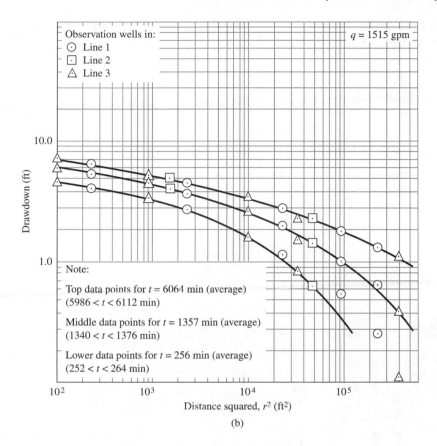

Observation wells in:
⊙ Line 1
▫ Line 2
△ Line 3

$q = 1515$ gpm

10.0

Drawdown (ft)

1.0

Note:

Top data points for $t = 6064$ min (average)
($5986 < t < 6112$ min)

Middle data points for $t = 1357$ min (average)
($1340 < t < 1376$ min)

Lower data points for $t = 256$ min (average)
($252 < t < 264$ min)

10^2 10^3 10^4 10^5

Distance squared, r^2 (ft²)

(b)

Figure 6.20 (*continued*)
(b) Plot of drawdown versus r^2 in a field pumping test (based on Ahmad, Lacroix, and Steinback, 1975)

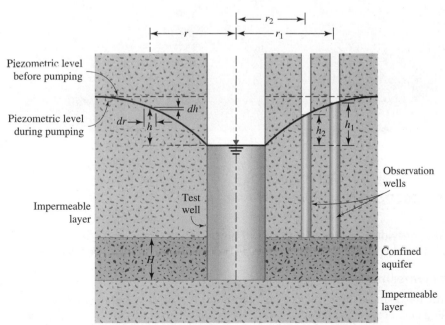

r_2

r r_1

Piezometric level before pumping

Piezometric level during pumping

dr h dh

h_2 h_1

Observation wells

Impermeable layer

Test well

H

Confined aquifer

Impermeable layer

Figure 6.21
Pumping test from a well penetrating the full depth in a confined aquifer

Because water can enter the test well only from the aquifer of thickness H, the steady state of discharge is

$$q = k \left(\frac{dh}{dr} \right) 2\pi r H \tag{6.40}$$

or

$$\int_{r_2}^{r_1} \frac{dr}{r} = \int_{h_2}^{h_1} \frac{2\pi k H}{q} \, dh$$

This gives the hydraulic conductivity in the direction of flow as

$$k = \frac{q \log_{10} \left(\frac{r_1}{r_2} \right)}{2.727 H (h_1 - h_2)} \tag{6.41}$$

Example 6.9

Consider the case of pumping from a well in an unconfined permeable layer underlain by an impermeable stratum (see Figure 6.20a). Given that

- $q = 0.74 \text{ m}^3/\text{min}$
- $h_1 = 6 \text{ m at } r_1 = 60 \text{ m}$
- $h_2 = 5.2 \text{ m at } r_2 = 30 \text{ m}$

calculate the hydraulic conductivity (in m/min) of the permeable layer.

Solution
From Eq. (6.38),

$$k = \frac{2.303 q \log_{10} \left(\frac{r_1}{r_2} \right)}{\pi (h_1^2 - h_2^2)} = \frac{(2.303)(0.74) \log_{10} \left(\frac{60}{30} \right)}{\pi (6^2 - 5.2^2)} = \textbf{0.018 ft/min} \qquad \blacksquare$$

6.12   In Situ *Hydraulic Conductivity of Compacted Clay Soils*

Daniel (1989) provided an excellent review of nine methods to estimate the *in situ* hydraulic conductivity of compacted clay layers. Three of these methods are described.

Boutwell Permeameter

A schematic diagram of the Boutwell permeameter is shown in Figure 6.22. A hole is first drilled and a casing is placed in it (Figure 6.22a). The casing is filled with water

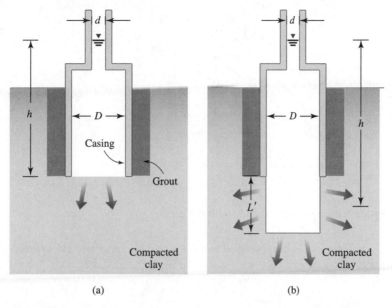

(a)   (b)

Figure 6.22 Permeability test with Boutwell permeameter

and a falling-head permeability test is conducted. Based on the test results, the hydraulic conductivity k_1 is calculated as

$$k_1 = \frac{\pi d^2}{\pi D(t_2 - t_1)} \ln\left(\frac{h_1}{h_2}\right)$$ (6.42)

where d = diameter of the standpipe
D = diameter of the casing
h_1 = head at time t_1
h_2 = head at time t_2

After the hydraulic conductivity is determined, the hole is deepened by augering, and the permeameter is reassembled as shown in Figure 6.22b. A falling-head hydraulic conductivity test is conducted again. The hydraulic conductivity is calculated as

$$k_2 = \frac{A'}{B'} \ln\left(\frac{h_1}{h_2}\right)$$ (6.43)

where

$$A' = d^2\left\{\ln\left[\frac{L'}{D} + \sqrt{1 + \left(\frac{L'}{D}\right)^2}\right]\right\}$$ (6.44)

$$B' = 8D\frac{L'}{D}(t_2 - t_1)\left\{1 - 0.562 \exp\left[-1.57\left(\frac{L'}{D}\right)\right]\right\}$$ (6.45)

The anisotropy with respect to permeability is determined by referring to Figure 6.23, which is a plot of k_2/k_1 versus m ($m = \sqrt{k_H/k_V}$) for various values of

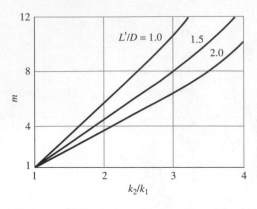

Figure 6.23 Variation of k_2/k_1 with m [Eq. (6.46)]

L'/D. Figure 6.23 can be used to determine m using the experimental values of k_2/k_1 and L'/D. The plots in this figure are determined from

$$\frac{k_2}{k_1} = \frac{\ln[(L'/D) + \sqrt{1 + (L'/D)^2}]}{\ln[(mL'/D) + \sqrt{1 + (mL'/D)^2}]} m \tag{6.46}$$

Once m is determined, we can calculate

$$k_H = mk_1 \tag{6.47}$$

and

$$k_V = \frac{k_1}{m} \tag{6.48}$$

Constant-Head Borehole Permeameter

Figure 6.24 shows a constant-head borehole permeameter. In this arrangement a constant head h is maintained by supplying water, and the rate of flow q is measured. The hydraulic conductivity can be calculated as

$$k = \frac{q}{r^2 \sqrt{R^2 - 1} [F_1 + (F_2/A'')]} \tag{6.49}$$

where

$$R = \frac{h}{r} \tag{6.50}$$

$$F_1 = \frac{4.117(1 - R^2)}{\ln(R + \sqrt{R^2 - 1}) - [1 - (1/R^2)]^{0.5}} \tag{6.51}$$

$$F_2 = \frac{4.280}{\ln(R + \sqrt{R^2 - 1})} \tag{6.52}$$

$$A'' = \frac{1}{2} \alpha r \tag{6.53}$$

Typical values of α range from 0.002 to 0.01 cm^{-1} for fine-grained soil.

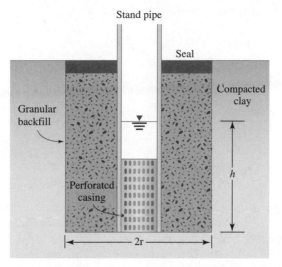

Figure 6.24
Borehole test with constant water level

Porous Probes

Porous probes (Figure 6.25) are pushed or driven into the soil. Constant- or falling-head permeability tests are performed. The hydraulic conductivity is calculated as follows:

The constant head is given by

$$k = \frac{q}{Fh} \tag{6.54}$$

The falling head is given by

$$k = \frac{\pi d^2/4}{F(t_2 - t_1)} \ln\left(\frac{h_1}{h_2}\right) \tag{6.55}$$

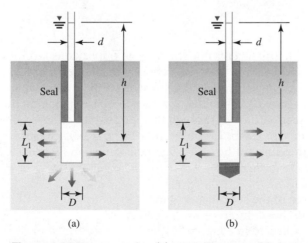

Figure 6.25 Porous probe: (a) test with permeable base; (b) test with impermeable base

For probes with permeable bases (Figure 6.25a),

$$F = \frac{2\pi L_1}{\ln[(L_1/D) + \sqrt{1 + (L_1/D)^2}]} \qquad (6.56)$$

For probes with impermeable bases (Figure 6.25b),

$$F = \frac{2\pi L_1}{\ln[(L_1/D) + \sqrt{1 + (L_1/D)^2}]} - 2.8D \qquad (6.57)$$

6.13   Summary and General Comments

In this chapter, we discussed Darcy's law, definition of hydraulic conductivity, laboratory determination of hydraulic conductivity and the empirical relations for it, and field determination of hydraulic conductivity of various types of soil. Hydraulic conductivity of various soil layers is highly variable. The empirical relations for hydraulic conductivity should be used as a general guide for all practical considerations. The accuracy of the values of k determined in the laboratory depends on several factors:

1. Temperature of the fluid
2. Viscosity of the fluid
3. Trapped air bubbles present in the soil specimen
4. Degree of saturation of the soil specimen
5. Migration of fines during testing
6. Duplication of field conditions in the laboratory

The hydraulic conductivity of saturated cohesive soils can also be determined by laboratory consolidation tests. (See Example 10.10.) The actual value of the hydraulic conductivity in the field may also be somewhat different than that obtained in the laboratory because of the nonhomogeneity of the soil. Hence, proper care should be taken in assessing the order of the magnitude of k for all design considerations.

Problems

6.1   A permeable soil layer is underlain by an impervious layer, as shown in Figure 6.26. With $k = 4.8 \times 10^{-3}$ cm/sec for the permeable layer, calculate the rate of seepage through it in m³/hr/m width if $H = 3$ m and $\alpha = 5°$.

6.2   Refer to Figure 6.27. Find the flow rate in m³/sec/m length (at right angles to the cross section shown) through the permeable soil layer. Given $H = 6$ m, $H_1 = 2.5$ m, $h = 2.8$ m, $L = 40$ m, $\alpha = 10°$, $k = 0.05$ cm/sec.

6.3   Refer to the constant-head arrangement shown in Figure 6.5. For a test, the following are given:
   • $L = 24$ in.
   • $A = $ area of the specimen $= 4$ in.^{2}
   • Constant head difference $= h = 30$ in.
   • Water collected in 3 min $= 25.1$ in.^{3}
   Calculate the hydraulic conductivity (in./min).

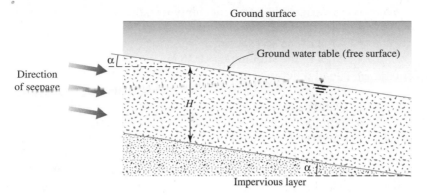

Figure 6.26

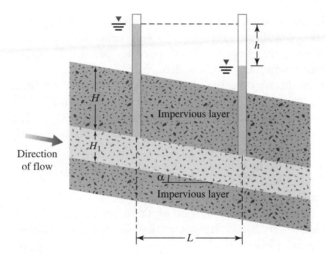

Figure 6.27

6.4   In a constant-head permeability test in the laboratory, the following are given: $L = 300$ mm and $A = 110$ cm^2. If the value of $k = 0.02$ cm/sec and a flow rate of 140 cm^3/min must be maintained through the soil, what is the head difference, h, across the specimen? Also, determine the discharge velocity under the test conditions.

6.5   Refer to Figure 6.5. For a constant-head permeability test in a sand, the following are given:
- $L = 400$ mm
- $A = 135$ cm^2
- $h = 450$ mm
- Water collected in 3 min = 640 cm^3
- Void ratio of sand = 0.54

Determine the
a. Hydraulic conductivity, k (cm/sec)
b. Seepage velocity

6.6   For a variable-head permeability test, the following are given:
- Length of the soil specimen = 20 in.
- Area of the soil specimen = 2.5 in.^{2}
- Area of the standpipe = 0.15 in.^{2}
- Head difference at time $t = 0$ is 30 in.
- Head difference at time $t = 8$ min is 16 in.

    a. Determine the hydraulic conductivity of the soil (in./min)
    b. What is the head difference at time $t = 6$ min?

6.7   For a variable-head test, the following are given: length of specimen = 380 mm; area of specimen = 6.5 cm^2; $k = 0.175$ cm/min. What should be the area of the standpipe for the head to drop from 650 cm to 300 cm in 8 min?

6.8   The hydraulic conductivity k of a soil is 10^{-6} cm/sec at a temperature of 28° C. Determine its absolute permeability at 20° C given that, at 20° C, $\gamma_w = 9.789$ kN/m^3 and $\eta = 1.005 \times 10^{-3}$ N.s/m^2 (Newton second per meter squared).

6.9   The hydraulic conductivity of a sand at a void ratio of 0.58 is 0.04 cm/sec. Estimate its hydraulic conductivity at a void ratio of 0.45. Use Eq. (6.24).

6.10   The following are given for a sand: porosity $(n) = 0.31$ and $k = 0.062$ cm/sec. Determine k when $n = 0.4$. Use Eq. (6.25).

6.11   The maximum dry density determined in the laboratory for a quartz sand is 1650 kg/m^3. In the field, if the relative of compaction is 90%, determine the hydraulic conductivity of the sand in the field compaction condition (given that k for the sand at the maximum dry density condition is 0.04 cm/sec and $G_s = 2.68$). Use Eq. (6.25).

6.12   For a sandy soil, the following are given:
- Maximum void ratio = 0.68
- Minimum void ratio = 0.42
- Hydraulic conductivity of sand at a relative density of 70% = 0.006 cm/sec

    Determine the hydraulic conductivity of the sand at a relative density of 32%. Use Eq. (6.24).

6.13   Repeat Problem 6.12 using Eq. (6.25).

6.14   For a normally consolidated clay, the following values are given:

Void ratio, e	k (cm/sec)
0.75	1.2×10^{-6}
1.2	2.8×10^{-6}

    Estimate the hydraulic conductivity of the clay at a void ratio $e = 0.6$. Use Eq. (6.28).

6.15   For a normally consolidated clay, the following values are given:

Void ratio, e	k (cm/sec)
0.95	0.2×10^{-6}
1.6	0.91×10^{-6}

    Determine the magnitude of k at a void ratio of 1.1. Use Eq. (6.28).

6.16   Figure 6.28 shows three layers of soil in a tube that is 100 mm × 100 mm in cross section. Water is supplied to maintain a constant head difference of

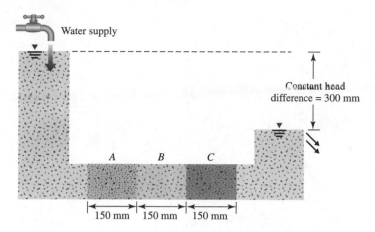

Figure 6.28

300 mm across the sample. The hydraulic conductivity of the soils in the direction of flow through them are as follows:

Soil	k (cm/sec)
A	10^{-2}
B	3×10^{-3}
C	4.9×10^{-4}

Find the rate of water supply in cm^3/hr.

6.17 For a clay soil, the following are given:
- Saturated unit weight = 121 lb/ft^3
- Specific gravity of soil solids (G_s) = 2.69
- Liquid limit = 46
- Plastic limit = 24
- Percent finer than 0.002 mm = 62

Estimate the hydraulic conductivity, k. Use Figure 6.10.

6.18 A layered soil is shown in Figure 6.29. Given that
- $H_1 = 1$ m $\quad k_1 = 10^{-4}$ cm/sec
- $H_2 = 1$ m $\quad k_2 = 2.8 \times 10^{-2}$ cm/sec
- $H_3 = 2$ m $\quad k_3 = 3.5 \times 10^{-5}$ cm/sec

Estimate the ratio of equivalent permeability, $k_{H(eq)}/k_{V(eq)}$.

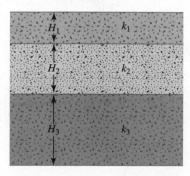

Figure 6.29

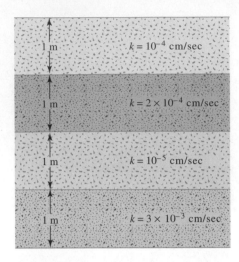

Figure 6.30

6.19   A layered soil is shown in Figure 6.30. Estimate the ratio of equivalent permeability, $k_{H(\text{eq})}/k_{V(\text{eq})}$.

References

AHMAD, S., LACROIX, Y., and STEINBACK, J. (1975). "Pumping Tests in an Unconfined Aquifer," *Proceedings,* Conference on *in situ* Measurement of Soil Properties, ASCE, Vol. 1, 1–21.

AL-TABBAA, A., and WOOD, D. M. (1987). "Some Measurements of the Permeability of Kaolin," *Geotechnique,* Vol. 37, 499–503.

AMER, A. M., and AWAD, A. A. (1974). "Permeability of Cohesionless Soils," *Journal of the Geotechnical Engineering Division,* ASCE, Vol. 100, No. GT12, 1309–1316.

BASAK, P. (1972). "Soil Structure and Its Effects on Hydraulic Conductivity," *Soil Science,* Vol. 114, No. 6, 417–422.

BENSON, C. H., and DANIEL, D. E. (1990). "Influence of Clods on Hydraulic Conductivity of Compacted Clay," *Journal of Geotechnical Engineering,* ASCE, Vol. 116, No. 8, 1231–1248.

CHAN, H. T., and KENNEY, T. C. (1973). "Laboratory Investigation of Permeability Ratio of New Liskeard Varved Soil," *Canadian Geotechnical Journal,* Vol. 10, No. 3, 453–472.

CHAPUIS, R. P., GILL, D. E., and BAASS, K. (1989). "Laboratory Permeability Tests on Sand: Influence of Compaction Method on Anisotropy," *Canadian Geotechnical Journal,* Vol. 26, 614–622.

DANIEL, D. E. (1989). "*In Situ* Hydraulic Conductivity Tests for Compacted Clay," *Journal of Geotechnical Engineering,* ASCE, Vol. 115, No. 9, 1205–1226.

DANIEL, D. E., and BENSON, C. H. (1990). "Water Content-Density Criteria for Compacted Soil Liners," *Journal of Geotechnical Engineering,* ASCE, Vol. 116, No. 12, 1811–1830.

DARCY, H. (1856). *Les Fontaines Publiques de la Ville de Dijon,* Dalmont, Paris.

ENVIRONMENTAL PROTECTION AGENCY (1989). *Requirements for Hazardous Waste Landfill Design, Construction, and Closure,* Publication no. EPA-625/4-89-022, Cincinnati, Ohio.

HANSBO, S. (1960). "Consolidation of Clay with Special Reference to Influence of Vertical Sand Drains," Swedish Geotechnical Institute, *Proc. No. 18,* 41–61.

HAZEN, A. (1930). "Water Supply," in *American Civil Engineers Handbook,* Wiley, New York.

KENNEY, T. C., LAU, D., and OFOEGBU, G. I. (1984). "Permeability of Compacted Granular Materials," *Canadian Geotechnical Journal,* Vol. 21, No. 4, 726–729.

KRUMBEIN, W. C., and MONK, G. D. (1943). "Permeability as a Function of the Size Parameters of Unconsolidated Sand," *Transactions,* AIMME (Petroleum Division), Vol. 151, 153–163.

MESRI, G., and OLSON, R. E. (1971). "Mechanism Controlling the Permeability of Clays," *Clay and Clay Minerals,* Vol. 19, 151–158.

MITCHELL, J. K. (1976). *Fundamentals of Soil Behavior,* Wiley, New York.

MITCHELL, J. K., HOOPER, D. R., and CAMPANELLA, R. G. (1965). "Permeability of Compacted Clay," *Journal of the Soil Mechanics and Foundations Division,* ASCE, Vol. 91, No. SM4, 41–65.

SAMARASINGHE, A. M., HUANG, Y. H., and DRNEVICH, V. P. (1982). "Permeability and Consolidation of Normally Consolidated Soils," *Journal of the Geotechnical Engineering Division,* ASCE, Vol. 108, No. GT6, 835–850.

SHAHABI, A. A., DAS, B. M., and TARQUIN, A. J. (1984). "An Empirical Relation for Coefficient of Permeability of Sand," *Proceedings,* Fourth Australia-New Zealand Conference on Geomechanics, Vol. 1, 54–57.

TAVENAS, F., JEAN, P., LEBLOND, F. T. P., and LEROUEIL, S. (1983). "The Permeability of Natural Soft Clays. Part II: Permeability Characteristics," *Canadian Geotechnical Journal,* Vol. 20, No. 4, 645–660.

U.S. DEPARTMENT OF NAVY (1971). "Design Manual — Soil Mechanics, Foundations, and Earth Structures," *NAVFAC DM-7,* U.S. Government Printing Office, Washington, D.C.

7

Seepage

In the preceding chapter, we considered some simple cases for which direct application of Darcy's law was required to calculate the flow of water through soil. In many instances, the flow of water through soil is not in one direction only, nor is it uniform over the entire area perpendicular to the flow. In such cases, the groundwater flow is generally calculated by the use of graphs referred to as *flow nets*. The concept of the flow net is based on *Laplace's equation of continuity*, which governs the steady flow condition for a given point in the soil mass. In the following sections of this chapter, the derivation of Laplace's equation of continuity will be presented along with its application to seepage problems.

7.1 *Laplace's Equation of Continuity*

To derive the Laplace differential equation of continuity, let us consider a single row of sheet piles that have been driven into a permeable soil layer, as shown in Figure 7.1a. The row of sheet piles is assumed to be impervious. The steady state flow of water from the upstream to the downstream side through the permeable layer is a two-dimensional flow. For flow at a point A, we consider an elemental soil block. The block has dimensions dx, dy, and dz (length dy is perpendicular to the plane of the paper); it is shown in an enlarged scale in Figure 7.1b. Let v_x and v_z be the components of the discharge velocity in the horizontal and vertical directions, respectively. The rate of flow of water into the elemental block in the horizontal direction is equal to $v_x\, dz\, dy$, and in the vertical direction it is $v_z\, dx\, dy$. The rates of outflow from the block in the horizontal and vertical directions are, respectively,

$$\left(v_x + \frac{\partial v_x}{\partial x}\, dx \right) dz\, dy$$

and

$$\left(v_z + \frac{\partial v_z}{\partial z}\, dz \right) dx\, dy$$

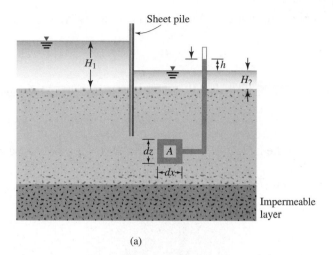

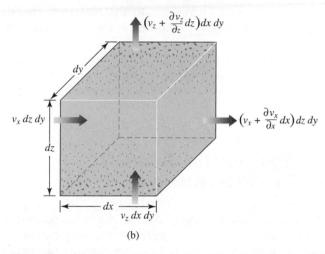

Figure 7.1 (a) Single-row sheet piles driven into permeable layer; (b) flow at A

Assuming that water is incompressible and that no volume change in the soil mass occurs, we know that the total rate of inflow should equal the total rate of outflow. Thus,

$$\left[\left(v_x + \frac{\partial v_x}{\partial x} dx \right) dz\, dy + \left(v_z + \frac{\partial v_z}{\partial z} dz \right) dx\, dy \right] - [v_x\, dz\, dy + v_z\, dx\, dy] = 0$$

or

$$\frac{\partial v_x}{\partial x} + \frac{\partial v_z}{\partial z} = 0 \tag{7.1}$$

With Darcy's law, the discharge velocities can be expressed as

$$v_x = k_x i_x = k_x \frac{\partial h}{\partial x} \tag{7.2}$$

and

$$v_z = k_z i_z = k_z \frac{\partial h}{\partial z} \tag{7.3}$$

where k_x and k_z are the hydraulic conductivities in the vertical and horizontal directions, respectively.

From Eqs. (7.1), (7.2), and (7.3), we can write

$$k_x \frac{\partial^2 h}{\partial x^2} + k_z \frac{\partial^2 h}{\partial z^2} = 0 \tag{7.4}$$

If the soil is isotropic with respect to the hydraulic conductivity — that is, $k_x = k_z$ — the preceding continuity equation for two-dimensional flow simplifies to

$$\frac{\partial^2 h}{\partial x^2} + \frac{\partial^2 h}{\partial z^2} = 0 \tag{7.5}$$

7.2   Continuity Equation for Solution of Simple Flow Problems

The continuity equation given in Eq. (7.5) can be used in solving some simple flow problems. To illustrate this, let us consider a one-dimensional flow problem, as shown in Figure 7.2, in which a constant head is maintained across a two-layered soil for the flow of water. The head difference between the top of soil layer no. 1 and the bottom of soil layer no. 2 is h_1. Because the flow is in only the z direction, the continuity equation [Eq. (7.5)] is simplified to the form

$$\frac{\partial^2 h}{\partial z^2} = 0 \tag{7.6}$$

or

$$h = A_1 z + A_2 \tag{7.7}$$

where A_1 and A_2 are constants.

To obtain A_1 and A_2 for flow through soil layer no. 1, we must know the boundary conditions, which are as follows:

- *Condition 1*: At $z = 0$, $h = h_1$.
- *Condition 2*: At $z = H_1$, $h = h_2$.

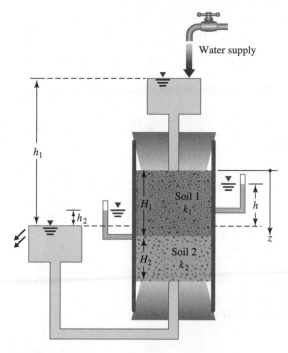

Figure 7.2 Flow through a two-layered soil

Combining Eq. (7.7) and condition 1 gives

$$A_2 = h_1 \qquad (7.8)$$

Similarly, combining Eq. (7.7) and condition 2 with Eq. (7.8) gives

$$h_2 = A_1 H_1 + h_1$$

or

$$A_1 = -\left(\frac{h_1 - h_2}{H_1}\right) \qquad (7.9)$$

Combining Eqs. (7.7), (7.8), and (7.9), we obtain

$$h = -\left(\frac{h_1 - h_2}{H_1}\right)z + h_1 \qquad (\text{for } 0 \le z \le H_1) \qquad (7.10)$$

For flow through soil layer no. 2, the boundary conditions are

- *Condition 1*: At $z - H_1, h - h_2$.
- *Condition 2*: At $z = H_1 + H_2, h = 0$.

From condition 1 and Eq. (7.7),

$$A_2 = h_2 - A_1 H_1 \qquad (7.11)$$

Also, from condition 2 and Eqs. (7.7) and (7.11),

$$0 = A_1(H_1 + H_2) + (h_2 - A_1H_1)$$
$$A_1H_1 + A_1H_2 + h_2 - A_1H_1 = 0$$

or

$$A_1 = -\frac{h_2}{H_2} \qquad (7.12)$$

So, from Eqs. (7.7), (7.11), and (7.12),

$$h = -\left(\frac{h_2}{H_2}\right)z + h_2\left(1 + \frac{H_1}{H_2}\right) \qquad \text{(for } H_1 \le z \le H_1 + H_2) \qquad (7.13)$$

At any given time, flow through soil layer no. 1 equals flow through soil layer no. 2, so

$$q = k_1\left(\frac{h_1 - h_2}{H_1}\right)A = k_2\left(\frac{h_2 - 0}{H_2}\right)A$$

where A = area of cross section of the soil
k_1 = hydraulic conductivity of soil layer no. 1
k_2 = hydraulic conductivity of soil layer no. 2

or

$$h_2 = \frac{h_1 k_1}{H_1\left(\dfrac{k_1}{H_1} + \dfrac{k_2}{H_2}\right)} \qquad (7.14)$$

Substituting Eq. (7.14) into Eq. (7.10), we obtain

$$h = h_1\left(1 - \frac{k_2 z}{k_1 H_2 + k_2 H_1}\right) \qquad \text{(for } 0 \le z \le H_1) \qquad (7.15)$$

Similarly, combining Eqs. (7.13) and (7.14) gives

$$h = h_1\left[\left(\frac{k_1}{k_1 H_2 + k_2 H_1}\right)(H_1 + H_2 - z)\right] \qquad \text{(for } H_1 \le z \le H_1 + H_2)$$

$$(7.16)$$

Example 7.1

Refer to Figure 7.2. Given that $H_1 = 300$ mm, $H_2 = 500$ mm, and $h_1 = 600$ mm, and that at $z = 200$ mm, $h = 500$ mm, determine h at $z = 600$ mm.

Solution
We know that $z = 200$ mm is located in soil layer no. 1, so Eq. (7.15) is valid. Thus,

$$h = h_1 \left(1 - \frac{k_2 z}{k_1 H_2 + k_2 H_1} \right)$$

$$500 = 600 \left[1 - \frac{k_2(200)}{k_1(500) + k_2(300)} \right]$$

or

$$\frac{k_1}{k_2} = \mathbf{1.8}$$

Because $z = 600$ mm is located in soil layer no. 2, Eq. (7.16) is valid, yielding

$$h = h_1 \left[\frac{1}{H_2 + \left(\dfrac{k_2}{k_1} \right) H_1} (H_1 + H_2 - z) \right]$$

or

$$h = 600 \left[\left(\frac{1}{500 + \dfrac{300}{1.8}} \right)(300 + 500 - 600) \right] = \mathbf{179.9\ mm} \qquad \blacksquare$$

7.3 *Flow Nets*

The continuity equation [Eq. (7.5)] in an isotropic medium represents two orthogonal families of curves — that is, the flow lines and the equipotential lines. A *flow line* is a line along which a water particle will travel from upstream to the downstream side in the permeable soil medium. An *equipotential line* is a line along which the potential head at all points is equal. Thus, if piezometers are placed at different points along an equipotential line, the water level will rise to the same elevation in all of them. Figure 7.3a demonstrates the definition of flow and equipotential lines for flow in the permeable soil layer around the row of sheet piles shown in Figure 7.1 (for $k_x = k_z = k$).

A combination of a number of flow lines and equipotential lines is called a *flow net*. As mentioned in the introduction, flow nets are constructed for the calculation of groundwater flow and the evaluation of heads in the media. To complete the graphic construction of a flow net, one must draw the flow and equipotential lines in such a way that

1. The equipotential lines intersect the flow lines at right angles.
2. The flow elements formed are approximate squares.

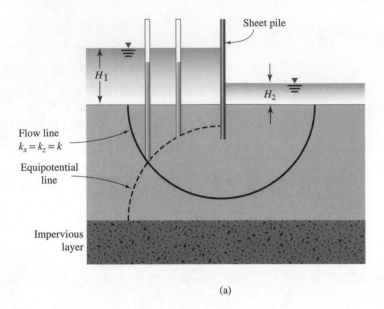

(a)

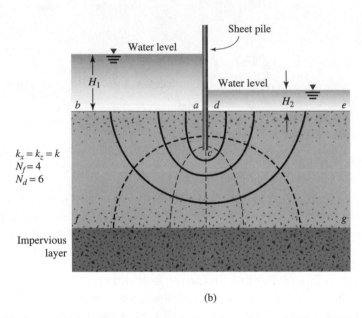

(b)

Figure 7.3 (a) Definition of flow lines and equipotential lines; (b) completed flow net

Figure 7.3b shows an example of a completed flow net. Two more examples of flow net in isotropic permeable layer are given in Figures 7.4 and 7.5. In these figures, N_f is the number of flow channels in the flow net, and N_d is the number of potential drops (defined later in this chapter).

Drawing a flow net takes several trials. While constructing the flow net, keep the boundary conditions in mind. For the flow net shown in Figure 7.3b, the following four boundary conditions apply:

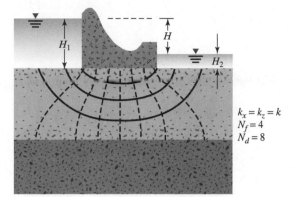

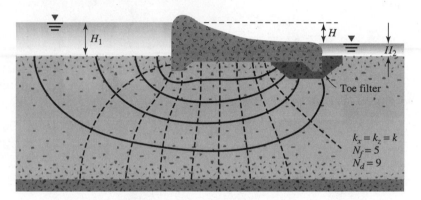

Figure 7.4 Flow net under a dam

Figure 7.5 Flow net under a dam with toe filter

1. The upstream and downstream surfaces of the permeable layer (lines *ab* and *de*) are equipotential lines.
2. Because *ab* and *de* are equipotential lines, all the flow lines intersect them at right angles.
3. The boundary of the impervious layer — that is, line *fg* — is a flow line, and so is the surface of the impervious sheet pile, line *acd*.
4. The equipotential lines intersect *acd* and *fg* at right angles.

7.4 Seepage Calculation from a Flow Net

In any flow net, the strip between any two adjacent flow lines is called a *flow channel*. Figure 7.6 shows a flow channel with the equipotential lines forming square elements. Let h_1, h_2, h_3, h_4, . . ., h_n be the piezometric levels corresponding to the equipotential lines. The rate of seepage through the flow channel per unit length (perpendicular to the vertical section through the permeable layer) can be calculated as follows: Because there is no flow across the flow lines,

$$\Delta q_1 = \Delta q_2 = \Delta q_3 = \cdots = \Delta q \tag{7.17}$$

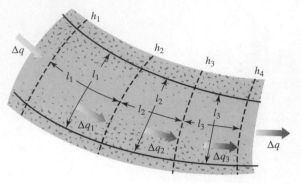

Figure 7.6
Seepage through a flow channel with square elements

From Darcy's law, the flow rate is equal to kiA. Thus, Eq. (7.17) can be written as

$$\Delta q = k\left(\frac{h_1 - h_2}{l_1}\right)l_1 = k\left(\frac{h_2 - h_3}{l_2}\right)l_2 = k\left(\frac{h_3 - h_4}{l_3}\right)l_3 = \cdots \qquad (7.18)$$

Eq. (7.18) shows that if the flow elements are drawn as approximate squares, the drop in the piezometric level between any two adjacent equipotential lines is the same. This is called the *potential drop*. Thus,

$$h_1 - h_2 = h_2 - h_3 = h_3 - h_4 = \cdots = \frac{H}{N_d} \qquad (7.19)$$

and

$$\Delta q = k\frac{H}{N_d} \qquad (7.20)$$

where H = head difference between the upstream and downstream sides
      N_d = number of potential drops

In Figure 7.3a, for any flow channel, $H = H_1 - H_2$ and $N_d = 6$.

If the number of flow channels in a flow net is equal to N_f, the total rate of flow through all the channels per unit length can be given by

$$q = k\frac{HN_f}{N_d} \qquad (7.21)$$

Although drawing square elements for a flow net is convenient, it is not always necessary. Alternatively, one can draw a rectangular mesh for a flow channel, as shown in Figure 7.7, provided that the width-to-length ratios for all the rectangular elements in the flow net are the same. In this case, Eq. (7.18) for rate of flow through the channel can be modified to

$$\Delta q = k\left(\frac{h_1 - h_2}{l_1}\right)b_1 = k\left(\frac{h_2 - h_3}{l_2}\right)b_2 = k\left(\frac{h_3 - h_4}{l_3}\right)b_3 = \cdots \qquad (7.22)$$

If $b_1/l_1 = b_2/l_2 = b_3/l_3 = \cdots = n$ (i.e., the elements are not square), Eqs. (7.20) and (7.21) can be modified to

$$\Delta q = kH\left(\frac{n}{N_d}\right) \qquad (7.23)$$

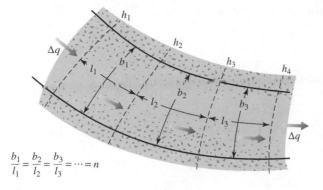

Figure 7.7
Seepage through a flow channel with rectangular elements

$$\frac{b_1}{l_1} = \frac{b_2}{l_2} = \frac{b_3}{l_3} = \cdots = n$$

or

$$q = kH\left(\frac{N_f}{N_d}\right)n \qquad (7.24)$$

Figure 7.8 shows a flow net for seepage around a single row of sheet piles. Note that flow channels 1 and 2 have square elements. Hence, the rate of flow through these two channels can be obtained from Eq. (7.20):

$$\Delta q_1 + \Delta q_2 = \frac{k}{N_d}H + \frac{k}{N_d}H = \frac{2kH}{N_d}$$

However, flow channel 3 has rectangular elements. These elements have a width-to-length ratio of about 0.38; hence, from Eq. (7.23)

$$\Delta q_3 = \frac{k}{N_d}H(0.38)$$

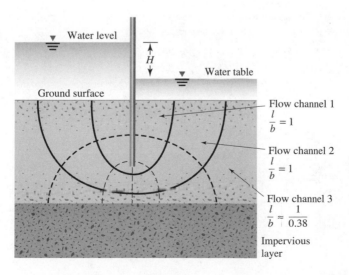

Figure 7.8
Flow net for seepage around a single row of sheet piles

So, the total rate of seepage can be given as

$$q = \Delta q_1 + \Delta q_2 + \Delta q_3 = 2.38\frac{kH}{N_d}$$

Example 7.2

A flow net for flow around a single row of sheet piles in a permeable soil layer is shown in Figure 7.9. Given that $k_x = k_z = k = 4.2 \times 10^{-3}$ cm/sec, determine

a. How high (above the ground surface) the water will rise if piezometers are placed at points a, b, c, and d

b. The rate of seepage through flow channel II per unit length (perpendicular to the section shown)

Solution

a. From Figure 7.9, we see that $N_f = 3$ and $N_d = 6$. The head difference between the upstream and downstream sides is 3.5 m, so the head loss for each drop is 3.5/6 = 0.583 m.

Point a is located on equipotential line 1, which means that the potential drop at a is 1×0.583 m. The water in the piezometer at a will rise to an elevation of $(5 - 0.583) = 4.417$ m **above the ground surface**.

Similarly, the piezometric levels for

$$b = (5 - 2 \times 0.583) = 3.834 \text{ m above the ground surface}$$

$$c = (5 - 5 \times 0.583) = 2.085 \text{ m above the ground surface}$$

$$d = (5 - 5 \times 0.583) = 2.085 \text{ m above the ground surface}$$

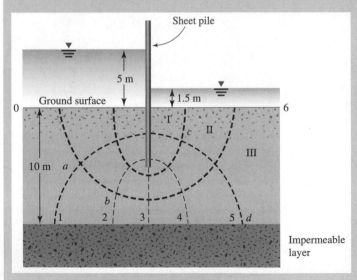

Figure 7.9

b. From Eq. (7.20),

$$\Delta q = k \frac{H}{N_d}$$

$$k = 4.2 \times 10^{-3} \text{ cm/sec} = 4.2 \times 10^{-5} \text{ m/sec}$$

$$\Delta q = (4.2 \times 10^{-5})(0.583) = 2.45 \times 10^{-5} \text{ m}^3\text{/sec/m}$$ ∎

7.5 *Flow Nets in Anisotropic Soil*

The flow net construction described thus far and the derived Eqs. (7.21) and (7.24) for seepage calculation have been based on the assumption that the soil is isotropic. However, in nature, most soils exhibit some degree of anisotropy. To account for soil anisotropy with respect to hydraulic conductivity, we must modify the flow net construction.

The differential equation of continuity for a two-dimensional flow [Eq. (7.4)] is

$$k_x \frac{\partial^2 h}{\partial x^2} + k_z \frac{\partial^2 h}{\partial z^2} = 0$$

For anisotropic soils, $k_x \neq k_z$. In this case, the equation represents two families of curves that do not meet at 90°. However, we can rewrite the preceding equation as

$$\frac{\partial^2 h}{(k_z/k_x) \partial x^2} + \frac{\partial^2 h}{\partial z^2} = 0 \tag{7.25}$$

Substituting $x' = \sqrt{k_z/k_x} \, x$, we can express Eq. (7.25) as

$$\frac{\partial^2 h}{\partial x'^2} + \frac{\partial^2 h}{\partial z^2} = 0 \tag{7.26}$$

Now Eq. (7.26) is in a form similar to that of Eq. (7.5), with x replaced by x', which is the new transformed coordinate. To construct the flow net, use the following procedure:

1. Adopt a vertical scale (that is, z axis) for drawing the cross section.
2. Adopt a horizontal scale (that is, x axis) such that horizontal scale = $\sqrt{k_z/k_x} \times$ vertical scale.
3. With scales adopted as in steps 1 and 2, plot the vertical section through the permeable layer parallel to the direction of flow.
4. Draw the flow net for the permeable layer on the section obtained from step 3, with flow lines intersecting equipotential lines at right angles and the elements as approximate squares.

The rate of seepage per unit length can be calculated by modifying Eq. (7.21) to

$$q = \sqrt{k_x k_z} \frac{H N_f}{N_d} \tag{7.27}$$

where H = total head loss

N_f and N_d = number of flow channels and potential drops, respectively (from flow net drawn in step 4)

Note that when flow nets are drawn in transformed sections (in anisotropic soils), the flow lines and the equipotential lines are orthogonal. However, when they are redrawn in a true section, these lines are not at right angles to each other. This fact is shown in Figure 7.10. In this figure, it is assumed that $k_x = 6k_z$. Figure 7.10a shows a flow element in a transformed section. The flow element has been redrawn in a true section in Figure 7.10b.

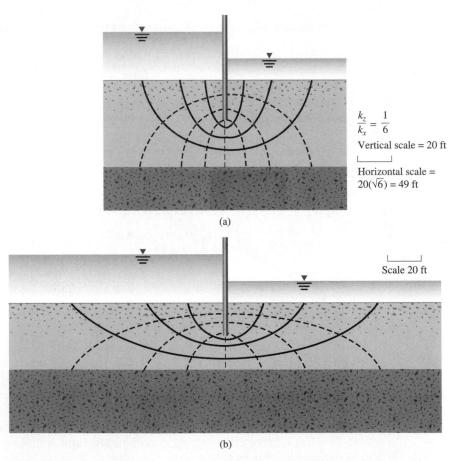

$$\frac{k_z}{k_x} = \frac{1}{6}$$

Vertical scale = 20 ft

Horizontal scale = $20(\sqrt{6}) = 49$ ft

(a)

Scale 20 ft

(b)

Figure 7.10 A flow element in anisotropic soil: (a) in transformed section; (b) in true section

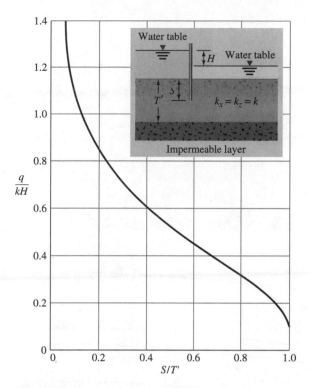

Figure 7.11 Plot of q/kH against S/T' for flow around a single row of sheet piles (after Harr, 1962)

7.6 Mathematical Solution for Seepage

The seepage under several simple hydraulic structures can be solved mathematically. Harr (1962) has analyzed many such conditions. Figure 7.11 shows a nondimensional plot for the rate of seepage around a single row of sheet piles. In a similar manner, Figure 7.12 is a nondimensional plot for the rate of seepage under a dam. In Figure 7.11, the depth of penetration of the sheet pile is S, and the thickness of the permeable soil layer is T'.

7.7 Uplift Pressure under Hydraulic Structures

Flow nets can be used to determine the uplift pressure at the base of a hydraulic structure. This general concept can be demonstrated by a simple example. Figure 7.13a shows a weir, the base of which is 2 m below the ground surface. The necessary flow net has also been drawn (assuming that $k_x = k_z = k$). The pressure distribution diagram at the base of the weir can be obtained from the equipotential lines as follows.

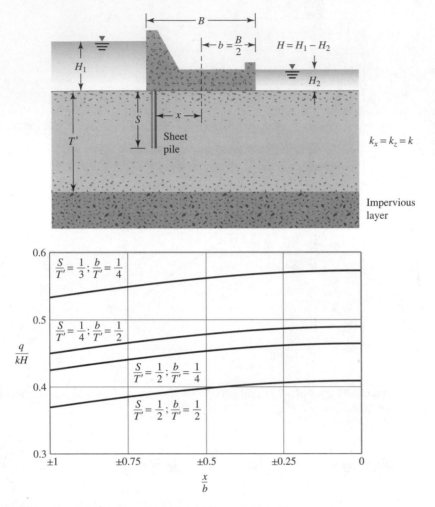

Figure 7.12 Seepage under a dam (after Harr, 1962)

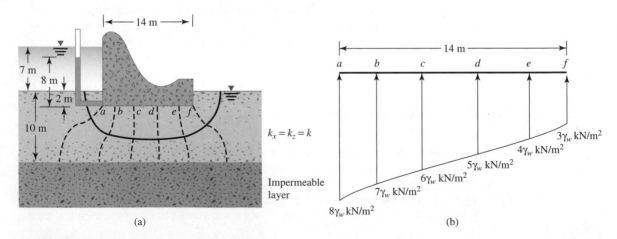

Figure 7.13 (a) A weir; (b) uplift force under a hydraulic structure

There are seven equipotential drops (N_d) in the flow net, and the difference in the water levels between the upstream and downstream sides is $H = 7$ m. The head loss for each potential drop is $H/7 = 7/7 = 1$ m. The uplift pressure at

$$a \text{ (left corner of the base)} = \text{(Pressure head at } a\text{)} \times (\gamma_w)$$
$$= [(7 + 2) - 1]\gamma_w = 8\gamma_w$$

Similarly, the uplift pressure at

$$b = [9 - (2)(1)]\gamma_w = 7\gamma_w$$

and at

$$f = [9 - (6)(1)]\gamma_w = 3\gamma_w$$

The uplift pressures have been plotted in Figure 7.13b. The uplift force per unit length measured along the axis of the weir can be calculated by finding the area of the pressure diagram.

7.8   *Seepage through an Earth Dam on an Impervious Base*

Figure 7.14 shows a homogeneous earth dam resting on an impervious base. Let the hydraulic conductivity of the compacted material of which the earth dam is made be equal to k. The free surface of the water passing through the dam is given by $abcd$. It is assumed that $a'bc$ is parabolic. The slope of the free surface can be assumed to be equal to the hydraulic gradient. It is also assumed that, because this hydraulic gradient is constant with depth (Dupuit, 1863),

$$i \simeq \frac{dz}{dx} \tag{7.28}$$

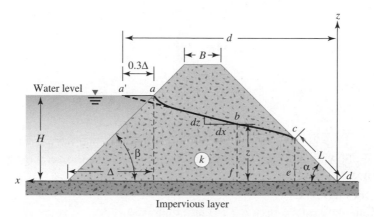

Figure 7.14 Flow through an earth dam constructed over an impervious base

Considering the triangle *cde,* we can give the rate of seepage per unit length of the dam (at right angles to the cross section shown in Figure 7.14) as

$$q = kiA$$

$$i = \frac{dz}{dx} = \tan \alpha$$

$$A = (\overline{ce})(1) = L \sin \alpha$$

So

$$q = k(\tan \alpha)(L \sin \alpha) = kL \tan \alpha \sin \alpha \qquad (7.29)$$

Again, the rate of seepage (per unit length of the dam) through the section *bf* is

$$q = kiA = k\left(\frac{dz}{dx}\right)(z \times 1) = kz\frac{dz}{dx} \qquad (7.30)$$

For continuous flow,

$$q_{\text{Eq. (7.29)}} = q_{\text{Eq. (7.30)}}$$

or

$$kz\frac{dz}{dx} = kL \tan \alpha \sin \alpha$$

or

$$\int_{z=L \sin \alpha}^{z=H} kz \, dz = \int_{x=L \cos \alpha}^{x=d} (kL \tan \alpha \sin \alpha) \, dx$$

$$\tfrac{1}{2}(H^2 - L^2 \sin^2 \alpha) = L \tan \alpha \sin \alpha(d - L \cos \alpha)$$

$$\frac{H^2}{2} - \frac{L^2 \sin^2 \alpha}{2} = Ld\left(\frac{\sin^2 \alpha}{\cos \alpha}\right) - L^2 \sin^2 \alpha$$

$$\frac{H^2 \cos \alpha}{2 \sin^2 \alpha} - \frac{L^2 \cos \alpha}{2} = Ld - L^2 \cos \alpha$$

or

$$L^2 \cos \alpha - 2Ld + \frac{H^2 \cos \alpha}{\sin^2 \alpha} = 0$$

$$L = \frac{d}{\cos \alpha} - \sqrt{\frac{d^2}{\cos^2 \alpha} - \frac{H^2}{\sin^2 \alpha}} \qquad (7.31)$$

Following is a step-by-step procedure to obtain the seepage rate q (per unit length of the dam):

1. Obtain α.
2. Calculate Δ (see Figure 7.14) and then 0.3Δ.

3. Calculate d.
4. With known values of α and d, calculate L from Eq. (7.31).
5. With known values of L, calculate q from Eq. (7.29).

Example 7.3

Refer to the earth dam shown in Figure 7.14. Given that $\beta = 45°$, $\alpha = 30°$, $B = 10$ ft, $H = 20$ ft, height of dam = 25 ft, and $k = 2 \times 10^{-4}$ ft/min, calculate the seepage rate q in ft³/day/ft length.

Solution
We know that $\beta = 45°$ and $\alpha = 30°$. Thus,

$$\Delta = \frac{H}{\tan \beta} = \frac{20}{\tan 45°} = 20 \text{ ft} \qquad 0.3\Delta = (0.3)(20) = 6 \text{ ft}$$

$$d = 0.3\Delta + \frac{(25 - 20)}{\tan \beta} + B + \frac{25}{\tan \alpha}$$

$$= 6 + \frac{(25 - 20)}{\tan 45°} + 10 + \frac{25}{\tan 30} = 64.3 \text{ ft}$$

From Eq. (7.31),

$$L = \frac{d}{\cos \alpha} - \sqrt{\frac{d^2}{\cos^2 \alpha} - \frac{H^2}{\sin^2 \alpha}}$$

$$= \frac{64.3}{\cos 30} - \sqrt{\left(\frac{64.3}{\cos 30}\right)^2 - \left(\frac{20}{\sin 30}\right)^2} = 11.7 \text{ ft}$$

From Eq. (7.29),

$$q = kL \tan \alpha \sin \alpha = (2 \times 10^{-4})(11.7)(\tan 30)(\sin 30)$$

$$= 6.754 \times 10^{-4} \text{ ft}^3/\text{min/ ft} = \textbf{0.973 ft}^3/\textbf{day/ft} \qquad \blacksquare$$

7.9   L. Casagrande's Solution for Seepage through an Earth Dam

Equation (7.31) is derived on the basis of Dupuit's assumption (i.e., $i \approx dz/dx$). It was shown by Casagrande (1932) that, when the downstream slope angle α in Figure 7.14 becomes greater than 30°, deviations from Dupuit's assumption become more noticeable. Thus (see Fig. 7.14), L. Casagrande (1932) suggested that

$$i = \frac{dz}{ds} = \sin \alpha \qquad (7.32)$$

where $ds = \sqrt{dx^2 + dz^2}$

So Eq. (7.29) can now be modified as

$$q = kiA = k\sin\alpha(L\sin\alpha) = kL\sin^2\alpha \tag{7.33}$$

Again,

$$q = kiA = k\left(\frac{dz}{ds}\right)(1 \times z) \tag{7.34}$$

Combining Eqs. (7.33) and (7.34) yields

$$\int_{L\sin\alpha}^{H} z\,dz = \int_{L}^{s} L\sin^2\alpha\,ds \tag{7.35}$$

where s = length of curve $a'bc$

$$\frac{1}{2}(H^2 - L^2\sin^2\alpha) = L\sin^2\alpha(s - L)$$

or

$$L = s - \sqrt{s^2 - \frac{H^2}{\sin^2\alpha}} \tag{7.36}$$

With about 4–5% error, we can write

$$s = \sqrt{d^2 + H^2} \tag{7.37}$$

Combining Eqs. (7.36) and (7.37) yields

$$L = \sqrt{d^2 + H^2} - \sqrt{d^2 - H^2\cot^2\alpha} \tag{7.38}$$

Once the magnitude of L is known, the rate of seepage can be calculated from Eq. (7.33) as

$$q = kL\sin^2\alpha$$

Example 7.4

Solve Example 7.3 using L. Casagrande's method.

Solution
From Example 7.3, $d = 64.3$ ft, $H = 20$ ft, and $\alpha = 30°$. So

$$L = \sqrt{d^2 + H^2} - \sqrt{d^2 - H^2\cot^2\alpha}$$
$$= \sqrt{(64.3)^2 + (20)^2} - \sqrt{(64.3)^2 - 20^2\cot^2 30} = 13.17 \text{ ft}$$
$$q = (2 \times 10^{-4})(13.17)(\sin^2 30) = 6.585 \times 10^{-4} \text{ ft}^3/\text{min/ft} = \mathbf{0.948 \text{ ft}^3/day/ft} \quad \blacksquare$$

7.10   *Summary*

In this chapter, we studied Laplace's equation of continuity and its application in solving problems related to seepage calculation. The continuity equation is the fundamental basis on which the concept of drawing flow nets is derived. Flow nets are very powerful tools for calculation of seepage as well as uplift pressure under various hydraulic structures.

Also discussed in this chapter (Sections 7.8 and 7.9) is the procedure to calculate seepage through an earth dam constructed over an impervious base. Section 7.8 derives the relationship for seepage based on Dupuit's assumption that the hydraulic gradient is constant with depth. An improved procedure (L. Casagrande's solution) for seepage calculation is provided in Section 7.9.

Problems

7.1   Refer to Figure 7.15. Given that
- H_1 = 6 m D = 3 m
- H_2 = 1.5 m D_1 = 6 m

draw a flow net. Calculate the seepage loss per meter length of the sheet pile (at a right angle to the cross section shown).

7.2   Draw a flow net for the single row of sheet piles driven into a permeable layer as shown in Figure 7.15. Given that
- H_1 = 6 m D = 2 m
- H_2 = 1 m D_1 = 5 m

calculate the seepage loss per meter length of the sheet pile (at right angles to the cross section shown).

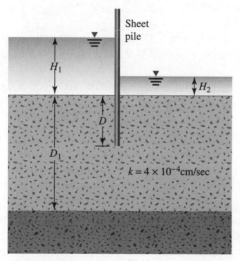

$k = 4 \times 10^{-4}$ cm/sec

Impermeable layer

Figure 7.15

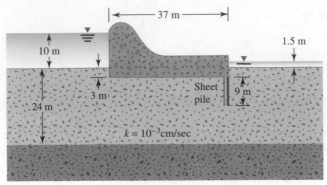

Figure 7.16

7.3 Draw a flow net for the weir shown in Figure 7.16. Calculate the rate of seepage under the weir.

7.4 For the flow net drawn in Problem 7.3, calculate the uplift force at the base of the weir per meter length (measured along the axis) of the structure.

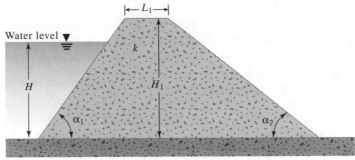

Figure 7.17

7.5 An earth dam is shown in Figure 7.17. Determine the seepage rate, q, in $m^3/day/m$ length. Given: $\alpha_1 = \alpha_2 = 45°$, $L_1 = 5$ m, $H = 10$ m, $H_1 = 13$ m, and $k = 2 \times 10^{-4}$ cm/sec.

7.6 Repeat Problem 7.5 with the following: $\alpha_1 = 28°$, $\alpha_2 = 35°$, $L_1 = 7$ m, $H = 8$ m, $H_1 = 12$ m, and $k = 1.5 \times 10^{-4}$ cm/sec.

7.7 Repeat Problem 7.5 using L. Casagrande's method (Section 7.9).

7.8 Repeat Problem 7.6 using L. Casagrande's method (Section 7.9).

References

CASAGRANDE, L. (1932). "Naeherungsmethoden zur Bestimmurg von Art und Menge der Sickerung durch geschuettete Daemme," Thesis, Technische Hochschule, Vienna.

DUPUIT, J. (1863). *Etudes Theoriques et Practiques sur le Mouvement des Eaux dans les Canaux Decouverts et a Travers les Terrains Permeables,* Dunod, Paris.

HARR, M. E. (1962). *Ground Water and Seepage,* McGraw-Hill, New York.

8

In Situ Stresses

As described in Chapter 3, soils are multiphase systems. In a given volume of soil, the solid particles are distributed randomly with void spaces between. The void spaces are continuous and are occupied by water and/or air. To analyze problems such as compressibility of soils, bearing capacity of foundations, stability of embankments, and lateral pressure on earth-retaining structures, we need to know the nature of the distribution of stress along a given cross section of the soil profile. We can begin the analysis by considering a saturated soil with no seepage.

8.1 Stresses in Saturated Soil without Seepage

Figure 8.1a shows a column of saturated soil mass with no seepage of water in any direction. The total stress at the elevation of point A can be obtained from the saturated unit weight of the soil and the unit weight of water above it. Thus,

$$\sigma = H\gamma_w + (H_A - H)\gamma_{\text{sat}} \tag{8.1}$$

where σ = total stress at the elevation of point A
    γ_w = unit weight of water
    γ_{sat} = saturated unit weight of the soil
    H = height of water table from the top of the soil column
    H_A = distance between point A and the water table

The total stress, σ, given by Eq. (8.1) can be divided into two parts:

1. A portion is carried by water in the continuous void spaces. This portion acts with equal intensity in all directions.
2. The rest of the total stress is carried by the soil solids at their points of contact. The sum of the vertical components of the forces developed at the points of contact of the solid particles per unit cross sectional area of the soil mass is called the *effective stress*.

This can be seen by drawing a wavy line, *a-a*, through point A that passes only through the points of contact of the solid particles. Let $P_1, P_2, P_3, \ldots, P_n$ be the forces that act at the points of contact of the soil particles (Figure 8.1b). The sum of the

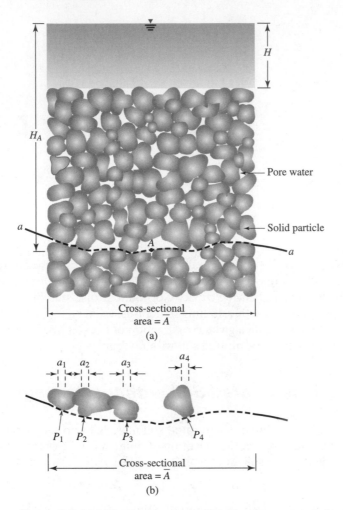

Figure 8.1 (a) Effective stress consideration for a saturated soil column without seepage; (b) forces acting at the points of contact of soil particles at the level of point A

vertical components of all such forces over the unit cross-sectional area is equal to the effective stress σ', or

$$\sigma' = \frac{P_{1(v)} + P_{2(v)} + P_{3(v)} + \cdots + P_{n(v)}}{\overline{A}} \tag{8.2}$$

where $P_{1(v)}, P_{2(v)}, P_{3(v)}, \ldots, P_{n(v)}$ are the vertical components of $P_1, P_2, P_3, \ldots, P_n$, respectively, and $\overline{A}$ is the cross-sectional area of the soil mass under consideration.

Again, if a_s is the cross-sectional area occupied by solid-to-solid contacts (that is, $a_s = a_1 + a_2 + a_3 + \cdots + a_n$), then the space occupied by water equals $(\overline{A} - a_s)$. So we can write

$$\sigma = \sigma' + \frac{u(\overline{A} - a_s)}{\overline{A}} = \sigma' + u(1 - a'_s) \tag{8.3}$$

where $u = H_A \gamma_w$ = pore water pressure (that is, the hydrostatic pressure at A)

$a_s' = a_s/A$ = fraction of unit cross-sectional area of the soil mass occupied by solid-to-solid contacts

The value of a_s' is extremely small and can be neglected for pressure ranges generally encountered in practical problems. Thus, Eq. (8.3) can be approximated by

$$\sigma = \sigma' + u \tag{8.4}$$

where u is also referred to as *neutral stress*. Substitution of Eq. (8.1) for σ in Eq. (8.4) gives

$$
\begin{aligned}
\sigma' &= [H\gamma_w + (H_A - H)\gamma_{\text{sat}}] - H_A\gamma_w \\
&= (H_A - H)(\gamma_{\text{sat}} - \gamma_w) \\
&= (\text{Height of the soil column}) \times \gamma' \tag{8.5}
\end{aligned}
$$

where $\gamma' = \gamma_{\text{sat}} - \gamma_w$ equals the submerged unit weight of soil. Thus, we can see that the effective stress at any point A is independent of the depth of water, H, above the submerged soil.

Figure 8.2a shows a layer of submerged soil in a tank where there is no seepage. Figures 8.2b through 8.2d show plots of the variations of the total stress, pore water pressure, and effective stress, respectively, with depth for a submerged layer of soil placed in a tank with no seepage.

The principle of effective stress [Eq. (8.4)] was first developed by Terzaghi (1925, 1936). Skempton (1960) extended the work of Terzaghi and proposed the relationship between total and effective stress in the form of Eq. (8.3).

In summary, effective stress is approximately the force per unit area carried by the soil skeleton. The effective stress in a soil mass controls its volume change and strength. Increasing the effective stress induces soil to move into a denser state of packing.

The effective stress principle is probably the most important concept in geotechnical engineering. The compressibility and shearing resistance of a soil depend to a great extent on the effective stress. Thus, the concept of effective stress is significant in solving geotechnical engineering problems, such as the lateral earth pressure on retaining structures, the load-bearing capacity and settlement of foundations, and the stability of earth slopes.

In Eq. (8.2), the effective stress, σ', is defined as the sum of the vertical components of all intergranular *contact* forces over a unit gross cross-sectional area. This definition is mostly true for granular soils; however, for fine-grained soils, intergranular contact may not physically be there, because the clay particles are surrounded by tightly held water film. In a more general sense, Eq. (8.3) can be rewritten as

$$\sigma = \sigma_{\text{ig}} + u(1 - a_s') - A' + R' \tag{8.6}$$

where σ_{ig} = intergranular stress

A' = electrical attractive force per unit cross-sectional area of soil

R' = electrical repulsive force per unit cross-sectional area of soil

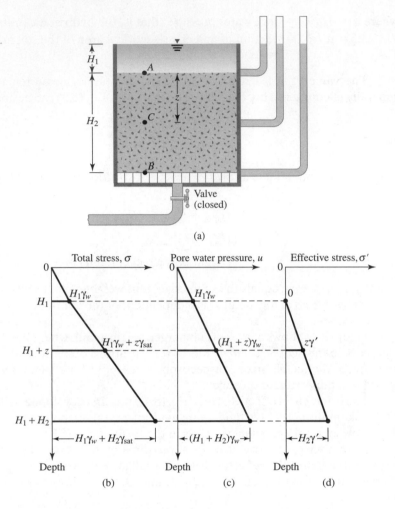

Figure 8.2 (a) Layer of soil in a tank where there is no seepage; variation of (b) total stress; (c) pore water pressure; (d) effective stress with depth for a submerged soil layer without seepage

For granular soils, silts, and clays of low plasticity, the magnitudes of A' and R' are small. Hence, for all practical purposes,

$$\sigma_{ig} = \sigma' \approx \sigma - u$$

However, if $A' - R'$ is large, then $\sigma_{ig} \neq \sigma'$. Such situations can be encountered in highly plastic, dispersed clay. Many interpretations have been made in the past to distinguish between the intergranular stress and effective stress. In any case, the effective stress principle is an excellent approximation used in solving engineering problems.

Example 8.1

A soil profile is shown in Figure 8.3. Calculate the total stress, pore water pressure, and effective stress at A, B, C, and D.

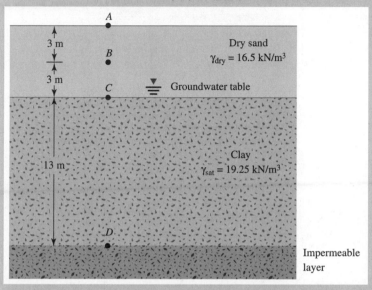

Figure 8.3 A soil profile for calculation of total stress, pore water pressure, and effective stress

Solution

At A
$$\text{total stress: } \sigma_A = 0$$
$$\text{pore water pressure: } u_A = 0$$
$$\text{effective stress: } \sigma'_A = 0$$

At B
$$\sigma_B = 3\gamma_{dry(sand)} = 3 \times 16.5 = 49.5 \text{ kN/m}^2$$
$$u_B = 0 \text{ kN/m}^2$$
$$\sigma_B' = 49.5 - 0 = 49.5 \text{ kN/m}^2$$

At C
$$\sigma_C = 6\gamma_{dry(sand)} = 6 \times 16.5 = 99 \text{ kN/m}^2$$
$$u_C = 0 \text{ kN/m}^2$$
$$\sigma_C' = 99 - 0 = 99 \text{ kN/m}^2$$

At D
$$\sigma_D = 6\gamma_{dry(sand)} + 13\gamma_{sat(clay)}$$
$$= 6 \times 16.5 + 13 \times 19.25$$
$$= 99 + 250.25 = 349.25 \text{ kN/m}^2$$
$$u_D = 13\gamma_w = 13 \times 9.81 = 127.53 \text{ kN/m}^2$$
$$\sigma_D' = 349.25 - 127.53 = 221.72 \text{ kN/m}^2$$

∎

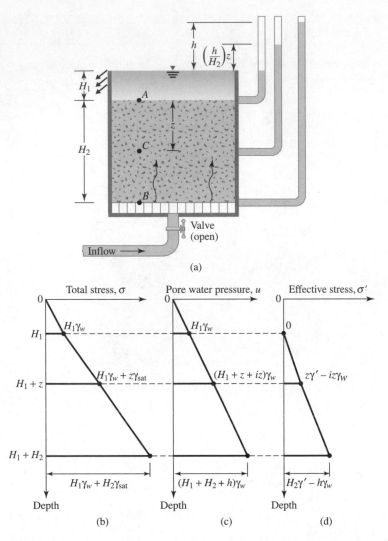

Figure 8.4 (a) Layer of soil in a tank with upward seepage; variation of (b) total stress; (c) pore water pressure; (d) effective stress with depth for a soil layer with upward seepage

8.2 Stresses in Saturated Soil with Upward Seepage

It water is seeping, the effective stress at any point in a soil mass will differ from that in the static case. It will increase or decrease, depending on the direction of seepage.

Figure 8.4a shows a layer of granular soil in a tank where upward seepage is caused by adding water through the valve at the bottom of the tank. The rate of water supply is kept constant. The loss of head caused by upward seepage between the levels of A and B is h. Keeping in mind that the total stress at any point in the soil mass is due solely to the weight of soil and water above it, we find that the effective stress calculations at points A and B are as follows:

At A

- Total stress: $\sigma_A = H_1\gamma_w$
- Pore water pressure: $u_A = H_1\gamma_w$
- Effective stress: $\sigma'_A = \sigma_A - u_A = 0$

At B

- Total stress: $\sigma_B = H_1\gamma_w + H_2\gamma_{sat}$
- Pore water pressure: $u_B = (H_1 + H_2 + h)\gamma_w$
- Effective stress: $\sigma'_B = \sigma_B - u_B$

$$= H_2(\gamma_{sat} - \gamma_w) - h\gamma_w$$
$$= H_2\gamma' - h\gamma_w$$

Similarly, the effective stress at a point C located at a depth z below the top of the soil surface can be calculated as follows:

At C

- Total stress: $\sigma_C = H_1\gamma_w + z\gamma_{sat}$
- Pore water pressure: $u_C = \left(H_1 + z + \dfrac{h}{H_2}z\right)\gamma_w$
- Effective stress: $\sigma'_C = \sigma_C - u_C$

$$= z(\gamma_{sat} - \gamma_w) - \frac{h}{H_2}z\gamma_w$$
$$= z\gamma' - \frac{h}{H_2}z\gamma_w$$

Note that h/H_2 is the hydraulic gradient i caused by the flow, and therefore,

$$\sigma'_C = z\gamma' - iz\gamma_w \tag{8.7}$$

The variations of total stress, pore water pressure, and effective stress with depth are plotted in Figures 8.4b through 8.4d, respectively. A comparison of Figures 8.2d and 8.3d shows that the effective stress at a point located at a depth z measured from the surface of a soil layer is reduced by an amount $iz\gamma_w$ because of upward seepage of water. If the rate of seepage and thereby the hydraulic gradient are gradually increased, a limiting condition will be reached, at which point

$$\sigma'_c = z\gamma' - i_{cr}z\gamma_w = 0 \tag{8.8}$$

where i_{cr} = critical hydraulic gradient (for zero effective stress).

Under such a situation, soil stability is lost. This situation is generally referred to as *boiling,* or a *quick condition.*

From Eq. (8.8),

$$i_{cr} = \frac{\gamma'}{\gamma_w} \tag{8.9}$$

For most soils, the value of i_{cr} varies from 0.9 to 1.1, with an average of 1.

Example 8.2

A 10-m thick layer of stiff saturated clay is underlain by a layer of sand (Figure 8.5). The sand is under artesian pressure. Calculate the maximum depth of cut H that can be made in the clay.

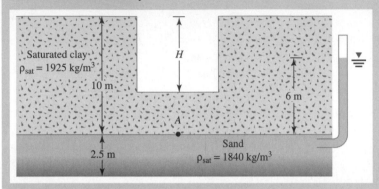

Figure 8.5

Solution
Given that

$$\rho_{sat(clay)} = 1925 \text{ kg/m}^3,$$

we have

$$\gamma_{sat(clay)} = \frac{1925 \times 9.81}{1000} = 18.88 \text{ kN/m}^3$$

Due to excavation, there will be unloading of the overburden pressure. Let the depth of the cut be H, at which point the bottom will heave. Let us consider the stability of point A at that time:

$$\sigma_A = (10 - H)\gamma_{sat(clay)}$$

$$u_A = 6\gamma_w$$

For heave to occur, σ_A' should be 0. So

$$\sigma_A - u_A = (10 - H)\gamma_{sat(clay)} - 6\gamma_w$$

or

$$(10 - H)18.88 - (6)9.81 = 0$$

$$H = \frac{(10)18.88 - (6)9.81}{18.88} = \textbf{6.88 m} \qquad \blacksquare$$

8.3   ## Stresses in Saturated Soil with Downward Seepage

The condition of downward seepage is shown in Figure 8.6a. The water level in the soil tank is held constant by adjusting the supply from the top and the outflow at the bottom.

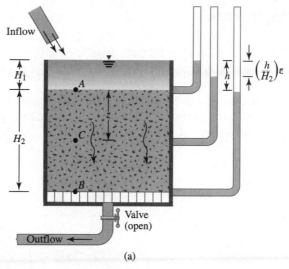

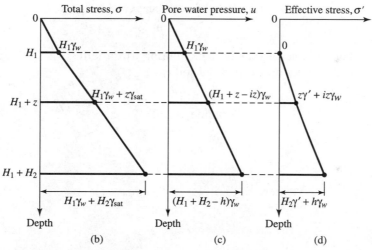

Figure 8.6 (a) Layer of soil in a tank with downward seepage; variation of (b) total stress; (c) pore water pressure; (d) effective stress with depth for a soil layer with downward seepage

The hydraulic gradient caused by the downward seepage equals $i = h/H_2$. The total stress, pore water pressure, and effective stress at any point C are, respectively,

$$\sigma_C = H_1\gamma_w + z\gamma_{sat}$$

$$u_C = (H_1 + z - iz)\gamma_w$$

$$\sigma'_C = (H_1\gamma_w + z\gamma_{sat}) - (H_1 + z - iz)\gamma_w$$

$$= z\gamma' + iz\gamma_w$$

The variations of total stress, pore water pressure, and effective stress with depth are also shown graphically in Figures 8.6b through 8.6d.

8.4	*Seepage Force*

The preceding section showed that the effect of seepage is to increase or decrease the effective stress at a point in a layer of soil. Often, expressing the seepage force per unit volume of soil is convenient.

In Figure 8.2, it was shown that, with no seepage, the effective stress at a depth z measured from the surface of the soil layer in the tank is equal to $z\gamma'$. Thus, the effective force on an area A is

$$P'_1 = z\gamma' A$$

(The direction of the force P'_1 is shown in Figure 8.7a.)

Again, if there is an upward seepage of water in the vertical direction through the same soil layer (Figure 8.4), the effective force on an area A at a depth z can be given by

$$P'_2 = (z\gamma' - iz\gamma_w)A$$

Hence, the decrease in the total force because of seepage is

$$P'_1 - P'_2 = iz\gamma_w A \tag{8.10}$$

The volume of the soil contributing to the effective force equals zA, so the seepage force per unit volume of soil is

$$\frac{P'_1 - P'_2}{(\text{Volume of soil})} = \frac{iz\gamma_w A}{zA} = i\gamma_w \tag{8.11}$$

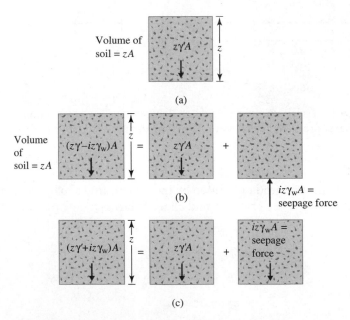

Figure 8.7 Force due to (a) no seepage; (b) upward seepage; (c) downward seepage on a volume of soil

The force per unit volume, $i\gamma_w$, for this case acts in the upward direction — that is, in the direction of flow. This upward force is demonstrated in Figure 8.7b. Similarly, for downward seepage, it can be shown that the seepage force in the downward direction per unit volume of soil is $i\gamma_w$ (Figure 8.7c).

From the preceding discussions, we can conclude that the seepage force per unit volume of soil is equal to $i\gamma_w$, and in isotropic soils the force acts in the same direction as the direction of flow. This statement is true for flow in any direction. Flow nets can be used to find the hydraulic gradient at any point and, thus, the seepage force per unit volume of soil.

This concept of seepage force can be effectively used to obtain the factor of safety against heave on the downstream side of a hydraulic structure. To see this, consider the case of flow around a sheet pile (Figure 8.8a). After conducting several model tests, Terzaghi (1922) concluded that heaving generally occurs within a distance of $D/2$ from the sheet piles (when D equals the depth of embedment of sheet piles into the permeable layer). Therefore, we need to investigate the stability of soil in a zone measuring D by $D/2$ in cross section, as shown in Figure 8.8a.

The factor of safety against heaving can be given by

$$FS = \frac{W'}{U} \tag{8.12}$$

where FS = factor of safety
$\quad W'$ = submerged weight of soil in the heave zone per unit length of
$\qquad$ sheet pile = $D(D/2)(\gamma_{sat} - \gamma_w) = (\frac{1}{2})D^2\gamma'$
$\quad U$ = uplifting force caused by seepage on the same volume of soil

From Eq. (8.11),

$$U = (\text{Soil volume}) \times (i_{av}\gamma_w) = \tfrac{1}{2}D^2 i_{av}\gamma_w$$

where i_{av} = average hydraulic gradient at the bottom of the block of soil.

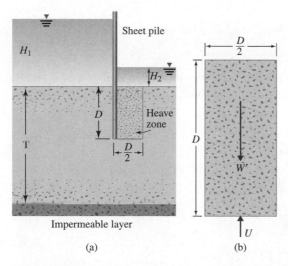

Figure 8.8 (a) Check for heaving on the downstream side for a row of sheet piles driven into a permeable layer; (b) enlargement of heave zone

Substituting the values of W' and U in Eq. (8.12), we can write

$$FS = \frac{\gamma'}{i_{av}\gamma_w} \tag{8.13}$$

For the case of *flow around a sheet pile* in a *homogeneous soil,* as shown in Figure 8.8, it can be demonstrated that

$$\frac{U}{0.5\gamma_w D(H_1 - H_2)} = C_o$$

where C_o is a function of D/T (see Table 8.1). Hence, from Eq. (8.12),

$$FS = \frac{W'}{U} = \frac{0.5D^2\gamma'}{0.5C_o\gamma_w D(H_1 - H_2)} = \frac{D\gamma'}{C_o\gamma_w D(H_1 - H_2)} \tag{8.13a}$$

Table 8.1 Variation of C_o with D/T

D/T	C_o
0.1	0.385
0.2	0.365
0.3	0.359
0.4	0.353
0.5	0.347
0.6	0.339
0.7	0.327
0.8	0.309
0.9	0.274

Example 8.3

Consider the upward flow of water through a layer of sand in a tank as shown in Figure 8.9. For the sand, the following are given: void ratio $(e) = 0.52$ and specific gravity of solids $= 2.67$.

a. Calculate the total stress, pore water pressure, and effective stress at points A and B.
b. What is the upward seepage force per unit volume of soil?

Solution
a. The saturated unit weight of sand is calculated as follows:

$$\gamma_{sat} = \frac{(G_s + e)\gamma_w}{1 + e} = \frac{(2.67 + 0.52)9.81}{1 + 0.52} = 20.59 \text{ kN/m}^3$$

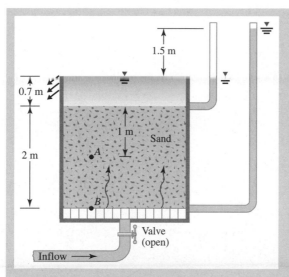

Figure 8.9 Upward flow of water through a layer of sand in a tank

Now, the following table can be prepared:

Point	Total stress, σ (kN/m²)	Pore water pressure, u (kN/m²)	Effective stress, $\sigma' = \sigma - u$ (kN/m²)
A	$0.7\gamma_w + 1\gamma_{sat} = (0.7)(9.81)$ $+ (1)(20.59) = \textbf{27.46}$	$\left[(1 + 0.7) + \left(\dfrac{1.5}{2}\right)(1)\right]\gamma_w$ $= (2.45)(9.81) = \textbf{24.03}$	3.43
B	$0.7\gamma_w + 2\gamma_{sat} = (0.7)(9.81)$ $+ (2)(20.59) = \textbf{48.05}$	$(2 + 0.7 + 1.5)\gamma_w$ $= (4.2)(9.81) = \textbf{41.2}$	6.85

b. Hydraulic gradient (i) = 1.5/2 = 0.75. Thus, the seepage force per unit volume can be calculated as

$$i\gamma_w = (0.75)(9.81) = \textbf{7.36 kN/m}^3 \qquad \blacksquare$$

Example 8.4

Figure 8.10 shows the flow net for seepage of water around a single row of sheet piles driven into a permeable layer. Calculate the factor of safety against downstream heave, given that γ_{sat} for the permeable layer = 17.7 kN/m³. (Note: thickness of permeable layer $T = 18$ m)

Solution
From the dimensions given in Figure 8.10, the soil prism to be considered is 6 m × 3 m in cross section.

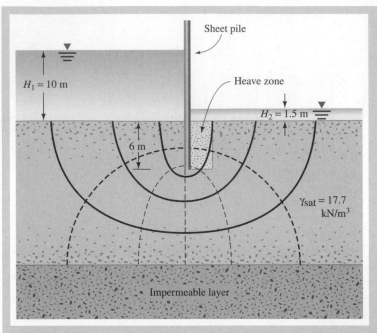

Figure 8.10 Flow net for seepage of water around sheet piles driven into permeable layer

The soil prism is drawn to an enlarged scale in Figure 8.11. By use of the flow net, we can calculate the head loss through the prism as follows:

- At b, the driving head $= \frac{3}{6}(H_1 - H_2)$
- At c, the driving head $\approx \frac{1.6}{6}(H_1 - H_2)$

Similarly, for other intermediate points along bc, the approximate driving heads have been calculated and are shown in Figure 8.11.

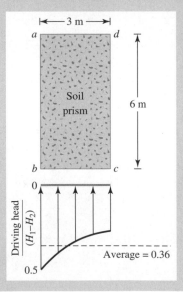

Figure 8.11 Soil prism — enlarged scale

The average value of the head loss in the prism is $0.36(H_1 - H_2)$, and the average hydraulic gradient is

$$i_{av} = \frac{0.36(H_1 - H_2)}{D}$$

Thus, the factor of safety [Eq. (8.13)] is

$$FS = \frac{\gamma'}{i_{av}\gamma_w} = \frac{\gamma' D}{0.36(H_1 - H_2)\gamma_w} = \frac{(17.7 - 9.81)\,6}{0.36(10 - 1.5) \times 9.81} = \mathbf{1.58}$$

Alternate solution

For this case, $D/T = 1/3$. From Table 8.1, for $D/T = 1/3$, the value of $C_o = 0.357$. Thus, from Eq. (8.13a),

$$FS = \frac{D\gamma'}{C_o\gamma_w(H_1 - H_2)} = \frac{(6)(17.7 - 9.81)}{(0.357)(9.81)(10 - 1.5)} = \mathbf{1.59} \qquad \blacksquare$$

8.5 Use of Filters to Increase the Factor of Safety against Heave

The factor of safety against heave as calculated in Example 8.4 is low. In practice, a minimum factor of safety of about 4 to 5 is required for the safety of the structure. Such a high factor of safety is recommended primarily because of the inaccuracies inherent in the analysis. One way to increase the factor of safety against heave is to use a *filter* in the downstream side of the sheet-pile structure (Figure 8.12a). A filter is a granular material with openings small enough to prevent the movement of the soil particles upon which it is placed and, at the same time, is pervious enough to offer little resistance to seepage through it. In Figure 8.12a, the thickness of the filter material is D_1. In this case, the factor of safety against heave can be calculated as follows (Figure 8.12b).

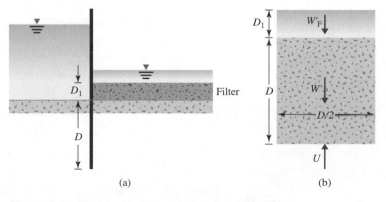

Figure 8.12 Factor of safety against heave, with a filter

The submerged weight of the soil and the filter in the heave zone per unit length of sheet pile = $W' + W'_F$, where

$$W' = (D)\left(\frac{D}{2}\right)(\gamma_{sat} - \gamma_w) = \frac{1}{2}D^2\gamma'$$

$$W'_F = (D_1)\left(\frac{D}{2}\right)(\gamma'_F) = \frac{1}{2}D_1 D\gamma'_F$$

in which γ'_F = effective unit weight of the filter.

The uplifting force caused by seepage on the same volume of soil is given by

$$U = \tfrac{1}{2}D^2 i_{av}\gamma_w$$

The preceding relationship was derived in Section 8.4.

The factor of safety against heave is thus

$$FS = \frac{W' + W'_F}{U} = \frac{\dfrac{1}{2}D^2\gamma' + \dfrac{1}{2}D_1 D\gamma'_F}{\dfrac{1}{2}D^2 i_{av}\gamma_w} = \frac{\gamma' + \left(\dfrac{D_1}{D}\right)\gamma'_F}{i_{av}\gamma_w} \tag{8.14}$$

8.6 *Selection of Filter Material*

It is extremely important that the filter material mentioned in Section 8.5 be chosen carefully, taking into consideration that the soil is to be protected. To describe the selection criteria of a filter, refer to Figure 8.13. Note that, in this figure, the soil to be protected is referred to as the *base material*. Terzaghi and Peck (1948) suggested the following criteria for selection of the filter material:

1. $\dfrac{D_{15(F)}}{D_{85(B)}} < 4$

2. $\dfrac{D_{15(F)}}{D_{15(B)}} > 4$

Here, $D_{15(F)}$, $D_{15(B)}$ = diameters through which 15% of the filter and base material, respectively, will pass

$D_{85(B)}$ = diameter through which 85% of the base material will pass

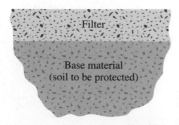

Figure 8.13 Definition of base material and filter material

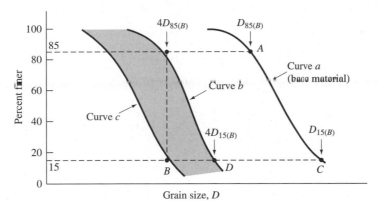

Figure 8.14 Filter selection criteria

The first criterion is for prevention of the movement of the soil particles of the base material (that is, the soil to be protected) through the filter.

The application of the filter selection criteria just described can be explained by using Figure 8.14, in which curve a is the grain-size distribution curve of the base material. From criterion 1, $D_{15(F)} < 4D_{85(B)}$. The abscissa of point A is $D_{85(B)}$, so the magnitude of $4D_{85(B)}$ can be calculated, and point B, whose abscissa is $4D_{85(B)}$, can be plotted. Similarly, from criterion 2, $D_{15(F)} > 4D_{15(B)}$. The abscissas of points C and D are $D_{15(B)}$ and $4D_{15(B)}$, respectively. The curves b and c are drawn, which are geometrically similar to curve a and are within the limits of points B and D. A soil whose grain-size curve falls within the bounds of curves b and c is a good filter material.

8.7 Capillary Rise in Soils

The continuous void spaces in soil can behave as bundles of capillary tubes of variable cross section. Because of surface tension force, water may rise above the phreatic surface.

Figure 8.15 shows the fundamental concept of the height of rise in a capillary tube. The height of rise of water in the capillary tube can be given by summing the forces in the vertical direction, or

$$\left(\frac{\pi}{4} d^2\right) h_c \gamma_w = \pi d T \cos \alpha$$

$$h_c = \frac{4T \cos \alpha}{d \gamma_w} \tag{8.15}$$

where T = surface tension (force/length)
α = angle of contact
d = diameter of capillary tube
γ_w = unit weight of water

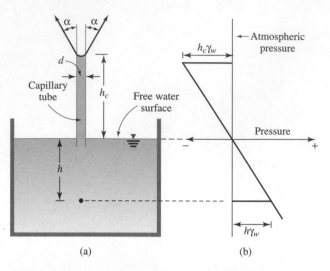

(a)   (b)

Figure 8.15 (a) Rise of water in the capillary tube; (b) pressure within the height of rise in the capillary tube (atmospheric pressure taken as datum)

For pure water and clean glass, $\alpha = 0$. Thus, Eq. (8.15) becomes

$$h_c = \frac{4T}{d\gamma_w} \tag{8.16}$$

For water, $T = 72$ mN/m. From Eq. (8.16), we see that the height of capillary rise

$$h_c \propto \frac{1}{d} \tag{8.17}$$

Thus, the smaller the capillary tube diameter, the larger the capillary rise. This fact is shown in Figure 8.16.

Although the concept of capillary rise as demonstrated for an ideal capillary tube can be applied to soils, one must realize that the capillary tubes formed in soils because of the continuity of voids have variable cross sections. The results of the

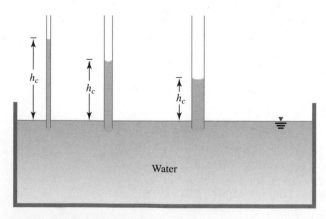

Figure 8.16 Nature of variation of capillary rise with diameter of capillary tube

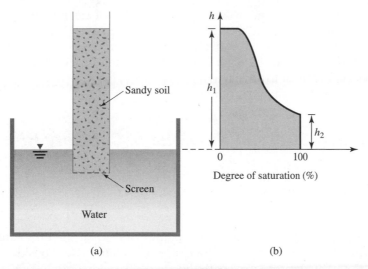

Figure 8.17 Capillary effect in sandy soil: (a) a soil column in contact with water; (b) variation of degree of saturation in the soil column

nonuniformity on capillary rise can be seen when a dry column of sandy soil is placed in contact with water (Figure 8.17). After the lapse of a given amount of time, the variation of the degree of saturation with the height of the soil column caused by capillary rise is approximately as shown in Figure 8.17b. The degree of saturation is about 100% up to a height of h_2, and this corresponds to the largest voids. Beyond the height h_2, water can occupy only the smaller voids; hence, the degree of saturation is less than 100%. The maximum height of capillary rise corresponds to the smallest voids. Hazen (1930) gave a formula for the approximation of the height of capillary rise in the form.

$$h_1 \, (\text{mm}) = \frac{C}{eD_{10}} \tag{8.18}$$

where D_{10} = effective size (mm)
  e = void ratio
  C = a constant that varies from 10 to 50 mm^2

Equation (8.18) has an approach similar to that of Eq. (8.17). With the decrease of D_{10}, the pore size in soil decreases, which causes higher capillary rise. Table 8.2 shows the approximate range of capillary rise that is encountered in various types of soils.

Table 8.2 Approximate Range of Capillary Rise in Soils

Soil type	Range of capillary rise	
	m	ft
Coarse sand	0.1–0.2	0.3–0.6
Fine sand	0.3–1.2	1–4
Silt	0.75–7.5	2.5–25
Clay	7.5–23	25–75

Capillary rise is important in the formation of some types of soils such as *caliche*, which can be found in the desert Southwest of the United States. Caliche is a mixture of sand, silt, and gravel bonded by calcareous deposits. These deposits are brought to the surface by a net upward migration of water by capillary action. The water evaporates in the high local temperature. Because of sparse rainfall, the carbonates are not washed out of the top soil layer.

8.8

Effective Stress in the Zone of Capillary Rise

The general relationship among total stress, effective stress, and pore water pressure was given in Eq. (8.4) as

$$\sigma = \sigma' + u$$

The pore water pressure u at a point in a layer of soil fully saturated by capillary rise is equal to $-\gamma_w h$ (h = height of the point under consideration measured from the groundwater table) with the atmospheric pressure taken as datum. If partial saturation is caused by capillary action, it can be approximated as

$$u = -\left(\frac{S}{100}\right)\gamma_w h \tag{8.19}$$

where S = degree of saturation, in percent.

Example 8.5

A soil profile is shown in Figure 8.18. Note the zone of capillary rise in the sand layer overlying clay. In this zone, the average degree of saturation and the moist unit weight are 60% and 17.6 kN/m³, respectively. Calculate and plot the variation of σ, u, and σ' with depth.

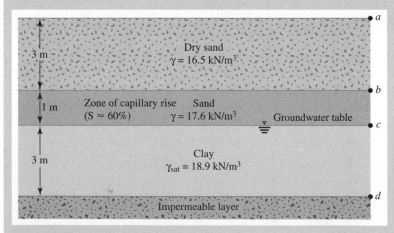

Figure 8.18 Soil profile with capillary rise

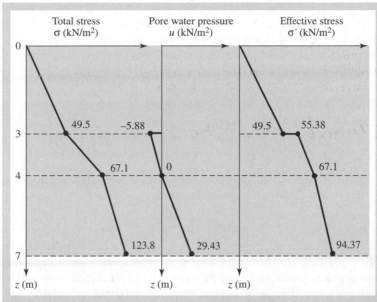

Figure 8.19 Plot of σ, u, and σ' with depth

Solution

The following table can be prepared:

Depth below ground surface (m)	Total stress, σ (kN/m^2)	Pore water pressure, u (kN/m^2)	Effective stress, σ' (kN/m^2)
0	0	0	0
3			
Immediately above the capillary zone	$(3)(16.5) = 49.5$	0	49.5
Just inside the capillary zone	$(3)(16.5) = 49.5$	$-(S\gamma_w)(1) =$ $-(0.6)(9.81)(1) = -5.88$	55.38
4	$(3)(16.5) + (1)(17.6) = 67.1$	0	67.1
7	$(3)(16.5) + (1)(17.6) +$ $(3)(18.9) = 123.8$	$3\gamma_w = (3)(9.81) = 29.43$	94.37

The plot of σ, u, and σ' with depth is shown in Figure 8.19. ∎

8.9 *Summary and General Comments*

The effective stress principle is probably the most important concept in geotechnical engineering. The compressibility and shearing resistance of a soil depend to a great extent on the effective stress. Thus, the concept of effective stress is significant in solving geotechnical engineering problems, such as the lateral earth pressure on retaining structures, the load-bearing capacity and settlement of foundations, and the stability of earth slopes.

In Eq. (8.2), the effective stress σ' is defined as the sum of the vertical components of all intergranular *contact* forces over a unit gross cross-sectional area. This definition is mostly true for granular soils; however, for fine-grained soils, intergranular contact may not physically be there, because the clay particles are surrounded by tightly held water film. In a more general sense, Eq. (8.3) can be rewritten as

$$\sigma = \sigma_{ig} + u(1 - a_s') - A' + R' \tag{8.20}$$

where σ_{ig} = intergranular stress
   A' = electrical attractive force per unit cross-sectional area of soil
   R' = electrical repulsive force per unit cross-sectional area of soil

For granular soils, silts, and clays of low plasticity, the magnitudes of A' and R' are small. Hence, for all practical purposes,

$$\sigma_{ig} = \sigma' \approx \sigma - u$$

However, if $A' - R'$ is large, then $\sigma_{ig} \neq \sigma'$. Such situations can be encountered in highly plastic, dispersed clay. Many interpretations have been made in the past to distinguish between the intergranular stress and effective stress. In any case, the effective stress principle is an excellent approximation used in solving engineering problems.

Problems

8.1   A soil profile is shown in Figure 8.20. Calculate the values of σ, u, and σ' at points A, B, C, and D. Plot the variation of σ, u, and σ' with depth. The following values are given:

Layer no.	Thickness (ft)	Unit weight (lb/ft³)
I	$H_1 = 5$	$\gamma_d = 112$
II	$H_2 = 6$	$\gamma_{sat} = 120$
III	$H_3 = 8$	$\gamma_{sat} = 125$

8.2   Repeat Problem 8.1 with the following data:

Layer no.	Thickness (ft)	Unit weight (lb/ft³)
I	$H_1 = 5$	$\gamma_d = 100$
II	$H_2 = 10$	$\gamma_{sat} = 116$
III	$H_3 = 9$	$\gamma_{sat} = 122$

8.3   Repeat Problem 8.1 with the following values:

Layer no.	Thickness (m)	Unit weight (kN/m³)
I	$H_1 = 3$	$\gamma_d = 15$
II	$H_2 = 4$	$\gamma_{sat} = 16$
III	$H_3 = 5$	$\gamma_{sat} = 18$

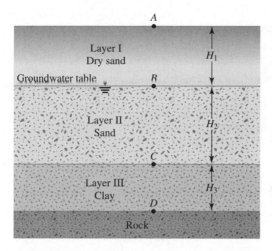

Figure 8.20

8.4 Repeat Problem 8.1 with the following data:

Layer no.	Thickness (m)	Soil parameters
I	$H_1 = 4$	$e = 0.4, G_s = 2.62$
II	$H_2 = 5$	$e = 0.60, G_s = 2.68$
III	$H_3 = 3$	$e = 0.81, G_s = 2.73$

8.5 Repeat Problem 8.1 with the following values:

Layer no.	Thickness (m)	Soil parameters
I	$H_1 = 4$	$e = 0.6, G_s = 2.65$
II	$H_2 = 3$	$e = 0.52, G_s = 2.68$
III	$H_3 = 1.5$	$w = 40\%, e = 1.1$

8.6 Plot the variation of total stress, pore water pressure, and effective stress with depth for the sand and clay layers shown in Figure 8.21 with $H_1 = 3$ m and $H_2 = 4$ m. Give numerical values.

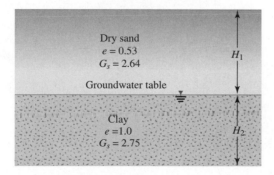

Figure 8.21

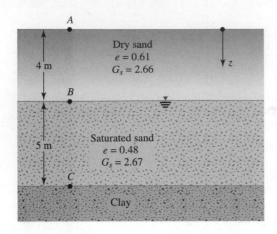

Figure 8.22

Figure 8.23

8.7 A soil profile is shown in Figure 8.22.
   a. Calculate the total stress, pore water pressure, and effective stress at A, B, and C.
   b. How high should the groundwater table rise so that the effective stress at C is 105 kN/m²?

8.8 A sand has $G_s = 2.68$. Calculate the hydraulic gradient that will cause boiling for $e = 0.38, 0.48, 0.6, 0.7$, and 0.8. Plot a graph for i_{cr} versus e.

8.9 An exploratory drill hole was made in a stiff saturated clay. (See Figure 8.23.) The sand layer underlying the clay was observed to be under artesian pressure. Water in the drill hole rose to a height of 12 ft above the top of the sand layer. If an open excavation is to be made in the clay, how deep can the excavation proceed before the bottom heaves?

8.10 A cut is made in a stiff saturated clay that is underlain by a layer of sand. (See Figure 8.24.) What should be the height of the water, h, in the cut so that the stability of the saturated clay is not lost?

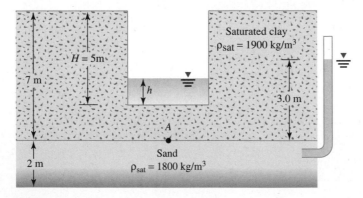

Figure 8.24

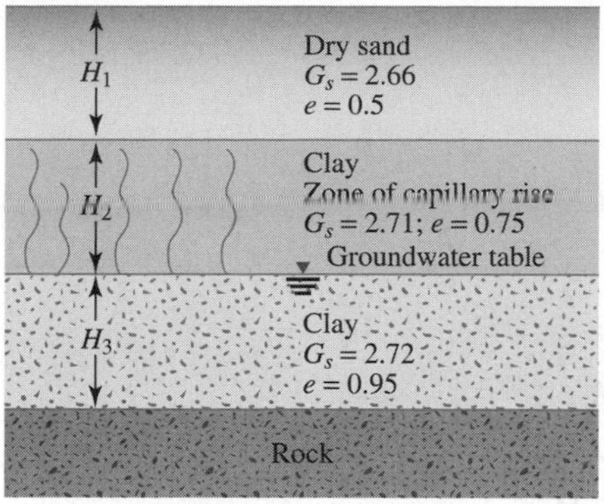

Figure 8.25

8.11 Refer to Figure 8.4a. Given that $H_1 = 1.5$ m, $H_2 = 2.5$ m, $h = 1.5$ m, void ratio of sand $(e) = 0.5$, specific gravity of soil solids $(G_s) = 2.68$, area of the tank = 0.6 m^2, and hydraulic conductivity of sand = 0.1 cm/sec,
   a. What is the rate of upward seepage?
   b. If $h = 1.5$ m, will boiling occur? Why?
   c. What should be the value of h to cause boiling?
8.12 Refer to Figure 8.4a. If $H_1 = 3$ ft, $H_2 = 4$ ft, $h = 1.5$ ft, $\gamma_{sat} = 120$ lb/ft^3, hydraulic conductivity of sand $(k) = 0.18$ cm/sec, and area of tank = 5.2 ft^2, what is the rate of upward seepage of water (ft^3/min)?
8.13 A soil profile is shown in Figure 8.25. Given: $H_1 = 6$ ft, $H_2 = 4$ ft, and $H_3 = 9$ ft, plot the variation of σ, u, and σ' with depth. Assume 50% saturation in the zone of capillary rise.
8.14 Repeat Problem 8.13 with $H_1 = 2$ m, $H_2 = 2.5$ m, and $H_3 = 3.5$ m.
8.15 Find the factor of safety against heave on the downstream side of the single-row sheet pile structure shown in Figure 7.9. (*Note:* The depth of penetration of sheet piles into the permeable layer is 5 m.) Assume $\gamma_{sat} = 19$ kN/m^3.
8.16 Repeat Problem 8.15 assuming that a filter 1.5 m thick is placed on the downstream side. We are given $\gamma_{sat(filter)} = 20$ kN/m^3.

References

HAZEN, A. (1930). "Water Supply," in *American Civil Engineering Handbook,* Wiley, New York.
SKEMPTON, A. W. (1960). "Correspondence," *Geotechnique,* Vol. 10, No. 4, 186.
TERZAGHI, K. (1922). "Der Grundbruch an Stauwerken und seine Verhütung," *Die Wasserkraft,* Vol. 17, 445–449.
TERZAGHI, K. (1925). *Erdbaumechanik auf Bodenphysikalischer Grundlage,* Dueticke, Vienna.
TERZAGHI, K. (1936). "Relation Between Soil Mechanics and Foundation Engineering: Presidential Address," *Proceedings,* First International Conference on Soil Mechanics and Foundation Engineering, Boston, Vol. 3, 13–18.
TERZAGHI, K., and PECK, R. B. (1948). *Soil Mechanics in Engineering Practice,* Wiley, New York.

9

Stresses in a Soil Mass

Construction of a foundation causes changes in the stress, usually a net increase. The net stress increase in the soil depends on the load per unit area to which the foundation is subjected, the depth below the foundation at which the stress estimation is desired, and other factors. It is necessary to estimate the net increase of vertical stress in soil that occurs as a result of the construction of a foundation so that settlement can be calculated. The settlement calculation procedure is discussed in more detail in Chapter 10. This chapter discusses the principles of estimation of vertical stress increase in soil caused by various types of loading, based on the theory of elasticity. Although natural soil deposits, in most cases, are not fully elastic, isotropic, or homogeneous materials, calculations for estimating increases in vertical stress yield fairly good results for practical work.

9.1 Normal and Shear Stresses on a Plane

Students in a soil mechanics course are familiar with the fundamental principles of the mechanics of deformable solids. This section is a brief review of the basic concepts of normal and shear stresses on a plane that can be found in any course on the mechanics of materials.

Figure 9.1a shows a two-dimensional soil element that is being subjected to normal and shear stresses ($\sigma_y > \sigma_x$). To determine the normal stress and the shear stress on a plane EF that makes an angle θ with the plane AB, we need to consider the free body diagram of EFB shown in Figure 9.1b. Let σ_n and τ_n be the normal stress and the shear stress, respectively, on the plane EF. From geometry, we know that

$$\overline{EB} = \overline{EF} \cos \theta \tag{9.1}$$

and

$$\overline{FB} = \overline{EF} \sin \theta \tag{9.2}$$

Summing the components of forces that act on the element in the direction of N and T, we have

$$\sigma_n(\overline{EF}) = \sigma_x(\overline{EF}) \sin^2 \theta + \sigma_y(\overline{EF}) \cos^2 \theta + 2\tau_{xy}(\overline{EF}) \sin \theta \cos \theta$$

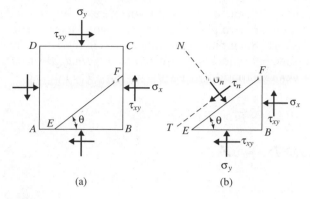

Figure 9.1 (a) A soil element with normal and shear stresses acting on it; (b) free body diagram of *EFB* as shown in (a)

or

$$\sigma_n = \sigma_x \sin^2 \theta + \sigma_y \cos^2 \theta + 2\tau_{xy} \sin \theta \cos \theta$$

or

$$\sigma_n = \frac{\sigma_y + \sigma_x}{2} + \frac{\sigma_y - \sigma_x}{2} \cos 2\theta + \tau_{xy} \sin 2\theta \qquad (9.3)$$

Again,

$$\tau_n(\overline{EF}) = -\sigma_x(\overline{EF}) \sin \theta \cos \theta + \sigma_y(\overline{EF}) \sin \theta \cos \theta$$
$$- \tau_{xy}(\overline{EF}) \cos^2 \theta + \tau_{xy}(\overline{EF}) \sin^2 \theta$$

or

$$\tau_n = \sigma_y \sin \theta \cos \theta - \sigma_x \sin \theta \cos \theta - \tau_{xy}(\cos^2 \theta - \sin^2 \theta)$$

or

$$\tau_n = \frac{\sigma_y - \sigma_x}{2} \sin 2\theta - \tau_{xy} \cos 2\theta \qquad (9.4)$$

From Eq. (9.4), we can see that we can choose the value of θ in such a way that τ_n will be equal to zero. Substituting $\tau_n = 0$, we get

$$\tan 2\theta = \frac{2\tau_{xy}}{\sigma_y - \sigma_x} \qquad (9.5)$$

For given values of τ_{xy}, σ_x, and σ_y, Eq. (9.5) will give two values of θ that are 90° apart. This means that there are two planes that are at right angles to each other on which the shear stress is zero. Such planes are called *principal planes*. The normal stresses that act on the principal planes are referred to as *principal stresses*. The values of principal stresses can be found by substituting Eq. (9.5) into Eq. (9.3), which yields

Major principal stress:

$$\sigma_n = \sigma_1 = \frac{\sigma_y + \sigma_x}{2} + \sqrt{\left[\frac{(\sigma_y - \sigma_x)}{2}\right]^2 + \tau_{xy}^2} \qquad (9.6)$$

Minor principal stress:

$$\sigma_n = \sigma_3 = \frac{\sigma_y + \sigma_x}{2} - \sqrt{\left[\frac{(\sigma_y - \sigma_x)}{2}\right]^2 + \tau_{xy}^2} \qquad (9.7)$$

The normal stress and shear stress that act on any plane can also be determined by plotting a Mohr's circle, as shown in Figure 9.2. The following sign conventions are used in Mohr's circles: compressive normal stresses are taken as positive, and shear stresses are considered positive if they act on opposite faces of the element in such a way that they tend to produce a counterclockwise rotation.

For plane AD of the soil element shown in Figure 9.1a, normal stress equals $+\sigma_x$ and shear stress equals $+\tau_{xy}$. For plane AB, normal stress equals $+\sigma_y$ and shear stress equals $-\tau_{xy}$.

The points R and M in Figure 9.2 represent the stress conditions on planes AD and AB, respectively. O is the point of intersection of the normal stress axis with the line RM. The circle $MNQRS$ drawn with O as the center and OR as the radius is the

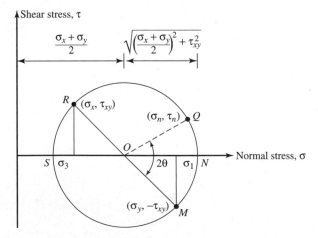

Figure 9.2 Principles of the Mohr's circle

Mohr's circle for the stress conditions considered. The radius of the Mohr's circle is equal to

$$\sqrt{\left[\frac{(\sigma_y - \sigma_x)}{2}\right]^2 + \tau_{xy}^2}$$

The stress on plane EF can be determined by moving an angle 2θ (which is twice the angle that the plane EF makes in a counterclockwise direction with plane AB in Figure 9.1a) in a counterclockwise direction from point M along the circumference of the Mohr's circle to reach point Q. The abscissa and ordinate of point Q, respectively, give the normal stress σ_n and the shear stress τ_n on plane EF.

Because the ordinates (that is, the shear stresses) of points N and S are zero, they represent the stresses on the principal planes. The abscissa of point N is equal to σ_1 [Eq. (9.6)], and the abscissa for point S is σ_3 [Eq. (9.7)].

As a special case, if the planes AB and AD were major and minor principal planes, the normal stress and the shear stress on plane EF could be found by substituting $\tau_{xy} = 0$. Equations (9.3) and (9.4) show that $\sigma_y = \sigma_1$ and $\sigma_x = \sigma_3$ (Figure 9.3a). Thus,

$$\sigma_n = \frac{\sigma_1 + \sigma_3}{2} + \frac{\sigma_1 - \sigma_3}{2} \cos 2\theta \qquad (9.8)$$

$$\tau_n = \frac{\sigma_1 - \sigma_3}{2} \sin 2\theta \qquad (9.9)$$

The Mohr's circle for such stress conditions is shown in Figure 9.3b. The abscissa and the ordinate of point Q give the normal stress and the shear stress, respectively, on the plane EF.

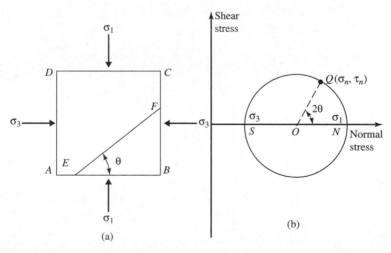

Figure 9.3 (a) Soil element with AB and AD as major and minor principal planes; (b) Mohr's circle for soil element shown in (a)

Example 9.1

A soil element is shown in Figure 9.4. The magnitudes of stresses are $\sigma_x = 120$ kN/ m^2, $\tau = 40$ kN/m^2, $\sigma_y = 300$ kN/m^2, and $\theta = 20°$. Determine

a. Magnitudes of the principal stresses
b. Normal and shear stresses on plane AB. Use Eqs. (9.3), (9.4), (9.6), and (9.7).

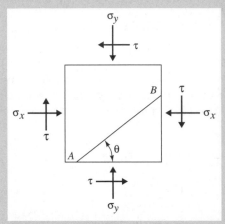

Figure 9.4 Soil element with stresses acting on it

Solution

a. From Eqs. (9.6) and (9.7),

$$\left.\begin{array}{c}\sigma_3 \\ \sigma_1\end{array}\right\} = \frac{\sigma_y + \sigma_x}{2} \pm \sqrt{\left[\frac{\sigma_y - \sigma_x}{2}\right]^2 + \tau_{xy}^2}$$

$$= \frac{300 + 120}{2} \pm \sqrt{\left[\frac{300 - 120}{2}\right]^2 + (-40)^2}$$

$$\sigma_1 = \mathbf{308.5 \ kN/m^2}$$

$$\sigma_3 = \mathbf{111.5 \ kN/m^2}$$

b. From Eq. (9.3),

$$\sigma_n = \frac{\sigma_y + \sigma_x}{2} + \frac{\sigma_y - \sigma_x}{2} \cos 2\theta + \tau \sin 2\theta$$

$$= \frac{300 + 120}{2} + \frac{300 - 120}{2} \cos (2 \times 20) + (-40) \sin (2 \times 20)$$

$$= \mathbf{252.23 \ kN/m^2}$$

From Eq. (7.4),

$$\tau_n = \frac{\sigma_y - \sigma_x}{2} \sin 2\theta - \tau \cos 2\theta$$

$$= \frac{300 \quad 120}{2} \sin (2 \times 20) - (-40) \cos (2 \times 20)$$

$$= \textbf{88.49 kN/m}^2 \qquad \blacksquare$$

9.2 *Stress Caused by a Point Load*

Boussinesq (1883) solved the problem of stresses produced at any point in a homogeneous, elastic, and isotropic medium as the result of a point load applied on the surface of an infinitely large half-space. According to Figure 9.5, Boussinesq's solution for normal stresses at a point caused by the point load P is

$$\Delta\sigma_x = \frac{P}{2\pi} \left\{ \frac{3x^2z}{L^5} - (1 - 2\mu)\left[\frac{x^2 - y^2}{Lr^2(L + z)} + \frac{y^2z}{L^3r^2} \right] \right\} \qquad (9.10)$$

$$\Delta\sigma_y = \frac{P}{2\pi} \left\{ \frac{2y^2z}{L^5} - (1 - 2\mu)\left[\frac{y^2 - x^2}{Lr^2(L + z)} + \frac{x^2z}{L^3r^2} \right] \right\} \qquad (9.11)$$

and

$$\Delta\sigma_z = \frac{3P}{2\pi} \frac{z^3}{L^5} = \frac{3P}{2\pi} \frac{z^3}{(r^2 + z^2)^{5/2}} \qquad (9.12)$$

where $r = \sqrt{x^2 + y^2}$
$L = \sqrt{x^2 + y^2 + z^2} = \sqrt{r^2 + z^2}$
μ = Poisson's ratio

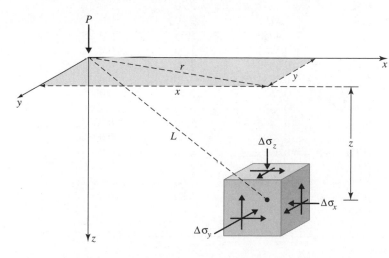

Figure 9.5
Stresses in an elastic medium caused by a point load

Note that Eqs. (9.10) and (9.11), which are the expressions for horizontal normal stresses, depend on the Poisson's ratio of the medium. However, the relationship for the vertical normal stress, $\Delta\sigma_z$, as given by Eq. (9.12), is independent of Poisson's ratio. The relationship for $\Delta\sigma_z$ can be rewritten as

$$\Delta\sigma_z = \frac{P}{z^2}\left\{\frac{3}{2\pi}\frac{1}{[(r/z)^2 + 1]^{5/2}}\right\} = \frac{P}{z^2}I_1 \tag{9.13}$$

where

$$I_1 = \frac{3}{2\pi}\frac{1}{[(r/z)^2 + 1]^{5/2}} \tag{9.14}$$

The variation of I_1 for various values of r/z is given in Table 9.1.

Table 9.1 Variation of I_1 for Various Values of r/z [Eq. (9.14)]

r/z	I_1	r/z	I_1
0	0.4775	0.75	0.1565
0.02	0.4770	0.80	0.1386
0.04	0.4765	0.85	0.1226
0.06	0.4723	0.90	0.1083
0.08	0.4699	0.95	0.0956
0.10	0.4657	1.00	0.0844
0.12	0.4607	1.20	0.0513
0.14	0.4548	1.40	0.0317
0.16	0.4482	1.60	0.0200
0.18	0.4409	1.80	0.0129
0.20	0.4329	2.00	0.0085
0.22	0.4242	2.20	0.0058
0.24	0.4151	2.40	0.0040
0.26	0.4050	2.60	0.0029
0.28	0.3954	2.80	0.0021
0.30	0.3849	3.00	0.0015
0.32	0.3742	3.20	0.0011
0.34	0.3632	3.40	0.00085
0.36	0.3521	3.60	0.00066
0.38	0.3408	3.80	0.00051
0.40	0.3294	4.00	0.00040
0.45	0.3011	4.20	0.00032
0.50	0.2733	4.40	0.00026
0.55	0.2466	4.60	0.00021
0.60	0.2214	4.80	0.00017
0.65	0.1978	5.00	0.00014
0.70	0.1762		

Example 9.2

Consider a point load $P = 5$ kN (Fig. 9.5). Calculate the vertical stress increase $(\Delta\sigma_z)$ at $z = 0, 2$ m, 4 m, 6 m, 10 m, and 20 m. Given $x = 3$ m and $y = 4$ m.

Solution

$$r = \sqrt{x^2 + y^2} = \sqrt{3^2 + 4^2} = 5 \text{ m}$$

The following table can now be prepared:

r (m)	z (m)	$\dfrac{r}{z}$	I_1	$\Delta\sigma_z = \left(\dfrac{P}{z^2}\right)I_1$ (kN/m²)
5	0	∞	0	0
	2	2.5	0.0034	0.0043
	4	1.25	0.0424	0.0133
	6	0.83	0.1295	0.0180
	10	0.5	0.2733	0.0137
	20	0.25	0.4103	0.0051

■

9.3 *Vertical Stress Caused by a Line Load*

Figure 9.6 shows a flexible line load of infinite length that has an intensity q/unit length on the surface of a semi-infinite soil mass. The vertical stress increase, $\Delta\sigma_z$, inside the soil mass can be determined by using the principles of the theory of elasticity, or

$$\Delta\sigma_z = \frac{2qz^3}{\pi(x^2 + z^2)^2} \tag{9.15}$$

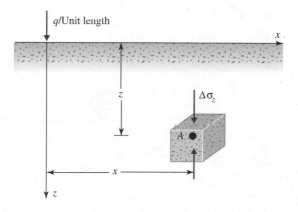

Figure 9.6
Line load over the surface of a
semi-infinite soil mass

Table 9.2 Variation of $\Delta\sigma_z/(q/z)$ with x/z [Eq. (9.16)]

x/z	$\Delta\sigma_z/(q/z)$	x/z	$\Delta\sigma_z/(q/z)$
0	0.637	1.3	0.088
0.1	0.624	1.4	0.073
0.2	0.589	1.5	0.060
0.3	0.536	1.6	0.050
0.4	0.473	1.7	0.042
0.5	0.407	1.8	0.035
0.6	0.344	1.9	0.030
0.7	0.287	2.0	0.025
0.8	0.237	2.2	0.019
0.9	0.194	2.4	0.014
1.0	0.159	2.6	0.011
1.1	0.130	2.8	0.008
1.2	0.107	3.0	0.006

This equation can be rewritten as

$$\Delta\sigma_z = \frac{2q}{\pi z[(x/z)^2 + 1]^2}$$

or

$$\frac{\Delta\sigma_z}{(q/z)} = \frac{2}{\pi[(x/z)^2 + 1]^2} \tag{9.16}$$

Note that Eq. (9.16) is in a nondimensional form. Using this equation, we can calculate the variation of $\Delta\sigma_z/(q/z)$ with x/z. This is given in Table 9.2. The value of $\Delta\sigma_z$ calculated by using Eq. (9.16) is the additional stress on soil caused by the line load. The value of $\Delta\sigma_z$ does not include the overburden pressure of the soil above point A.

Example 9.3

Figure 9.7a shows two line loads and a point load acting at the ground surface. Determine the increase in vertical stress at point A, which is located at a depth of 1.5 m.

Solution
Referring to Figures 9.7b through 9.7d, we find that

$$\Delta\sigma_z = \Delta\sigma_{z(1)} + \Delta\sigma_{z(2)} + \Delta\sigma_{z(3)}$$

$$= \underbrace{\frac{2q_1 z^3}{\pi(x_1^2 + z^2)^2}}_{\text{Eq. (9.15)}} + \underbrace{\frac{2q_2 z^3}{\pi(x_2^2 + z^2)^2}}_{\text{Eq. (9.15)}} + \underbrace{\frac{3P}{2\pi}\frac{z^3}{(r^2 + z^2)^{5/2}}}_{\text{Eq. (9.12)}}$$

$$= \frac{(2)(15)(1.5)^3}{\pi[(2)^2 + (1.5)^2]^2} + \frac{(2)(10)(1.5)^3}{\pi[(4)^2 + (1.5)^2]^2} + \frac{(3)(30)}{(2)(\pi)}\frac{(1.5)^3}{\{[(3)^2 + (4)^2] + (1.5)^2\}^{5/2}}$$

$$= 0.825 + 0.065 + 0.012 = \mathbf{0.902\ kN/m^2} \qquad \blacksquare$$

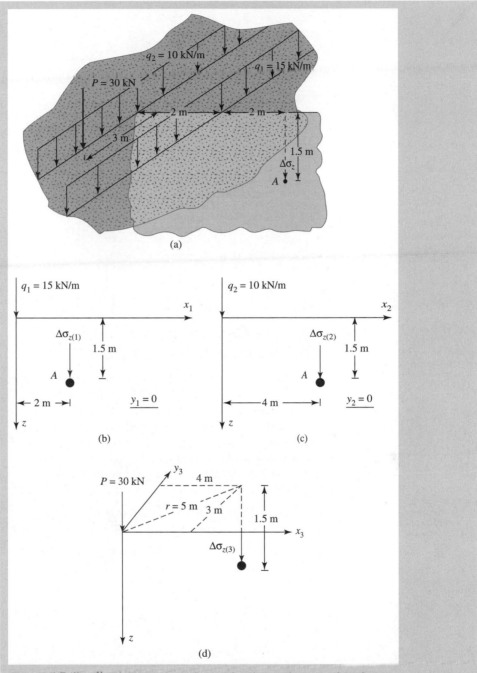

Figure 9.7 Two line loads and a point load acting at the ground surface

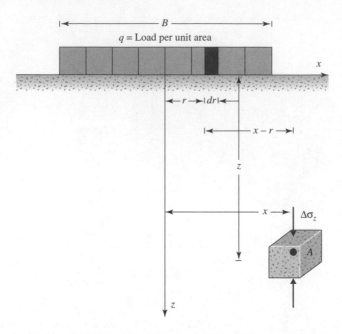

Figure 9.8
Vertical stress caused by a
flexible strip load

9.4   *Vertical Stress Caused by a Strip Load (Finite Width and Infinite Length)*

The fundamental equation for the vertical stress increase at a point in a soil mass as the result of a line load (Section 9.3) can be used to determine the vertical stress at a point caused by a flexible strip load of width B. (See Figure 9.8.) Let the load per unit area of the strip shown in Figure 9.8 be equal to q. If we consider an elemental strip of width dr, the load per unit length of this strip is equal to $q\,dr$. This elemental strip can be treated as a line load. Equation (9.15) gives the vertical stress increase $d\sigma_z$ at point A inside the soil mass caused by this elemental strip load. To calculate the vertical stress increase, we need to substitute $q\,dr$ for q and $(x - r)$ for x. So

$$d\sigma_z = \frac{2(q\,dr)z^3}{\pi[(x - r)^2 + z^2]^2} \tag{9.17}$$

The total increase in the vertical stress ($\Delta\sigma_z$) at point A caused by the entire strip load of width B can be determined by integration of Eq. (9.17) with limits of r from $-B/2$ to $+B/2$, or

$$\Delta\sigma_z = \int d\sigma_z = \int_{-B/2}^{+B/2} \left(\frac{2q}{\pi}\right)\left\{\frac{z^3}{[(x - r)^2 + z^2]^2}\right\} dr$$

$$= \frac{q}{\pi}\left\{\tan^{-1}\left[\frac{z}{x - (B/2)}\right] - \tan^{-1}\left[\frac{z}{x + (B/2)}\right]\right. \tag{9.18}$$

$$\left. - \frac{Bz[x^2 - z^2 - (B^2/4)]}{[x^2 + z^2 - (B^2/4)]^2 + B^2z^2}\right\}$$

Table 9.3 shows the variation of $\Delta\sigma_z/q$ with $2z/B$ for $2x/B$. This table can be used conveniently for the calculation of vertical stress at a point caused by a flexible strip load.

Table 9.3 Variation of $\Delta\sigma_z/q$ with $2z/B$ and $2x/B$ [Eq. (9.18)]

2z/B	2x/B										
	0.0	0.1	0.2	0.3	0.4	0.5	0.6	0.7	0.8	0.9	1.0
0.00	1.000	1.000	1.000	1.000	1.000	1.000	1.000	1.000	1.000	1.000	0.000
0.10	1.000	1.000	0.999	0.999	0.999	0.998	0.997	0.993	0.980	0.909	0.500
0.20	0.997	0.997	0.996	0.995	0.992	0.988	0.979	0.959	0.909	0.775	0.500
0.30	0.990	0.989	0.987	0.984	0.978	0.967	0.947	0.908	0.833	0.697	0.499
0.40	0.977	0.976	0.973	0.966	0.955	0.937	0.906	0.855	0.773	0.651	0.498
0.50	0.959	0.958	0.953	0.943	0.927	0.902	0.864	0.808	0.727	0.620	0.497
0.60	0.937	0.935	0.928	0.915	0.896	0.866	0.825	0.767	0.691	0.598	0.495
0.70	0.910	0.908	0.899	0.885	0.863	0.831	0.788	0.732	0.662	0.581	0.492
0.80	0.881	0.878	0.869	0.853	0.829	0.797	0.755	0.701	0.638	0.566	0.489
0.90	0.850	0.847	0.837	0.821	0.797	0.765	0.724	0.675	0.617	0.552	0.485
1.00	0.818	0.815	0.805	0.789	0.766	0.735	0.696	0.650	0.598	0.540	0.480
1.10	0.787	0.783	0.774	0.758	0.735	0.706	0.670	0.628	0.580	0.529	0.474
1.20	0.755	0.752	0.743	0.728	0.707	0.679	0.646	0.607	0.564	0.517	0.468
1.30	0.725	0.722	0.714	0.699	0.679	0.654	0.623	0.588	0.548	0.506	0.462
1.40	0.696	0.693	0.685	0.672	0.653	0.630	0.602	0.569	0.534	0.495	0.455
1.50	0.668	0.666	0.658	0.646	0.629	0.607	0.581	0.552	0.519	0.484	0.448
1.60	0.642	0.639	0.633	0.621	0.605	0.586	0.562	0.535	0.506	0.474	0.440
1.70	0.617	0.615	0.608	0.598	0.583	0.565	0.544	0.519	0.492	0.463	0.433
1.80	0.593	0.591	0.585	0.576	0.563	0.546	0.526	0.504	0.479	0.453	0.425
1.90	0.571	0.569	0.564	0.555	0.543	0.528	0.510	0.489	0.467	0.443	0.417
2.00	0.550	0.548	0.543	0.535	0.524	0.510	0.494	0.475	0.455	0.433	0.409
2.10	0.530	0.529	0.524	0.517	0.507	0.494	0.479	0.462	0.443	0.423	0.401
2.20	0.511	0.510	0.506	0.499	0.490	0.479	0.465	0.449	0.432	0.413	0.393
2.30	0.494	0.493	0.489	0.483	0.474	0.464	0.451	0.437	0.421	0.404	0.385
2.40	0.477	0.476	0.473	0.467	0.460	0.450	0.438	0.425	0.410	0.395	0.378
2.50	0.462	0.461	0.458	0.452	0.445	0.436	0.426	0.414	0.400	0.386	0.370
2.60	0.447	0.446	0.443	0.439	0.432	0.424	0.414	0.403	0.390	0.377	0.363
2.70	0.433	0.432	0.430	0.425	0.419	0.412	0.403	0.393	0.381	0.369	0.355
2.80	0.420	0.419	0.417	0.413	0.407	0.400	0.392	0.383	0.372	0.360	0.348
2.90	0.408	0.407	0.405	0.401	0.396	0.389	0.382	0.373	0.363	0.352	0.341
3.00	0.396	0.395	0.393	0.390	0.385	0.379	0.372	0.364	0.355	0.345	0.334
3.10	0.385	0.384	0.382	0.379	0.375	0.369	0.363	0.355	0.347	0.337	0.327
3.20	0.374	0.373	0.372	0.369	0.365	0.360	0.354	0.347	0.339	0.330	0.321
3.30	0.364	0.363	0.362	0.359	0.355	0.351	0.345	0.339	0.331	0.323	0.315
3.40	0.354	0.354	0.352	0.350	0.346	0.342	0.337	0.331	0.324	0.316	0.308
3.50	0.345	0.345	0.343	0.341	0.338	0.334	0.329	0.323	0.317	0.310	0.302
3.60	0.337	0.336	0.335	0.333	0.330	0.326	0.321	0.316	0.310	0.304	0.297
3.70	0.328	0.328	0.327	0.325	0.322	0.318	0.314	0.309	0.304	0.298	0.291
3.80	0.320	0.320	0.319	0.317	0.315	0.311	0.307	0.303	0.297	0.292	0.285
3.90	0.313	0.313	0.312	0.310	0.307	0.304	0.301	0.296	0.291	0.286	0.280
4.00	0.306	0.305	0.304	0.303	0.301	0.298	0.294	0.290	0.285	0.280	0.275
4.10	0.299	0.299	0.298	0.296	0.294	0.291	0.288	0.284	0.280	0.275	0.270
4.20	0.292	0.292	0.291	0.290	0.288	0.285	0.282	0.278	0.274	0.270	0.265
4.30	0.286	0.286	0.285	0.283	0.282	0.279	0.276	0.273	0.269	0.265	0.260
4.40	0.280	0.280	0.279	0.278	0.276	0.274	0.271	0.268	0.264	0.260	0.256
4.50	0.274	0.274	0.273	0.272	0.270	0.268	0.266	0.263	0.259	0.255	0.251
4.60	0.268	0.268	0.268	0.266	0.265	0.263	0.260	0.258	0.254	0.251	0.247
4.70	0.263	0.263	0.262	0.261	0.260	0.258	0.255	0.253	0.250	0.246	0.243
4.80	0.258	0.258	0.257	0.256	0.255	0.253	0.251	0.248	0.245	0.242	0.239
4.90	0.253	0.253	0.252	0.251	0.250	0.248	0.246	0.244	0.241	0.238	0.235
5.00	0.248	0.248	0.247	0.246	0.245	0.244	0.242	0.239	0.237	0.234	0.231

(continued)

Table 9.3 (*continued*)

2z/B	2x/B									
	1.1	1.2	1.3	1.4	1.5	1.6	1.7	1.8	1.9	2.0
0.00	0.000	0.000	0.000	0.000	0.000	0.000	0.000	0.000	0.000	0.000
0.10	0.091	0.020	0.007	0.003	0.002	0.001	0.001	0.000	0.000	0.000
0.20	0.225	0.091	0.040	0.020	0.011	0.007	0.004	0.003	0.002	0.002
0.30	0.301	0.165	0.090	0.052	0.031	0.020	0.013	0.009	0.007	0.005
0.40	0.346	0.224	0.141	0.090	0.059	0.040	0.027	0.020	0.014	0.011
0.50	0.373	0.267	0.185	0.128	0.089	0.063	0.046	0.034	0.025	0.019
0.60	0.391	0.298	0.222	0.163	0.120	0.088	0.066	0.050	0.038	0.030
0.70	0.403	0.321	0.250	0.193	0.148	0.113	0.087	0.068	0.053	0.042
0.80	0.411	0.338	0.273	0.218	0.173	0.137	0.108	0.086	0.069	0.056
0.90	0.416	0.351	0.291	0.239	0.195	0.158	0.128	0.104	0.085	0.070
1.00	0.419	0.360	0.305	0.256	0.214	0.177	0.147	0.122	0.101	0.084
1.10	0.420	0.366	0.316	0.271	0.230	0.194	0.164	0.138	0.116	0.098
1.20	0.419	0.371	0.325	0.282	0.243	0.209	0.178	0.152	0.130	0.111
1.30	0.417	0.373	0.331	0.291	0.254	0.221	0.191	0.166	0.143	0.123
1.40	0.414	0.374	0.335	0.298	0.263	0.232	0.203	0.177	0.155	0.135
1.50	0.411	0.374	0.338	0.303	0.271	0.240	0.213	0.188	0.165	0.146
1.60	0.407	0.373	0.339	0.307	0.276	0.248	0.221	0.197	0.175	0.155
1.70	0.402	0.370	0.339	0.309	0.281	0.254	0.228	0.205	0.183	0.164
1.80	0.396	0.368	0.339	0.311	0.284	0.258	0.234	0.212	0.191	0.172
1.90	0.391	0.364	0.338	0.312	0.286	0.262	0.239	0.217	0.197	0.179
2.00	0.385	0.360	0.336	0.311	0.288	0.265	0.243	0.222	0.203	0.185
2.10	0.379	0.356	0.333	0.311	0.288	0.267	0.246	0.226	0.208	0.190
2.20	0.373	0.352	0.330	0.309	0.288	0.268	0.248	0.229	0.212	0.195
2.30	0.366	0.347	0.327	0.307	0.288	0.268	0.250	0.232	0.215	0.199
2.40	0.360	0.342	0.323	0.305	0.287	0.268	0.251	0.234	0.217	0.202
2.50	0.354	0.337	0.320	0.302	0.285	0.268	0.251	0.235	0.220	0.205
2.60	0.347	0.332	0.316	0.299	0.283	0.267	0.251	0.236	0.221	0.207
2.70	0.341	0.327	0.312	0.296	0.281	0.266	0.251	0.236	0.222	0.208
2.80	0.335	0.321	0.307	0.293	0.279	0.265	0.250	0.236	0.223	0.210
2.90	0.329	0.316	0.303	0.290	0.276	0.263	0.249	0.236	0.223	0.211
3.00	0.323	0.311	0.299	0.286	0.274	0.261	0.248	0.236	0.223	0.211
3.10	0.317	0.306	0.294	0.283	0.271	0.259	0.247	0.235	0.223	0.212
3.20	0.311	0.301	0.290	0.279	0.268	0.256	0.245	0.234	0.223	0.212
3.30	0.305	0.296	0.286	0.275	0.265	0.254	0.243	0.232	0.222	0.211
3.40	0.300	0.291	0.281	0.271	0.261	0.251	0.241	0.231	0.221	0.211
3.50	0.294	0.286	0.277	0.268	0.258	0.249	0.239	0.229	0.220	0.210
3.60	0.289	0.281	0.273	0.264	0.255	0.246	0.237	0.228	0.218	0.209
3.70	0.284	0.276	0.268	0.260	0.252	0.243	0.235	0.226	0.217	0.208
3.80	0.279	0.272	0.264	0.256	0.249	0.240	0.232	0.224	0.216	0.207
3.90	0.274	0.267	0.260	0.253	0.245	0.238	0.230	0.222	0.214	0.206
4.00	0.269	0.263	0.256	0.249	0.242	0.235	0.227	0.220	0.212	0.205
4.10	0.264	0.258	0.252	0.246	0.239	0.232	0.225	0.218	0.211	0.203
4.20	0.260	0.254	0.248	0.242	0.236	0.229	0.222	0.216	0.209	0.202
4.30	0.255	0.250	0.244	0.239	0.233	0.226	0.220	0.213	0.207	0.200
4.40	0.251	0.246	0.241	0.235	0.229	0.224	0.217	0.211	0.205	0.199
4.50	0.247	0.242	0.237	0.232	0.226	0.221	0.215	0.209	0.203	0.197
4.60	0.243	0.238	0.234	0.229	0.223	0.218	0.212	0.207	0.201	0.195
4.70	0.239	0.235	0.230	0.225	0.220	0.215	0.210	0.205	0.199	0.194
4.80	0.235	0.231	0.227	0.222	0.217	0.213	0.208	0.202	0.197	0.192
4.90	0.231	0.227	0.223	0.219	0.215	0.210	0.205	0.200	0.195	0.190
5.00	0.227	0.224	0.220	0.216	0.212	0.207	0.203	0.198	0.193	0.188

Example 9.4

With reference to Figure 9.8, we are given $q = 200$ kN/m^2, $B = 6$ m, and $z = 3$ m. Determine the vertical stress increase at $x = \pm9, \pm6, \pm3$, and 0 m. Plot a graph of $\Delta\sigma_z$ against x.

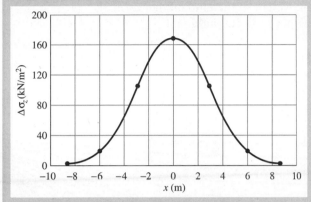

Figure 9.9
Plot of $\Delta\sigma_z$ against distance x

Solution

The following table can be made:

x(m)	$2x/B$	$2z/B$	$\Delta\sigma_z/q$[a]	$\Delta\sigma_z$[b] (kN/m^2)
±9	±3	1	0.017	3.4
±6	±2	1	0.084	16.8
±3	±1	1	0.480	96.0
0	0	1	0.818	163.6

[a] From Table 9.3
[b] $q = 200$ kN/m^2

The plot of $\Delta\sigma_z$ against x is given in Figure 9.9. ∎

9.5 *Vertical Stress Due to Embankment Loading*

Figure 9.10 shows the cross section of an embankment of height H. For this two-dimensional loading condition the vertical stress increase may be expressed as

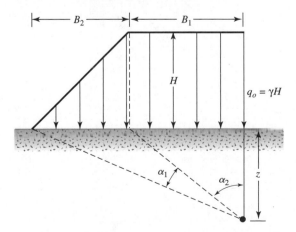

Figure 9.10
Embankment loading

$$\Delta\sigma_z = \frac{q_o}{\pi}\left[\left(\frac{B_1 + B_2}{B_2}\right)(\alpha_1 + \alpha_2) - \frac{B_1}{B_2}(\alpha_2)\right] \tag{9.19}$$

where $q_o = \gamma H$
   γ = unit weight of the embankment soil
   H = height of the embankment

$$\alpha_1\,(\text{radians}) = \tan^{-1}\left(\frac{B_1 + B_2}{z}\right) - \tan^{-1}\left(\frac{B_1}{z}\right) \tag{9.20}$$

$$\alpha_2 = \tan^{-1}\left(\frac{B_1}{z}\right) \tag{9.21}$$

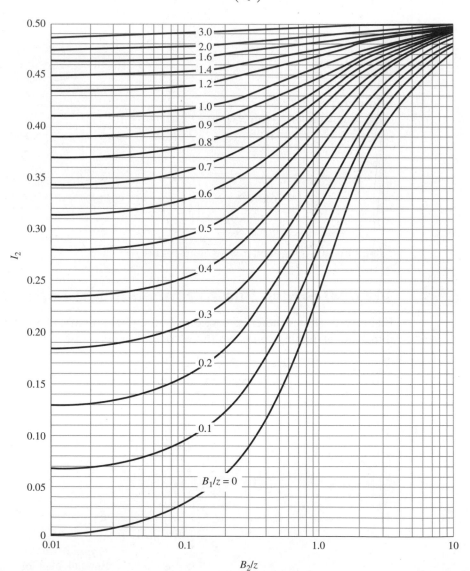

Figure 9.11
Osterberg's chart
for determination
of vertical stress
due to embank-
ment loading

For a detailed derivation of the equation, see Das (1997). A simplified form of Eq. (9.19) is

$$\Delta\sigma_z = q_o I_2 \qquad (9.22)$$

where I_2 = a function of B_1/z and B_2/z.

The variation of I_2 with B_1/z and B_2/z is shown in Figure 9.11 (Osterberg, 1957).

Example 9.5

An embankment is shown in Figure 9.12a. Determine the stress increase under the embankment at points A_1 and A_2.

Solution

$$\gamma H = (17.5)(7) = 122.5\ \text{kN/m}^2$$

Stress Increase at A_1

The left side of Figure 9.12 indicates that $B_1 = 2.5$ m and $B_2 = 14$ m. So

$$\frac{B_1}{z} = \frac{2.5}{5} = 0.5;\ \frac{B_2}{z} = \frac{14}{5} = 2.8$$

According to Figure 9.11, in this case, $I_2 = 0.445$. Because the two sides in Figure 9.12b are symmetrical, the value of I_2 for the right side will also be 0.445. So

$$\Delta\sigma_z = \Delta\sigma_{z(1)} + \Delta\sigma_{z(2)} = q_o[I_{2(\text{Left})} + I_{2(\text{Right})}]$$

$$= 122.5[0.445 + 0.445] = \mathbf{109.03\ kN/m^2}$$

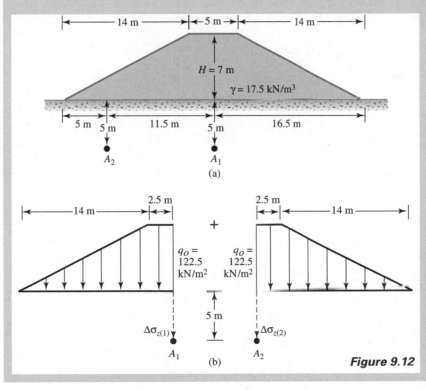

Figure 9.12

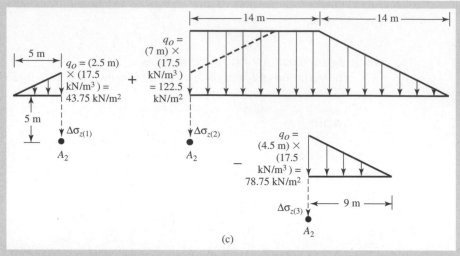

Figure 9.12 (*continued*)

Stress Increase at A_2

Refer to Figure 9.12c. For the left side, $B_2 = 5$ m and $B_1 = 0$. So

$$\frac{B_2}{z} = \frac{5}{5} = 1; \quad \frac{B_1}{z} = \frac{0}{5} = 0$$

According to Figure 9.11, for these values of B_2/z and B_1/z, $I_2 = 0.25$. So

$$\Delta\sigma_{z(1)} = 43.75(0.25) = 10.94 \text{ kN/m}^2$$

For the middle section,

$$\frac{B_2}{z} = \frac{14}{5} = 2.8; \quad \frac{B_1}{z} = \frac{14}{5} = 2.8$$

Thus, $I_2 = 0.495$. So

$$\Delta\sigma_{z(2)} = 0.495(122.5) = 60.64 \text{ kN/m}^2$$

For the right side

$$\frac{B_2}{z} = \frac{9}{5} = 1.8; \quad \frac{B_1}{z} = \frac{0}{5} = 0$$

and $I_2 = 0.335$. So

$$\Delta\sigma_{z(3)} = (78.75)(0.335) = 26.38 \text{ kN/m}^2$$

Total stress increase at point A_2 is

$$\Delta\sigma_z = \Delta\sigma_{z(1)} + \Delta\sigma_{z(2)} - \Delta\sigma_{z(3)} = 10.94 + 60.64 - 26.38 = \textbf{45.2 kN/m}^2 \quad \blacksquare$$

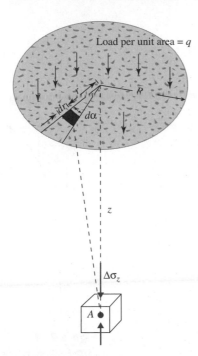

Figure 9.13
Vertical stress below the center of a uniformly loaded
flexible circular area

9.6   *Vertical Stress Below the Center of a Uniformly Loaded Circular Area*

Using Boussinesq's solution for vertical stress $\Delta\sigma_z$ caused by a point load [Eq. (9.12)], one can also develop an expression for the vertical stress below the center of a uniformly loaded flexible circular area.

From Figure 9.13, let the intensity of pressure on the circular area of radius R be equal to q. The total load on the elemental area (shaded in the figure) is equal to $qr\,dr\,d\alpha$. The vertical stress, $d\sigma_z$ at point A caused by the load on the elemental area (which may be assumed to be a concentrated load) can be obtained from Eq. (9.12):

$$d\sigma_z = \frac{3(qr\,dr\,d\alpha)}{2\pi}\frac{z^3}{(r^2 + z^2)^{5/2}} \tag{9.23}$$

The increase in the stress at point A caused by the entire loaded area can be found by integrating Eq. (9.23):

$$\Delta\sigma_z = \int d\sigma_z = \int_{\alpha=0}^{\alpha=2\pi}\int_{r=0}^{r=R}\frac{3q}{2\pi}\frac{z^3 r}{(r^2 + z^2)^{5/2}}\,dr\,d\alpha$$

So

$$\Delta\sigma_z = q\left\{1 - \frac{1}{[(R/z)^2 + 1]^{3/2}}\right\} \tag{9.24}$$

Table 9.4 Variation of $\Delta\sigma_z/q$ with z/R [Eq. (9.24)]

z/R	$\Delta\sigma_z$
0	1
0.02	0.9999
0.05	0.9998
0.10	0.9990
0.2	0.9925
0.4	0.9488
0.5	0.9106
0.8	0.7562
1.0	0.6465
1.5	0.4240
2.0	0.2845
2.5	0.1996
3.0	0.1436
4.0	0.0869
5.0	0.0571

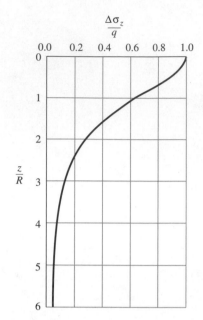

Figure 9.14
Stress under the center of a uniformly loaded flexible circular area

The variation of $\Delta\sigma_z/q$ with z/R as obtained from Eq. (9.24) is given in Table 9.4. A plot of this is also shown in Figure 9.14. The value of $\Delta\sigma_z$ decreases rapidly with depth, and at $z = 5R$, it is about 6% of q, which is the intensity of pressure at the ground surface.

9.7 Vertical Stress at any Point Below a Uniformly Loaded Circular Area

A detailed tabulation for calculation of vertical stress below a uniformly loaded flexible circular area was given by Ahlvin and Ulery (1962). Referring to Figure 9.15, we find that $\Delta\sigma_z$ at any point A located at a depth z at any distance r from the center of the loaded area can be given as

$$\Delta\sigma_z = q(A' + B') \tag{9.25}$$

where A' and B' are functions of z/R and r/R. (See Tables 9.5 and 9.6 on pages 244 and 245.)

9.8 Vertical Stress Caused by a Rectangularly Loaded Area

Boussinesq's solution can also be used to calculate the vertical stress increase below a flexible rectangular loaded area, as shown in Figure 9.16. The loaded area is located at the ground surface and has length L and width B. The uniformly distributed

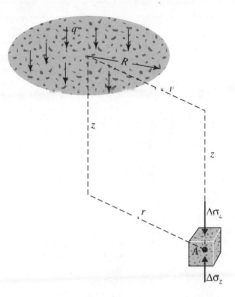

Figure 9.15 Vertical stress at any point below a uniformly loaded circular area

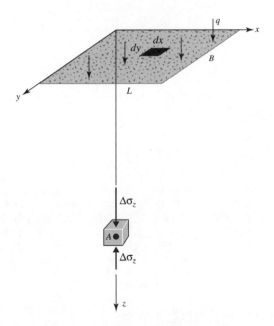

Figure 9.16 Vertical stress below the corner of a uniformly loaded flexible rectangular area

load per unit area is equal to q. To determine the increase in the vertical stress $(\Delta\sigma_z)$ at point A, which is located at depth z below the corner of the rectangular area, we need to consider a small elemental area $dx\,dy$ of the rectangle. (This is shown in Figure 9.16.) The load on this elemental area can be given by

$$dq = q\,dx\,dy \qquad (9.26)$$

Table 9.5 Variation of A' with z/R and r/R*

									r/R
z/R	0	0.2	0.4	0.6	0.8	1	1.2	1.5	2
0	1.0	1.0	1.0	1.0	1.0	0.5	0	0	0
0.1	0.90050	0.89748	0.88679	0.86126	0.78797	0.43015	0.09645	0.02787	0.00856
0.2	0.80388	0.79824	0.77884	0.73483	0.63014	0.38269	0.15433	0.05251	0.01680
0.3	0.71265	0.70518	0.68316	0.62690	0.52081	0.34375	0.17964	0.07199	0.02440
0.4	0.62861	0.62015	0.59241	0.53767	0.44329	0.31048	0.18709	0.08593	0.03118
0.5	0.55279	0.54403	0.51622	0.46448	0.38390	0.28156	0.18556	0.09499	0.03701
0.6	0.48550	0.47691	0.45078	0.40427	0.33676	0.25588	0.17952	0.10010	
0.7	0.42654	0.41874	0.39491	0.35428	0.29833	0.21727	0.17124	0.10228	0.04558
0.8	0.37531	0.36832	0.34729	0.31243	0.26581	0.21297	0.16206	0.10236	
0.9	0.33104	0.32492	0.30669	0.27707	0.23832	0.19488	0.15253	0.10094	
1	0.29289	0.28763	0.27005	0.24697	0.21468	0.17868	0.14329	0.09849	0.05185
1.2	0.23178	0.22795	0.21662	0.19890	0.17626	0.15101	0.12570	0.09192	0.05260
1.5	0.16795	0.16552	0.15877	0.14804	0.13436	0.11892	0.10296	0.08048	0.05116
2	0.10557	0.10453	0.10140	0.09647	0.09011	0.08269	0.07471	0.06275	0.04496
2.5	0.07152	0.07098	0.06947	0.06698	0.06373	0.05974	0.05555	0.04880	0.03787
3	0.05132	0.05101	0.05022	0.04886	0.04707	0.04487	0.04241	0.03839	0.03150
4	0.02986	0.02976	0.02907	0.02802	0.02832	0.02749	0.02651	0.02490	0.02193
5	0.01942	0.01938				0.01835			0.01573
6	0.01361					0.01307			0.01168
7	0.01005					0.00976			0.00894
8	0.00772					0.00755			0.00703
9	0.00612					0.00600			0.00566
10								0.00477	0.00465

*After Ahlvin and Ulery (1962)

Table 9.6 Variation of B' with z/R and r/R*

									r/R
z/R	0	0.2	0.4	0.6	0.8	1	1.2	1.5	2
0	0	0	0	0	0	0	0	0	0
0.1	0.09852	0.10140	0.11138	0.13424	0.18796	0.05388	−0.07899	−0.02672	−0.00845
0.2	0.18857	0.19306	0.20772	0.23524	0.25983	0.08513	−0.07759	−0.04448	−0.01593
0.3	0.26362	0.26787	0.28018	0.29483	0.27257	0.10757	−0.04316	−0.04999	−0.02166
0.4	0.32016	0.32259	0.32748	0.32273	0.26925	0.12404	−0.00766	−0.04535	−0.02522
0.5	0.35777	0.35752	0.35323	0.33106	0.26236	0.13591	0.02165	−0.03455	−0.02651
0.6	0.37831	0.37531	0.36308	0.32822	0.25411	0.14440	0.04457	−0.02101	
0.7	0.38487	0.37962	0.36072	0.31929	0.24638	0.14986	0.06209	−0.00702	−0.02329
0.8	0.38091	0.37408	0.35133	0.30699	0.23779	0.15292	0.07530	0.00614	
0.9	0.36962	0.36275	0.33734	0.29299	0.22891	0.15404	0.08507	0.01795	
1	0.35355	0.34553	0.32075	0.27819	0.21978	0.15355	0.09210	0.02814	−0.01005
1.2	0.31485	0.30730	0.28481	0.24836	0.20113	0.14915	0.10002	0.04378	0.00023
1.5	0.25602	0.25025	0.23338	0.20694	0.17368	0.13732	0.10193	0.05745	0.01385
2	0.17889	0.18144	0.16644	0.15198	0.13375	0.11331	0.09254	0.06371	0.02836
2.5	0.12807	0.12633	0.12126	0.11327	0.10298	0.09130	0.07869	0.06022	0.03429
3	0.09487	0.09394	0.09099	0.08635	0.08033	0.07325	0.06551	0.05354	0.03511
4	0.05707	0.05666	0.05562	0.05383	0.05145	0.04773	0.04532	0.03995	0.03066
5	0.03772	0.03760				0.03384			0.02474
6	0.02666					0.02468			0.01968
7	0.01980					0.01868			0.01577
8	0.01526					0.01459			0.01279
9	0.01212					0.01170			0.01054
.10								0.00924	0.00879

*Source: From "Tabulated Values for Determining the Complete Pattern of Stresses, Strains, and Deflections Beneath a Uniform Circular Load on a Homogeneous Half Space," by R. G Ahlvin and H. H. Ulery. In Highway Research Bulletin 342, Transportation Research Board, National Research Council, Washington, DC, 1962.

Table 9.5 (*continued*)

3	4	5	6	7	8	10	12	14
0	0	0	0	0	0	0	0	0
0.00211	0.00084	0.00042						
0.00419	0.00167	0.00083	0.00048	0.00030	0.00020			
0.00622	0.00250							
0.01013	0.00407	0.00209	0.00118	0.00071	0.00053	0.00025	0.00014	0.00009
0.01742	0.00761	0.00393	0.00226	0.00143	0.00097	0.00050	0.00029	0.00018
0.01935	0.00871	0.00459	0.00269	0.00171	0.00115			
0.02142	0.01013	0.00548	0.00325	0.00210	0.00141	0.00073	0.00043	0.00027
0.02221	0.01160	0.00659	0.00399	0.00264	0.00180	0.00094	0.00056	0.00036
0.02143	0.01221	0.00732	0.00463	0.00308	0.00214	0.00115	0.00068	0.00043
0.01980	0.01220	0.00770	0.00505	0.00346	0.00242	0.00132	0.00079	0.00051
0.01592	0.01109	0.00768	0.00536	0.00384	0.00282	0.00160	0.00099	0.00065
0.01249	0.00949	0.00708	0.00527	0.00394	0.00298	0.00179	0.00113	0.00075
0.00983	0.00795	0.00628	0.00492	0.00384	0.00299	0.00188	0.00124	0.00084
0.00784	0.00661	0.00548	0.00445	0.00360	0.00291	0.00193	0.00130	0.00091
0.00635	0.00554	0.00472	0.00398	0.00332	0.00276	0.00189	0.00134	0.00094
0.00520	0.00466	0.00409	0.00353	0.00301	0.00256	0.00184	0.00133	0.00096
0.00438	0.00397	0.00352	0.00326	0.00273	0.00241			

Table 9.6 (*continued*)

3	4	5	6	7	8	10	12	14
0	0	0	0	0	0	0	0	0
−0.00210	−0.00084	−0.00042						
−0.00412	−0.00166	−0.00083	−0.00024	−0.00015	−0.00010			
−0.00599	−0.00245							
−0.00991	−0.00388	−0.00199	−0.00116	−0.00073	−0.00049	−0.00025	−0.00014	−0.00009
−0.01115	−0.00608	−0.00344	−0.00210	−0.00135	−0.00092	−0.00048	−0.00028	−0.00018
−0.00995	−0.00632	−0.00378	−0.00236	−0.00156	−0.00107			
−0.00669	−0.00600	−0.00401	−0.00265	−0.00181	−0.00126	−0.00068	−0.00040	−0.00026
0.00028	−0.00410	−0.00371	−0.00278	−0.00202	−0.00148	−0.00084	−0.00050	−0.00033
0.00661	−0.00130	−0.00271	−0.00250	−0.00201	−0.00156	−0.00094	−0.00059	−0.00039
0.01112	0.00157	−0.00134	−0.00192	−0.00179	−0.00151	−0.00099	−0.00065	−0.00046
0.01515	0.00595	0.00155	−0.00029	−0.00094	−0.00109	−0.00094	−0.00068	−0.00050
0.01522	0.00810	0.00371	0.00132	0.00013	−0.00043	−0.00070	−0.00061	−0.00049
0.01380	0.00867	0.00496	0.00254	0.00110	0.00028	−0.00037	−0.00047	−0.00045
0.01204	0.00842	0.00547	0.00332	0.00185	0.00093	−0.00002	−0.00029	−0.00037
0.01034	0.00779	0.00554	0.00372	0.00236	0.00141	0.00035	−0.00008	−0.00025
0.00888	0.00705	0.00533	0.00386	0.00265	0.00178	0.00066	0.00012	−0.00012
0.00764	0.00631	0.00501	0.00382	0.00281	0.00199			

The increase in the stress ($d\sigma_z$) at point A caused by the load dq can be determined by using Eq. (9.12). However, we need to replace P with $dq = q \, dx \, dy$ and r^2 with $x^2 + y^2$. Thus,

$$d\sigma_z = \frac{3q \, dx \, dy \, z^3}{2\pi(x^2 + y^2 + z^2)^{5/2}} \tag{9.27}$$

The increase in the stress, at point A caused by the entire loaded area can now be determined by integrating the preceding equation. We obtain

$$\Delta\sigma_z = \int d\sigma_z = \int_{y=0}^{B} \int_{x=0}^{L} \frac{3qz^3(dx \, dy)}{2\pi(x^2 + y^2 + z^2)^{5/2}} = qI_3 \tag{9.28}$$

where

$$I_3 = \frac{1}{4\pi}\left[\frac{2mn\sqrt{m^2 + n^2 + 1}}{m^2 + n^2 + m^2n^2 + 1}\left(\frac{m^2 + n^2 + 2}{m^2 + n^2 + 1}\right) + \tan^{-1}\left(\frac{2mn\sqrt{m^2 + n^2 + 1}}{m^2 + n^2 - m^2n^2 + 1}\right)\right] \tag{9.29}$$

$$m = \frac{B}{z} \tag{9.30}$$

$$n = \frac{L}{z} \tag{9.31}$$

The variation of I_3 with m and n is shown in Table 9.7.

Table 9.7 Variation of I_3 with m and n [Eq. (9.29)]

										m
n	0.1	0.2	0.3	0.4	0.5	0.6	0.7	0.8	0.9	1.0
0.1	0.0047	0.0092	0.0132	0.0168	0.0198	0.0222	0.0242	0.0258	0.0270	0.0279
0.2	0.0092	0.0179	0.0259	0.0328	0.0387	0.0435	0.0474	0.0504	0.0528	0.0547
0.3	0.0132	0.0259	0.0374	0.0474	0.0559	0.0629	0.0686	0.0731	0.0766	0.0794
0.4	0.0168	0.0328	0.0474	0.0602	0.0711	0.0801	0.0873	0.0931	0.0977	0.1013
0.5	0.0198	0.0387	0.0559	0.0711	0.0840	0.0947	0.1034	0.1104	0.1158	0.1202
0.6	0.0222	0.0435	0.0629	0.0801	0.0947	0.1069	0.1168	0.1247	0.1311	0.1361
0.7	0.0242	0.0474	0.0686	0.0873	0.1034	0.1169	0.1277	0.1365	0.1436	0.1491
0.8	0.0258	0.0504	0.0731	0.0931	0.1104	0.1247	0.1365	0.1461	0.1537	0.1598
0.9	0.0270	0.0528	0.0766	0.0977	0.1158	0.1311	0.1436	0.1537	0.1619	0.1684
1.0	0.0279	0.0547	0.0794	0.1013	0.1202	0.1361	0.1491	0.1598	0.1684	0.1752
1.2	0.0293	0.0573	0.0832	0.1063	0.1263	0.1431	0.1570	0.1684	0.1777	0.1851
1.4	0.0301	0.0589	0.0856	0.1094	0.1300	0.1475	0.1620	0.1739	0.1836	0.1914
1.6	0.0306	0.0599	0.0871	0.1114	0.1324	0.1503	0.1652	0.1774	0.1874	0.1955
1.8	0.0309	0.0606	0.0880	0.1126	0.1340	0.1521	0.1672	0.1797	0.1899	0.1981
2.0	0.0311	0.0610	0.0887	0.1134	0.1350	0.1533	0.1686	0.1812	0.1915	0.1999
2.5	0.0314	0.0616	0.0895	0.1145	0.1363	0.1548	0.1704	0.1832	0.1938	0.2024
3.0	0.0315	0.0618	0.0898	0.1150	0.1368	0.1555	0.1711	0.1841	0.1947	0.2034
4.0	0.0316	0.0619	0.0901	0.1153	0.1372	0.1560	0.1717	0.1847	0.1954	0.2042
5.0	0.0316	0.0620	0.0901	0.1154	0.1374	0.1561	0.1719	0.1849	0.1956	0.2044
6.0	0.0316	0.0620	0.0902	0.1154	0.1374	0.1562	0.1719	0.1850	0.1957	0.2045

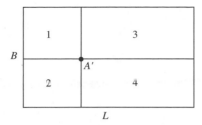

Figure 9.17 Increase of stress at any point below a rectangularly loaded flexible area

The increase in the stress at any point below a rectangularly loaded area can be found by using Eq. (9.28). This can be explained by reference to Figure 9.17. Let us determine the stress at a point below point A' at depth z. The loaded area can be divided into four rectangles as shown. The point A' is the corner common to all four rectangles. The increase in the stress at depth z below point A' due to each rectangular area can now be calculated by using Eq. (9.28). The total stress increase caused by the entire loaded area can be given by

$$\Delta\sigma_z = q[I_{3(1)} + I_{3(2)} + I_{3(3)} + I_{3(4)}] \tag{9.32}$$

where $I_{3(1)}, I_{3(2)}, I_{3(3)},$ and $I_{3(4)}$ = values of I_3 for rectangles 1, 2, 3, and 4, respectively.

Table 9.7 (*continued*)

1.2	1.4	1.6	1.8	2.0	2.5	3.0	4.0	5.0	6.0
0.0293	0.0301	0.0306	0.0309	0.0311	0.0314	0.0315	0.0316	0.0316	0.0316
0.0573	0.0589	0.0599	0.0606	0.0610	0.0616	0.0618	0.0619	0.0620	0.0620
0.0832	0.0856	0.0871	0.0880	0.0887	0.0895	0.0898	0.0901	0.0901	0.0902
0.1063	0.1094	0.1114	0.1126	0.1134	0.1145	0.1150	0.1153	0.1154	0.1154
0.1263	0.1300	0.1324	0.1340	0.1350	0.1363	0.1368	0.1372	0.1374	0.1374
0.1431	0.1475	0.1503	0.1521	0.1533	0.1548	0.1555	0.1560	0.1561	0.1562
0.1570	0.1620	0.1652	0.1672	0.1686	0.1704	0.1711	0.1717	0.1719	0.1719
0.1684	0.1739	0.1774	0.1797	0.1812	0.1832	0.1841	0.1847	0.1849	0.1850
0.1777	0.1836	0.1874	0.1899	0.1915	0.1938	0.1947	0.1954	0.1956	0.1957
0.1851	0.1914	0.1955	0.1981	0.1999	0.2024	0.2034	0.2042	0.2044	0.2045
0.1958	0.2028	0.2073	0.2103	0.2124	0.2151	0.2163	0.2172	0.2175	0.2176
0.2028	0.2102	0.2151	0.2184	0.2206	0.2236	0.2250	0.2260	0.2263	0.2264
0.2073	0.2151	0.2203	0.2237	0.2261	0.2294	0.2309	0.2320	0.2323	0.2325
0.2103	0.2183	0.2237	0.2274	0.2299	0.2333	0.2350	0.2362	0.2366	0.2367
0.2124	0.2206	0.2261	0.2299	0.2325	0.2361	0.2378	0.2391	0.2395	0.2397
0.2151	0.2236	0.2294	0.2333	0.2361	0.2401	0.2420	0.2434	0.2439	0.2441
0.2163	0.2250	0.2309	0.2350	0.2378	0.2420	0.2439	0.2455	0.2461	0.2463
0.2172	0.2260	0.2320	0.2362	0.2391	0.2434	0.2455	0.2472	0.2479	0.2481
0.2175	0.2263	0.2324	0.2366	0.2395	0.2439	0.2460	0.2479	0.2486	0.2489
0.2176	0.2264	0.2325	0.2367	0.2397	0.2441	0.2463	0.2482	0.2489	0.2492

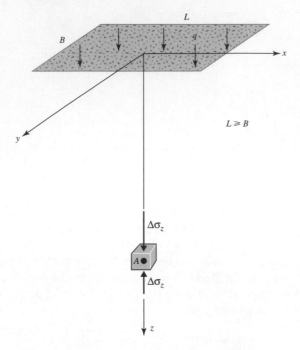

Figure 9.18 Vertical stress below the center of a uniformly loaded flexible rectangular area

In most cases the vertical stress increase below the center of a rectangular area (Figure 9.18) is important. This stress increase can be given by the relationship

$$\Delta\sigma_z = qI_4 \tag{9.33}$$

where

$$I_4 = \frac{2}{\pi}\left[\frac{m_1 n_1}{\sqrt{1 + m_1^2 + n_1^2}}\frac{1 + m_1^2 + 2n_1^2}{(1 + n_1^2)(m_1^2 + n_1^2)}\right] + \sin^{-1}\frac{m_1}{\sqrt{m_1^2 + n_1^2}\sqrt{1 + n_1^2}} \tag{9.34}$$

$$m_1 = \frac{L}{B} \tag{9.35}$$

$$n_1 = \frac{z}{b} \tag{9.36}$$

$$b = \frac{B}{2} \tag{9.37}$$

The variation of I_4 with m_1 and n_1 is given in Table 9.8.

Table 9.8 Variation of I_4 with m_1 and n_1 [Eq. (9.34)]

n_1	m_1									
	1	**2**	**3**	**4**	**5**	**6**	**7**	**8**	**9**	**10**
0.20	0.994	0.997	0.997	0.997	0.997	0.997	0.997	0.997	0.997	0.997
0.40	0.960	0.976	0.977	0.977	0.977	0.977	0.977	0.977	0.977	0.977
0.60	0.892	0.932	0.936	0.936	0.937	0.937	0.937	0.937	0.937	0.937
0.80	0.800	0.870	0.878	0.880	0.881	0.881	0.881	0.881	0.881	0.881
1.00	0.701	0.800	0.814	0.817	0.818	0.818	0.818	0.818	0.818	0.818
1.20	0.606	0.727	0.748	0.753	0.754	0.755	0.755	0.755	0.755	0.755
1.40	0.522	0.658	0.685	0.692	0.694	0.695	0.695	0.696	0.696	0.696
1.60	0.449	0.593	0.627	0.636	0.639	0.640	0.641	0.641	0.641	0.642
1.80	0.388	0.534	0.573	0.585	0.590	0.591	0.592	0.592	0.593	0.593
2.00	0.336	0.481	0.525	0.540	0.545	0.547	0.548	0.549	0.549	0.549
3.00	0.179	0.293	0.348	0.373	0.384	0.389	0.392	0.393	0.394	0.395
4.00	0.108	0.190	0.241	0.269	0.285	0.293	0.298	0.301	0.302	0.303
5.00	0.072	0.131	0.174	0.202	0.219	0.229	0.236	0.240	0.242	0.244
6.00	0.051	0.095	0.130	0.155	0.172	0.184	0.192	0.197	0.200	0.202
7.00	0.038	0.072	0.100	0.122	0.139	0.150	0.158	0.164	0.168	0.171
8.00	0.029	0.056	0.079	0.098	0.113	0.125	0.133	0.139	0.144	0.147
9.00	0.023	0.045	0.064	0.081	0.094	0.105	0.113	0.119	0.124	0.128
10.00	0.019	0.037	0.053	0.067	0.079	0.089	0.097	0.103	0.108	0.112

Example 9.6

The flexible area shown in Figure 9.19 is uniformly loaded. Given that $q = 150 \text{ kN/m}^2$, determine the vertical stress increase at point A.

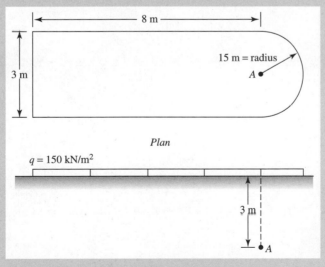

Figure 9.19 Uniformly loaded flexible area

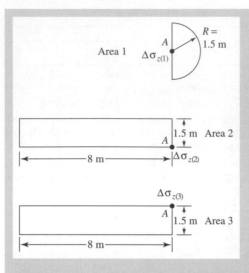

Figure 9.20
Division of uniformly loaded flexible area into three parts

Solution

The flexible area shown in Figure 9.19 is divided into three parts in Figure 9.20. At A,

$$\Delta\sigma_z = \Delta\sigma_{z(1)} + \Delta\sigma_{z(2)} + \Delta\sigma_{z(3)}$$

From Eq. (9.24),

$$\Delta\sigma_{z(1)} = \left(\frac{1}{2}\right)q\left\{1 - \frac{1}{[(R/z)^2 + 1]^{3/2}}\right\}$$

We know that $R = 1.5$ m, $z = 3$ m, and $q = 150$ kN/m², so

$$\Delta\sigma_{z(1)} = \frac{150}{2}\left\{1 - \frac{1}{[(1.5/3)^2 + 1]^{3/2}}\right\} = 21.3 \text{ kN/m}^2$$

We can see that $\Delta\sigma_{z(2)} = \Delta\sigma_{z(3)}$. From Eqs. (9.30) and (9.31),

$$m = \frac{1.5}{3} = 0.5$$

$$n = \frac{8}{3} = 2.67$$

From Table 9.7, for $m = 0.5$ and $n = 2.67$, the magnitude of $I_3 = 0.1365$. Thus, from Eq. (9.28),

$$\Delta\sigma_{z(2)} = \Delta\sigma_{z(3)} = qI_3 = (150)(0.1365) = 20.48 \text{ kN/m}^2$$

so

$$\Delta\sigma_z = 21.3 + 20.48 + 20.48 = \textbf{62.26 kN/m}^2$$

∎

Influence Chart for Vertical Pressure

Equation (9.24) can be rearranged and written in the form

$$\frac{R}{z} = \sqrt{\left(1 - \frac{\Delta\sigma_z}{q}\right)^{-2/3} - 1}$$

(9.38)

Note that R/z and $\Delta\sigma_z/q$ in this equation are nondimensional quantities. The values of R/z that correspond to various pressure ratios are given in Table 9.9.

Table 9.9 Values of R/z for Various Pressure Ratios [Eq. (9.38)]

$\Delta\sigma_z/q$	R/z	$\Delta\sigma_z/q$	R/z
0	0	0.55	0.8384
0.05	0.1865	0.60	0.9176
0.10	0.2698	0.65	1.0067
0.15	0.3383	0.70	1.1097
0.20	0.4005	0.75	1.2328
0.25	0.4598	0.80	1.3871
0.30	0.5181	0.85	1.5943
0.35	0.5768	0.90	1.9084
0.40	0.6370	0.95	2.5232
0.45	0.6997	1.00	∞
0.50	0.7664		

Using the values of R/z obtained from Eq. (9.38) for various pressure ratios, Newmark (1942) presented an influence chart that can be used to determine the vertical pressure at any point below a uniformly loaded flexible area of any shape.

Figure 9.21 shows an influence chart that has been constructed by drawing concentric circles. The radii of the circles are equal to the R/z values corresponding to $\Delta\sigma_z/q = 0, 0.1, 0.2, \ldots, 1$. (*Note:* For $\Delta\sigma_z/q = 0$, $R/z = 0$, and for $\Delta\sigma_z/q = 1$, $R/z = \infty$, so nine circles are shown.) The unit length for plotting the circles is $\overline{AB}$. The circles are divided by several equally spaced radial lines. The influence value of the chart is given by $1/N$, where N is equal to the number of elements in the chart. In Figure 9.21, there are 200 elements; hence, the influence value is 0.005.

The procedure for obtaining vertical pressure at any point below a loaded area is as follows:

1. Determine the depth z below the uniformly loaded area at which the stress increase is required.
2. Plot the plan of the loaded area with a scale of z equal to the unit length of the chart ($\overline{AB}$).
3. Place the plan (plotted in step 2) on the influence chart in such a way that the point below which the stress is to be determined is located at the center of the chart.
4. Count the number of elements (M) of the chart enclosed by the plan of the loaded area.

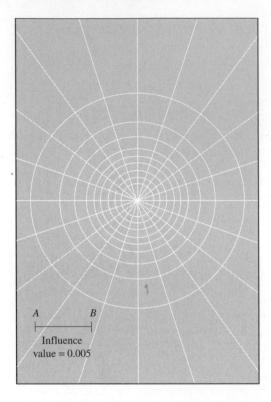

A

B

Influence
value = 0.005

Figure 9.21 Influence chart for vertical
pressure based on Boussinesq's theory
(after Newmark, 1942)

The increase in the pressure at the point under consideration is given by

$$\Delta\sigma_z = (IV)qM \qquad (9.39)$$

where IV = influence value
q = pressure on the loaded area

Example 9.7

The cross section and plan of a column footing are shown in Figure 9.22. Find the increase in vertical stress produced by the column footing at point A.

Solution
Point A is located at à depth 3 m below the bottom of the footing. The plan of the square footing has been replotted to a scale of $\overline{AB}$ = 3 m and placed on the influence chart (Figure 9.23) in such a way that point A on the plan falls directly over the center of the chart. The number of elements inside the outline of the plan is about 48.5. Hence,

$$\Delta\sigma_z = (IV)qM = 0.005\left(\frac{660}{3 \times 3}\right)48.5 = \textbf{17.78 kN/m}^2 \qquad \blacksquare$$

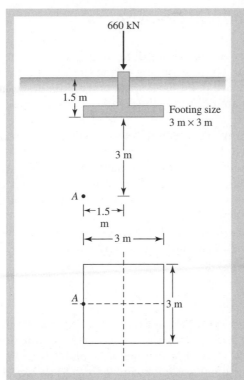

Figure 9.22 Cross section and plan of a column footing

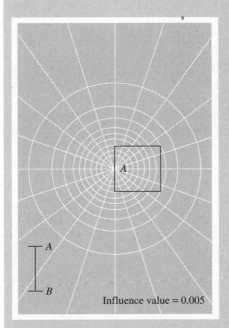

Figure 9.23 Determination of stress at a point by use of Newmark's influence chart

9.10   *Summary and General Comments*

This chapter presents the relationships for determining vertical stress at a point due to the application of various types of loading on the surface of a soil mass. The types of loading considered here are point, line, strip, embankment, circular, and rectangular. These relationships are derived by integration of Boussinesq's equation for a point load.

The equations and graphs presented in this chapter are based entirely on the principles of the theory of elasticity; however, one must realize the limitations of these theories when they are applied to a soil medium. This is because soil deposits, in general, are not homogeneous, perfectly elastic, and isotropic. Hence, some deviations from the theoretical stress calculations can be expected in the field. Only a limited number of field observations are available in the literature at the present time. On the basis of these results, it appears that one could expect a difference of ± 25 to 30% between theoretical estimates and actual field values.

Problems

9.1–9.5   For the soil elements shown in Figures 9.24–28, determine the maximum and minimum principal stresses. Also determine the normal and shear stresses on plane *AB*. [*Note:* For Problems 9.1 and 9.2 use Eqs. (9.3), (9.4), (9.6), and (9.7); for Problems 9.3, 9.4, and 9.5 use Mohr's circle.]

9.6   Point loads having magnitudes of 15 kN, 20 kN, and 30 kN act at *A, B,* and *C,* respectively (Figure 9.29). Determine the increase in vertical stress below point *D* at a depth of 5 m.

9.7   Refer to Figure 9.30. Determine the stress increase, $\Delta\sigma_z$, at *A*, given the following data:

$$q_1 = 75 \text{ kN/m} \qquad x_1 = 2 \text{ m} \qquad z = 1.5 \text{ m}$$
$$q_2 = 0 \qquad\qquad x_2 = 1 \text{m}$$

9.8   Repeat Problem 9.7 with the following values:

$$q_1 = 0 \qquad\qquad x_1 = 5 \text{ ft} \qquad z = 5 \text{ ft}$$
$$q_2 = 300 \text{ lb/ft} \qquad x_2 = 3 \text{ ft}$$

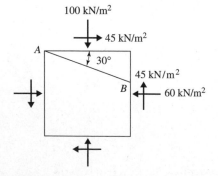

Figure 9.24 Soil element for Problem 9.1

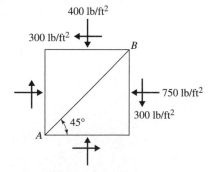

Figure 9.25 Soil elememt for Problem 9.2

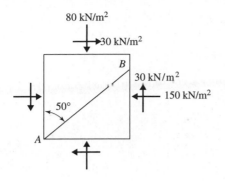

Figure 9.26 Soil element for Problem 9.3

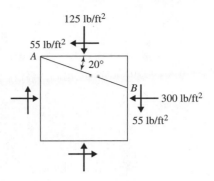

Figure 9.27 Soil element for Problem 9.4

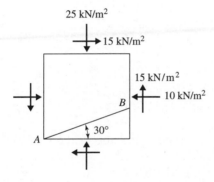

Figure 9.28 Soil element for Problem 9.5

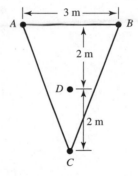

Figure 9.29

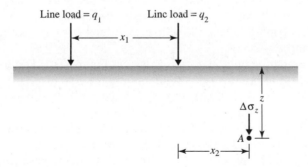

Figure 9.30

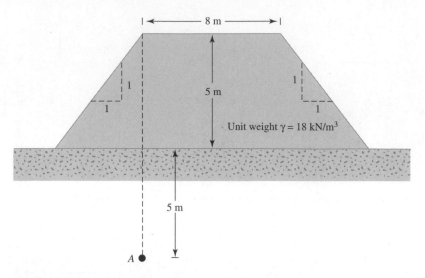

Figure 9.31

9.9 Repeat Problem 9.7 with the following values:
$$q_1 = 100 \text{ kN/m} \qquad x_1 = 3 \text{ m} \qquad z = 2\text{m}$$
$$q_2 = 200 \text{ kN/m} \qquad x_2 = 2 \text{ m}$$

9.10 Refer to Figure 9.6. The magnitude of the line load q is 2500 lb/ft. Calculate and plot the variation of the vertical stress increase $\Delta\sigma_z$ between the limits of $x = -10$ ft and $x = +10$ ft, given that $z = 5$ ft.

9.11 Refer to Figure 9.30. Given that $q_1 = 10$ kN/m, $x_1 = 3$ m, $x_2 = 2$ m, and $z = 1$ m, if the vertical stress increase at point A due to the loading is 3 kN/m^2, determine the magnitude of q_2.

9.12 Refer to Figure 9.8. Given that $B = 12$ ft, $q = 350$ lb/ft^2, $x = 9$ ft, and $z = 5$ ft, determine the vertical stress increase, $\Delta\sigma_z$, at point A.

9.13 Repeat Problem 9.12 using the following values: $q = 7000$ kN/m^2, $B = 2$ m, $x = 2$ m, and $z = 2.5$ m.

9.14 An earth embankment diagram is shown in Figure 9.31. Determine the stress increase at point A due to the embankment load.

9.15 Figure 9.32 shows an embankment load for a silty clay soil layer. Determine the vertical stress increase at points A, B, and C.

9.16 Consider a circularly loaded flexible area on the ground surface. Given that the radius of the circular area is $(R) = 4$ m and the uniformly distributed load is $q = 200$ kN/m^2, calculate the vertical stress increase $\Delta\sigma_z$ at a point located 5 m (z) below the ground surface (immediately below the center of the circular area).

9.17 Consider a circularly loaded flexible area on the ground surface. Given that the radius of the circular area $(R) = 6$ ft and that the uniformly distributed load $(q) = 4200$ lb/ft^2, calculate the vertical stress increase $\Delta\sigma_z$ at points 1.5, 3, 6, 9, and 12 ft below the ground surface (immediately below the center of the circular area).

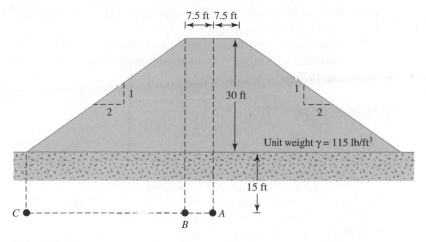

7.5 ft 7.5 ft

1

2

30 ft

1

2

Unit weight $\gamma = 115$ lb/ft^3

15 ft

C

B

A

Figure 9.32

9.18 Figure 9.15 shows a flexible circular area of radius $R = 4$ m. The uniformly distributed load on the circular area is 300 kN/m^2. Calculate the vertical stress increase at $r = 0, 0.8, 1.6, 4, 6$, and 8 m, and $z = 4.8$ m.

9.19 Refer to Figure 9.33. The circular flexible area is uniformly loaded. Given $q = 320$ kN/m^2 and using Newmark's chart, determine the vertical stress increase $\Delta\sigma_z$ at point A.

9.20 The plan of a flexible rectangular loaded area is shown in Figure 9.34. The uniformly distributed load on the flexible area, q is 90 kN/m^2. Determine the vertical stress increase, $\Delta\sigma_z$, at a depth of $z = 2$ m below
  a. Point A
  b. Point B
  c. Point C

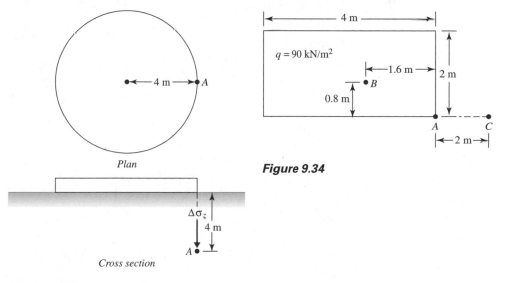

Plan

$\Delta\sigma_z$

4 m

A

Cross section

Figure 9.33

4 m

$q = 90$ kN/m^2

1.6 m

2 m

B

0.8 m

A

C

2 m

Figure 9.34

9.21 Repeat Problem 9.20. Use Newmark's influence chart for vertical pressure distribution.

9.22 Refer to the uniformly loaded rectangular area shown in Figure 9.34. Estimate the stress below the center of the area at a depth of 3.5 m. Use Eq. (9.33).

References

AHLVIN, R. G., and ULERY, H. H. (1962). "Tabulated Values for Determining the Complete Pattern of Stresses, Strains, and Deflections Beneath a Uniform Circular Load on a Homogeneous Half Space," in *Highway Research Bulletin 342,* Transportation Research Board, National Research Council, Washington, D.C., 1–13.

BOUSSINESQ, J. (1883). *Application des Potentials à L'Etude de L'Equilibre et du Mouvement des Solides Elastiques,* Gauthier-Villars, Paris.

DAS, B. (1997). *Advanced Soil Mechanics,* 2nd ed., Taylor and Francis, Washington, D.C.

NEWMARK, N. M. (1942). "Influence Charts for Computation of Stresses in Elastic Soil," University of Illinois Engineering Experiment Station, *Bulletin No. 338.*

OSTERBERG, J. O. (1957). "Influence Values for Vertical Stresses in Semi-Infinite Mass Due to Embankment Loading," *Proceedings,* Fourth International Conference on Soil Mechanics and Foundation Engineering, London, Vol. 1, 393–396.

10

Compressibility of Soil

A stress increase caused by the construction of foundations or other loads compresses soil layers. The compression is caused by (a) deformation of soil particles, (b) relocations of soil particles, and (c) expulsion of water or air from the void spaces. In general, the soil settlement caused by loads may be divided into three broad categories:

1. *Immediate settlement* (or *elastic settlement*), which is caused by the elastic deformation of dry soil and of moist and saturated soils without any change in the moisture content. Immediate settlement calculations are generally based on equations derived from the theory of elasticity.
2. *Primary consolidation settlement,* which is the result of a volume change in saturated cohesive soils because of expulsion of the water that occupies the void spaces.
3. *Secondary consolidation settlement,* which is observed in saturated cohesive soils and is the result of the plastic adjustment of soil fabrics. It is an additional form of compression that occurs at constant effective stress.

This chapter presents the fundamental principles for estimating the immediate and consolidation settlements of soil layers under superimposed loadings.

The total settlement of a foundation can then be given as

$$S_T = S_c + S_s + S_e$$

where S_T = total settlement
S_c = primary consolidation settlement
S_s = secondary consolidation settlement
S_e = immediate settlement

When foundations are constructed on very compressible clays, the consolidation settlement can be several times greater than the immediate settlement.

IMMEDIATE SETTLEMENT

10.1 Contact Pressure and Settlement Profile

Immediate, or elastic, settlement of foundations (S_e) occurs directly after the application of a load, without a change in the moisture content of the soil. The magnitude

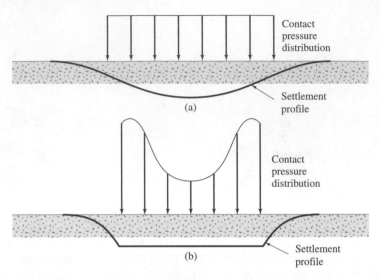

Figure 10.1 Immediate settlement profile and contact pressure in clay: (a) flexible foundation; (b) rigid foundation

of the contact settlement will depend on the flexibility of the foundation and the type of material on which it is resting.

In Chapter 9, the relationships for determining the increase in stress (which causes elastic settlement) due to the application of line load, strip load, embankment load, circular load, and rectangular load were based on the following assumptions:

a. The load is applied at the ground surface,
b. The loaded area is *flexible,* and
c. The soil medium is homogeneous, elastic, isotropic, and extends to a great depth.

In general, foundations are not perfectly flexible and are embedded at a certain depth below the ground surface. It is instructive, however, to evaluate the distribution of the contact pressure under a foundation along with the settlement profile under idealized conditions. Figure 10.1a shows a *perfectly flexible* foundation resting on an elastic material such as saturated clay. If the foundation is subjected to a uniformly distributed load, the contact pressure will be uniform and the foundation will experience a sagging profile. On the other hand, if we consider a *perfectly rigid* foundation resting on the ground surface subjected to a uniformly distributed load, the contact pressure and foundation settlement profile will be as shown in Figure 10.1b: the foundation will undergo a uniform settlement and the contact pressure will be redistributed.

The settlement profile and contact pressure distribution described above are true for soils in which the modulus of elasticity is fairly constant with depth. In the case of cohesionless sand, the modulus of elasticity increases with depth. Additionally, there is a lack of lateral confinement on the edge of the foundation at the ground surface. The sand at the edge of a flexible foundation is pushed outward, and the deflection curve of the foundation takes a concave downward shape. The distributions of contact pressure and the settlement profiles of a flexible and a rigid founda-

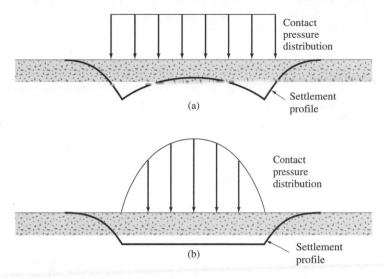

Figure 10.2 Immediate settlement profile and contact pressure in sand: (a) flexible foundation; (b) rigid foundation

tion resting on sand and subjected to uniform loading are shown in Figs. 10.2a and 10.2b, respectively.

10.2   *Relations for Immediate Settlement Calculation*

Immediate settlement for foundations that rest on elastic material (of infinite thickness) can be calculated from equations derived by using the principles of the theory of elasticity. They are of the form

$$S_e = \Delta\sigma B \frac{1 - \mu_s^2}{E_s} I_\rho \tag{10.1}$$

where S_e = immediate settlement
$\Delta\sigma$ = net pressure applied
B = width of the foundation ($=$ diameter of circular foundation)
μ_s = Poisson's ratio of soil
E_s = modulus of elasticity of soil
I_ρ = nondimensional influence factor

Schleicher (1926) expressed the influence factor for the corner of a flexible rectangular footing as

$$I_\rho = \frac{1}{\pi}\left[m_1 \ln\left(\frac{1 + \sqrt{m_1^2 + 1}}{m_1} \right) + \ln(m_1 + \sqrt{m_1^2 + 1}) \right] \tag{10.2}$$

where m_1 = length of the foundation divided by the width of the foundation.

Table 10.1 Influence Factors for Foundations [Eq. (10.2)]

Shape	m_1	I_p Flexible Center	I_p Flexible Corner	Rigid
Circle	—	1.00	0.64	0.79
Rectangle	1	1.12	0.56	0.88
	1.5	1.36	0.68	1.07
	2	1.53	0.77	1.21
	3	1.78	0.89	1.42
	5	2.10	1.05	1.70
	10	2.54	1.27	2.10
	20	2.99	1.49	2.46
	50	3.57	1.8	3.0
	100	4.01	2.0	3.43

Table 10.1 gives the influence factors for rigid and flexible foundations. Representative values of the modulus of elasticity and Poisson's ratio for different types of soils are given in Tables 10.2 and 10.3, respectively.

Note that Eq. (10.1) is based on the assumption that the pressure $\Delta\sigma$ is applied at the ground surface. In practice, foundations are placed at a certain depth below the ground surface. Deeper foundation embedment tends to reduce the magnitude of the foundation settlement S_e. However, if Eq. (10.1) is used to calculate settlement, it results in a conservative estimate.

Table 10.2 Representative Values of the Modulus of Elasticity of Soil

Soil type	E_s kN/m²	E_s lb/in.²
Soft clay	1,800–3,500	250–500
Hard clay	6,000–14,000	850–2,000
Loose sand	10,000–28,000	1,500–4,000
Dense sand	35,000–70,000	5,000–10,000

Table 10.3 Representative Values of Poisson's Ratio

Type of soil	Poisson's ratio, μ_s
Loose sand	0.2–0.4
Medium sand	0.25–0.4
Dense sand	0.3–0.45
Silty sand	0.2–0.4
Soft clay	0.15–0.25
Medium clay	0.2–0.5

Example 10.1

Estimate the immediate settlement of a column footing 1.5 m in diameter that is constructed on an unsaturated clay layer, given that the total load carried by the column footing = 150 kN, E_s = 7000 kN/m², and μ_s = 0.25. Assume the footing to be rigid.

Solution

Using Eq. (10.1), we have

$$S_e = \Delta\sigma B \frac{1 - \mu_s^2}{E_s} I_\rho$$

$$\Delta\sigma = \frac{(150)}{\dfrac{\pi}{4}(1.5)^2} = 84.9 \text{ kN/m}^2$$

From Table 10.1, for a circular rigid foundation, I_ρ = 0.79, so

$$S_e = (84.9)(1.5)\left[\frac{1 - 0.25^2}{7000}\right](0.79) = 0.0135 \text{ m} = \mathbf{13.5 \ mm}$$ ∎

10.3 *Improved Relationship for Immediate Settlement*

Mayne and Poulos (1999) recently presented an improved relationship for calculating the immediate settlement of foundations. This relationship takes into account the rigidity of the foundation, the depth of embedment of the foundation, the increase in the modulus of elasticity of soil with depth, and the location of rigid layers at limited depth. In order to use this relationship, one needs to determine the equivalent diameter of a rectangular foundation, which is

$$B_e = \sqrt{\frac{4BL}{\pi}} \tag{10.3a}$$

where B = width of foundation
$\quad\quad L$ = length of foundation

For circular foundations,

$$B_e = B \tag{10.3b}$$

where B = diameter of foundation.

Figure 10.3 shows a foundation having an equivalent diameter of B_e located at a depth D_f below the ground surface. Let the thickness of the foundation be t and the modulus of elasticity of the foundation material be E_f. A rigid layer is located at a depth h below the bottom of the foundation. The modulus of elasticity of the compressible soil layer can be given as

$$E_s = E_o + kz \tag{10.4}$$

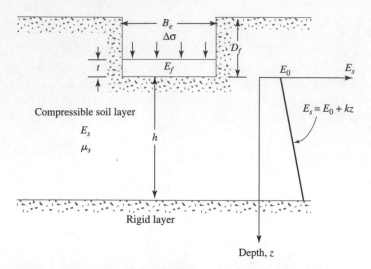

Figure 10.3 Improved relationship for immediate settlement

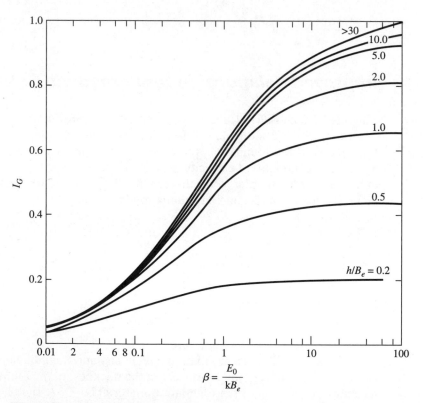

Figure 10.4 Variation of I_G with β

With the preceding parameters defined, the immediate settlement can be given as

$$S_e = \frac{\Delta\sigma B_e I_G I_F I_E}{E_o}(1 - \mu_s^2) \tag{10.5}$$

where I_G = influence factor for the variation of E_s with depth = $f(E_o, k, B_e, \text{ and } h)$
I_F = foundation rigidity correction factor
I_E = foundation embedment correction factor

Figure 10.4 shows the variation of I_G with $\beta = E_o/kB_e$ and h/B_e. The foundation rigidity correction factor can be expressed as

$$I_F = \frac{\pi}{4} + \frac{1}{4.6 + 10\left(\dfrac{E_f}{E_o + \dfrac{B_e}{2}k}\right)\left(\dfrac{2t}{B_e}\right)^3} \tag{10.6}$$

Similarly, the embedment correction factor is

$$I_E = 1 - \frac{1}{3.5\exp(1.22\mu_s - 0.4)\left(\dfrac{B_e}{D_f} + 1.6\right)} \tag{10.7}$$

Figures 10.5 and 10.6 show the variations of I_F and I_E expressed by Eqs. (10.6) and (10.7).

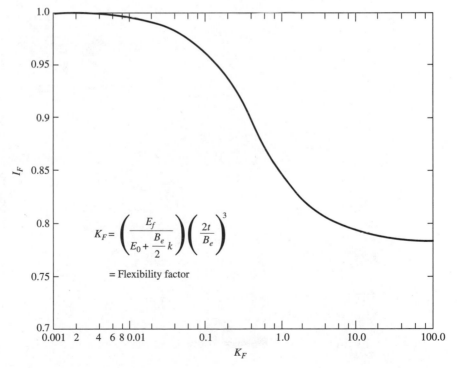

Figure 10.5 Variation of rigidity correction factor, I_F, with flexibility factor, K_F. [Eq. (10.6)]

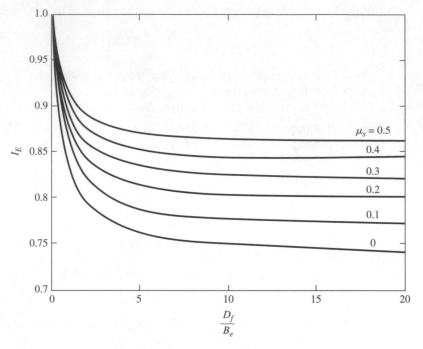

Figure 10.6 Variation of embedment correction factor, I_E [Eq. (10.7)]

Example 10.2

Refer to Figure 10.3. For a shallow foundation supported by a silty clay, the following are given:

    Length $= L = 1.5$ m
    Width $= B = 1$ m
    Depth of foundation $= D_f = 1$ m
    Thickness of foundation $= t = 0.23$ m
    Load per unit area $= \Delta\sigma = 190$ kN/m^2
    $E_f = 15 \times 10^6$ kN/m^2

The silty clay soil had the following properties:

    $h = 2$m
    $\mu_s = 0.3$
    $E_o = 9000$ kN/m^2
    $k = 500$ kN/m^2/m

Estimate the immediate settlement of the foundation.

Solution

From Eq. (10.3a), the equivalent diameter is

$$B_e = \sqrt{\frac{4BL}{\pi}} = \sqrt{\frac{(4)(1.5)(1)}{\pi}} = 1.38 \text{ m}$$

$$\Delta\sigma = 190 \text{ kN/m}^2$$

$$\beta = \frac{E_o}{kB_e} = \frac{9000}{(500)(1.38)} = 13.04$$

$$\frac{h}{B_e} = \frac{2}{1.38} = 1.45$$

From Figure 10.4, for $\beta = 13.04$ and $h/B_e = 1.45$, the value of $I_G \approx 0.74$. Thus, from Eq. (10.6),

$$I_F = \frac{\pi}{4} + \cfrac{1}{4.6 + 10\left(\cfrac{E_f}{E_o + \cfrac{B_e}{2}k}\right)\left(\cfrac{2t}{B_e}\right)^3}$$

$$= \frac{\pi}{4} + \cfrac{1}{4.6 + 10\left[\cfrac{15 \times 10^6}{9000 + \left(\cfrac{1.38}{2}\right)(500)}\right]\left[\cfrac{(2)(0.23)}{1.38}\right]^3} = 0.787$$

From Eq. (10.7),

$$I_E = 1 - \cfrac{1}{3.5 \exp(1.22\mu_s - 0.4)\left(\cfrac{B_e}{D_f} + 1.6\right)}$$

$$= 1 - \cfrac{1}{3.5 \exp[(1.22)(0.3) - 0.4]\left(\cfrac{1.38}{1} + 1.6\right)} = 0.907$$

From Eq. (10.5),

$$S_e = \frac{\Delta\sigma B_e I_G I_F I_E}{E_o}(1 - \mu_s^2) = \frac{(190)(1.38)(0.74)(0.787)(0.907)}{9000}(1 - 0.3^2)$$

$$= 0.014 \text{ m} \approx \textbf{14 mm}$$ ∎

CONSOLIDATION SETTLEMENT

10.4   *Fundamentals of Consolidation*

When a saturated soil layer is subjected to a stress increase, the pore water pressure is suddenly increased. In sandy soils that are highly permeable, the drainage caused by the increase in the pore water pressure is completed immediately. Pore water drainage is accompanied by a reduction in the volume of the soil mass, which results in settlement. Because of rapid drainage of the pore water in sandy soils, immediate settlement and consolidation occur simultaneously.

When a saturated compressible clay layer is subjected to a stress increase, elastic settlement occurs immediately. Because the hydraulic conductivity of clay is significantly smaller than that of sand, the excess pore water pressure generated by loading gradually dissipates over a long period. Thus, the associated volume change (that is, the consolidation) in the clay may continue long after the immediate settlement. The settlement caused by consolidation in clay may be several times greater than the immediate settlement.

The time-dependent deformation of saturated clayey soil can best be understood by considering a simple model that consists of a cylinder with a spring at its center. Let the inside area of the cross section of the cylinder be equal to A. The cylinder is filled with water and has a frictionless watertight piston and valve as shown in Figure 10.7a. At this time, if we place a load P on the piston (Figure 10.7b) and keep the valve closed, the entire load will be taken by the water in the cylinder because water is *incompressible*. The spring will not go through any deformation. The excess hydrostatic pressure at this time can be given as

$$\Delta u = \frac{P}{A} \tag{10.8}$$

This value can be observed in the pressure gauge attached to the cylinder.

In general, we can write

$$P = P_s + P_w \tag{10.9}$$

where P_s = load carried by the spring and P_w = load carried by the water.

From the preceding discussion, we can see that when the valve is closed after the placement of the load P,

$$P_s = 0 \quad \text{and} \quad P_w = P$$

Now, if the valve is opened, the water will flow outward (Figure 10.7c). This flow will be accompanied by a reduction of the excess hydrostatic pressure and an increase in the compression of the spring. So, at this time, Eq. (10.9) will hold. However,

$$P_s > 0 \quad \text{and} \quad P_w < P \quad \text{(that is, } \Delta u < P/A)$$

After some time, the excess hydrostatic pressure will become zero and the system will reach a state of equilibrium, as shown in Figure 10.7d. Now we can write

$$P_s = P \quad \text{and} \quad P_w = 0$$

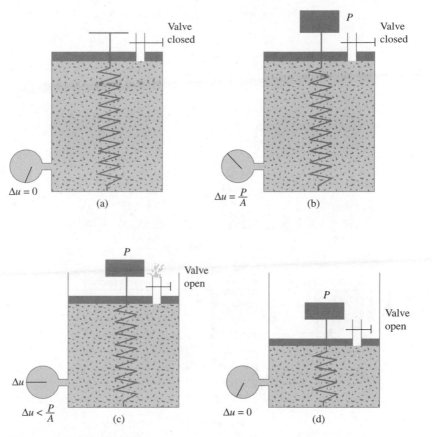

Figure 10.7 Spring-cylinder model

and

$$P = P_s + P_w$$

With this in mind, we can analyze the strain of a saturated clay layer subjected to a stress increase (Figure 10.8a). Consider the case where a layer of saturated clay of thickness H that is confined between two layers of sand is being subjected to an instantaneous increase of *total stress* of $\Delta\sigma$. This incremental total stress will be transmitted to the pore water and the soil solids. This means that the total stress, $\Delta\sigma$, will be divided in some proportion between effective stress and pore water pressure. The behavior of the effective stress change will be similar to that of the spring in Figure 10.7, and the behavior of the pore water pressure change will be similar to that of the excess hydrostatic pressure in Figure 10.7. From the principle of effective stress (Chapter 8), it follows that

$$\Delta\sigma = \Delta\sigma' + \Delta u \tag{10.10}$$

where $\Delta\sigma'$ = increase in the effective stress
Δu = increase in the pore water pressure

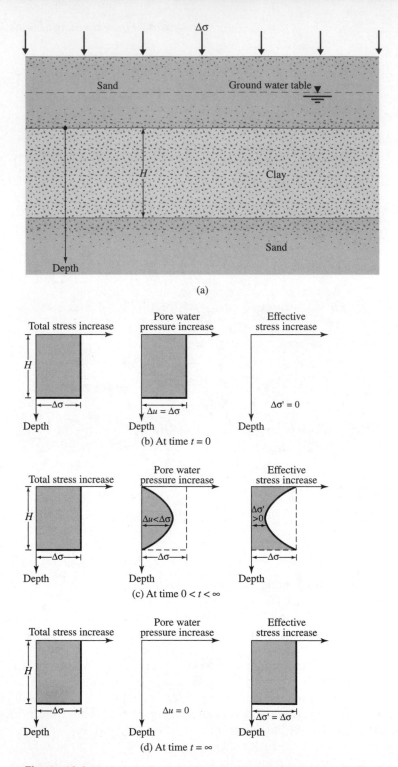

Figure 10.8 Variation of total stress, pore water pressure, and effective stress in a clay layer drained at top and bottom as the result of an added stress, $\Delta\sigma$

Because clay has a very low hydraulic conductivity and water is incompressible as compared with the soil skeleton, at time $t = 0$, the entire incremental stress, $\Delta\sigma$, will be carried by water ($\Delta\sigma = \Delta u$) at all depths (Figure 10.8b). None will be carried by the soil skeleton — that is, incremental effective stress ($\Delta\sigma'$) = 0.

After the application of incremental stress, Δu, to the clay layer, the water in the void spaces will start to be squeezed out and will drain in both directions into the sand layers. By this process, the excess pore water pressure at any depth in the clay layer will gradually decrease, and the stress carried by the soil solids (effective stress) will increase. Thus, at time $0 < t < \infty$,

$$\Delta\sigma = \Delta\sigma' + \Delta u \qquad (\Delta\sigma' > 0 \quad \text{and} \quad \Delta u < \Delta\sigma)$$

However, the magnitudes of $\Delta\sigma'$ and Δu at various depths will change (Figure 10.8c), depending on the minimum distance of the drainage path to either the top or bottom sand layer.

Theoretically, at time $t = \infty$, the entire excess pore water pressure would be dissipated by drainage from all points of the clay layer; thus, $\Delta u = 0$. Now the total stress increase, $\Delta\sigma$, will be carried by the soil structure (Figure 10.8d). Hence,

$$\Delta\sigma = \Delta\sigma'$$

This gradual process of drainage under an additional load application and the associated transfer of excess pore water pressure to effective stress cause the time-dependent settlement in the clay soil layer.

10.5 One-Dimensional Laboratory Consolidation Test

The one-dimensional consolidation testing procedure was first suggested by Terzaghi. This test is performed in a consolidometer (sometimes referred to as an *oedometer*). The schematic diagram of a consolidometer is shown in Figure 10.9a. Figure 10.9b. shows a photograph of a consolidometer. The soil specimen is placed inside a metal ring with two porous stones, one at the top of the specimen and another at the bottom. The specimens are usually 64 mm (≈ 2.5 in.) in diameter and 25 mm. (≈ 1 in.) thick. The load on the specimen is applied through a lever arm, and compression is measured by a micrometer dial gauge. The specimen is kept under water during the test. Each load is usually kept for 24 hours. After that, the load is usually doubled, which doubles the pressure on the specimen, and the compression measurement is continued. At the end of the test, the dry weight of the test specimen is determined. Figure 10.9c shows a consolidation test in progress (right-hand side).

The general shape of the plot of deformation of the specimen against time for a given load increment is shown in Figure 10.10. From the plot, we can observe three distinct stages, which may be described as follows:

Stage I   Initial compression, which is caused mostly by preloading.

Stage II   Primary consolidation, during which excess pore water pressure is gradually transferred into effective stress because of the expulsion of pore water.

Stage III   Secondary consolidation, which occurs after complete dissipation of the excess pore water pressure, when some deformation of the specimen takes place because of the plastic readjustment of soil fabric.

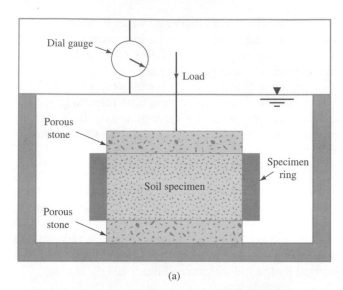

(a)

(b)

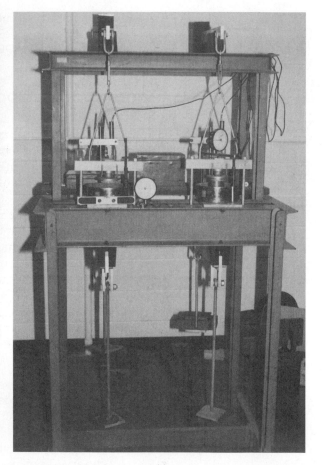

(c)

Figure 10.9

(a) Schematic diagram of a consolidometer;
(b) photograph of a consolidometer; (c) a consolidation test in progress (right-hand side)

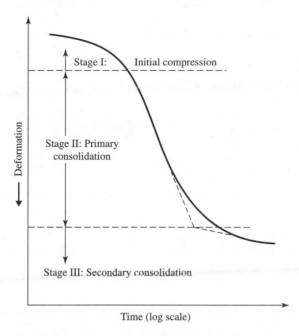

Stage I: Initial compression

Stage II: Primary
consolidation

Stage III: Secondary consolidation

Time (log scale)

Figure 10.10
Time–deformation plot during
consolidation for a given load
increment

10.6   *Void Ratio–Pressure Plots*

After the time–deformation plots for various loadings are obtained in the laboratory, it is necessary to study the change in the void ratio of the specimen with pressure. Following is a step-by-step procedure for doing so:

1. Calculate the height of solids, H_s, in the soil specimen (Figure 10.11) using the
equation

$$H_s = \frac{W_s}{AG_s\gamma_w} = \frac{M_s}{AG_s\rho_w} \tag{10.11}$$

where W_s = dry weight of the specimen
   M_s = dry mass of the specimen
   A = area of the specimen
   G_s = specific gravity of soil solids
   γ_w = unit weight of water
   ρ_w = density of water

2. Calculate the initial height of voids as

$$H_v = H - H_s \tag{10.12}$$

where H = initial height of the specimen.

3. Calculate the initial void ratio, e_O, of the specimen, using the equation

$$e_O = \frac{V_v}{V_s} = \frac{H_v}{H_s}\frac{A}{A} = \frac{H_v}{H_s} \tag{10.13}$$

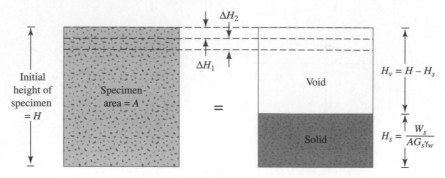

Figure 10.11 Change of height of specimen in one-dimensional consolidation test

4. For the first incremental loading, σ_1 (total load/unit area of specimen), which causes a deformation ΔH_1, calculate the change in the void ratio as

$$\Delta e_1 = \frac{\Delta H_1}{H_s} \qquad (10.14)$$

(ΔH_1 is obtained from the initial and the final dial readings for the loading).

It is important to note that, at the end of consolidation, total stress σ_1 is equal to effective stress σ_1'.

5. Calculate the new void ratio after consolidation caused by the pressure increment as

$$e_1 = e_O - \Delta e_1 \qquad (10.15)$$

For the next loading, σ_2 (*note:* σ_2 equals the cumulative load per unit area of specimen), which causes additional deformation ΔH_2, the void ratio at the end of consolidation can be calculated as

$$e_2 = e_1 - \frac{\Delta H_2}{H_s} \qquad (10.16)$$

At this time, σ_2 = effective stress, σ_2'. Proceeding in a similar manner, one can obtain the void ratios at the end of the consolidation for all load increments.

The effective stress σ' and the corresponding void ratios (e) at the end of consolidation are plotted on semilogarithmic graph paper. The typical shape of such a plot is shown in Figure 10.12.

10.7   *Normally Consolidated and Overconsolidated Clays*

Figure 10.12 shows that the upper part of the e–log σ' plot is somewhat curved with a flat slope, followed by a linear relationship for the void ratio with log σ' having a steeper slope. This phenomenon can be explained in the following manner:

A soil in the field at some depth has been subjected to a certain maximum effective past pressure in its geologic history. This maximum effective past pressure may be equal to or less than the existing effective overburden pressure at the time of sampling. The reduction of effective pressure in the field may be caused by natural

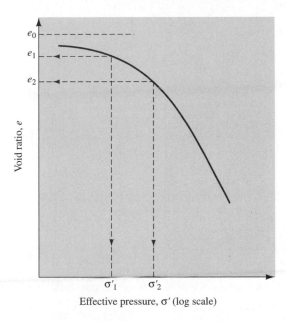

Figure 10.12
Typical plot of e against log σ'

geologic processes or human processes. During the soil sampling, the existing effective overburden pressure is also released, which results in some expansion. When this specimen is subjected to a consolidation test, a small amount of compression (that is, a small change in void ratio) will occur when the effective pressure applied is less than the maximum effective overburden pressure in the field to which the soil has been subjected in the past. When the effective pressure on the specimen becomes greater than the maximum effective past pressure, the change in the void ratio is much larger, and the e–log σ' relationship is practically linear with a steeper slope.

This relationship can be verified in the laboratory by loading the specimen to exceed the maximum effective overburden pressure, and then unloading and reloading again. The e–log σ' plot for such cases is shown in Figure 10.13, in which cd represents unloading and dfg represents the reloading process.

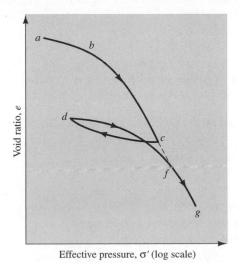

Figure 10.13
Plot of e against log σ' showing loading, unloading, and reloading branches

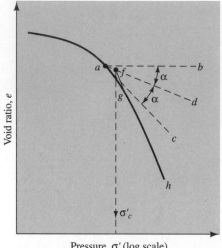

Figure 10.14 Graphic procedure for determining preconsolidation pressure

This leads us to the two basic definitions of clay based on stress history:

1. *Normally consolidated,* whose present effective overburden pressure is the maximum pressure that the soil was subjected to in the past.
2. *Overconsolidated,* whose present effective overburden pressure is less than that which the soil experienced in the past. The maximum effective past pressure is called the *preconsolidation pressure.*

Casagrande (1936) suggested a simple graphic construction to determine the preconsolidation pressure σ_c' from the laboratory e–$\log \sigma'$ plot. The procedure is as follows (see Figure 10.14):

1. By visual observation, establish point *a*, at which the e–$\log \sigma'$ plot has a minimum radius of curvature.
2. Draw a horizontal line *ab*.
3. Draw the line *ac* tangent at *a*.
4. Draw the line *ad*, which is the bisector of the angle *bac*.
5. Project the straight-line portion *gh* of the e–$\log \sigma'$ plot back to intersect line *ad* at *f*. The abscissa of point *f* is the preconsolidation pressure, σ_c'.

The overconsolidation ratio (*OCR*) for a soil can now be defined as

$$OCR = \frac{\sigma_c'}{\sigma'} \tag{10.17}$$

where σ_c' = preconsolidation pressure of a specimen
σ' = present effective vertical pressure

Effect of Disturbance on Void Ratio–Pressure Relationship

A soil specimen will be remolded when it is subjected to some degree of disturbance. This remolding will result in some deviation of the e–log σ' plot as observed in the laboratory from the actual behavior in the field. The field e–log σ' plot can be reconstructed from the laboratory test results in the manner described in this section (Terzaghi and Peck, 1967).

Normally Consolidated Clay of Low to Medium Plasticity (Figure 10.15)

1. In Figure 10.15, curve 2 is the laboratory e–log σ' plot. From this plot, determine the preconsolidation pressure $(\sigma'_c) = \sigma'_O$ (that is, the present effective overburden pressure). Knowing where $\sigma'_c = \sigma'_O$, draw vertical line ab.
2. Calculate the void ratio in the field, e_O [Section 10.6, Eq. (10.13)]. Draw horizontal line cd.
3. Calculate $0.4e_O$ and draw line ef. (*Note: f* is the point of intersection of the line with curve 2.)
4. Join points f and g. Note that g is the point of intersection of lines ab and cd. This is the *virgin compression curve*.

It is important to point out that if a soil is completely remolded, the general position of the e–log σ' plot will be as represented by curve 3.

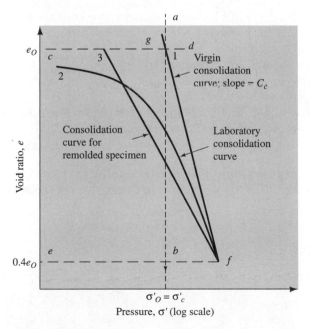

Figure 10.15 Consolidation characteristics of normally consolidated clay of low to medium sensitivity

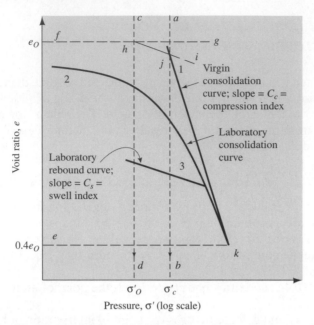

Figure 10.16 Consolidation characteristics of overconsolidated clay of low to medium sensitivity

Overconsolidated Clay of Low to Medium Plasticity (Figure 10.16)

1. In Figure 10.16, curve 2 is the laboratory e–log σ' plot (loading), and curve 3 is the laboratory unloading, or rebound, curve. From curve 2, determine the preconsolidation pressure σ'_c. Draw the vertical line *ab*.
2. Determine the field effective overburden pressure σ'_O. Draw vertical line *cd*.
3. Determine the void ratio in the field, e_O. Draw the horizontal line *fg*. The point of intersection of lines *fg* and *cd* is *h*.
4. Draw a line *hi*, which is parallel to curve 3 (which is practically a straight line). The point of intersection of lines *hi* and *ab* is *j*.
5. Join points *j* and *k*. Point *k* is on curve 2, and its ordinate is $0.4e_O$.

The field consolidation plot will take a path *hjk*. The recompression path in the field is *hj* and is parallel to the laboratory rebound curve (Schmertmann, 1953).

Example 10.3

Following are the results of a laboratory consolidation test on a soil specimen obtained from the field: Dry mass of specimen = 128 g, height of specimen at the beginning of the test = 2.54 cm, G_s = 2.75, and area of the specimen = 30.68 cm².

Effective pressure, σ' (ton/ft²)	Final height of specimen at the end of consolidation (cm)
0	2.540
0.5	2.488
1	2.465
2	2.431
4	2.389
8	2.324
16	2.225
32	2.115

Make necessary calculations and draw an e vs. log σ' curve.

Solution

From Eq. (10.11)

$$H_s = \frac{W_s}{AG_s\gamma_w} = \frac{M_s}{AG_s\rho_w} = \frac{128 \text{ g}}{(30.68 \text{ cm}^2)(2.75)(1 \text{ g/cm}^3)} = 1.52 \text{ cm}$$

Now the following table can be prepared:

Effective pressure, σ' (ton/ft²)	Height at the end of consolidation, H (cm)	$H_v = H - H_s$ (cm)	$e = H_v/H_s$
0	2.540	1.02	0.671
0.5	2.488	0.968	0.637
1	2.465	0.945	0.622
2	2.431	0.911	0.599
4	2.389	0.869	0.572
8	2.324	0.804	0.529
16	2.225	0.705	0.464
32	2.115	0.595	0.390

The e vs. log σ' plot is shown in Figure 10.17. ∎

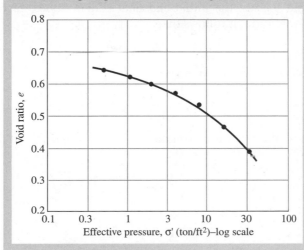

Figure 10.17
Variation of void ratio with effective pressure

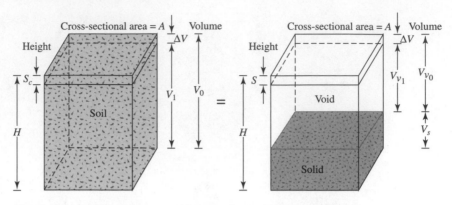

Figure 10.18 Settlement caused by one-dimensional consolidation

Calculation of Settlement from One-Dimensional Primary Consolidation

With the knowledge gained from the analysis of consolidation test results, we can now proceed to calculate the probable settlement caused by primary consolidation in the field, assuming one-dimensional consolidation.

Let us consider a saturated clay layer of thickness H and cross-sectional area A under an existing average effective overburden pressure σ'_O. Because of an increase of effective pressure, $\Delta\sigma'$, let the primary settlement be S_c. Thus, the change in volume (Figure 10.18) can be given by

$$\Delta V = V_0 - V_1 = HA - (H - S_c)A = S_c A \tag{10.18}$$

where V_0 and V_1 are the initial and final volumes, respectively. However, the change in the total volume is equal to the change in the volume of voids, ΔV_v. Hence,

$$\Delta V = S_c A = V_{v0} - V_{v1} = \Delta V_v \tag{10.19}$$

where V_{v0} and V_{v1} are the initial and final void volumes, respectively. From the definition of void ratio, it follows that

$$\Delta V_v = \Delta e V_s \tag{10.20}$$

where Δe = change of void ratio. But

$$V_s = \frac{V_0}{1 + e_O} = \frac{AH}{1 + e_O} \tag{10.21}$$

where e_O = initial void ratio at volume V_0. Thus, from Eqs. (10.18) through (10.21),

$$\Delta V = S_c A = \Delta e V_s = \frac{AH}{1 + e_O}\Delta e$$

or

$$S_c = H\frac{\Delta e}{1 + e_O} \tag{10.22}$$

For normally consolidated clays that exhibit a linear e–log σ' relationship (see Figure 10.15),

$$\Delta e = C_c[\log(\sigma'_O + \Delta\sigma') - \log\sigma'_O] \tag{10.23}$$

where C_c = slope of the e–log σ' plot and is defined as the compression index. Substitution of Eq. (10.23) into Eq. (10.22) gives

$$S_c = \frac{C_c H}{1 + e_O}\log\left(\frac{\sigma'_O + \Delta\sigma'}{\sigma'_O}\right) \tag{10.24}$$

In overconsolidated clays (see Figure 10.16), for $\sigma'_O + \Delta\sigma' \le \sigma'_c$, field e–log σ' variation will be along the line hj, the slope of which will be approximately equal to that for the laboratory rebound curve. The slope of the rebound curve C_s is referred to as the *swell index;* so

$$\Delta e = C_s[\log(\sigma'_O + \Delta\sigma') - \log\sigma'_O] \tag{10.25}$$

From Eqs. (8.15) and (8.18), we obtain

$$S_c = \frac{C_s H}{1 + e_O}\log\left(\frac{\sigma'_O + \Delta\sigma'}{\sigma'_O}\right) \tag{10.26}$$

If $\sigma'_O + \Delta\sigma' > \sigma'_c$, then

$$S_c = \frac{C_s H}{1 + e_O}\log\frac{\sigma'_c}{\sigma'_O} + \frac{C_c H}{1 + e_O}\log\left(\frac{\sigma'_O + \Delta\sigma'}{\sigma'_c}\right) \tag{10.27}$$

However, if the e–log σ' curve is given, one can simply pick Δe off the plot for the appropriate range of pressures. This number may be substituted into Eq. (10.22) for the calculation of settlement, S_c.

10.10 Compression Index (C_c) and Swell Index (C_s)

The compression index for the calculation of field settlement caused by consolidation can be determined by graphic construction (as shown in Figure 10.15) after one obtains the laboratory test results for void ratio and pressure.

Skempton (1944) suggested the following empirical expression for the compression index for undisturbed clays:

$$C_c = 0.009(LL - 10) \tag{10.28}$$

where LL = liquid limit.

Several other correlations for the compression index are also available. They have been developed by tests on various clays. Some of these correlations are given in Table 10.4.

Table 10.4 Correlations for Compression Index, C_c*

Equation	Reference	Region of applicability
$C_c = 0.007(LL - 7)$	Skempton (1944)	Remolded clays
$C_c = 0.01w_N$		Chicago clays
$C_c = 1.15(e_O - 0.27)$	Nishida (1956)	All clays
$C_c = 0.30(e_O - 0.27)$	Hough (1957)	Inorganic cohesive soil: silt, silty clay, clay
$C_c = 0.0115w_N$		Organic soils, peats, organic silt, and clay
$C_c = 0.0046(LL - 9)$		Brazilian clays
$C_c = 0.75(e_O - 0.5)$		Soils with low plasticity
$C_c = 0.208e_O + 0.0083$		Chicago clays
$C_c = 0.156e_O + 0.0107$		All clays

* After Rendon-Herrero (1980)
Note: $e_O = $ *in situ* void ratio; $w_N = $ *in situ* water content.

On the basis of observations on several natural clays, Rendon-Herrero (1983) gave the relationship for the compression index in the form

$$C_c = 0.141 G_s^{1.2} \left(\frac{1 + e_O}{G_s} \right)^{2.38} \tag{10.29}$$

Nagaraj and Murty (1985) expressed the compression index as

$$C_c = 0.2343 \left[\frac{LL\,(\%)}{100} \right] G_s \tag{10.30}$$

The swell index is appreciably smaller in magnitude than the compression index and can generally be determined from laboratory tests. In most cases,

$$C_s \simeq \tfrac{1}{5} \text{ to } \tfrac{1}{10} C_c \tag{10.31}$$

The swell index was expressed by Nagaraj and Murty (1985) as

$$C_s = 0.0463 \left[\frac{LL\,(\%)}{100} \right] G_s \tag{10.32}$$

Example 10.4

A soil profile is shown in Figure 10.19. If a uniformly distributed load, $\Delta\sigma$, is applied at the ground surface, what is the settlement of the clay layer caused by primary consolidation if

    a. The clay is normally consolidated
    b. The preconsolidation pressure $(\sigma_c') = 190 \text{ kN/m}^2$
    c. $\sigma_c' = 170 \text{ kN/m}^2$

Use $C_s \approx \tfrac{1}{6} C_c$.

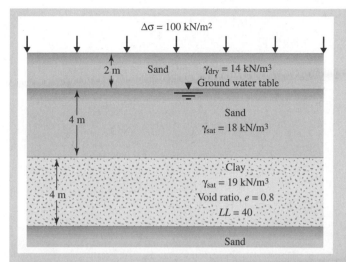

Figure 10.19

Solution

a. The average effective stress at the middle of the clay layer is

$$\sigma'_O = 2\gamma_{\text{dry}} + 4[\gamma_{\text{sat(sand)}} - \gamma_w] + \tfrac{4}{2}[\gamma_{\text{sat(clay)}} - \gamma_w]$$

or

$$\sigma'_O = (2)(14) + 4(18 - 9.81) + 2(19 - 9.81) = 79.14 \text{ kN/m}^2$$

From Eq. (10.24),

$$S_c = \frac{C_c H}{1 + e_O} \log\left(\frac{\sigma'_O + \Delta\sigma'}{\sigma'_O}\right)$$

From Eq. (10.28),

$$C_c = 0.009(LL - 10) - 0.009(40 - 10) = 0.27$$

So

$$S_c = \frac{(0.27)(4)}{1 + 0.8} \log\left(\frac{79.14 + 100}{79.14}\right) = 0.213 \text{ m} = \mathbf{213 \text{ mm}}$$

b. $\sigma'_O + \Delta\sigma' = 79.14 + 100 = 179.14 \text{ kN/m}^2$
$\sigma'_c = 190 \text{ kN/m}^2$

Because $\sigma'_O + \Delta\sigma' > \sigma'_c$, use Eq. (10.26) to get

$$S_c = \frac{C_s H}{1 + e_O} \log\left(\frac{\sigma'_O + \Delta\sigma'}{\sigma'_O}\right)$$

$$C_s = \frac{C_c}{6} = \frac{0.27}{6} = 0.045$$

$$S_c = \frac{(0.045)(4)}{1 + 0.8} \log\left(\frac{79.14 + 100}{79.14}\right) = 0.036 \text{ m} = \mathbf{36 \text{ mm}}$$

c.
$$\sigma'_O = 79.14 \text{ kN/m}^2$$
$$\sigma'_O + \Delta\sigma' = 179.14 \text{ kN/m}^2$$
$$\sigma'_c = 170 \text{ kN/m}^2$$

Because $\sigma'_O < \sigma'_c < \sigma'_O + \Delta\sigma'$, use Eq. (10.27),

$$S_c = \frac{C_s H}{1 + e_o} \log \frac{\sigma'_c}{\sigma'_O} + \frac{C_c H}{1 + e_o} \log\left(\frac{\sigma'_O + \Delta\sigma'}{\sigma'_c}\right)$$

$$= \frac{(0.045)(4)}{1.8} \log\left(\frac{170}{79.14}\right) + \frac{(0.27)(4)}{1.8} \log\left(\frac{179.14}{170}\right) = 0.0468 \text{ m}$$

$$= \textbf{46.8 mm} \qquad \blacksquare$$

Example 10.5

The laboratory consolidation data for an undisturbed clay specimen are as follows:

$$e_1 = 1.1 \qquad \sigma'_1 = 95 \text{ kN/m}^2$$
$$e_2 = 0.9 \qquad \sigma'_2 = 475 \text{ kN/m}^2$$

What will be the void ratio for a pressure of 600 kN/m²? (*Note:* $\sigma'_c < 95$ kN/m².)

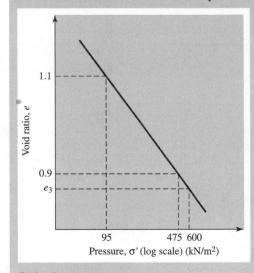

Figure 10.20

Solution
From Figure 10.20,

$$C_c = \frac{e_1 - e_2}{\log \sigma'_2 - \log \sigma'_1} = \frac{1.1 - 0.9}{\log 475 - \log 95} = 0.286$$

$$e_1 - e_3 = C_c(\log 600 - \log 95)$$

$$e_3 = e_1 - C_c \log \frac{600}{95}$$

$$= 1.1 - 0.286 \log \frac{600}{95} = \textbf{0.87} \qquad \blacksquare$$

10.11   *Secondary Consolidation Settlement*

Section 10.5 showed that at the end of primary consolidation (that is, after complete dissipation of excess pore water pressure) some settlement is observed because of the plastic adjustment of soil fabrics. This stage of consolidation is called *secondary consolidation*. During secondary consolidation the plot of deformation against the log of time is practically linear (see Figure 10.10). The variation of the void ratio, *e,* with time *t* for a given load increment will be similar to that shown in Figure 10.10. This variation is shown in Figure 10.21. From Figure 10.21, the secondary compression index can be defined as

$$C_\alpha = \frac{\Delta e}{\log t_2 - \log t_1} = \frac{\Delta e}{\log(t_2/t_1)} \qquad (10.33)$$

where C_α = secondary compression index
Δe = change of void ratio
t_1, t_2 = time

The magnitude of the secondary consolidation can be calculated as

$$S_s = C'_\alpha H \log\left(\frac{t_2}{t_1}\right) \qquad (10.34)$$

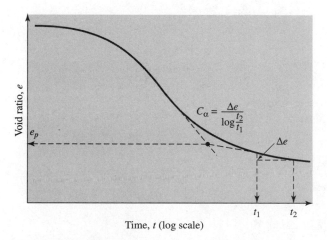

Figure 10.21 Variation of *e* with log *t* under a given load increment and definition of secondary consolidation index

where

$$C'_\alpha = \frac{C_\alpha}{1 + e_p}$$ (10.35)

e_p = void ratio at the end of primary consolidation (see Figure 10.21)
H = thickness of clay layer

The general magnitudes of C'_α as observed in various natural deposits are as follows:

- Overconsolidated clays = 0.001 or less
- Normally consolidated clays = 0.005 to 0.03
- Organic soil = 0.04 or more

Secondary consolidation settlement is more important than primary consolidation in organic and highly compressible inorganic soils. In overconsolidated inorganic clays, the secondary compression index is very small and of less practical significance.

Example 10.6

For a normally consolidated clay layer in the field, the following values are given:

- thickness of clay layer = 8.5 ft
- Void ratio (e_O) = 0.8
- Compression index (C_c) = 0.28
- Average effective pressure on the clay layer (σ'_O) = 2650 lb/ft^2
- $\Delta\sigma'$ = 970 lb/ft^2
- Secondary compression index (C_α) = 0.02

What is the total consolidation settlement of the clay layer five years after the completion of primary consolidation settlement? (*Note:* Time for completion of primary settlement = 1.5 years.)

Solution
From Eq. (10.35),

$$C'_\alpha = \frac{C_\alpha}{1 + e_p}$$

The value of e_p can be calculated as

$$e_p = e_O - \Delta e_{\text{primary}}$$

Combining Eqs. (10.22) and (10.23), we find that

$$\Delta e = C_c \log\left(\frac{\sigma_O' + \Delta\sigma'}{\sigma_O'}\right) = 0.28 \log\left(\frac{2650 + 970}{2650}\right)$$

$$= 0.038$$

Primary consolidation, $S_c = \dfrac{\Delta e H}{1 + e_O} = \dfrac{(0.038)(8.5 \times 12)}{1 + 0.8} = 2.15$ in.

It is given that $e_O = 0.8$, and thus,

$$e_p = 0.8 - 0.038 = 0.762$$

Hence,

$$C_\alpha' = \frac{0.02}{1 + 0.762} = 0.011$$

From Eq. (10.34),

$$S_s = C_\alpha' H \log\left(\frac{t_2}{t_1}\right) = (0.011)(8.5 \times 12) \log\left(\frac{5}{1.5}\right) \approx 0.59 \text{ in.}$$

Total consolidation settlement = primary consolidation (S_c) + secondary settlement (S_s). So

$$\text{total consolidation settlement} = 2.15 + 0.59 = \mathbf{2.74 \text{ in.}} \qquad \blacksquare$$

10.12 Time Rate of Consolidation

The total settlement caused by primary consolidation resulting from an increase in the stress on a soil layer can be calculated by the use of one of the three equations — (10.24), (10.26), or (10.27)— given in Section 10.9. However, they do not provide any information regarding the rate of primary consolidation. Terzaghi (1925) proposed the first theory to consider the rate of one-dimensional consolidation for saturated clay soils. The mathematical derivations are based on the following six assumptions (also see Taylor, 1948):

1. The clay–water system is homogeneous.
2. Saturation is complete.
3. Compressibility of water is negligible.
4. Compressibility of soil grains is negligible (but soil grains rearrange).
5. The flow of water is in one direction only (that is, in the direction of compression).
6. Darcy's law is valid.

Figure 10.22a shows a layer of clay of thickness $2H_{dr}$ that is located between two highly permeable sand layers. If the clay layer is subjected to an increased pressure of $\Delta\sigma$, the pore water pressure at any point A in the clay layer will increase. For one-dimensional consolidation, water will be squeezed out in the vertical direction toward the sand layer.

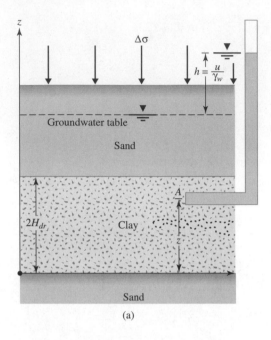

(a)

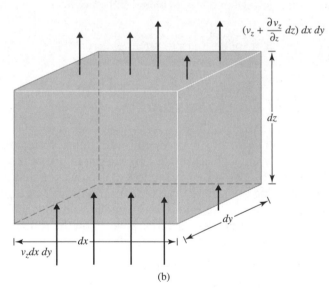

(b)

Figure 10.22
(a) Clay layer undergoing consolidation; (b) flow of water at A during consolidation

Figure 10.22b shows the flow of water through a prismatic element at A. For the soil element shown,

$$\frac{\text{Rate of outflow}}{\text{of water}} - \frac{\text{Rate of inflow}}{\text{of water}} = \frac{\text{Rate of}}{\text{volume change}}$$

Thus,

$$\left(v_z + \frac{\partial v_z}{\partial z} dz \right) dx\, dy - v_z\, dx\, dy = \frac{\partial V}{\partial t}$$

where V = volume of the soil element
  v_z = velocity of flow in z direction

or

$$\frac{\partial v_z}{\partial z} dx\, dy\, dz = \frac{\partial V}{\partial t} \tag{10.36}$$

Using Darcy's law, we have

$$v_z = ki = -k\frac{\partial h}{\partial z} = -\frac{k}{\gamma_w}\frac{\partial u}{\partial z} \tag{10.37}$$

where u = excess pore water pressure caused by the increase of stress.
  From Eqs. (10.36) and (10.37),

$$-\frac{k}{\gamma_w}\frac{\partial^2 u}{\partial z^2} = \frac{1}{dx\, dy\, dz}\frac{\partial V}{\partial t} \tag{10.38}$$

During consolidation, the rate of change in the volume of the soil element is equal to the rate of change in the volume of voids. Thus,

$$\frac{\partial V}{\partial t} = \frac{\partial V_v}{\partial t} = \frac{\partial(V_s + eV_s)}{\partial t} = \frac{\partial V_s}{\partial t} + V_s\frac{\partial e}{\partial t} + e\frac{\partial V_s}{\partial t} \tag{10.39}$$

where V_s = volume of soil solids
  V_v = volume of voids

But (assuming that soil solids are incompressible)

$$\frac{\partial V_s}{\partial t} = 0$$

and

$$V_s = \frac{V}{1 + e_O} = \frac{dx\, dy\, dz}{1 + e_O}$$

Substitution for $\partial V_s/\partial t$ and V_s in Eq. (10.39) yields

$$\frac{\partial V}{\partial t} = \frac{dx\, dy\, dz}{1 + e_O}\frac{\partial e}{\partial t} \tag{10.40}$$

where e_O = initial void ratio.
  Combining Eqs. (10.38) and (10.40) gives

$$-\frac{k}{\gamma_w}\frac{\partial^2 u}{\partial z^2} = \frac{1}{1 + e_O}\frac{\partial e}{\partial t} \tag{10.41}$$

The change in the void ratio is caused by the increase of effective stress (i.e., a decrease of excess pore water pressure). Assuming that they are linearly related, we have

$$\partial e = a_v\partial(\Delta\sigma') = -a_v\partial u \tag{10.42}$$

where $\partial(\Delta\sigma')$ = change in effective pressure
  a_v = coefficient of compressibility (a_v can be considered constant for a narrow range of pressure increase)

Combining Eqs. (10.41) and (10.42) gives

$$-\frac{k}{\gamma_w}\frac{\partial^2 u}{\partial z^2} = -\frac{a_v}{1+e_O}\frac{\partial u}{\partial t} = -m_v\frac{\partial u}{\partial t}$$

where

$$m_v = \text{coefficient of volume compressibility} = a_v/(1+e_O) \qquad (10.43)$$

or,

$$\frac{\partial u}{\partial t} = c_v\frac{\partial^2 u}{\partial z^2} \qquad (10.44)$$

where

$$c_v = \text{coefficient of consolidation} = k/(\gamma_w m_v) \qquad (10.45)$$

Thus,

$$c_v = \frac{k}{\gamma_w m_v} = \frac{k}{\gamma_w\left(\dfrac{a_v}{1+e_O}\right)} \qquad (10.46)$$

Eq. (10.44) is the basic differential equation of Terzaghi's consolidation theory and can be solved with the following boundary conditions:

$$z = 0, \quad u = 0$$
$$z = 2H_{dr}, \quad u = 0$$
$$t = 0, \quad u = u_O$$

The solution yields

$$u = \sum_{m=0}^{m=\infty}\left[\frac{2u_O}{M}\sin\left(\frac{Mz}{H_{dr}}\right)\right]e^{-M^2 T_v} \qquad (10.47)$$

where m = an integer
$M = (\pi/2)(2m+1)$
u_O = initial excess pore water pressure

$$T_v = \frac{c_v t}{H_{dr}^2} = \text{time factor} \qquad (10.48)$$

The time factor is a nondimensional number.

Because consolidation progresses by the dissipation of excess pore water pressure, the degree of consolidation at a distance z at any time t is

$$U_z = \frac{u_O - u_z}{u_O} = 1 - \frac{u_z}{u_O} \qquad (10.49)$$

where u_z = excess pore water pressure at time t.

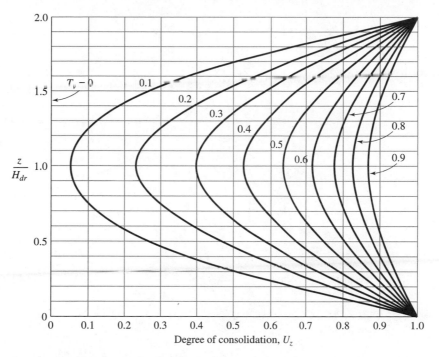

Figure 10.23 Variation of U_z with T_v and z/H_{dr}

Equations (10.47) and (10.49) can be combined to obtain the degree of consolidation at any depth z. This is shown in Figure 10.23.

The average degree of consolidation for the entire depth of the clay layer at any time t can be written from Eq. (10.49) as

$$U = \frac{S_{c(t)}}{S_c} = 1 - \frac{\left(\dfrac{1}{2H_{dr}}\right) \displaystyle\int_0^{2H_{dr}} u_z\, dz}{u_O} \qquad (10.50)$$

where U = average degree of consolidation
$S_{c(t)}$ = settlement of the layer at time t
S_c = ultimate settlement of the layer from primary consolidation

Substitution of the expression for excess pore water pressure u_z given in Eq. (10.47) into Eq. (10.50) gives

$$U = 1 - \sum_{m=0}^{m=\infty} \frac{2}{M^2} e^{-M^2 T_v} \qquad (10.51)$$

The variation in the average degree of consolidation with the nondimensional time factor, T_v, is given in Figure 10.24, which represents the case where u_O is the same for the entire depth of the consolidating layer.

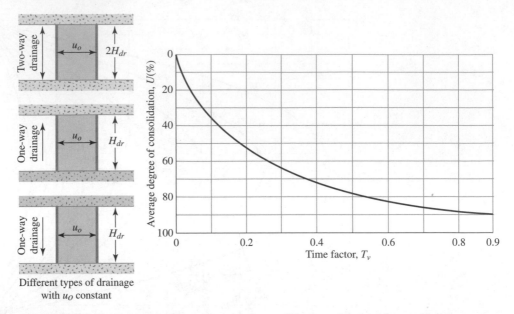

Figure 10.24 Variation of average degree of consolidation with time factor, T_v (u_O constant with depth)

The values of the time factor and their corresponding average degrees of consolidation for the case presented in Figure 10.24 may also be approximated by the following simple relationship:

$$\text{For } U = 0 \text{ to } 60\%, \quad T_v = \frac{\pi}{4}\left(\frac{U\%}{100}\right)^2 \tag{10.52}$$

$$\text{For } U > 60\%, \quad T_v = 1.781 - 0.933 \log(100 - U\%) \tag{10.53}$$

Table 10.5 gives the variation of T_v with U on the basis of Eqs. (10.52) and (10.53).

10.13 Coefficient of Consolidation

The coefficient of consolidation c_v generally decreases as the liquid limit of soil increases. The range of variation of c_v for a given liquid limit of soil is wide.

For a given load increment on a specimen, two graphical methods are commonly used for determining c_v from laboratory one-dimensional consolidation tests. The first is the *logarithm-of-time method* proposed by Casagrande and Fadum (1940), and the other is the *square-root-of-time method* given by Taylor (1942). More recently, at least two other methods were proposed. They are the *hyperbola method* (Sridharan and Prakash, 1985) and the *early stage log-t method* (Robinson and Allam, 1996). The general procedures for obtaining c_v by these methods are described in this section.

Table 10.5 Variation of T_v with U

U (%)	T_v	U (%)	T_v	U (%)	T_v
0	0	34	0.0907	68	0.377
1	0.00008	35	0.0962	69	0.390
2	0.0003	36	0.102	70	0.403
3	0.00071	37	0.107	71	0.417
4	0.00126	38	0.113	72	0.431
5	0.00196	39	0.119	73	0.446
6	0.00283	40	0.126	74	0.461
7	0.00385	41	0.132	75	0.477
8	0.00502	42	0.138	76	0.493
9	0.00636	43	0.145	77	0.511
10	0.00785	44	0.152	78	0.529
11	0.0095	45	0.159	79	0.547
12	0.0113	46	0.166	80	0.567
13	0.0133	47	0.173	81	0.588
14	0.0154	48	0.181	82	0.610
15	0.0177	49	0.188	83	0.633
16	0.0201	50	0.197	84	0.658
17	0.0227	51	0.204	85	0.684
18	0.0254	52	0.212	86	0.712
19	0.0283	53	0.221	87	0.742
20	0.0314	54	0.230	88	0.774
21	0.0346	55	0.239	89	0.809
22	0.0380	56	0.248	90	0.848
23	0.0415	57	0.257	91	0.891
24	0.0452	58	0.267	92	0.938
25	0.0491	59	0.276	93	0.993
26	0.0531	60	0.286	94	1.055
27	0.0572	61	0.297	95	1.129
28	0.0615	62	0.307	96	1.219
29	0.0660	63	0.318	97	1.336
30	0.0707	64	0.329	98	1.500
31	0.0754	65	0.304	99	1.781
32	0.0803	66	0.352	100	∞
33	0.0855	67	0.364		

Logarithm-of-Time Method

For a given incremental loading of the laboratory test, the specimen deformation against log-of-time plot is shown in Figure 10.25. The following constructions are needed to determine c_v:

1. Extend the straight-line portions of primary and secondary consolidations to intersect at A. The ordinate of A is represented by d_{100} — that is, the deformation at the end of 100% primary consolidation.
2. The initial curved portion of the plot of deformation versus log t is approximated to be a parabola on the natural scale. Select times t_1 and t_2 on the curved portion such that $t_2 = 4t_1$. Let the difference of specimen deformation during time $(t_2 - t_1)$ be equal to x.

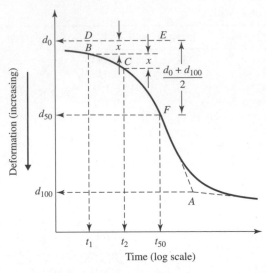

Figure 10.25 Logarithm-of-time method for determining coefficient of consolidation

3. Draw a horizontal line DE such that the vertical distance BD is equal to x. The deformation corresponding to the line DE is d_0 (that is, deformation at 0% consolidation).
4. The ordinate of point F on the consolidation curve represents the deformation at 50% primary consolidation, and its abscissa represents the corresponding time (t_{50}).
5. For 50% average degree of consolidation, $T_v = 0.197$ (see Table 10.5), so,

$$T_{50} = \frac{c_v t_{50}}{H_{dr}^2}$$

or

$$c_v = \frac{0.197 H_{dr}^2}{t_{50}} \tag{10.54}$$

where H_{dr} = average longest drainage path during consolidation.

For specimens drained at both top and bottom, H_{dr} equals one-half the average height of the specimen during consolidation. For specimens drained on only one side, H_{dr} equals the average height of the specimen during consolidation.

Square-Root-of-Time Method

In the square-root-of-time method, a plot of deformation against the square root of time is made for the incremental loading (Figure 10.26). Other graphic constructions required are as follows:

1. Draw a line AB through the early portion of the curve.
2. Draw a line AC such that $\overline{OC} = 1.15\overline{OB}$. The abscissa of point D, which is the intersection of AC and the consolidation curve, gives the square root of time for 90% consolidation ($\sqrt{t_{90}}$).

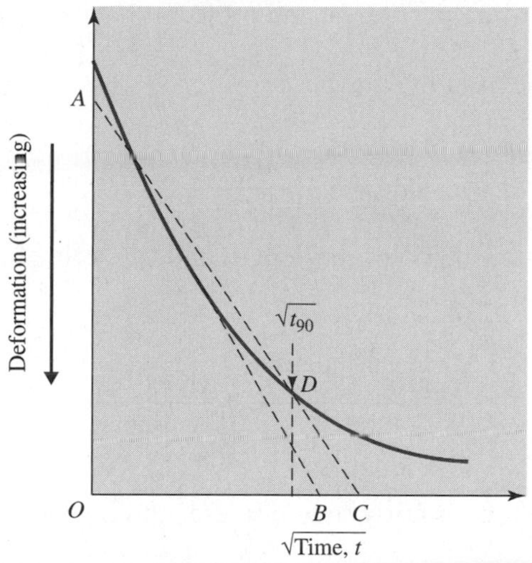

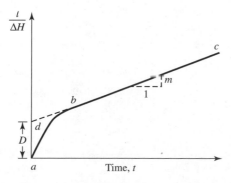

Figure 10.26
Square-root-of-time fitting method

3. For 90% consolidation, $T_{90} = 0.848$ (see Table 10.5), so

$$T_{90} = 0.848 = \frac{c_v t_{90}}{H_{dr}^2}$$

or

$$c_v = \frac{0.848 H_{dr}^2}{t_{90}} \tag{10.55}$$

H_{dr} in Eq. (10.55) is determined in a manner similar to that in the logarithm-of-time method.

Hyperbola Method

In the hyperbola method, the following procedure is recommended for the determination of c_v:

1. Obtain the time t and the specimen deformation (ΔH) from the laboratory consolidation test.
2. Plot the graph of $t/\Delta H$ against t as shown in Figure 10.27.

Figure 10.27
Hyperbola method for determination of c_v

3. Identify the straight-line portion *bc* and project it back to point *d.* Determine the intercept *D.*
4. Determine the slope *m* of the line *bc.*
5. Calculate c_v as

$$c_v = 0.3\left(\frac{mH_{dr}^2}{D}\right) \tag{10.56}$$

Note that because the unit of D is time/length and the unit of m is (time/length)/ time = 1/length, the unit of c_v is

$$\frac{\left(\dfrac{1}{length}\right)(length)^2}{\left(\dfrac{time}{length}\right)} = \frac{(length)^2}{time}$$

The hyperbola method is fairly simple to use, and it gives good results for $U = 60\%$ to 90%.

Early Stage log-t Method

The early stage log-*t* method, an extension of the logarithm-of-time method,.is based on specimen deformation against log-of-time plot as shown in Figure 10.28. According-ing to this method, follow steps 2 and 3 described for the logarithm-of-time method to determine d_0. Draw a horizontal line *DE* through d_0. Then draw a tangent through the point of inflection, *F.* The tangent intersects line *DE* at point *G.* Determine the time *t* corresponding to *G,* which is the time at $U = 22.14\%$. So

$$c_v = \frac{0.0385H_{dr}^2}{t_{22.14}} \tag{10.57}$$

In most cases, for a given soil and pressure range, the magnitude of c_v determined by using the *logarithm-of-time method* provides *lowest value.* The *highest value* is ob-tained from the *early stage log-t method.* The primary reason is because the early stage log-*t* method uses the earlier part of the consolidation curve, whereas the logarithm-of-time method uses the lower portion of the consolidation curve. When the lower

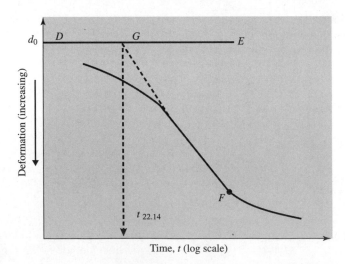

Figure 10.28
Early stage log-*t* method

portion of the consolidation curve is taken into account, the effect of secondary consolidation plays a role in the magnitude of c_v. This fact is demonstrated for several soils in Table 10.6.

Several investigators have also reported that the c_v value obtained from the field is substantially higher than that obtained from laboratory tests conducted by using conventional testing methods (that is, logarithm-of-time and square-root-of-time methods). Hence, the early stage log-t method may provide a more realistic value of fieldwork.

Table 10.6 Comparison of c_v Obtained from Various Methods*

Soil	Range of pressure σ' (kN/m²)	$c_v \times 10^4$ cm²/sec		
		Logarithm-of-time method	Square-root-of-time method	Early stage log t method
Red earth	25–50	4.63	5.45	6.12
	50–100	6.43	7.98	9.00
	100–200	7.32	9.99	11.43
	200–400	8.14	10.90	12.56
	400–800	8.10	11.99	12.80
Brown soil	25–50	3.81	4.45	5.42
	50–100	3.02	3.77	3.80
	100–200	2.86	3.40	3.52
	200–400	2.09	2.21	2.74
	400–800	1.30	1.45	1.36
Black cotton soil	25–50	5.07	6.55	9.73
	50–100	3.06	3.69	4.78
	100–200	2.00	2.50	3.45
	200–400	1.15	1.57	2.03
	400–800	0.56	0.64	0.79
Illite	25–50	1.66	2.25	2.50
	50–100	1.34	3.13	3.32
	100–200	2.20	3.18	3.65
	200–400	3.15	4.59	5.14
	400–800	4.15	5.82	6.45
Bentonite	25–50	0.063	0.130	0.162
	50–100	0.046	0.100	0.130
	100–200	0.044	0.052	0.081
	200–400	0.021	0.022	0.040
	400–800	0.015	0.017	0.022
Chicago clay (Taylor, 1948)	12.5–25	25.10	45.50	46.00
	25–50	20.10	23.90	31.50
	50–100	13.70	17.40	20.20
	100–200	3.18	4.71	4.97
	200–400	4.56	4.40	4.91
	400–800	6.05	6.44	7.41
	800–1600	7.09	8.62	9.09

*After a table from "Determination of Coefficient of Consolidation from Early Stage of Log t Plot," by R. G. Robinson and M. M. Allam, 1996, *Geotechnical Testing Journal, 19*(3) pp. 316–320. Copyright © 1996 American Society for Testing and Materials. Reprinted with permission.

Example 10.7

The time required for 50% consolidation of a 25-mm-thick clay layer (drained at both top and bottom) in the laboratory is 2 min. 20 sec. How long (in days) will it take for a 3-m-thick clay layer of the same clay in the field under the same pressure increment to reach 50% consolidation? In the field, there is a rock layer at the bottom of the clay.

Solution

$$T_{50} = \frac{c_v t_{\text{lab}}}{H_{dr(\text{lab})}^2} = \frac{c_v t_{\text{field}}}{H_{dr(\text{field})}^2}$$

or

$$\frac{t_{\text{lab}}}{H_{dr(\text{lab})}^2} = \frac{t_{\text{field}}}{H_{dr(\text{field})}^2}$$

$$\frac{140 \text{ sec}}{\left(\dfrac{0.025 \text{ m}}{2}\right)^2} = \frac{t_{\text{field}}}{(3 \text{ m})^2}$$

$$t_{\text{field}} = 8{,}064{,}000 \text{ sec} = \textbf{93.33 days} \qquad \blacksquare$$

Example 10.8

Refer to Example 10.7. How long (in days) will it take in the field for 30% primary consolidation to occur? Use Eq. (10.52).

Solution
From Eq. (10.52),

$$\frac{c_v t_{\text{field}}}{H_{dr(\text{lab})}^2} = T_v \propto U^2$$

So

$$t \propto U^2$$

$$\frac{t_1}{t_2} = \frac{U_1^2}{U_2^2}$$

or

$$\frac{93.33 \text{ days}}{t_2} = \frac{50^2}{30^2}$$

$$t_2 = \textbf{33.6 days} \qquad \blacksquare$$

Example 10.9

A 3-in.-thick layer (double drainage) of saturated clay under a surcharge loading underwent 90% primary consolidation in 75 days. Find the coefficient of consolidation of clay for the pressure range.

Solution

$$T_{90} = \frac{c_v t_{90}}{H_{dr}^2}$$

Because the clay layer has two-way drainage, $H_{dr} = 3 \text{ m}/2 = 1.5 \text{ m}$. Also, $T_{90} = 0.848$ (see Table 10.5). So

$$0.848 = \frac{c_v(75 \times 24 \times 60 \times 60)}{(1.5 \times 100)^2}$$

$$c_v = \frac{0.848 \times 2.25 \times 10^4}{75 \times 24 \times 60 \times 60} = \textbf{0.00294 cm}^2/\textbf{sec} \qquad \blacksquare$$

Example 10.10

For a normally consolidated laboratory clay specimen drained on both sides, the following are given:

$$\sigma'_O = 3000 \text{ lb/ft}^2 \qquad e = e_O = 1.1$$
$$\sigma'_O + \Delta\sigma' = 6000 \text{ lb/ft}^2 \qquad e = 0.9$$

Thickness of clay specimen = 1 in.
Time for 50% consolidation = 2 min

a. Determine the hydraulic conductivity (ft/min) of the clay for the loading range.
b. How long (in days) will it take for a 6-ft clay layer in the field (drained on one side) to reach 60% consolidation?

Solution
Part A
The coefficient of compressibility is

$$m_v = \frac{a_v}{1 + e_{av}} = \frac{\left(\dfrac{\Delta e}{\Delta\sigma'}\right)}{1 + e_{av}}$$

$$\Delta e = 1.1 - 0.9 = 0.2$$

$$\Delta\sigma' = 6000 - 3000 = 3000 \text{ lb/ft}^2$$

$$e_{av} = \frac{1.1 + 0.9}{2} = 1.0$$

So

$$m_v = \frac{\dfrac{0.2}{3000}}{1 + 1.0} = 3.33 \times 10^{-5}\, \text{ft}^2/\text{lb}$$

From Table 10.5, for $U = 50\%$, $T_v = 0.197$; thus,

$$c_v = \frac{(0.197)\left(\dfrac{1}{2 \times 12}\right)^2}{2} = 1.71 \times 10^{-4}\, \text{ft}^2/\text{min}$$

$$k = c_v m_v \gamma_w = (1.71 \times 10^{-4}\, \text{ft}^2/\text{min})(3.33 \times 10^{-5}\, \text{ft}^2/\text{lb})(62.4\, \text{lb/ft}^3)$$

$$= \mathbf{3.55 \times 10^{-7}\, ft/min}$$

Part B

$$T_{60} = \frac{c_v t_{60}}{H_{dr}^2}$$

$$t_{60} = \frac{T_{60} H_{dr}^2}{c_v}$$

From Table 10.5, for $U = 60\%$ and $T_{60} = 0.286$,

$$t_{60} = \frac{(0.286)(6)^2}{1.71 \times 10^{-4}} = 60{,}211\ \text{min} = \mathbf{41.8\ days} \qquad \blacksquare$$

10.14 Calculation of Consolidation Settlement under a Foundation

Chapter 9 showed that the increase in the vertical stress in soil caused by a load applied over a limited area decreases with depth z measured from the ground surface downward. Hence to estimate the one-dimensional settlement of a foundation, we can use Eq. (10.24), (10.26), or (10.27). However, the increase of effective stress, $\Delta\sigma'$, in these equations should be the average increase in the pressure below the center of the foundation. The values can be determined by using the procedure described in Chapter 9.

Assuming that the pressure increase varies parabolically, using Simpson's rule, we can estimate the value of $\Delta\sigma'_{av}$ as

$$\Delta\sigma'_{av} = \frac{\Delta\sigma'_t + 4\Delta\sigma'_m + \Delta\sigma'_b}{6} \qquad (10.58)$$

where $\Delta\sigma'_t$, $\Delta\sigma'_m$, and $\Delta\sigma'_b$ represent the increase in the effective pressure at the top, middle, and bottom of the layer, respectively.

Example 10.11

Calculate the settlement of the 10-ft-thick clay layer (Figure 10.29) that will result from the load carried by a 5-ft-square footing. The clay is normally consolidated. Use the weighted average method [Eq. (10.58)] to calculate the average increase of effective pressure in the clay layer.

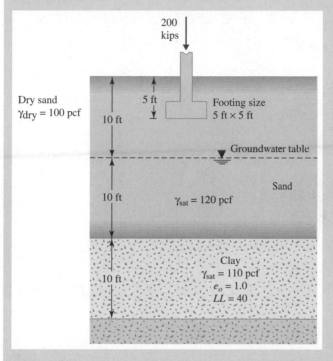

Figure 10.29

Solution
For normally consolidated clay, from Eq. (10.24),

$$S_e = \frac{C_c H}{1 + e_O} \log \frac{\sigma'_O + \Delta\sigma'_{av}}{\sigma'_O}$$

where

$C_c = 0.009(LL - 10) = 0.009(40 - 10) = 0.27$

$H = 10 \times 12 = 120$ in.

$e_O = 1.0$

$\sigma'_O = 10 \text{ ft} \times \gamma_{dry(sand)} + 10 \text{ ft}[\gamma_{sat(sand)} - 62.4] + \frac{10}{2}[\gamma_{sat(clay)} - 62.4]$

$= 10 \times 100 + 10(120 - 62.4) + 5(110 - 62.4)$

$= 1814 \text{ lb/ft}^2$

From Eq. (10.58),

$$\Delta\sigma'_{av} = \frac{\Delta\sigma'_t + 4\Delta\sigma'_m + \Delta\sigma'_b}{6}$$

$\Delta\sigma'_t, \Delta\sigma'_m,$ and $\Delta\sigma'_b$ below the center of the footing can be obtained from Eq. (9.33). Now we can prepare the following table (*note: $L/B = 5/5 = 1$*):

m_1	z (ft)	$b = B/2$ (ft)	$n_1 = z/b$	q (kip/ft^2)	I_4	$\Delta\sigma' = qI_4$ (kip/ft^2)
1	15	2.5	6	$\frac{200}{5 \times 5} = 8$	0.051	$0.408 = \Delta\sigma'_t$
1	20	2.5	8	8	0.029	$0.232 = \Delta\sigma'_m$
1	25	2.5	10	8	0.019	$0.152 = \Delta\sigma'_b$

So

$$\Delta\sigma'_{av} = \frac{0.408 + (4)(0.232) + 0.152}{6} = 0.248 \text{ kip/ft}^2 = 248 \text{ lb/ft}^2$$

Hence,

$$S_c = \frac{(0.27)(120)}{1 + 1} \log \frac{1814 + 248}{1814} \approx \textbf{0.9 in.} \qquad \blacksquare$$

10.15   *Methods for Accelerating Consolidation Settlement*

In many instances, *sand drains* and *prefabricated vertical drains* are used in the field to accelerate consolidation settlement in soft, normally consolidated clay layers and to achieve precompression before the construction of a desired foundation. Sand drains are constructed by drilling holes through the clay layer(s) in the field at regular intervals. The holes are then backfilled with sand. This can be achieved by several means, such as (a) rotary drilling and then backfilling with sand; (b) drilling by continuous flight auger with hollow stem and backfilling with sand (through the hollow stem); and (c) driving hollow steel piles. The soil inside the pile is then jetted out, and the hole is backfilled with sand. Figure 10.30 shows a schematic diagram of sand drains. After backfilling the drill holes with sand, a surcharge is applied at the ground surface. This surcharge will increase the pore water pressure in the clay. The excess pore water pressure in the clay will be dissipated by drainage — both vertically and radially to the sand drains — which accelerates settlement of the clay layer. In Figure 10.30a, note that the radius of the sand drains is r_w. Figure 10.30b shows the plan of the layout of the sand drains. The effective zone from which the radial drainage will be directed toward a given sand drain is approximately cylindrical, with a diameter of d_e. The surcharge that needs to be applied at the ground surface and the length of time it has to be maintained to achieve the desired degree of consolidation will be a function of r_w, d_e, and other soil parameters. Figure 10.31 shows a sand drain installation in progress.

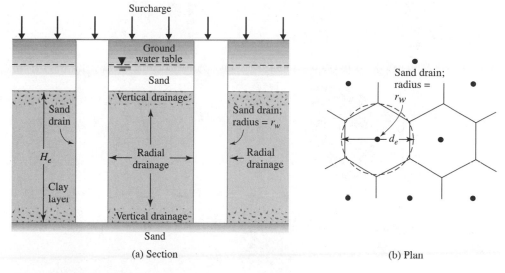

(a) Section

(b) Plan

Figure 10.30 Sand drains

Figure 10.31
Sand drain installation in
progress (courtesy of E. C. Shin,
University of Inchon,
South Korea)

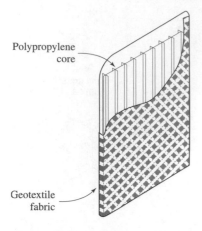

Figure 10.32 Prefabricated vertical drain (PVD)

Figure 10.33
Intallation of PVDs in
progress (courtesy of E. C.
Shin, University of Inchon,
South Korea)

Prefabricated vertical drains (PVDs), which are also referred to as *wick* or *strip drains,* were originally developed as a substitute for the commonly used sand drain. With the advent of materials science, these drains are manufactured from synthetic polymers such as polypropylene and high-density polyethylene. PVDs are normally manufactured with a corrugated or channeled synthetic core enclosed by a geotextile filter, as shown schematically in Figure 10.32. Installation rates reported in the literature are on the order of 0.1 to 0.3 m/s, excluding equipment mobilization and setup time. PVDs have been used extensively in the past for expedient consolidation of low permeability soils under surface surcharge. The main advantage of PVDs over sand drains is that they do not require drilling and, thus, installation is much faster. Figure 10.33 shows the installation of PVDs in the field.

10.16   Summary and General Comments

In this chapter we discussed the fundamental concepts and theories for estimating elastic and consolidation (primary and secondary) settlement. Elastic settlement of a foundation is primarily a function of the size and rigidity of the foundation, the modulus of elasticity, the Poisson's ratio of the soil, and the intensity of load on the foundation.

Consolidation is a time-dependent process of settlement of saturated clay layers located below the ground water table by extrusion of excess water pressure generated by application of load on the foundation. Total consolidation settlement of a clay foundation is a function of compression index (C_c), swell index (C_s), initial void ratio, (e_o) and the average stress increase in the clay layer. The degree of consolidation for a given soil layer at a certain time after the load application depends on its coefficient of consolidation (c_v) and also on the length of the minimum drainage path. Installation of sand drains and wick drains helps reduce the time for accomplishing the desired degree of consolidation for a given construction project.

There are several case histories in the literature for which the fundamental principles of soil compressibility have been used to predict and compare the actual total settlement and the time rate of settlement of soil profiles under superimposed loading. In some cases, the actual and predicted maximum settlements agree remarkably; in many others, the predicted settlements deviate to a large extent from the actual settlements observed. The disagreement in the latter cases may have several causes:

1. Improper evaluation of soil properties
2. Nonhomogeneity and irregularity of soil profiles
3. Error in the evaluation of the net stress increase with depth, which induces settlement

The variation between the predicted and observed time rate of settlement may also be due to

a. Improper evaluation of c_v (see Section 10.13)
b. Presence of irregular sandy seams within the clay layer, which reduces the length of the maximum drainage path, H_{dr}

Problems

10.1 Estimate the immediate settlement of a column footing 4.5 ft in diameter that is constructed on an unsaturated clay layer. The column carries a load of 20 tons, and it is given that $E_s = 1500$ lb/in.^{2} and $\mu_s = 0.25$. Assume the footing to be rigid. [Use Eq. (10.1).]

10.2 Refer to Figure 10.3. For a square foundation measuring 3 m $\times$ 3 m in plan supported by a layer of sand and given that $D_f = 1.5$ m, $t = 0.25$ m, $E_o = 16,000$ kN/m^2, $k = 400$ kN/m^2/m, $\mu_s = 0.3$, $h = 20$ m, $E_f = 15 \times 10^6$ kN/m^2, and $\Delta\sigma = 100$ kN/m^2, calculate the immediate settlement.

10.3 Following are the results of a consolidation test:

e	Pressure, σ' (ton/ft^2)
1.1	0.25
1.085	0.5
1.055	1.0
1.01	2.0
0.94	4.0
0.79	8.0
0.63	16.0

a. Plot the e–log σ' curve.
b. Using Casagrande's method, determine the preconsolidation pressure.
c. Calculate the compression index C_c from the laboratory e–log σ' curve.

10.4   Repeat Problem 10.3, using the following values:

e	Pressure, σ' (kN/m²)
1.21	25
1.195	50
1.15	100
1.06	200
0.98	400
0.925	500

10.5   A soil profile is shown in Figure 10.34. The uniformly distributed load on the ground surface is $\Delta\sigma$. Estimate the primary settlement of the normally consolidated clay layer, given that
$H_1 = 4$ ft, $H_2 = 6$ ft, $H_3 = 4$ ft
For sand, $e = 0.58$, $G_s = 2.67$
For clay, $e = 1.1$, $G_s = 2.72$, $LL = 45$
$\Delta\sigma = 1800$ lb/ft²

10.6   Repeat Problem 10.5, using the following data:
$H_1 = 2.5$ m, $H_2 = 2.5$ m, $H_3 = 3$ m
For sand, $e = 0.64$, $G_s = 2.65$
For clay, $e = 0.9$, $G_s = 2.75$, $LL = 55$
$\Delta\sigma = 100$ kN/m²

10.7   Repeat Problem 10.5, using the following data:
$\Delta\sigma = 90$ kN/m²
$H_1 = 2$ m, $H_2 = 2$ m, $H_3 = 1.5$ m
For sand, $\gamma_{dry} = 14.6$ kN/m³, $\gamma_{sat} = 17.3$ kN/m³
For clay, $\gamma_{sat} = 19.3$ kN/m³, $LL = 38$, $e = 0.75$

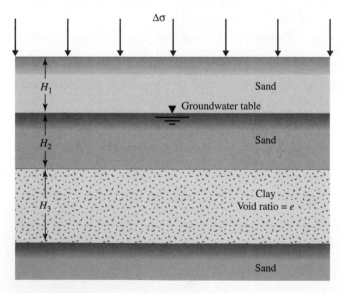

Figure 10.34

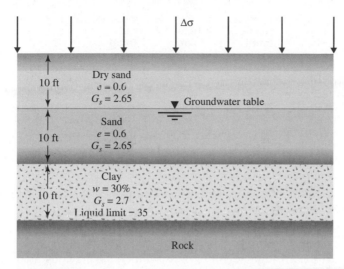

Figure 10.35

10.8 A soil profile is shown in Figure 10.35. The preconsolidation pressure of the clay is 3400 lb/ft². Estimate the primary consolidation settlement that will take place as the result of a surcharge equal to 2200 lb/ft². Assume $C_s = \frac{1}{5}C_c$.

10.9 Refer to Problem 10.6. Given that $c_v = 2.8 \times 10^{-6}$ m²/min, how long will it take for 60% primary consolidation to take place?

10.10 The coordinates of two points on a virgin compression curve are as follows:
$$e_1 = 1.82 \qquad \sigma_1' = 200 \text{ kN/m}^2$$
$$e_2 = 1.54 \qquad \sigma_2' = 400 \text{ kN/m}^2$$
a. Determine the coefficient of volume compressibility for the pressure range stated above.
b. Given that $c_v = 0.003$ cm²/sec, determine k in cm/sec corresponding to the average void ratio.

10.11 For the virgin curve stated in Problem 10.10, what would be the effective pressure σ' corresponding to $e = 1.7$?

10.12 For the virgin curve stated in Problem 10.10, what would be the void ratio corresponding to an effective pressure σ' that is equal to 500 kN/m²?

10.13 Following are the relationships of e and σ' for a clay soil:

e	σ' (ton/ft²)
1.0	0.2
0.97	0.5
0.85	1.8
0.75	3.2

For this clay soil in the field, the following values are given: $H = 4.5$ ft, $\sigma_0' = 0.7$ ton/ft², and $\sigma_0' + \Delta\sigma' = 2$ ton/ft². Calculate the expected settlement caused by primary consolidation.

10.14 During a laboratory consolidation test, the time and dial gauge readings obtained from an increase in pressure on the specimen from 50 to 100 kN/m² are given in the following table:

Time (min)	Dial gauge reading (mm)	Time (min)	Dial gauge reading (mm)
0	3.98	16.0	4.57
0.10	4.08	30.0	4.74
0.25	4.10	60.0	4.92
0.60	4.13	120.0	5.08
1.0	4.17	240.0	5.21
2.0	4.22	480.0	5.28
4.0	4.30	960.0	5.33
8.0	4.42	1440.0	5.39

   a. Find the time for 50% primary consolidation (t_{50}) using the logarithm-of-time method.
   b. Find the time for 90% primary consolidation (t_{90}) using the square-root-of-time method.
   c. If the average height of the specimen during consolidation caused by this incremental loading was 22 mm and it was drained at both the top and the bottom, calculate the coefficient of consolidation using t_{50} and t_{90} obtained from parts (a) and (b).

10.15 Refer to the laboratory test results given in Problem 10.14. Using the hyperbola method, determine c_v. The average height of the specimen during consolidation was 22 mm, and it was drained at the top and bottom.

10.16 The time for 50% consolidation of a 25-mm-thick clay layer (drained at top and bottom) in the laboratory is 150 sec. How long (in days) will it take for a 3-m-thick layer of the same clay in the field under the same pressure increment to reach 50% consolidation? There is a rock layer at the bottom of the clay in the field.

10.17 For a normally consolidated clay, the following values are given:
$$\sigma'_O = 2 \text{ ton/ft}^2 \qquad e = e_o = 1.21$$
$$\sigma'_O + \Delta\sigma' = 4 \text{ ton/ft}^2 \qquad e = 0.96$$
The hydraulic conductivity k of the clay for the preceding loading range is 1.8×10^{-4} ft/day.
   a. How long (in days) will it take for a 9-ft-thick clay layer (drained on one side) in the field to reach 60% consolidation?
   b. What is the settlement at that time (i.e., at 60% consolidation)?

10.18 A 10-ft-thick layer (two-way drainage) of saturated clay under a surcharge loading underwent 90% primary consolidation in 100 days.
   a. Find the coefficient of consolidation of clay for the pressure range.
   b. For a 1-in-thick undisturbed clay specimen, how long will it take to undergo 90% consolidation in the laboratory for a similar consolidation pressure range? The laboratory tests's specimen will have two-way drainage.

10.19 Laboratory tests on a 25-mm-thick clay specimen drained at the top only show 50% consolidation takes place in 11 min.
   a. How long will it take for a similar clay layer in the field, 4 m thick and drained at the top and bottom, to undergo 50% consolidation?
   b. Find the time required for the clay layer in the field, as described in part (a), to reach 70% consolidation.

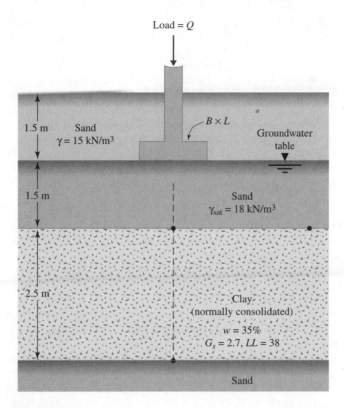

Figure 10.36

10.20 For a laboratory consolidation test on a clay specimen (drained on both sides), the following results were obtained:

Thickness of the clay soil = 25 mm

$\sigma_1' = 50$ kN/m^2 $e_1 = 0.92$

$\sigma_2' = 120$ kN/m^2 $e_2 = 0.78$

Time for 50% consolidation = 2.5 min

Determine the hydraulic conductivity of the clay for the loading range.

10.21 Refer to Figure 10.36. Given that $B = 1.5$ m, $L = 2.5$ m, and $Q = 120$ kN, calculate the primary consolidation settlement of the foundation.

10.22 Redo Problem 10.21 with $B = 1$ m, $L = 3$ m, and $Q = 110$ kN.

References

CASAGRANDE, A. (1936). "Determination of the Preconsolidation Load and Its Practical Significance," *Proceedings,* 1st International Conference on Soil Mechanics and Foundation Engineering, Cambridge, Mass., Vol. 3, 60–64.

CASAGRANDE, A. and FADUM, R. E. (1940). "Notes on Soil Testing for Engineering Purposes," Harvard University Graduate School of Engineering Publication No. 8.

HOUGH, B. K. (1957). *Basic Soils Engineering,* Ronald Press, New York.

MAYNE, P. W., and POULOS, H. G. (1999). "Approximate Displacement Influence Factors for Elastic Shallow Foundations," *Journal of Geotechnical and Geoenvironmental Engineering,* ASCE, Vol. 125, No. 6, 453–460.

NAGARAJ, T., and MURTY, B. R. S. (1985). "Prediction of the Preconsolidation Pressure and Recompression Index of Soils," *Geotechnical Testing Journal,* Vol. 8, No. 4, 199–202.

NISHIDA, Y. (1956). "A Brief Note on Compression Index of Soils," *Journal of the Soil Mechanics and Foundations Division,* ASCE, Vol. 82, No. SM3, 1027-1–1027-14.

RENDON-HERRERO, O. (1983). "Universal Compression Index Equation," *Discussion, Journal of Geotechnical Engineering,* ASCE, Vol. 109, No. 10, 1349.

RENDON-HERRERO, O. (1980). "Universal Compression Index Equation," *Journal of the Geotechnical Engineering Division,* ASCE, Vol. 106, No. GT11, 1179–1200.

ROBINSON, R. G., and ALLAM, M. M. (1996). "Determination of Coefficient of Consolidation from Early Stage of log *t* Plot," *Geotechnical Testing Journal,* ASTM, Vol. 19, No. 3, 316–320.

SCHLEICHER, F. (1926). "Zur Theorie des Baugrundes," *Bauingenieur,* Vol. 7, 931–935, 949–952.

SCHMERTMANN, J. H. (1953). "Undisturbed Consolidation Behavior of Clay," *Transactions, ASCE,* Vol. 120, 1201.

SKEMPTON, A. W. (1944). "Notes on the Compressibility of Clays," *Quarterly Journal of the Geological Society of London,* Vol. 100, 119–135.

SRIDHARAN, A., and PRAKASH, K. (1985). "Improved Rectangular Hyperbola Method for the Determination of Coefficient of Consolidation," *Geotechnical Testing Journal,* ASTM, Vol. 8, No. 1, 37–40.

TAYLOR, D. W. (1942). "Research on Consolidation of Clays," *Serial No. 82,* Department of Civil and Sanitary Engineering, Massachusetts Institute of Technology, Cambridge, Mass.

TAYLOR, D. W. (1948). *Fundamentals of Soil Mechanics,* Wiley, New York.

TERZAGHI, K. (1925). *Erdbaumechanik auf Bodenphysikalischer Grundlager,* Deuticke, Vienna.

TERZAGHI, K., and PECK, R. B. (1967). *Soil Mechanics in Engineering Practice,* 2nd ed., Wiley, New York.

11

Shear Strength of Soil

The *shear strength* of a soil mass is the internal resistance per unit area that the soil mass can offer to resist failure and sliding along any plane inside it. One must understand the nature of shearing resistance in order to analyze soil stability problems such as bearing capacity, slope stability, and lateral pressure on earthretaining structures.

11.1   Mohr–Coulomb Failure Criterion

Mohr (1900) presented a theory for rupture in materials that contended that a material fails because of a critical combination of normal stress and shearing stress, and not from either maximum normal or shear stress alone. Thus, the functional relationship between normal stress and shear stress on a failure plane can be expressed in the following form:

$$\tau_f = f(\sigma) \tag{11.1}$$

The failure envelope defined by Eq. (11.1) is a curved line. For most soil mechanics problems, it is sufficient to approximate the shear stress on the failure plane as a linear function of the normal stress (Coulomb, 1776). This linear function can be written as

$$\tau_f = c + \sigma \tan \phi \tag{11.2}$$

where c = cohesion
    ϕ = angle of internal friction
    σ = normal stress on the failure plane
    τ_f = shear strength

The preceding equation is called the *Mohr–Coulomb failure criterion*.

In saturated soil, the total normal stress at a point is the sum of the effective stress (σ') and pore water pressure (u), or

$$\sigma = \sigma' + u$$

Table 11.1 Typical Values of Drained Angle of Friction for Sands and Silts

Soil type	ϕ' (deg)
Sand: Rounded grains	
Loose	27–30
Medium	30–35
Dense	35–38
Sand: Angular grains	
Loose	30–35
Medium	35–40
Dense	40–45
Gravel with some sand	34–48
Silts	26–35

The effective stress σ' is carried by the soil solids. The Mohr–Coulomb failure criterion, expressed in terms of effective stress, will be of the form

$$\tau_f = c' + \sigma' \tan \phi' \tag{11.3}$$

where c' = cohesion and ϕ' = friction angle, based on effective stress.

Thus, Eqs. (11.2) and (11.3) are expressions of shear strength based on total stress and effective stress. The value of c' for sand and inorganic silt is 0. For normally consolidated clays, c' can be approximated at 0. Overconsolidated clays have values of c' that are greater than 0. The angle of friction, ϕ', is sometimes referred to as the *drained angle of friction*. Typical values of ϕ' for some granular soils are given in Table 11.1.

The significance of Eq. (11.3) can be explained by referring to Fig. 11.1, which shows an elemental soil mass. Let the effective normal stress and the shear stress on

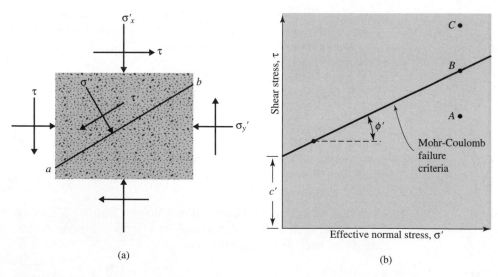

(a)

(b)

Figure 11.1 Mohr–Coulomb failure criterion

the plane *ab* be σ' and τ, respectively. Figure 11.1b shows the plot of the failure envelope defined by Eq. (11.3). If the magnitudes of σ' and τ on plane *ab* are such that they plot as point *A* in Figure 11.1b, shear failure will not occur along the plane. If the effective normal stress and the shear stress on plane *ab* plot as point *B* (which falls on the failure envelope), shear failure will occur along that plane. A state of stress on a plane represented by point *C* cannot exist, because it plots above the failure envelope, and shear failure in a soil would have occurred already.

11.2 Inclination of the Plane of Failure Caused by Shear

As stated by the Mohr–Coulomb failure criterion, failure from shear will occur when the shear stress on a plane reaches a value given by Eq. (11.3). To determine the inclination of the failure plane with the major principal plane, refer to Figure 11.2, where σ_1' and σ_3' are, respectively, the major and minor effective principal stresses. The failure plane *EF* makes an angle θ with the major principal plane. To determine the angle θ and the relationship between σ_1' and σ_3', refer to Figure 11.3, which is a plot of the Mohr's circle for the state of stress shown in Figure 11.2 (see Chapter 9). In Figure 11.3, *fgh* is the failure envelope defined by the relationship $\tau_f = c' + \sigma' \tan \phi'$. The radial line *ab* defines the major principal plane (*CD* in Figure 11.2), and the radial line *ad* defines the failure plane (*EF* in Figure 11.2). It can be shown that $\angle bad = 2\theta = 90 + \phi'$, or

$$\theta = 45 + \frac{\phi'}{2} \tag{11.4}$$

Again, from Figure 11.3,

$$\frac{\overline{ad}}{\overline{fa}} = \sin \phi' \tag{11.5}$$

$$\overline{fa} = \overline{fO} + \overline{Oa} = c' \cot \phi' + \frac{\sigma_1' + \sigma_3'}{2} \tag{11.6a}$$

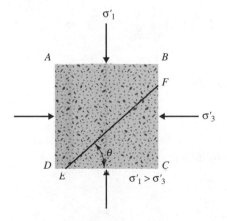

Figure 11.2 Inclination of failure plane in soil with major principal plane

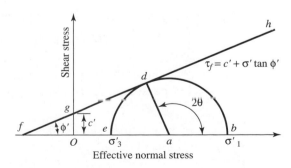

Figure 11.3 Mohr's circle and failure envelope

Also,

$$\overline{ad} = \frac{\sigma_1' - \sigma_3'}{2} \qquad (11.6b)$$

Substituting Eqs. (11.6a) and (11.6b) into Eq. (11.5), we obtain

$$\sin \phi' = \frac{\dfrac{\sigma_1' - \sigma_3'}{2}}{c' \cot \phi' + \dfrac{\sigma_1' + \sigma_3'}{2}}$$

or

$$\sigma_1' = \sigma_3'\left(\frac{1 + \sin \phi'}{1 - \sin \phi'}\right) + 2c\left(\frac{\cos \phi'}{1 - \sin \phi'}\right) \qquad (11.7)$$

However,

$$\frac{1 + \sin \phi'}{1 - \sin \phi'} = \tan^2\left(45 + \frac{\phi'}{2}\right)$$

and

$$\frac{\cos \phi'}{1 - \sin \phi'} = \tan\left(45 + \frac{\phi'}{2}\right)$$

Thus,

$$\sigma_1' = \sigma_3' \tan^2\left(45 + \frac{\phi'}{2}\right) + 2c' \tan\left(45 + \frac{\phi'}{2}\right) \qquad (11.8)$$

An expression similar to Eq. (11.8) could also be derived using Eq. (11.2) (that is, total stress parameters c and ϕ), or

$$\sigma_1 = \sigma_3 \tan^2\left(45 + \frac{\phi}{2}\right) + 2c \tan\left(45 + \frac{\phi}{2}\right) \qquad (11.9)$$

11.3   *Laboratory Test for Determination of Shear Strength Parameters*

There are several laboratory methods now available to determine the shear strength parameters (i.e., c, ϕ, c', ϕ') of various soil specimens in the laboratory. They are as follows:

a. Direct shear test
b. Triaxial test
c. Direct simple shear test
d. Plane strain triaxial test
e. Torsional ring shear test

The direct shear test and the triaxial test are the two commonly used techniques for determining the shear strength parameters. These two tests will be described in detail in the sections that follow.

11.4   *Direct Shear Test*

The direct shear test is the oldest and simplest form of shear test arrangement. A diagram of the direct shear test apparatus is shown in Figure 11.4. The test equipment consists of a metal shear box in which the soil specimen is placed. The soil specimens may be square or circular in plan. The size of the specimens generally used is about 51 mm × 51 mm or 102 mm × 102 mm (2 in. × 2 in. or 4 in. × 4 in.) across and about 25 mm (1 in.) high. The box is split horizontally into halves. Normal force on the specimen is applied from the top of the shear box. The normal stress on the specimens can be as great as 1050 kN/m² (150 lb/in.²). Shear force is applied by moving one-half of the box relative to the other to cause failure in the soil specimen.

Depending on the equipment, the shear test can be either stress controlled or strain controlled. In stress-controlled tests, the shear force is applied in equal increments until the specimen fails. The failure occurs along the plane of split of the shear box. After the application of each incremental load, the shear displacement of the top half of the box is measured by a horizontal dial gauge. The change in the height of the specimen (and thus the volume change of the specimen) during the test can be obtained from the readings of a dial gauge that measures the vertical movement of the upper loading plate.

In strain-controlled tests, a constant rate of shear displacement is applied to one-half of the box by a motor that acts through gears. The constant rate of shear displacement is measured by a horizontal dial gauge. The resisting shear force of the soil corresponding to any shear displacement can be measured by a horizontal proving ring or load cell. The volume change of the specimen during the test is obtained

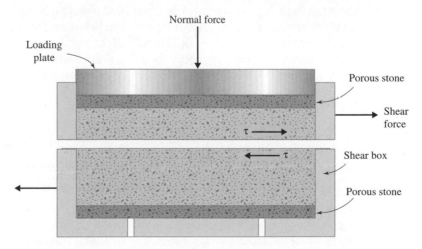

Figure 11.4 Diagram of direct shear test arrangement

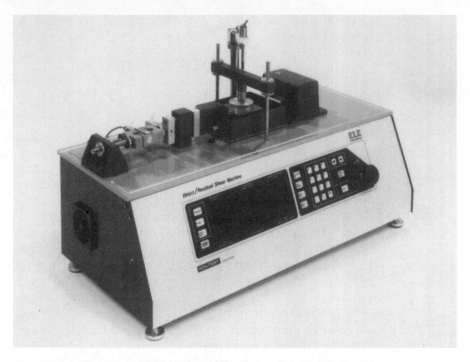

Figure 11.5 Strain-controlled direct shear test equipment (courtesy of Soiltest, Inc., Lake Bluff, Illinois)

in a manner similar to that in the stress-controlled tests. Figure 11.5 shows a photograph of strain-controlled direct shear test equipment.

The advantage of the strain-controlled tests is that in the case of dense sand, peak shear resistance (that is, at failure) as well as lesser shear resistance (that is, at a point after failure called *ultimate strength*) can be observed and plotted. In stress-controlled tests, only the peak shear resistance can be observed and plotted. Note that the peak shear resistance in stress-controlled tests can be only approximated because failure occurs at a stress level somewhere between the prefailure load increment and the failure load increment. Nevertheless, compared with strain-controlled tests, stress-controlled tests probably model real field situations better.

For a given test, the normal stress can be calculated as

$$\sigma = \text{Normal stress} = \frac{\text{Normal force}}{\text{Cross-sectional area of the specimen}} \tag{11.10}$$

The resisting shear stress for any shear displacement can be calculated as

$$\tau = \text{Shear stress} = \frac{\text{Resisting shear force}}{\text{Cross-sectional area of the specimen}} \tag{11.11}$$

Figure 11.6 shows a typical plot of shear stress and change in the height of the specimen against shear displacement for dry loose and dense sands. These observations were obtained from a strain-controlled test. The following generalizations can

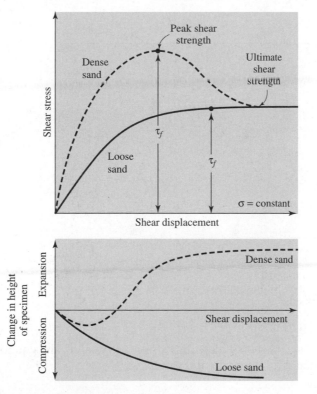

Figure 11.6 Plot of shear stress and change in height of specimen against shear displacement for loose and dense dry sand (direct shear test)

be developed from Figure 11.6 regarding the variation of resisting shear stress with shear displacement:

1. In loose sand, the resisting shear stress increases with shear displacement until a failure shear stress of τ_f is reached. After that, the shear resistance remains approximately constant for any further increase in the shear displacement.
2. In dense sand, the resisting shear stress increases with shear displacement until it reaches a failure stress of τ_f. This τ_f is called the *peak shear strength*. After failure stress is attained, the resisting shear stress gradually decreases as shear displacement increases until it finally reaches a constant value called the *ultimate shear strength*.

It is important to note that, in dry sand,

$$\sigma = \sigma'$$

and

$$c' = 0$$

Direct shear tests are repeated on similar specimens at various normal stresses. The normal stresses and the corresponding values of τ_f obtained from a number of tests are plotted on a graph from which the shear strength parameters are determined.

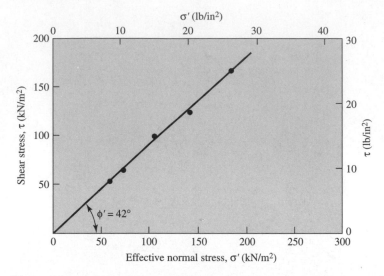

Figure 11.7 Determination of shear strength parameters for a dry sand using the results of direct shear tests

Figure 11.7 shows such a plot for tests on a dry sand. The equation for the average line obtained from experimental results is

$$\tau_f = \sigma' \tan \phi' \tag{11.12}$$

So, the friction angle can be determined as follows:

$$\phi' = \tan^{-1}\left(\frac{\tau_f}{\sigma'}\right)$$

It is important to note that *in situ* cemented sands may show a c' intercept.

11.5 Drained Direct Shear Test on Saturated Sand and Clay

In the direct shear test arrangement, the shear box that contains the soil specimen is generally kept inside a container that can be filled with water to saturate the specimen. A *drained test* is made on a saturated soil specimen by keeping the rate of loading slow enough so that the excess pore water pressure generated in the soil is completely dissipated by drainage. Pore water from the specimen is drained through two porous stones. (See Figure 11.4.)

Because the hydraulic conductivity of sand is high, the excess pore water pressure generated due to loading (normal and shear) is dissipated quickly. Hence, for

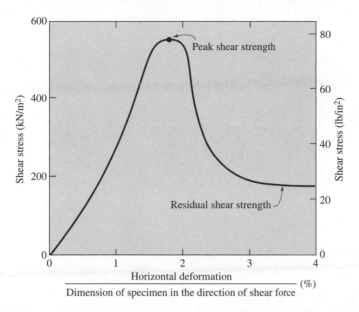

Figure 11.8 Results of a drained direct shear test on an overconsolidated clay. *Note:* Residual shear strength in clay is similar to ultimate shear strength in sand — see Figure 11.6

an ordinary loading rate, essentially full drainage conditions exist. The friction angle, ϕ', obtained from a drained direct shear test of saturated sand will be the same as that for a similar specimen of dry sand.

The hydraulic conductivity of clay is very small compared with that of sand. When a normal load is applied to a clay soil specimen, a sufficient length of time must elapse for full consolidation — that is, for dissipation of excess pore water pressure. For this reason, the shearing load must be applied very slowly. The test may last from two to five days. Figure 11.8 shows the results of a drained direct shear test on an overconsolidated clay. Figure 11.9 shows the plot of τ_f against σ' obtained from a number

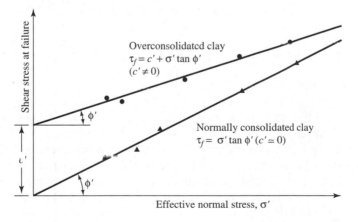

Figure 11.9 Failure envelope for clay obtained from drained direct shear tests

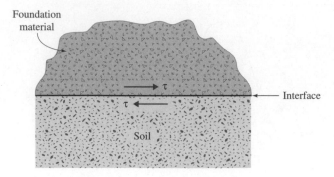

Figure 11.10 Interface of a foundation material and soil

of drained direct shear tests on a normally consolidated clay and an overconsolidated clay. Note that the value of $c' \simeq 0$ for a normally consolidated clay.

11.6 General Comments on Direct Shear Test

The direct shear test is simple to perform, but it has some inherent shortcomings. The reliability of the results may be questioned because the soil is not allowed to fail along the weakest plane but is forced to fail along the plane of split of the shear box. Also, the shear stress distribution over the shear surface of the specimen is not uniform. Despite these shortcomings, the direct shear test is the simplest and most economical for a dry or saturated sandy soil.

In many foundation design problems, one must determine the angle of friction between the soil and the material in which the foundation is constructed (Figure 11.10). The foundation material may be concrete, steel, or wood. The shear strength along the surface of contact of the soil and the foundation can be given as

$$\tau_f = c'_a + \sigma' \tan \delta \qquad (11.13)$$

where c'_a = adhesion
δ = effective angle of friction between the soil and the foundation material

Note that the preceding equation is similar in form to Eq. (11.3). The shear strength parameters between a soil and a foundation material can be conveniently determined by a direct shear test. This is a great advantage of the direct shear test. The foundation material can be placed in the bottom part of the direct shear test box and then the soil can be placed above it (that is, in the top part of the box), as shown in Figure 11.11, and the test can be conducted in the usual manner.

Figure 11.12 shows the results of direct shear tests conducted in this manner with a quartz sand and concrete wood, and steel as foundation materials, with $\sigma' = 100 \text{ kN/m}^2 \ (14.5 \text{ lb/in.}^2)$.

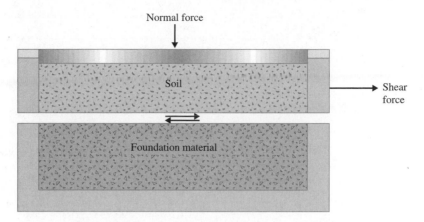

Figure 11.11 Direct shear test to determine interface friction angle

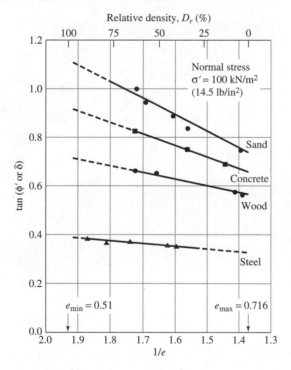

Figure 11.12 Variation of tan ϕ' and tan δ with $1/e$. [*Note:* e = void ratio, σ' = 100 kN/m² (14.5 lb/in.²), quartz sand (after Acar, Durgunoglu, and Tumay, 1982)]

Example 11.1

Following are the results of four drained direct shear tests on an *overconsolidated clay:*

Diameter of specimen = 50 mm

Height of specimen = 25 mm

Test no.	Normal force, N (N)	Shear force at failure, S_{peak} (N)	Residual shear force, $S_{residual}$ (N)[a]
1	150	157.5	44.2
2	250	199.9	56.6
3	350	257.6	102.9
4	550	363.4	144.5

[a]See Figure 11.8

Determine the relationships for *peak shear strength* (τ_f) and *residual shear strength* (τ_r).

Solution

Area of the specimen $(A) = (\pi/4)\left(\dfrac{50}{1000}\right)^2 = 0.0019634$ m^2. Now the following table can be prepared:

Test no.	Normal force, N (N)	Normal stress, σ' (kN/m^2)	Peak shear force, S_{peak} (N)	$\tau_f = \dfrac{S_{peak}}{A}$ (kN/m^2)	Residual shear force, $S_{residual}$ (N)	$\tau_r = \dfrac{S_{residual}}{A}$ (kN/m^2)
1	150	76.4	157.5	80.2	44.2	22.5
2	250	127.3	199.9	101.8	56.6	28.8
3	350	178.3	257.6	131.2	102.9	52.4
4	550	280.1	363.4	185.1	144.5	73.6

The variations of τ_f and τ_r with σ' are plotted in Figure 11.13. From the plots, we find that

Peak strength: $\tau_f(\text{kN/m}^2) = \mathbf{40 + \sigma' \tan 27}$

Residual strength: $\tau_r(\text{kN/m}^2) = \boldsymbol{\sigma' \tan 14.6}$

(*Note:* For all *overconsolidated clays,* the residual shear strength can be expressed as

$$\tau_r = \sigma' \tan \phi'_r$$

where ϕ'_r = effective residual friction angle.)   ∎

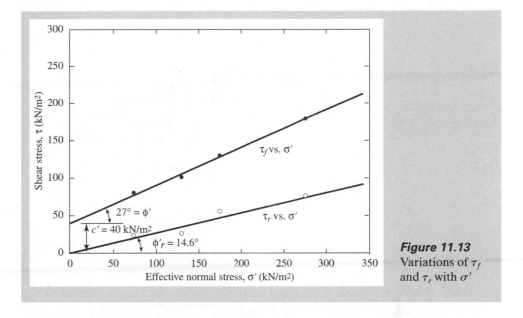

Figure 11.13
Variations of τ_f and τ_r with σ'

11.7 Triaxial Shear Test (General)

The triaxial shear test is one of the most reliable methods available for determining shear strength parameters. It is widely used for research and conventional testing. A diagram of the triaxial test layout is shown in Figure 11.14.

In this test, a soil specimen about 36 mm (1.4 in.) in diameter and 76 mm (3 in.) long is generally used. The specimen is encased by a thin rubber membrane and placed inside a plastic cylindrical chamber that is usually filled with water or glycerine. The specimen is subjected to a confining pressure by compression of the fluid in the chamber. (*Note:* Air is sometimes used as a compression medium.) To cause shear failure in the specimen, one must apply axial stress through a vertical loading ram (sometimes called *deviator stress*). This stress can be applied in one of two ways:

1. Application of dead weights or hydraulic pressure in equal increments until the specimen fails. (Axial deformation of the specimen resulting from the load applied through the ram is measured by a dial gauge.)
2. Application of axial deformation at a constant rate by means of a geared or hydraulic loading press. This is a strain-controlled test.

The axial load applied by the loading ram corresponding to a given axial deformation is measured by a proving ring or load cell attached to the ram.

Connections to measure drainage into or out of the specimen, or to measure pressure in the pore water (as per the test conditions), are also provided. The following three standard types of triaxial tests are generally conducted:

1. Consolidated-drained test or drained test (CD test)
2. Consolidated-undrained test (CU test)
3. Unconsolidated-undrained test or undrained test (UU test)

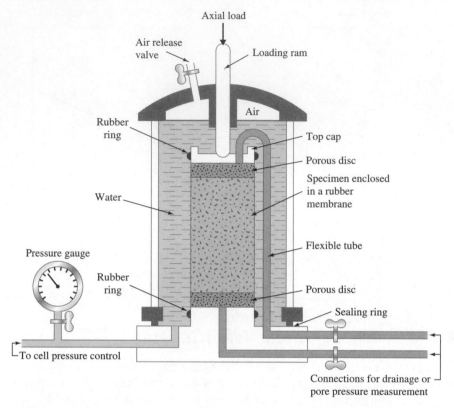

Figure 11.14 Diagram of triaxial test equipment (after Bishop and Bjerrum, 1960)

The general procedures and implications for each of the tests in *saturated soils* are described in the following sections.

11.8 *Consolidated-Drained Triaxial Test*

In the CD test, the saturated specimen is first subjected to an all around confining pressure, σ_3, by compression of the chamber fluid (Figure 11.15a). As confining pressure is applied, the pore water pressure of the specimen increases by u_c (if drainage is prevented). This increase in the pore water pressure can be expressed as a nondimensional parameter in the form

$$B = \frac{u_c}{\sigma_3} \tag{11.14}$$

where B = Skempton's pore pressure parameter (Skempton, 1954).

For saturated soft soils, B is approximately equal to 1; however, for saturated stiff soils, the magnitude of B can be less than 1. Black and Lee (1973) gave the theo-

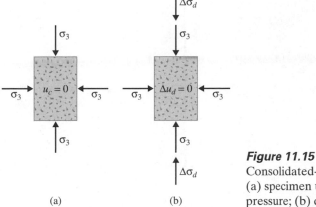

Figure 11.15
Consolidated-drained triaxial test:
(a) specimen under chamber confining
pressure; (b) deviator stress application

Table 11.2 Theoretical Values of B at Complete Saturation

Type of soil	Theoretical value
Normally consolidated soft clay	0.9998
Lightly overconsolidated soft clays and silts	0.9988
Overconsolidated stiff clays and sands	0.9877
Very dense sands and very stiff clays at high confining pressures	0.9130

retical values of B for various soils at complete saturation. These values are listed in Table 11.2.

Now, if the connection to drainage is opened, dissipation of the excess pore water pressure, and thus consolidation, will occur. With time, u_c will become equal to 0. In saturated soil, the change in the volume of the specimen (ΔV_c) that takes place during consolidation can be obtained from the volume of pore water drained (Figure 11.16a). Next, the deviator stress, $\Delta\sigma_d$, on the specimen is increased very slowly (Figure 11.15b). The drainage connection is kept open, and the slow rate of deviator stress application allows complete dissipation of any pore water pressure that developed as a result ($\Delta u_d = 0$).

A typical plot of the variation of deviator stress against strain in loose sand and normally consolidated clay is shown in Figure 11.16b. Figure 11.16c shows a similar plot for dense sand and overconsolidated clay. The volume change, ΔV_d, of specimens that occurs because of the application of deviator stress in various soils is also shown in Figures 11.16d and 11.16e.

Because the pore water pressure developed during the test is completely dissipated, we have

$$\text{total and effective confining stress} = \sigma_3 = \sigma_3'$$

and

$$\text{total and effective axial stress at failure} = \sigma_3 + (\Delta\sigma_d)_f = \sigma_1 = \sigma_1'$$

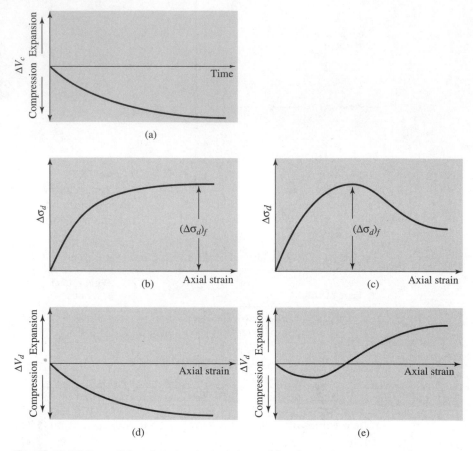

Figure 11.16 Consolidated-drained triaxial test: (a) volume change of specimen caused by chamber confining pressure; (b) plot of deviator stress against strain in the vertical direction for loose sand and normally consolidated clay; (c) plot of deviator stress against strain in the vertical direction for dense sand and overconsolidated clay; (d) volume change in loose sand and normally consolidated clay during deviator stress application; (e) volume change in dense sand and overconsolidated clay during deviator stress application

In a triaxial test, σ_1' is the major principal effective stress at failure and σ_3' is the minor principal effective stress at failure.

Several tests on similar specimens can be conducted by varying the confining pressure. With the major and minor principal stresses at failure for each test the Mohr's circles can be drawn and the failure envelopes can be obtained. Figure 11.17 shows the type of effective stress failure envelope obtained for tests on sand and normally consolidated clay. The coordinates of the point of tangency of the failure envelope with a Mohr's circle (that is, point A) give the stresses (normal and shear) on the failure plane of that test specimen.

Overconsolidation results when a clay is initially consolidated under an all-around chamber pressure of $\sigma_c \, (= \sigma_c')$ and is allowed to swell by reducing the chamber pressure to $\sigma_3 \, (= \sigma_3')$. The failure envelope obtained from drained triaxial tests of such overconsolidated clay specimens shows two distinct branches (*ab* and *bc* in

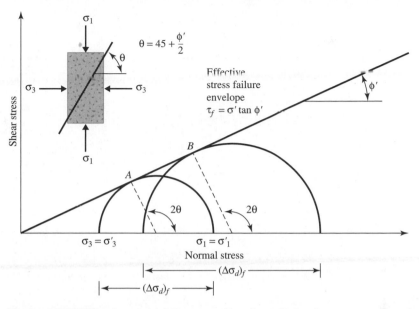

Figure 11.17 Effective stress failure envelope from drained tests on sand and normally consolidated clay

Figure 11.18). The portion ab has a flatter slope with a cohesion intercept, and the shear strength equation for this branch can be written as

$$\tau_f = c' + \sigma' \tan \phi'_1 \tag{11.15}$$

The portion bc of the failure envelope represents a normally consolidated stage of soil and follows the equation $\tau_f = \sigma' \tan \phi'$.

A consolidated-drained triaxial test on a clayey soil may take several days to complete. This amount of time is required because deviator stress must be applied very slowly to ensure full drainage from the soil specimen. For this reason, the CD type of triaxial test is uncommon.

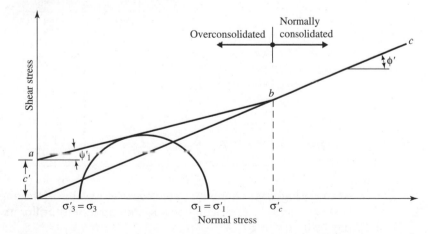

Figure 11.18 Effective stress failure envelope for overconsolidated clay

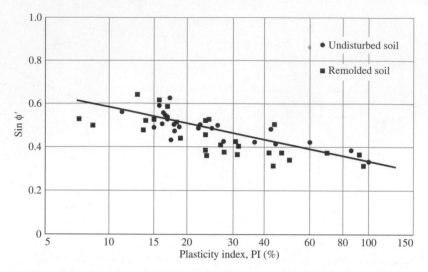

Figure 11.19 Variation of sin ϕ' with plasticity index for a number of soils (after Kenney, 1959)

Comments on Drained and Residual Friction Angles of Clay

The drained angle of friction, ϕ', generally decreases with the plasticity index of soil. This fact is illustrated in Figure 11.19 for a number of clays from data reported by Kenney (1959). Although the data are considerably scattered, the general pattern seems to hold. In Figure 11.8, the residual shear strength of clay soil is defined. Also in Example 11.1, the procedure to calculate residual friction angle ϕ'_r is shown.

Skempton (1964) provided the results of the variation of the residual angle of friction, ϕ'_r, of a number of clayey soils with the clay-size fraction (≤ 2 μm) present. The following table shows a summary of these results:

Soil	Clay-size fraction (%)	Residual friction angle, ϕ_r (deg)
Selset	17.7	29.8
Wiener Tegel	22.8	25.1
Jackfield	35.4	19.1
Oxford clay	41.9	16.3
Jari	46.5	18.6
London clay	54.9	16.3
Walton's Wood	67	13.2
Weser-Elbe	63.2	9.3
Little Belt	77.2	11.2
Biotite	100	7.5

At a very high clay content, ϕ'_r approaches the value of the angle of sliding friction for sheet minerals. For highly plastic sodium montmorillonites, the magnitude of ϕ'_r may be as low as 3 to 4°.

Example 11.2

For a normally consolidated clay, the results of a drained triaxial test are as follows:

Chamber confining pressure = 16 lb/in.^{2}

Deviator stress at failure = 25 lb/in.^{2}

a. Find the angle of friction, ϕ'.
b. Determine the angle θ that the failure plane makes with the major principal plane.

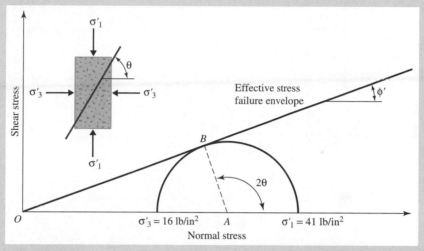

Figure 11.20 Mohr's circle and failure envelope for a normally consolidated clay

Solution

For a normally consolidated soil, the failure envelope equation is

$$\tau_f = \sigma' \tan \phi' \qquad (\text{since } c' = 0)$$

For the triaxial test, the effective major and minor principal stresses at failure are

$$\sigma_1' = \sigma_1 = \sigma_3 + (\Delta\sigma_d)_f = 16 + 25 = 41 \text{ lb/in.}^2$$

and

$$\sigma_3' = \sigma_3 = 16 \text{ lb/in.}^2$$

a. The Mohr's circle and the failure envelope are shown in Figure 11.20, from which we can write

$$\sin \phi' = \frac{AB}{OA} = \frac{\left(\dfrac{\sigma_1' - \sigma_3'}{2}\right)}{\left(\dfrac{\sigma_1' + \sigma_3'}{2}\right)}$$

or

$$\sin \phi' = \frac{\sigma_1' - \sigma_3'}{\sigma_1' + \sigma_3'} = \frac{41 - 16}{41 + 16} = 0.438$$

$$\phi' = \mathbf{26°}$$

b. $\theta = 45 + \dfrac{\phi'}{2} = 45° + \dfrac{26}{2} = \mathbf{58°}$ ∎

Example 11.3

Refer to Example 11.2.

    a. Find the normal stress σ' and the shear stress τ_f on the failure plane.
    b. Determine the effective normal stress on the plane of maximum shear stress.

Solution

    a. From Eqs. (9.8) and (9.9),

$$\sigma' \text{ (on the failure plane)} = \frac{\sigma_1' + \sigma_3'}{2} + \frac{\sigma_1' - \sigma_3'}{2} \cos 2\theta$$

and

$$\tau_f = \frac{\sigma_1' - \sigma_3'}{2} \sin 2\theta$$

Substituting the values of $\sigma_1' = 41$ lb/in.^{2}, $\sigma_3' = 16$ lb/in.^{2}, and $\theta = 58°$ into the preceding equations, we get

$$\sigma' = \frac{41 + 16}{2} + \frac{41 - 16}{2} \cos(2 \times 58) = \mathbf{23.0 \ lb/in.^2}$$

and

$$\tau_f = \frac{41 - 16}{2} \sin(2 \times 58) = \mathbf{11.2 \ lb/in.^2}$$

b. From Eq. (9.9), it can be seen that the maximum shear stress will occur on the plane with $\theta = 45°$. From Eq. (9.8),

$$\sigma' = \frac{\sigma_1' + \sigma_3'}{2} + \frac{\sigma_1' - \sigma_3'}{2} \cos 2\theta$$

Substituting $\theta = 45°$ into the preceding equation gives

$$\sigma' = \frac{41 + 16}{2} + \frac{41 - 16}{2} \cos 90 = \mathbf{28.5 \ lb/in.^2}$$ ∎

Example 11.4

The equation of the effective stress failure envelope for normally consolidated clayey soil is $\tau_f = \sigma' \tan 25°$. A drained triaxial test was conducted with the same soil at a chamber confining pressure of 80 kN/m^2. Calculate the deviator stress at failure.

Solution

For normally consolidated clay, $c' = 0$. Thus, from Eq. (11.8),

$$\sigma'_1 = \sigma'_3 \tan^2\left(45 + \frac{\phi'}{2}\right)$$

$$\phi' = 25°$$

$$\sigma'_1 = 80 \tan^2\left(45 + \frac{25}{2}\right) = 197 \text{ kN/m}^2$$

so

$$(\Delta\sigma_d)_f = \sigma'_1 - \sigma'_3 = 197 - 80 = \mathbf{117 \text{ kN/m}^2}$$ ∎

Example 11.5

The results of two drained triaxial tests on a saturated clay are as follows:

$$\text{Specimen I:} \qquad \sigma_3 = \sigma'_3 = 70 \text{ kN/m}^2$$

$$(\Delta\sigma_d)_f = 130 \text{ kN/m}^2$$

$$\text{Specimen II:} \qquad \sigma_3 = \sigma'_3 = 160 \text{ kN/m}^2$$

$$(\Delta\sigma_d)_f = 223.5 \text{ kN/m}^2$$

Determine the shear strength parameters c' and ϕ'.

Solution

For specimen I, the principal stresses at failure are

$$\sigma'_3 = \sigma_3 = 70 \text{ kN/m}^2$$

and

$$\sigma'_1 = \sigma_1 = \sigma_3 + (\Delta\sigma_d)_f = 70 + 130 = 200 \text{ kN/m}^2$$

Similarly, the principal stresses at failure for specimen II are

$$\sigma'_3 = \sigma_3 = 160 \text{ kN/m}^2$$

and

$$\sigma'_1 = \sigma_1 = \sigma_3 + (\Delta\sigma_d)_f = 160 + 223.5 = 383.5 \text{ kN/m}^2$$

Using the relationship given by Eq. (11.8), we get

$$\sigma_1' = \sigma_3' \tan^2\left(45 + \frac{\phi'}{2}\right) + 2c' \tan\left(45 + \frac{\phi'}{2}\right)$$

Thus, for specimen I,

$$200 = 70 \tan^2\left(45 + \frac{\phi'}{2}\right) + 2c' \tan\left(45 + \frac{\phi'}{2}\right)$$

and for specimen II,

$$383.5 = 160 \tan^2\left(45 + \frac{\phi'}{2}\right) + 2c' \tan\left(45 + \frac{\phi'}{2}\right)$$

Solving the two preceding equations, we obtain

$$\phi' = 20° \qquad c' = \textbf{20 kN/m}^2 \qquad \blacksquare$$

11.9   *Consolidated-Undrained Triaxial Test*

The consolidated-undrained test is the most common type of triaxial test. In this test, the saturated soil specimen is first consolidated by an all-around chamber fluid pressure, σ_3, that results in drainage (Figures 11.21a and 11.21b). After the pore water pressure generated by the application of confining pressure is dissipated, the deviator stress, $\Delta\sigma_d$, on the specimen is increased to cause shear failure (Figure 11.21c). During this phase of the test, the drainage line from the specimen is kept closed. Because drainage is not permitted, the pore water pressure, Δu_d, will increase. During the test, simultaneous measurements of $\Delta\sigma_d$ and Δu_d are made. The increase in the pore water pressure, Δu_d, can be expressed in a nondimensional form as

$$\overline{A} = \frac{\Delta u_d}{\Delta\sigma_d} \tag{11.16}$$

where $\overline{A}$ = Skempton's pore pressure parameter (Skempton, 1954).

The general patterns of variation of $\Delta\sigma_d$ and Δu_d with axial strain for sand and clay soils are shown in Figures 11.21d through 11.21g. In loose sand and normally consolidated clay, the pore water pressure increases with strain. In dense sand and overconsolidated clay, the pore water pressure increases with strain to a certain limit, beyond which it decreases and becomes negative (with respect to the atmospheric pressure). This decrease is because of a tendency of the soil to dilate.

Unlike the consolidated-drained test, the total and effective principal stresses are not the same in the consolidated-undrained test. Because the pore water pressure at failure is measured in this test, the principal stresses may be analyzed as follows:

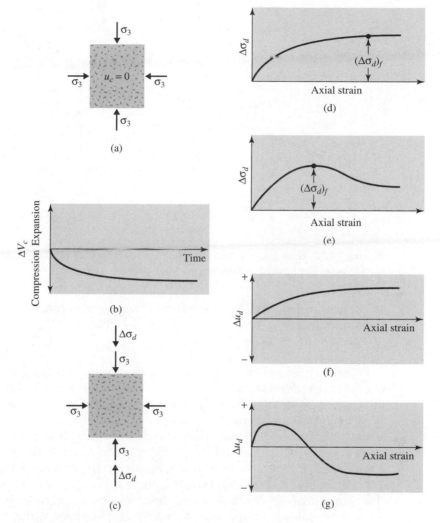

Figure 11.21 Consolidated undrained test: (a) specimen under chamber confining pressure; (b) volume change in specimen caused by confining pressure; (c) deviator stress application; (d) deviator stress against axial strain for loose sand and normally consolidated clay; (e) deviator stress against axial strain for dense sand and overconsolidated clay; (f) variation of pore water pressure with axial strain for loose sand and normally consolidated clay; (g) variation of pore water pressure with axial strain for dense sand and overconsolidated clay

— Major principal stress at failure (total): $\qquad \sigma_3 + (\Delta\sigma_d)_f = \sigma_1$

— Major principal stress at failure (effective): $\sigma_1 - (\Delta u_d)_f = \sigma_1'$

— Minor principal stress at failure (total): $\qquad \sigma_3$

— Minor principal stress at failure (effective): $\sigma_3 - (\Delta u_d)_f = \sigma_3'$

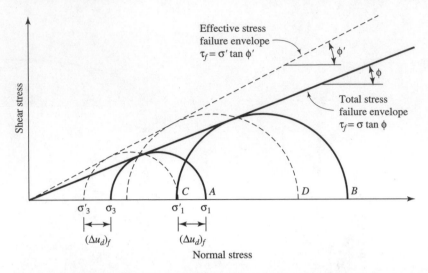

Figure 11.22 Total and effective stress failure envelopes for consolidated undrained triaxial tests. (*Note:* The figure assumes that no back pressure is applied.)

In these equations, $(\Delta u_d)_f$ = pore water pressure at failure. The preceding derivations show that

$$\sigma_1 - \sigma_3 = \sigma'_1 - \sigma'_3$$

Tests on several similar specimens with varying confining pressures may be conducted to determine the shear strength parameters. Figure 11.22 shows the total and effective stress Mohr's circles at failure obtained from consolidated-undrained triaxial tests in sand and normally consolidated clay. Note that A and B are two total stress Mohr's circles obtained from two tests. C and D are the effective stress Mohr's circles corresponding to total stress circles A and B, respectively. The diameters of circles A and C are the same; similarly, the diameters of circles B and D are the same.

In Figure 11.22, the total stress failure envelope can be obtained by drawing a line that touches all the total stress Mohr's circles. For sand and normally consolidated clays, this will be approximately a straight line passing through the origin and may be expressed by the equation

$$\tau_f = \sigma \tan \phi \qquad (11.17)$$

where σ = total stress

ϕ = the angle that the total stress failure envelope makes with the normal stress axis, also known as the *consolidated-undrained angle of shearing resistance*

Equation (11.17) is seldom used for practical considerations.

Again referring to Figure 11.22, we see that the failure envelope that is tangent to all the effective stress Mohr's circles can be represented by the equation $\tau_f = \sigma' \tan \phi'$, which is the same as that obtained from consolidated-drained tests (see Figure 11.17).

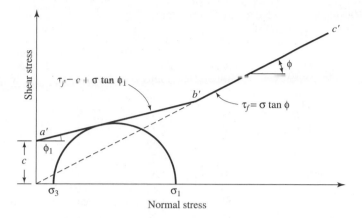

Figure 11.23 Total stress failure envelope obtained from consolidated-undrained tests in over-consolidated clay

In overconsolidated clays, the total stress failure envelope obtained from consolidated-undrained tests will take the shape shown in Figure 11.23. The straight line $a'b'$ is represented by the equation

$$\tau_f = c + \sigma \tan \phi_1 \tag{11.18}$$

and the straight line $b'c'$ follows the relationship given by Eq. (11.17). The effective stress failure envelope drawn from the effective stress Mohr's circles will be similar to that shown in Figure 11.23.

Consolidated-drained tests on clay soils take considerable time. For this reason, consolidated-undrained tests can be conducted on such soils with pore pressure measurements to obtain the drained shear strength parameters. Because drainage is not allowed in these tests during the application of deviator stress, they can be performed quickly.

Skempton's pore water pressure parameter $\overline{A}$ was defined in Eq. (11.16). At failure, the parameter $\overline{A}$ can be written as

$$\overline{A} = \overline{A}_f = \frac{(\Delta u_d)_f}{(\Delta \sigma_d)_f} \tag{11.19}$$

The general range of $\overline{A}_f$ values in most clay soils is as follows:

- Normally consolidated clays: 0.5 to 1
- Overconsolidated clays: −0.5 to 0

Table 11.3 gives the values of $\overline{A}_f$ for some normally consolidated clays as obtained by the Norwegian Geotechnical Institute.

Laboratory triaxial tests of Bjerrum and Simons (1960) on Oslo clay, Weald clay, and London clay showed that $\overline{A}_f$ becomes approximately zero at an overconsolidation value of about 3 or 4.

Table 11.3 Triaxial Test Results for Some Normally Consolidated Clays
Obtained by the Norwegian Geotechnical Institute*

Location	Liquid limit	Plastic limit	Liquidity index	Sensitivity[a]	Drained friction angle, ϕ' (deg)	$\overline{A}_f$
Seven Sisters, Canada	127	35	0.28		19	0.72
Sarpborg	69	28	0.68	5	25.5	1.03
Lilla Edet, Sweden	68	30	1.32	50	26	1.10
Fredrikstad	59	22	0.58	5	28.5	0.87
Fredrikstad	57	22	0.63	6	27	1.00
Lilla Edet, Sweden	63	30	1.58	50	23	1.02
Göta River, Sweden	60	27	1.30	12	28.5	1.05
Göta River, Sweden	60	30	1.50	40	24	1.05
Oslo	48	25	0.87	4	31.5	1.00
Trondheim	36	20	0.50	2	34	0.75
Drammen	33	18	1.08	8	28	1.18

*After Bjerrum and Simons (1960)
[a] See Section 11.15 for the definition of sensitivity.

Example 11.6

A consolidated-undrained test on a normally consolidated clay yielded the following results:

$\sigma_3 = 12$ lb/in.^{2}

Deviator stress, $(\Delta\sigma_d)_f = 9.1$ lb/in.^{2}

Pore pressure, $(\Delta u_d)_f = 6.8$ lb/in.^{2}

Calculate the consolidated-undrained friction angle and the consolidated-drained friction angle.

Solution

$$\sigma_3 = 12 \text{ lb/in.}^2$$

$$\sigma_1 = \sigma_3 + (\Delta\sigma_d)_f = 12 + 9.1 = 21.1 \text{ lb/in.}^2$$

From Eq. (11.9), for normally consolidated clay with $c = 0$,

$$\sigma_1 = \sigma_3 \tan^2\left(45 + \frac{\phi}{2}\right)$$

$$21.1 = 12 \tan^2\left(45 + \frac{\phi}{2}\right)$$

$$\phi = 2\left[\tan^{-1}\left(\frac{21.1}{12}\right)^{0.5} - 45\right] = \mathbf{16°}$$

Again,

$$\sigma_3' = \sigma_3 - (\Delta u_d)_f = 12 - 6.8 = 5.2 \text{ lb/in.}^2$$

$$\sigma_1' = \sigma_1 - (\Delta u_d)_f = 21.1 - 6.8 = 14.3 \text{ lb/in.}^2$$

From Eq. (11.8), for normally consolidated clay with $c' = 0$,

$$\sigma_1' = \sigma_3' \tan^2\left(45 + \frac{\phi'}{2}\right)$$

$$14.3 = 5.2 \tan^2\left(45 + \frac{\phi'}{2}\right)$$

$$\phi' = 2\left[\tan^{-1}\left(\frac{14.3}{5.2}\right)^{0.5} - 45\right] = \mathbf{27.8°}$$

∎

11.10 *Unconsolidated-Undrained Triaxial Test*

In unconsolidated-undrained tests, drainage from the soil specimen is not permitted during the application of chamber pressure σ_3. The test specimen is sheared to failure by the application of deviator stress, $\Delta\sigma_d$, and drainage is prevented. Because drainage is not allowed at any stage, the test can be performed quickly. Because of the application of chamber confining pressure σ_3, the pore water pressure in the soil specimen will increase by u_c. A further increase in the pore water pressure (Δu_d) will occur because of the deviator stress application. Hence, the total pore water pressure u in the specimen at any stage of deviator stress application can be given as

$$u = u_c + \Delta u_d \tag{11.20}$$

From Eqs. (11.14) and (11.16), $u_c = B\sigma_3$ and $\Delta u_d = \overline{A}\Delta\sigma_d$, so

$$u = B\sigma_3 + \overline{A}\Delta\sigma_d = B\sigma_3 + \overline{A}(\sigma_1 - \sigma_3) \tag{11.21}$$

This test is usually conducted on clay specimens and depends on a very important strength concept for cohesive soils if the soil is fully saturated. The added axial stress at failure $(\Delta\sigma_d)_f$ is practically the same regardless of the chamber confining pressure. This property is shown in Figure 11.24. The failure envelope for the total stress Mohr's circles becomes a horizontal line and hence is called a $\phi = 0$ condition. From Eq. (11.9) with $\phi = 0$, we get

$$\tau_f = c = c_u \tag{11.22}$$

where c_u is the undrained shear strength and is equal to the radius of the Mohr's circles. Note that the $\phi = 0$ concept is applicable to only saturated clays and silts.

The reason for obtaining the same added axial stress $(\Delta\sigma_d)_f$ regardless of the confining pressure can be explained as follows. If a clay specimen (no. I) is consoli-

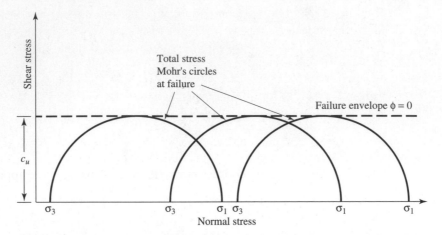

Figure 11.24 Total stress Mohr's circles and failure envelope ($\phi = 0$) obtained from unconsolidated-undrained triaxial tests on fully saturated cohesive soil

dated at a chamber pressure σ_3 and then sheared to failure without drainage, the total stress conditions at failure can be represented by the Mohr's circle P in Figure 11.25. The pore pressure developed in the specimen at failure is equal to $(\Delta u_d)_f$. Thus, the major and minor principal effective stresses at failure are, respectively,

$$\sigma_1' = [\sigma_3 + (\Delta\sigma_d)_f] - (\Delta u_d)_f = \sigma_1 - (\Delta u_d)_f$$

and

$$\sigma_3' = \sigma_3 - (\Delta u_d)_f$$

Q is the effective stress Mohr's circle drawn with the preceding principal stresses. Note that the diameters of circles P and Q are the same.

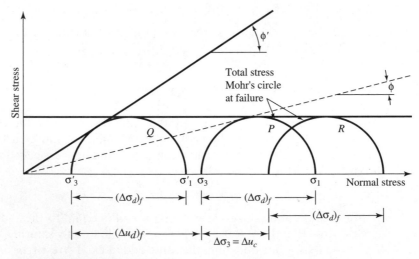

Figure 11.25 The $\phi = 0$ concept

Now let us consider another similar clay specimen (no. II) that has been consolidated under a chamber pressure σ_3 with initial pore pressure equal to zero. If the chamber pressure is increased by $\Delta\sigma_3$ without drainage, the pore water pressure will increase by an amount Δu_c. For saturated soils under isotropic stresses, the pore water pressure increase is equal to the total stress increase, so $\Delta u_c = \Delta\sigma_3$ ($B = 1$). At this time, the effective confining pressure is equal to $\sigma_3 + \Delta\sigma_3 - \Delta u_c = \sigma_3 + \Delta\sigma_3 - \Delta\sigma_3 = \sigma_3$. This is the same as the effective confining pressure of specimen no. I before the application of deviator stress. Hence, if specimen no. II is sheared to failure by increasing the axial stress, it should fail at the same deviator stress $(\Delta\sigma_d)_f$ that was obtained for specimen no. I. The total stress Mohr's circle at failure will be R (see Figure 11.25). The added pore pressure increase caused by the application of $(\Delta\sigma_d)_f$ will be $(\Delta u_d)_f$.

At failure, the minor principal effective stress is

$$[(\sigma_3 + \Delta\sigma_3)] - [\Delta u_c + (\Delta u_d)_f] = \sigma_3 - (\Delta u_d)_f = \sigma_3'$$

and the major principal effective stress is

$$[\sigma_3 + \Delta\sigma_3 + (\Delta\sigma_d)_f] - [\Delta u_c + (\Delta u_d)_f] = [\sigma_3 + (\Delta\sigma_d)_f] - (\Delta u_d)_f$$
$$= \sigma_1 - (\Delta u_d)_f = \sigma_1'$$

Thus, the effective stress Mohr's circle will still be Q because strength is a function of effective stress. Note that the diameters of circles P, Q, and R are all the same.

Any value of $\Delta\sigma_3$ could have been chosen for testing specimen no. II. In any case, the deviator stress $(\Delta\sigma_d)_f$ to cause failure would have been the same as long as the soil was fully saturated and fully undrained during both stages of the test.

11.11 *Unconfined Compression Test on Saturated Clay*

The unconfined compression test is a special type of unconsolidated-undrained test that is commonly used for clay specimens. In this test, the confining pressure σ_3 is 0. An axial load is rapidly applied to the specimen to cause failure. At failure, the total minor principal stress is zero and the total major principal stress is σ_1 (Figure 11.26).

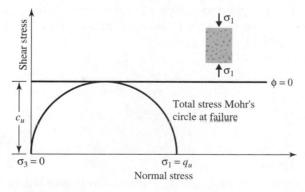

Figure 11.26 Unconfined compression test

Table 11.4 General Relationship of Consistency and Unconfined Compression Strength of Clays

Consistency	q_u	
	kN/m²	ton/ft²
Very soft	0–25	0–0.25
Soft	25–50	0.25–0.5
Medium	50–100	0.5–1
Stiff	100–200	1–2
Very stiff	200–400	2–4
Hard	>400	>4

Because the undrained shear strength is independent of the confining pressure as long as the soil is fully saturated and fully undrained, we have

$$\tau_f = \frac{\sigma_1}{2} = \frac{q_u}{2} = c_u \qquad (11.23)$$

where q_u is the *unconfined compression strength*. Table 11.4 gives the approximate consistencies of clays on the basis of their unconfined compression strength. A photograph of unconfined compression test equipment is shown in Figure 11.27.

Theoretically, for similar saturated clay specimens, the unconfined compression tests and the unconsolidated-undrained triaxial tests should yield the same values of c_u. In practice, however, unconfined compression tests on saturated clays yield slightly lower values of c_u than those obtained from unconsolidated-undrained tests.

11.12   *Stress Path*

Results of triaxial tests can be represented by diagrams called *stress paths*. A stress path is a line that connects a series of points, each of which represents a successive stress state experienced by a soil specimen during the progress of a test. There are several ways in which a stress path can be drawn. This section covers one of them.

Lambe (1964) suggested a type of stress path representation that plots q' against p' (where p' and q' are the coordinates of the top of the Mohr's circle). Thus, relationships for p' and q' are as follows:

$$p' = \frac{\sigma_1' + \sigma_3'}{2} \qquad (11.24)$$

$$q' = \frac{\sigma_1' - \sigma_3'}{2} \qquad (11.25)$$

This type of stress path plot can be explained with the aid of Figure 11.28. Let us consider a normally consolidated clay specimen subjected to an isotropically

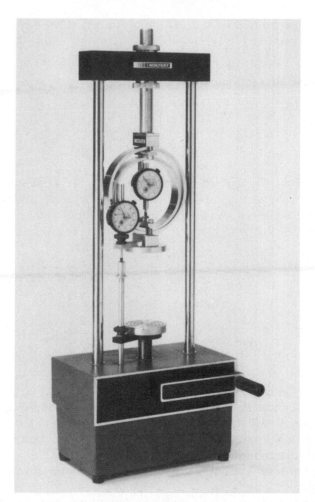

Figure 11.27
Unconfined compression test
equipment (courtesy of Soiltest,
Inc., Lake Bluff, Illinois)

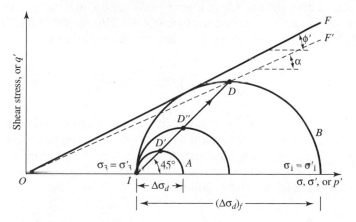

Figure 11.28 Stress path — plot of q' against p' for a consolidated-drained triaxial test on a normally consolidated clay

consolidated-drained triaxial test. At the beginning of the application of deviator stress, $\sigma_1' = \sigma_3' = \sigma_3$, so

$$p' = \frac{\sigma_3' + \sigma_3'}{2} = \sigma_3' = \sigma_3 \qquad (11.26)$$

and

$$q' = \frac{\sigma_3' - \sigma_3'}{2} = 0 \qquad (11.27)$$

For this condition, p' and q' will plot as a point (that is, I in Figure 11.28). At some other time during deviator stress application, $\sigma_1' = \sigma_3' + \Delta\sigma_d = \sigma_3 + \Delta\sigma_d; \sigma_3' = \sigma_3$. The Mohr's circle marked A in Figure 11.28 corresponds to this state of stress on the soil specimen. The values of p' and q' for this stress condition are

$$p' = \frac{\sigma_1' + \sigma_3'}{2} = \frac{(\sigma_3' + \Delta\sigma_d) + \sigma_3'}{2} = \sigma_3' + \frac{\Delta\sigma_d}{2} = \sigma_3 + \frac{\Delta\sigma_d}{2} \qquad (11.28)$$

and

$$q' = \frac{(\sigma_3' + \Delta\sigma_d) - \sigma_3'}{2} = \frac{\Delta\sigma_d}{2} \qquad (11.29)$$

If these values of p' and q' were plotted in Figure 11.28, they would be represented by point D' at the top of the Mohr's circle. So, if the values of p' and q' at various stages of the deviator stress application are plotted and these points are joined, a straight line like ID will result. The straight line ID is referred to as the *stress path* in a q'-p' plot for a consolidated-drained triaxial test. Note that the line ID makes an angle of 45° with the horizontal. Point D represents the failure condition of the soil specimen in the test. Also, we can see that Mohr's circle B represents the failure stress condition.

For normally consolidated clays, the failure envelope can be given by $\tau_f = \sigma' \tan \phi'$. This is the line OF in Figure 11.28. (See also Figure 11.17.) A modified failure envelope can now be defined by line OF'. This modified line is commonly called the K_f line. The equation of the K_f line can be expressed as

$$q' = p' \tan \alpha \qquad (11.30)$$

where α = the angle that the modified failure envelope makes with the horizontal.

The relationship between the angles ϕ' and α can be determined by referring to Figure 11.29, in which, for clarity, the Mohr's circle at failure (that is, circle B) and lines OF and OF' as shown in Figure 11.28 have been redrawn. Note that O' is the center of the Mohr's circle at failure. Now,

$$\frac{DO'}{OO'} = \tan \alpha$$

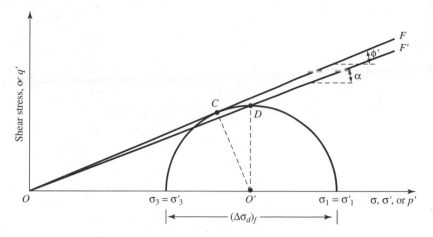

Figure 11.29 Relationship between ϕ' and α

and thus, we obtain

$$\tan \alpha = \frac{\dfrac{\sigma_1' - \sigma_3'}{2}}{\dfrac{\sigma_1' + \sigma_3'}{2}} = \frac{\sigma_1' - \sigma_3'}{\sigma_1' + \sigma_3'} \tag{11.31}$$

Again,

$$\frac{CO'}{OO'} = \sin \phi'$$

or

$$\sin \phi' = \frac{\dfrac{\sigma_1' - \sigma_3'}{2}}{\dfrac{\sigma_1' + \sigma_3'}{2}} = \frac{\sigma_1' - \sigma_3'}{\sigma_1' + \sigma_3'} \tag{11.32}$$

Comparing Eqs. (11.31) and (11.32), we see that

$$\sin \phi' = \tan \alpha \tag{11.33}$$

or

$$\phi' = \sin^{-1}(\tan \alpha) \tag{11.34}$$

Figure 11.30 shows a q'-p' plot for a normally consolidated clay specimen subjected to an isotropically consolidated-undrained triaxial test. At the beginning of the

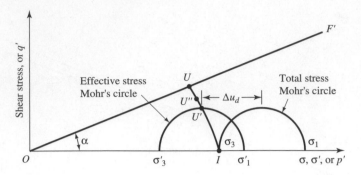

Figure 11.30 Stress path — plot of q' against p' for a consolidated-undrained triaxial test on a normally consolidated clay

application of deviator stress, $\sigma_1' = \sigma_3' = \sigma_3$. Hence, $p' = \sigma_3'$ and $q' = 0$. This relationship is represented by point I. At some other stage of the deviator stress application,

$$\sigma_1' = \sigma_3 + \Delta\sigma_d - \Delta u_d$$

and

$$\sigma_3' = \sigma_3 - \Delta u_d$$

So

$$p' = \frac{\sigma_1' + \sigma_3'}{2} = \sigma_3 + \frac{\Delta\sigma_d}{2} - \Delta u_d \tag{11.35}$$

and

$$q' = \frac{\sigma_1' - \sigma_3'}{2} = \frac{\Delta\sigma_d}{2} \tag{11.36}$$

The preceding values of p' and q' will plot as point U' in Figure 11.30. Points such as U'' represent values of p' and q' as the test progresses. At failure of the soil specimen,

$$p' = \sigma_3 + \frac{(\Delta\sigma_d)_f}{2} - (\Delta u_d)_f \tag{11.37}$$

and

$$q' = \frac{(\Delta\sigma_d)_f}{2} \tag{11.38}$$

The values of p' and q' given by Eqs. (11.37) and (11.38) will plot as point U. Hence, the effective stress path for a consolidated-undrained test can be given by the curve $IU'U$. Note that point U will fall on the modified failure envelope, OF' (see Figure 11.29), which is inclined at an angle α to the horizontal. Lambe (1964) proposed a technique to evaluate the elastic and consolidation settlements of foundations on clay soils by using the stress paths determined in this manner.

Example 11.7

For a normally consolidated clay, the failure envelope is given by the equation $\tau_f = \sigma' \tan \phi'$. The corresponding modified failure envelope (q'-p' plot) is given by Eq. (11.30) as $q' = p' \tan \alpha$. In a similar manner, if the failure envelope is $\tau_f = c' + \sigma' \tan \phi'$, the corresponding modified failure envelope is a q'-p' plot that can be expressed as $q' = m + p' \tan \alpha$. Express α as a function of ϕ', and give m as a function of c' and ϕ'.

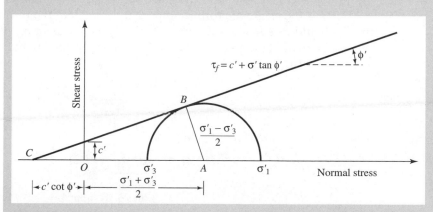

Figure 11.31 Derivation of α as a function of ϕ' and m as a function of c' and ϕ'

Solution
From Figure 11.31,

$$\sin \phi' = \frac{AB}{AC} = \frac{AB}{CO + OA} = \frac{\left(\dfrac{\sigma'_1 - \sigma'_3}{2}\right)}{c' \cot \phi' + \left(\dfrac{\sigma'_1 + \sigma'_3}{2}\right)}$$

so

$$\frac{\sigma'_1 - \sigma'_3}{2} = c' \cos \phi' + \left(\frac{\sigma'_1 + \sigma'_3}{2}\right) \sin \phi' \qquad \text{(a)}$$

or

$$q' = m + p' \tan \alpha \qquad \text{(b)}$$

Comparing Eqs. (a) and (b), we find that

$$m = c' \cos \phi'$$

and

$$\tan \alpha = \sin \phi'$$

or

$$\alpha = \tan^{-1}(\sin \phi') \qquad \blacksquare$$

11.13   *Vane Shear Test*

Fairly reliable results for the undrained shear strength, c_u ($\phi = 0$ concept), of very soft to medium cohesive soils may be obtained directly from vane shear tests. The shear vane usually consists of four thin, equal-sized steel plates welded to a steel torque rod (Figure 11.32). First, the vane is pushed into the soil. Then torque is applied at the top of the torque rod to rotate the vane at a uniform speed. A cylinder of soil of height h and diameter d will resist the torque until the soil fails. The undrained shear strength of the soil can be calculated as follows.

If T is the maximum torque applied at the head of the torque rod to cause failure, it should be equal to the sum of the resisting moment of the shear force along the side surface of the soil cylinder (M_s) and the resisting moment of the shear force at each end (M_e) (Figure 11.33):

$$T = M_s + \underbrace{M_e + M_e}_{\text{Two ends}} \tag{11.39}$$

The resisting moment can be given as

$$M_s = \underbrace{(\pi dh)c_u}_{\substack{\text{Surface} \\ \text{area}}} \; \underbrace{(d/2)}_{\substack{\text{Moment} \\ \text{arm}}} \tag{11.40}$$

where d = diameter of the shear vane
$\quad\quad h$ = height of the shear vane

For the calculation of M_e, investigators have assumed several types of distribution of shear strength mobilization at the ends of the soil cylinder:

1. *Triangular.* Shear strength mobilization is c_u at the periphery of the soil cylinder and decreases linearly to zero at the center.
2. *Uniform.* Shear strength mobilization is constant (that is, c_u) from the periphery to the center of the soil cylinder.
3. *Parabolic.* Shear strength mobilization is c_u at the periphery of the soil cylinder and decreases parabolically to zero at the center.

These variations in shear strength mobilization are shown in Figure 11.33b. In general, the torque, T, at failure can be expressed as

$$T = \pi c_u \left[\frac{d^2 h}{2} + \beta \frac{d^3}{4} \right] \tag{11.41}$$

or

$$c_u = \frac{T}{\pi \left[\dfrac{d^2 h}{2} + \beta \dfrac{d^3}{4} \right]} \tag{11.42}$$

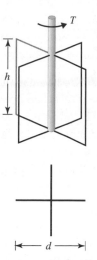

Figure 11.32
Diagram of vane shear
test equipment

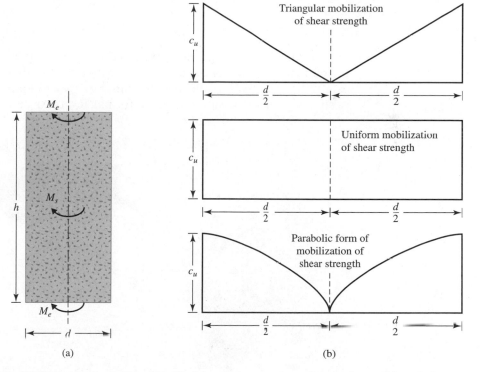

Figure 11.33 Derivation of Eq. (11.42): (a) resisting moment of shear force; (b) variations
in shear strength mobilization

where $\beta = \frac{1}{2}$ for triangular mobilization of undrained shear strength

$\beta = \frac{2}{3}$ for uniform mobilization of undrained shear strength

$\beta = \frac{3}{5}$ for parabolic mobilization of undrained shear strength

Note that Eq. (11.42) is usually referred to as *Calding's equation.*

Vane shear tests can be conducted in the laboratory and in the field during soil exploration. The laboratory shear vane has dimensions of about 13 mm ($\frac{1}{2}$ in.) in diameter and 25 mm (1 in.) in height. Figure 11.34 shows a photograph of laboratory vane shear test equipment. Figure 11.35 shows the field vanes recommended by ASTM (1994). Table 11.5 gives the ASTM recommended dimensions of field vanes.

According to ASTM (1994), if $h/d = 2$, then

$$c_u \, (\text{kN/m}^2) = \frac{T \, (\text{N} \cdot \text{m})}{(366 \times 10^{-8}) d^3} \atop \underset{(\text{cm})}{\uparrow} \qquad (11.43)$$

and

$$c_u \, (\text{lb/ft}^2) = \frac{T \, (\text{lb} \cdot \text{ft})}{0.0021 d^3} \atop \underset{(\text{in.})}{\uparrow} \qquad (11.44)$$

In the field, where considerable variation in the undrained shear strength can be found with depth, vane shear tests are extremely useful. In a short period, one can establish a reasonable pattern of the change of c_u with depth. However, if the clay deposit at a given site is more or less uniform, a few unconsolidated-undrained triaxial

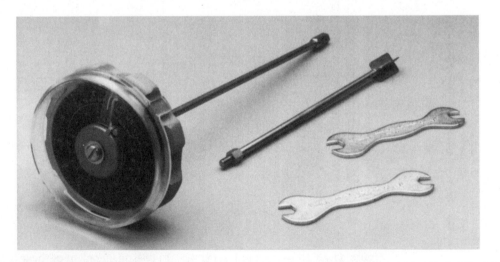

Figure 11.34 Laboratory vane shear test device (courtesy of Soiltest, Inc., Lake Bluff, Illinois)

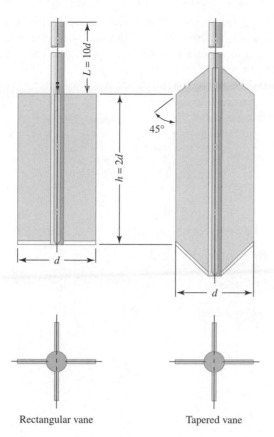

Rectangular vane   Tapered vane

Figure 11.35 Geometry of field vanes [*Source:* From *Annual Book of ASTM Standards,* 04.08, p. 346. Copyright © 1994 American Society for Testing and Materials. Reprinted with permission]

Table 11.5 Recommended Dimensions of Field Vanes*[a]

Casing size	Diameter, mm (in.)	Height, mm (In.)	Thickness of blade, mm (in.)	Diameter of rod mm (in.)
AX	38.1 ($1\frac{1}{2}$)	76.2 (3)	1.6 ($\frac{1}{16}$)	12.7 ($\frac{1}{2}$)
BX	50.8 (2)	101.6 (4)	1.6 ($\frac{1}{16}$)	12.7 ($\frac{1}{2}$)
NX	63.5 ($2\frac{1}{2}$)	127.0 (5)	3.2 ($\frac{1}{8}$)	12.7 ($\frac{1}{2}$)
101.6 mm (4 in.)[b]	92.1 ($3\frac{5}{8}$)	184.1 ($7\frac{1}{4}$)	3.2 ($\frac{1}{8}$)	12.7 ($\frac{1}{2}$)

*After ASTM, 1994
[a]Selection of vane size is directly related to the consistency of the soil being tested; that is, the softer the soil, the larger the vane diameter should be.
[b]Inside diameter

tests on undisturbed specimens will allow a reasonable estimation of soil parameters for design work. Vane shear tests are also limited by the strength of soils in which they can be used. The undrained shear strength obtained from a vane shear test also depends on the rate of application of torque T.

Bjerrum (1974) showed that as the plasticity of soils increases, c_u obtained from vane shear tests may give results that are unsafe for foundation design. For this reason, he suggested the correction

$$c_{u(\text{design})} = \lambda c_{u(\text{vane shear})} \tag{11.45}$$

where

$$\lambda = \text{correction factor} = 1.7 - 0.54 \log(PI) \tag{11.46}$$

$$PI = \text{plasticity index}$$

More recently, Morris and Williams (1994) gave the correlations of λ as

$$\lambda = 1.18e^{-0.08(PI)} + 0.57 \qquad (\text{for } PI > 5) \tag{11.47}$$

and

$$\lambda = 7.01e^{-0.08(LL)} + 0.57 \qquad (\text{for } LL > 20) \tag{11.48}$$

where LL = liquid limit (%).

11.14 Other Methods for Determining Undrained Shear Strength

A modified form of the vane shear test apparatus is the *Torvane* (Figure 11.36), which is a handheld device with a calibrated spring. This instrument can be used for determining c_u for tube specimens collected from the field during soil exploration, and it can be used in the field. The Torvane is pushed into the soil and then rotated until the soil fails. The undrained shear strength can be read at the top of the calibrated dial.

Figure 11.37 shows a *pocket penetrometer,* which is pushed directly into the soil. The unconfined compression strength (q_u) is measured by a calibrated spring. This device can be used both in the laboratory and in the field.

11.15 Sensitivity and Thixotropy of Clay

For many naturally deposited clay soils, the unconfined compression strength is greatly reduced when the soils are tested after remolding without any change in

Figure 11.36
Torvane (courtesy of
Soiltest, Inc., Lake Bluff,
Illinois)

Figure 11.37
Pocket penetrometer
(courtesy of Soiltest, Inc.,
Lake Bluff, Illinois)

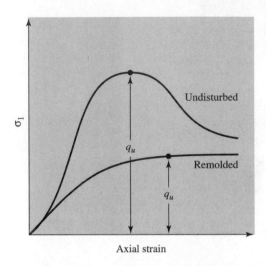

Figure 11.38 Unconfined compression strength for undisturbed and remolded clay

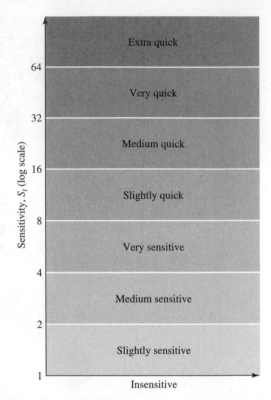

Figure 11.39 Classification of clays based on sensitivity

the moisture content, as shown in Figure 11.38. This property of clay soils is called *sensitivity*. The degree of sensitivity may be defined as the ratio of the unconfined compression strength in an undisturbed state to that in a remolded state, or

$$S_t = \frac{q_{u(\text{undisturbed})}}{q_{u(\text{remolded})}} \qquad (11.49)$$

The sensitivity ratio of most clays ranges from about 1 to 8; however, highly flocculent marine clay deposits may have sensitivity ratios ranging from about 10 to 80. Some clays turn to viscous fluids upon remolding. These clays are found mostly in the previously glaciated areas of North America and Scandinavia. Such clays are referred to as *quick* clays. Rosenqvist (1953) classified clays on the basis of their sensitivity. This general classification is shown in Figure 11.39.

The loss of strength of clay soils from remolding is caused primarily by the destruction of the clay particle structure that was developed during the original process of sedimentation.

If, however, after remolding, a soil specimen is kept in an undisturbed state (that is, without any change in the moisture content), it will continue to gain strength with time. This phenomenon is referred to as *thixotropy.* Thixotropy is a time-dependent, reversible process in which materials under constant composition and volume soften when remolded. This loss of strength is gradually regained with time when the materials are allowed to rest. This phenomenon is illustrated in Figure 11.40a.

Most soils, however, are partially thixotropic — that is, part of the strength loss caused by remolding is never regained with time. The nature of the strength-time variation for partially thixotropic materials is shown in Figure 11.40b. For soils, the difference between the undisturbed strength and the strength after thixotropic hardening can be attributed to the destruction of the clay-particle structure that was developed during the original process of sedimentation.

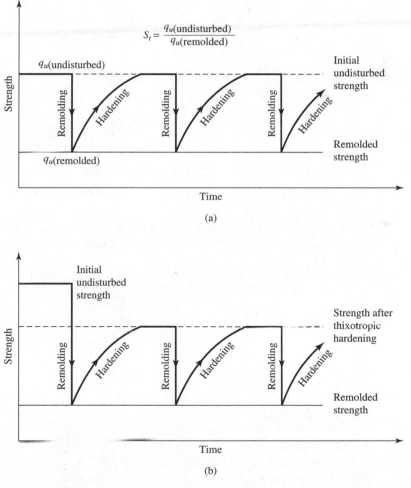

Figure 11.40 Behavior of (a) thixotropic material; (b) partially thixotropic material

Table 11.6 Empirical Equations Related to c_u and σ'_O

Reference	Relationship	Remarks
Skempton (1957)	$\dfrac{c_{u(\text{VST})}}{\sigma'_O} = 0.11 + 0.0037(PI)$ PI = plasticity index (%) $c_{u(\text{VST})}$ = undrained shear strength from vane shear test	For normally consolidated clay
Chandler (1988)	$\dfrac{c_{u(\text{VST})}}{\sigma'_c} = 0.11 + 0.0037(PI)$ σ'_c = preconsolidation pressure	Can be used in overconsolidated soil; accuracy $\pm 25\%$; not valid for sensitive and fissured clays
Jamiolkowski et al. (1985)	$\dfrac{c_u}{\sigma'_c} = 0.23 \pm 0.04$	For lightly overconsolidated clays
Mesri (1989)	$\dfrac{c_u}{\sigma'_O} = 0.22$	
Ladd et al. (1977)	$\dfrac{\left(\dfrac{c_u}{\sigma'_O}\right)_{\text{overconsolidated}}}{\left(\dfrac{c_u}{\sigma'_O}\right)_{\text{normally consolidated}}} = (OCR)^{0.8}$ OCR = overconsolidation ratio	

11.16 Empirical Relationships between Undrained Cohesion (c_u) and Effective Overburden Pressure (σ'_O)

Several empirical relationships can be observed between c_u and the effective overburden pressure (σ'_O) in the field. Some of these relationships are summarized in Table 11.6.

The overconsolidation ratio was defined in Chapter 10 as

$$OCR = \frac{\sigma'_c}{\sigma'_O} \tag{11.50}$$

where σ'_c = preconsolidation pressure.

Example 10.8

A soil profile is shown in Figure 11.41. The clay is normally consolidated. Its liquid limit is 60 and its plastic limit is 25. Estimate the unconfined compression strength of the clay at a depth of 10 m measured from the ground surface. Use Skempton's relationship from Table 11.6 and Eqs. (11.45) and (11.46).

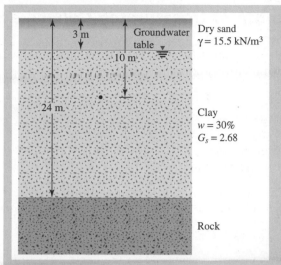

Figure 11.41

Solution

For the saturated clay layer, the void ratio is

$$e = wG_s = (2.68)(0.3) = 0.8$$

The effective unit weight is

$$\gamma'_{clay} = \left(\frac{G_s - 1}{1 + e}\right)\gamma_w = \frac{(2.68 - 1)(9.81)}{1 + 0.8} = 9.16 \text{ kN/m}^3$$

The effective stress at a depth of 10 m from the ground surface is

$$\sigma'_O = 3\gamma_{sand} + 7\gamma'_{clay} = (3)(15.5) + (7)(9.16)$$
$$= 110.62 \text{ kN/m}^2$$

From Table 11.6,

$$\frac{c_{u(VST)}}{\sigma'_O} = 0.11 + 0.0037(PI)$$

$$\frac{c_{u(VST)}}{110.62} = 0.11 + 0.0037(60 - 25)$$

and

$$c_{u(VST)} = 26.49 \text{ kN/m}^2$$

From Eqs. (11.45) and (11.46), we get

$$c_u = \lambda c_{u(VST)}$$
$$= [1.7 - 0.54 \log(PI)]c_{u(VST)}$$
$$= [1.7 - 0.54 \log(60 - 25)]26.49 = 22.95 \text{ kN/m}^2$$

So the unconfined compression strength is

$$q_u = 2c_u = (2)(22.95) = \textbf{45.9 kN/m}^2$$

■

11.17   *Shear Strength of Unsaturated Cohesive Soils*

The equation relating total stress, effective stress, and pore water pressure for unsaturated soils, can be expressed as

$$\sigma' = \sigma - u_a + \chi(u_a - u_w) \tag{11.51}$$

where σ' = effective stress
  σ = total stress
  u_a = pore air pressure
  u_w = pore water pressure

When the expression for σ' is substituted into the shear strength equation [Eq. (11.3)], which is based on effective stress parameters, we get

$$\tau_f = c' + [\sigma - u_a + \chi(u_a - u_w)]\tan \phi' \tag{11.52}$$

The values of χ depend primarily on the degree of saturation. With ordinary triaxial equipment used for laboratory testing, it is not possible to determine accurately the effective stresses in unsaturated soil specimens, so the common practice is to conduct undrained triaxial tests on unsaturated specimens and measure only the total stress. Figure 11.42 shows a total stress failure envelope obtained from a number of undrained triaxial tests conducted with a given initial degree of saturation. The failure envelope is generally curved. Higher confining pressure causes higher compression of the air in void spaces; thus, the solubility of void air in void water is increased. For design purposes, the curved envelope is sometimes approximated as a straight line, as shown in Figure 11.42, with an equation as follows:

$$\tau_f = c + \sigma \tan \phi \tag{11.53}$$

(Note that c and ϕ in the preceding equation are empirical constants.)

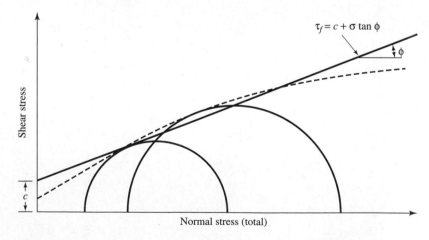

Figure 11.42 Total stress failure envelope for unsaturated cohesive soils

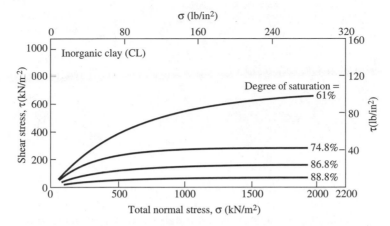

Figure 11.43 Variation of the total stress failure envelope with change of initial degree of saturation obtained from undrained tests of an inorganic clay (after Casagrande and Hirschfeld, 1960)

Figure 11.43 shows the variation of the total stress envelopes with change of the initial degree of saturation obtained from undrained tests on an inorganic clay. Note that for these tests the specimens were prepared with approximately the same initial dry unit weight of about $16.7 \, kN/m^3$ ($106 \, lb/m^3$). For a given total normal stress, the shear stress needed to cause failure decreases as the degree of saturation increases. When the degree of saturation reaches 100%, the total stress failure envelope becomes a horizontal line that is the same as with the $\phi = 0$ concept.

In practical cases where a cohesive soil deposit may become saturated because of rainfall or a rise in the groundwater table, the strength of partially saturated clay should not be used for design considerations. Instead, the unsaturated soil specimens collected from the field must be saturated in the laboratory and the undrained strength determined.

11.18 Summary and General Comments

In this chapter, the shear strengths of grandular and cohesive soils were examined. Laboratory procedures for determining the shear strength parameters were described.

In textbooks, determination of the shear strength parameters of cohesive soils appears to be fairly simple. However, in practice, the proper choice of these parameters for design and stability checks of various earth, earth-retaining, and earth-supported structures is very difficult and requires experience and an appropriate theoretical background in geotechnical engineering. In this chapter, three types of strength parameters (*consolidated-drained, consolidated-undrained, and unconsolidated-undrained*) were introduced. Their use depends on drainage conditions.

Consolidated-drained strength parameters can be used to determine the long-term stability of structures such as earth embankments and cut slopes. Consolidated-undrained shear strength parameters can be used to study stability problems relating to cases where the soil initially is fully consolidated and then there is rapid loading.

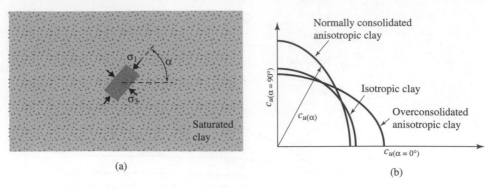

Figure 11.44 Strength anisotropy in clay

An excellent example of this is the stability of slopes of earth dams after rapid draw-down. The unconsolidated-undrained shear strength of clays can be used to evaluate the end-of-construction stability of saturated cohesive soils with the assumption that the load caused by construction has been applied rapidly and there has been little time for drainage to take place. The bearing capacity of foundations on soft saturated clays and the stability of the base of embankments on soft clays are examples of this condition.

The unconsolidated-undrained shear strength of some saturated clays can vary depending on the direction of load application; this is referred to as *anisotropy with respect to strength.* Anisotropy is primarily caused by the nature of the deposition of the cohesive soils, and subsequent consolidation makes the clay particles orient per-pendicular to the direction of the major principal stress. Parallel orientation of the clay particles can cause the strength of clay to vary with direction. Figure 11.44a shows an element of saturated clay in a deposit with the major principal stress mak-ing an angle α with respect to the horizontal. For anisotropic clays, the magnitude of c_u will be a function of α. For normally consolidated clays, $c_{u(\alpha=90°)} > c_{u(\alpha=0°)}$; for over-consolidated clays, $c_{u(\alpha=90°)} < c_{u(\alpha=0°)}$. Figure 11.44b shows the directional variation for $c_{u(\alpha)}$. The anisotropy with respect to strength for clays can have an important ef-fect on the load-bearing capacity of foundations and the stability of earth embank-ments because the direction of the major principal stress along the potential failure surfaces changes.

The sensitivity of clays was discussed in Section 11.15. It is imperative that sen-sitive clay deposits are properly identified. For instance, when machine foundations (which are subjected to vibratory loading) are constructed over sensitive clays, the clay may substantially lose its load-bearing capacity, and failure may occur.

Problems

11.1   A direct shear test was conducted on a specimen of dry sand with a normal stress of 200 kN/m². Failure occurred at a shear stress of 175 kN/m². The size of the specimen tested was 75 mm × 75 mm × 30 mm (height). Determine the angle of friction, ϕ'. For a normal stress of 150 kN/m², what shear force would be required to cause failure in the specimen?

11.2 The size of a sand specimen in a direct shear test was 50 mm $\times$ 50 mm $\times$ 30 mm (height). It is known that, for the sand, tan $\phi' = 0.65/e$ (where e = void ratio) and the specific gravity of solids, $G_s = 2.68$. During the test a normal stress of 150 kN/m^2 was applied. Failure occurred at a shear stress of 110 kN/m^2. What was the mass of the sand specimen?

11.3 Following are the results of four drained direct shear tests on a normally consolidated clay:

Size of specimen = 60 mm $\times$ 60 mm
Height of specimen = 30 mm

Test no.	Normal force (N)	Shear force at failure (N)
1	200	155
2	300	230
3	400	310
4	500	385

Draw a graph for the shear stress at failure against the normal stress and determine the drained angle of friction (ϕ') from the graph.

11.4 Following are the results of four drained direct shear tests on a normally consolidated clay:

Specimen size: diameter of specimen = 2 in.
height of specimen = 1 in.

Test no.	Normal force (lb)	Shear force at failure (lb)
1	60	37.5
2	90	55
3	110	70
4	125	80

Draw a graph for shear stress at failure against the normal stress and determine the drained angle of friction (ϕ') from the graph.

11.5 The equation of the effective stress failure envelope for a loose sandy soil was obtained from a direct shear test as $\tau_f = \sigma' \tan 30°$. A drained triaxial test was conducted with the same soil at a chamber confining pressure of 10 lb/in.^{2}

a. Calculate the deviator stress at failure.

b. Estimate the angle that the failure plane makes with the major principal plane.

c. Determine the normal stress and shear stress (when the specimen failed) on a plane that makes an angle of 30° with the major principal plane. Also, explain why the specimen did not fail along the plane during the test.

11.6 The relationship between the relative density D_r and the angle of friction, ϕ', of a sand can be given as $\phi'° = 28 + 0.18 D_r$ (D_r is in %). A drained triaxial test on the same sand was conducted with a chamber confining pressure of 120 kN/m^2. The relative density of compaction was 65%. Calculate the major principal stress at failure.

11.7 For a normally consolidated clay, the results of a drained triaxial test are as follows:

Chamber confining pressure = 15 lb/in.^{2}
Deviator stress at failure = 34 lb/in.^{2}

Determine the soil friction angle, ϕ'.

11.8 For a normally consolidated clay, it is given that $\phi' = 24°$. In a drained triaxial test, the specimen failed at a deviator stress of 175 kN/m^2. What was the chamber confining pressure, σ'_3?

11.9 For a normally consolidated clay, it is given that $\phi' = 28°$. In a drained triaxial test, the specimen failed at a deviator stress at 30 lb/in.^{2} What was the chamber confining pressure, σ'_3?

11.10 A consolidated-drained triaxial test was conducted on a normally consolidated clay. The results were as follows:

$\sigma_3 = 250$ kN/m^2
$(\Delta\sigma_d)_f = 275$ kN/m^2

Determine the following:

a. Angle of friction, ϕ'
b. Angle θ that the failure plane makes with the major principal plane
c. Normal stress, σ', and shear stress, τ_f, on the failure plane

11.11 The results of two drained triaxial tests on a saturated clay are as follows:

Specimen I: chamber confining pressure = 70 kN/m^2
                      deviator stress at failure = 215 kN/m^2
Specimen II: chamber confining pressure = 120 kN/m^2
                      deviator stress at failure = 260 kN/m^2

Calculate the shear strength parameters of the soil.

11.12 If a specimen of clay described in Problem 11.11 is tested in a triaxial apparatus with a chamber confining pressure of 200 kN/m^2, what will be the major principal stress at failure? Assume full drained condition during the test.

11.13 A sandy soil has a drained angle of friction of 35°. In a drained triaxial test on the same soil, the deviator stress at failure is 2.69 ton/ft^2. What is the chamber confining pressure?

11.14 A deposit of sand is shown in Figure 11.45. Find the maximum shear resistance in kN/m^2 along a horizontal plane located 10 m below the ground surface.

11.15 A consolidated-undrained test on a normally consolidated clay yielded the following results:

$\sigma_3 = 15$ lb/in.^{2}
Deviator stress, $(\Delta\sigma_d)_f = 11$ lb/in.^{2}
Pore pressure, $(\Delta u_d)_f = 7.2$ lb/in.^{2}

Calculate the consolidated-undrained friction angle and the drained friction angle.

11.16 Repeat Problem 11.15, using the following values:

$\sigma_3 = 140$ kN/m^2
$(\Delta\sigma_d)_f = 125$ kN/m^2
$(\Delta u_d)_f = 75$ kN/m^2

Ground water table

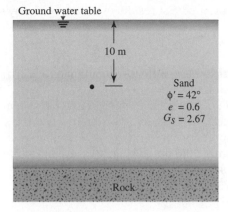

10 m

Sand
$\phi' = 42°$
$e = 0.6$
$G_S = 2.67$

Rock

Figure 11.45

11.17 The shear strength of a normally consolidated clay can be given by the equation $\tau_f = \sigma' \tan 31°$. A consolidated-undrained test was conducted on the clay. Following are the results of the test:

Chamber confining pressure = 112 kN/m²
Deviator stress at failure = 100 kN/m²

Determine
a. The consolidated-undrained friction angle, ϕ
b. The pore water pressure developed in the clay specimen at failure

11.18 For the clay specimen described in Problem 11.17, what would have been the deviator stress at failure if a drained test would have been conducted with the same chamber confining pressure (i.e., $\sigma_3 = 112$ kN/m²)?

11.19 A silty sand has a consolidated-undrained friction angle of 22° and a drained friction angle of 32° ($c' = 0$). If a consolidated-undrained test on such a soil is conducted at a chamber confining pressure of 1.2 ton/ft², what will be the major principal stress (total) at failure? Also, calculate the pore pressure that will be generated in the soil specimen at failure.

11.20 Repeat Problem 11.19, using the following values:
$\phi = 19°$
$\phi' = 28°$
$\sigma_3 = 85$ kN/m²

11.21 The following are the results of a consolidated-undrained triaxial test in a clay:

Specimen no.	σ_3 (kN/m²)	σ_1 at failure (kN/m²)
I	192	375
II	384	636

Draw the total stress Mohr's circles and determine the shear strength parameters for consolidated undrained conditions (i.e., ϕ and c).

11.22 The consolidated-undrained test results of a saturated clay specimen are as follows:

$$\sigma_3 = 97 \text{ kN/m}^2$$
$$\sigma_1 \text{ at failure} = 197 \text{ kN/m}^2$$

What will be the axial stress at failure if a similar specimen is subjected to an unconfined compression test?

11.23 The friction angle, ϕ', of a normally consolidated clay specimen collected during field exploration was determined from drained triaxial tests to be 25°. The unconfined compression strength, q_u, of a similar specimen was found to be 100 kN/m². Determine the pore water pressure at failure for the unconfined compression test.

11.24 Repeat Problem 11.23 using the following values:

$$\phi' = 23°$$
$$q_u = 120 \text{ kN/m}^2$$

11.25 The results of two consolidated-drained triaxial tests on a clayey soil are as follows:

Test no.	σ_3' (lb/in.²)	$\sigma_{1(failure)}'$ (lb/in.²)
1	27	73
2	12	48

Use the failure envelope equation given in Example 11.7 — that is, $q' = m + p' \tan \alpha$. (Do not plot the graph.)

a. Find m and α.
b. Find c' and ϕ'.

11.26 A 15-m-thick normally consolidated clay layer is shown in Figure 11.46. The plasticity index of the clay is 18. Estimate the undrained cohesion as would be determined from a vane shear test at a depth of 6 m below the ground surface. Use Skempton's equation in Table 11.6.

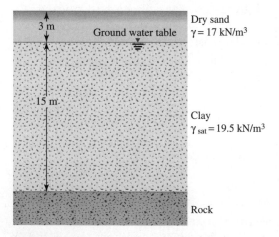

Figure 11.46

References

ACAR, Y. B., DURGUNOGLU, H. T., and TUMAY, M. T. (1982). "Interface Properties of Sand," *Journal of the Geotechnical Engineering Division,* ASCE, Vol. 108, No. GT4, 648–654.

AMERICAN SOCIETY FOR TESTING AND MATERIALS (1994). *Annual Book of ASTM Standards,* Vol. 04.08, Philadelphia, Pa.

BISHOP, A. W., and BJERRUM, L. (1960). "The Relevance of the Triaxial Test to the Solution of Stability Problems," *Proceedings,* Research Conference on Shear Strength of Cohesive Soils, ASCE, 437–501.

BJERRUM, L. (1974). "Problems of Soil Mechanics and Construction on Soft Clays," Norwegian Geotechnical Institute, *Publication No. 110,* Oslo.

BJERRUM, L., and SIMONS, N. E. (1960). "Compression of Shear Strength Characteristics of Normally Consolidated Clay," *Proceedings,* Research Conference on Shear Strength of Cohesive Soils, ASCE, 711–726.

BLACK, D. K., and LEE, K. L. (1973). "Saturating Laboratory Samples by Back Pressure," *Journal of the Soil Mechanics and Foundations Division,* ASCE, Vol. 99, No. SM1, 75–93.

CASAGRANDE, A., and HIRSCHFELD, R. C. (1960). "Stress Deformation and Strength Characteristics of a Clay Compacted to a Constant Dry Unit Weight," *Proceedings,* Research Conference on Shear Strength of Cohesive Soils, ASCE, 359–417.

CHANDLER, R. J. (1988). "The *in situ* Measurement of the Undrained Shear Strength of Clays Using the Field Vane," *STP 1014, Vane Shear Strength Testing in Soils: Field and Laboratory Studies,* ASTM, 13–44.

COULOMB, C. A. (1776). "Essai sur une application des regles de Maximums et Minimis á quelques Problèmes de Statique, relatifs á l'Architecture," *Memoires de Mathematique et de Physique,* Présentés, á l'Academie Royale des Sciences, Paris, Vol. 3, 38.

JAMIOLKOWSKI, M., LADD, C. C., GERMAINE, J. T., and LANCELLOTTA, R. (1985). "New Developments in Field and Laboratory Testing of Soils," *Proceedings,* XIth International Conference on Soil Mechanics and Foundation Engineering, San Francisco, Vol. 1, 57–153.

KENNEY, T. C. (1959). "Discussion," *Proceedings,* ASCE, Vol. 85, No. SM3, 67–79.

LADD, C. C., FOOTE, R., ISHIHARA, K., SCHLOSSER, F., and POULOS, H. G. (1977). "Stress Deformation and Strength Characteristics," *Proceedings,* 9th International Conference on Soil Mechanics and Foundation Engineering, Tokyo, Vol. 2, 421–494.

LAMBE, T. W. (1964). "Methods of Estimating Settlement," *Journal of the Soil Mechanics and Foundations Division,* ASCE, Vol. 90, No. SM5, 47–74.

MESRI, G. (1989). "A Re-evaluation of $s_{u(\text{mob})} \approx 0.22\,\sigma_p$ Using Laboratory Shear Tests," *Canadian Geotechnical Journal,* Vol. 26, No. 1, 162–164.

MOHR, O. (1900). "Welche Umstände Bedingen die Elastizitätsgrenze und den Bruch eines Materiales?" *Zeitschrift des Vereines Deutscher Ingenieure,* Vol. 44, 1524–1530, 1572–1577.

MORRIS, P. M., and WILLIAMS, D. J. (1994). "Effective Stress Vane Shear Strength Correction Factor Correlations," *Canadian Geotechnical Journal,* Vol. 31, No. 3, 335–342.

ROSENQVIST, I. TH. (1953). "Considerations on the Sensitivity of Norwegian Quick Clays," *Geotechnique,* Vol. 3, No. 5, 195–200.

SKEMPTON, A. W. (1954). "The Pore Water Coefficients A and B," *Geotechnique,* Vol. 4, 143–147.

SKEMPTON, A. W. (1957). "Discussion: The Planning and Design of New Hong Kong Airport," *Proceedings,* Institute of Civil Engineers, London, Vol. 7, 305–307.

SKEMPTON, A. W. (1964). "Long-Term Stability of Clay Slopes," *Geotechnique,* Vol. 14, 77.

12

Lateral Earth Pressure: At-Rest, Rankine, and Coulomb

Retaining structures such as retaining walls, basement walls, and bulkheads are commonly encountered in foundation engineering as they support slopes of earth masses. Proper design and construction of these structures require a thorough knowledge of the lateral forces that act between the retaining structures and the soil masses being retained. These lateral forces are caused by lateral earth pressure. This chapter is devoted to the study of the various earth pressure theories.

12.1   At-Rest, Active, and Passive Pressures

Consider a mass of soil shown in Figure. 12.1a. The mass is bounded by a *frictionless wall* of height AB. A soil element located at a depth z is subjected to a vertical effective pressure σ'_o and a horizontal effective pressure σ'_h. There are no shear stresses on the vertical and horizontal planes of the soil element. Let us define the ratio of σ'_h to σ'_o as a nondimensional quantity K, or

$$K = \frac{\sigma'_h}{\sigma'_o} \tag{12.1}$$

Now, three possible cases may arise concerning the retaining wall: and they are described

Case 1. If the wall AB is static — that is, if it does not move either to the right or to the left of its initial position — the soil mass will be in a state of *static equilibrium*. In that case, σ'_h is referred to as the *at-rest earth pressure*, or

$$K = K_o = \frac{\sigma'_h}{\sigma'_o} \tag{12.2}$$

where K_o = at-rest earth pressure coefficient.

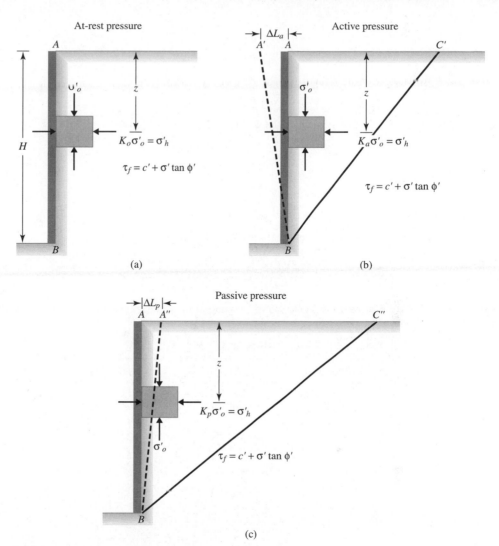

Figure 12.1 Definition of at-rest, active, and passive pressures (Note: Wall AB is frictionless)

Case 2. If the frictionless wall rotates sufficiently about its bottom to a position of $A'B$ (Figure 12.1b), then a triangular soil mass ABC' adjacent to the wall will reach a state of *plastic equilibrium* and will fail sliding down the plane BC'. At this time, the horizontal effective stress, $\sigma'_h = \sigma'_a$, will be referred to as *active pressure*. Now,

$$K = K_a = \frac{\sigma'_h}{\sigma'_o} = \frac{\sigma'_a}{\sigma'_o} \tag{12.3}$$

where K_a = active earth pressure coefficient.

Case 3. If the frictionless wall rotates sufficiently about its bottom to a position $A''B$ (Figure 12.1c), then a triangular soil mass ABC'' will reach a state of *plastic*

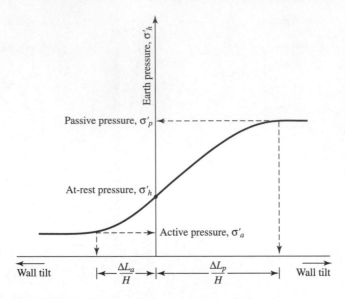

Figure 12.2 Variation of the magnitude of lateral earth pressure with wall tilt

Table 12.1 Typical Values of $\Delta L_a/H$ and $\Delta L_p/H$

Soil type	$\Delta L_a/H$	$\Delta L_p/H$
Loose sand	0.001–0.002	0.01
Dense sand	0.0005–0.001	0.005
Soft clay	0.02	0.04
Stiff clay	0.01	0.02

equilibrium and will fail sliding upward along the plane BC''. The horizontal effective stress at this time will be $\sigma'_h = \sigma'_p$, the so-called *passive pressure*. In this case,

$$K = K_p = \frac{\sigma'_h}{\sigma'_o} = \frac{\sigma'_p}{\sigma'_o} \tag{12.4}$$

where K_p = passive earth pressure coefficient

Figure 12.2 shows the nature of variation of lateral earth pressure with the wall tilt. Typical values of $\Delta L_a/H$ ($\Delta L_a = A'A$ in Figure 12.1b) and $\Delta L_p/H$ ($\Delta L_p = A''A$ in Figure 12.1c) for attaining the active and passive states in various soils are given in Table 12.1.

AT-REST LATERAL EARTH PRESSURE

12.2

Earth Pressure at Rest

The fundamental concept of earth pressure at rest was discussed in the preceding section. In order to define the earth pressure coefficient K_o at rest, we refer to Fig-

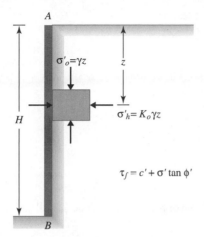

Figure 12.3 Earth pressure at rest

ure 12.3, which shows a wall AB retaining a dry soil with a unit weight of γ. The wall is static. At a depth z,

$$\text{Vertical effective stress} = \sigma'_o = \gamma z$$

$$\text{Horizontal effective stress} = \sigma'_h = K_o \gamma z$$

So

$$K_o = \frac{\sigma'_h}{\sigma'_o} = \text{at-rest earth pressure coefficient}$$

For coarse-grained soils, the coefficient of earth pressure at rest can be estimated by using the empirical relationship (Jaky, 1944)

$$K_o = 1 - \sin \phi' \tag{12.5}$$

where ϕ' = drained friction angle.

While designing a wall that may be subjected to lateral earth pressure at rest, one must take care in evaluating the value of K_o. Sherif, Fang, and Sherif (1984), on the basis of their laboratory tests, showed that Jaky's equation for K_o [Eq. (12.5)] gives good results when the backfill is loose sand. However, for a dense sand backfill, Eq. (12.5) may grossly underestimate the lateral earth pressure at rest. This underestimation results because of the process of compaction of backfill. For this reason, they recommended the design relationship

$$K_o = (1 - \sin \phi) + \left[\frac{\gamma_d}{\gamma_{d(min)}} - 1 \right] 5.5 \tag{12.6}$$

where γ_d = actual compacted dry unit weight of the sand behind the wall
   $\gamma_{d(min)}$ = dry unit weight of the sand in the loosest state (Chapter 2)

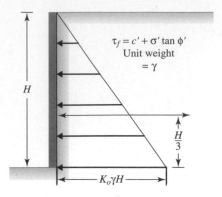

Figure 12.4 Distribution of lateral earth pressure at-rest on a wall

For fine-grained, normally consolidated soils, Massarsch (1979) suggested the following equation for K_o:

$$K_o = 0.44 + 0.42 \left[\frac{PI\,(\%)}{100} \right] \qquad (12.7)$$

For overconsolidated clays, the coefficient of earth pressure at rest can be approximated as

$$K_{o(\text{overconsolidated})} = K_{o(\text{normally consolidated})} \sqrt{OCR} \qquad (12.8)$$

where OCR = overconsolidation ratio. The overconsolidation ratio was defined in Chapter 10 as

$$OCR = \frac{\text{Preconsolidation pressure, } \sigma_c'}{\text{Present effective overburden pressure, } \sigma_o'} \qquad (12.9)$$

Figure 12.4 shows the distribution of lateral earth pressure at rest on a wall of height H retaining a dry soil having a unit weight of γ. The total force per unit length of the wall, P_o, is equal to the area of the pressure diagram, so

$$P_o = \tfrac{1}{2} K_o \gamma H^2 \qquad (12.10)$$

12.3   *Earth Pressure at Rest for Partially Submerged Soil*

Figure 12.5a shows a wall of height H. The groundwater table is located at a depth H_1 below the ground surface, and there is no compensating water on the other side of the wall. For $z \leq H_1$, the lateral earth pressure at rest can be given as $\sigma_h' = K_o \gamma z$. The variation of σ_h' with depth is shown by triangle ACE in Figure 12.5a. However, for $z \geq H_1$ (i.e., below the groundwater table), the pressure on the wall is found from the effective stress and pore water pressure components via the equation

$$\text{effective vertical pressure} = \sigma_o' = \gamma H_1 + \gamma'(z - H_1) \qquad (12.11)$$

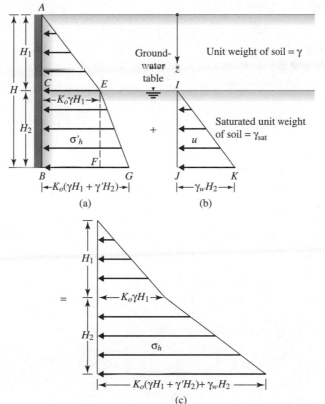

Figure 12.5
Distribution of earth
pressure at rest for partially
submerged soil

where $\gamma' = \gamma_{\mathrm{sat}} - \gamma_w$ = the effective unit weight of soil. So the effective lateral pressure at rest is

$$\sigma_h' = K_o\sigma_o' - K_o[\gamma H_1 + \gamma'(z - H_1)] \qquad (12.12)$$

The variation of σ_h' with depth is shown by $CEGB$ in Figure 12.5a. Again the lateral pressure from pore water is

$$u = \gamma_w(z - H_1) \qquad (12.13)$$

The variation of u with depth is shown in Figure 12.5b.

Hence, the total lateral pressure from earth and water at any depth $z \geq H_1$ is equal to

$$\sigma_h = \sigma_h' + u$$

$$= K_o[\gamma H_1 + \gamma'(z - H_1)] + \gamma_w(z - H_1) \qquad (12.14)$$

The force per unit length of the wall can be found from the sum of the areas of the pressure diagrams in Figures 12.5a and 12.5b and is equal to (Figure 12.5c)

$$P_o = \underbrace{\tfrac{1}{2}K_o\gamma H_1^2}_{\substack{\text{Area} \\ ACE}} + \underbrace{K_o\gamma H_1 H_2}_{\substack{\text{Area} \\ CEFB}} + \underbrace{\tfrac{1}{2}(K_o\gamma' + \gamma_w)H_2^2}_{\substack{\text{Areas} \\ EFG \text{ and } IJK}} \qquad (12.15)$$

Example 12.1

Figure 12.6a shows a 15-ft-high retaining wall. The wall is restrained from yielding. Calculate the lateral force P_o per unit length of the wall. Also, determine the location of the resultant force.

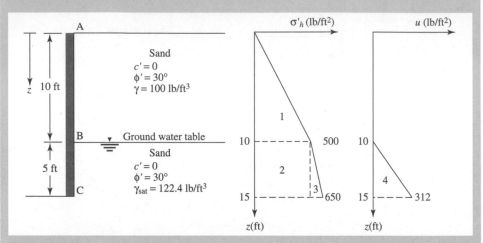

Figure 12.6

Solution

$K_o = 1 - \sin \phi' = 1 - \sin 30 = 0.5$

At $z = 0$: $\sigma'_o = 0$; $\sigma'_h = 0$; $u = 0$

At $z = 10$ ft: $\sigma'_o = (10)(100) = 1000 \text{ lb/ft}^2$

$\sigma'_h = K_o\sigma'_o = (0.5)(1000) = 500 \text{ lb/ft}^2$

$u = 0$

At $z = 15$ ft: $\sigma'_o = (10)(100) + (5)(122.4 - 62.4) = 1300 \text{ lb/ft}^2$

$\sigma'_h = K_o\sigma'_o = (0.5)(1300) = 650 \text{ lb/ft}^2$

$u = (5)(\gamma_w) = (5)(62.4) = 312 \text{ lb/ft}^2$

The variations of σ'_h and u with depth are shown in Figures 12.6b and 12.6c.

Lateral force P_o = Area 1 + Area 2 + Area 3 + Area 4

or

$$P_o = \left(\frac{1}{2}\right)(10)(500) + (5)(500) + \left(\frac{1}{2}\right)(5)(150) + \left(\frac{1}{2}\right)(5)(312)$$

$$= 2500 + 2500 + 375 + 780 = 6155 \text{ lb/ft}$$

The location of the resultant, measured from the bottom of the wall, is

$$\bar{z} = \frac{\Sigma \text{ moment of pressure diagram about } C}{P_o}$$

or

$$\bar{z} = \frac{(2500)\left(5 + \dfrac{10}{3}\right) + (2500)\left(\dfrac{5}{2}\right) + (375)\left(\dfrac{5}{3}\right) + (780)\left(\dfrac{5}{3}\right)}{6155} = \textbf{4.71 ft} \quad \blacksquare$$

12.4 *Lateral Pressure on Retaining Walls from Surcharges—Based on Theory of Elasticity*

Point-Load Surcharge

The equations for normal stresses inside a homogeneous, elastic, and isotropic medium produced from a point load on the surface were given in Chapter 9 [Eqs. (9.10), (9.11) and 9.12].

We now apply Eq. (9.10) to determine the lateral pressure on a retaining wall caused by the concentrated point load Q placed at the surface of the backfill as shown in Figure 12.7a. If the load Q is placed on the plane of the section shown, we can substitute $y = 0$ in Eq. (9.10). Also, assuming that $\mu = 0.5$, we can write

$$\sigma_h' = \frac{Q}{2\pi}\left(\frac{3x^2 z}{L^5}\right) \tag{12.16}$$

where $L = \sqrt{x^2 + z^2}$. Substituting $x = mH$ and $z = nH$ into Eq. (12.16), we have

$$\sigma_h' = \frac{3Q}{2\pi H^2}\frac{m^2 n}{(m^2 + n^2)^{5/2}} \tag{12.17}$$

The horizontal stress expressed by Eq. (12.17) does not include the restraining effect of the wall. This expression was investigated by Gerber (1929) and Spangler (1938) with large-scale tests. On the basis of the experimental findings, Eq. (12.17) has been modified as follows to agree with the real conditions:

For $m > 0.4$,

$$\sigma_h' = \frac{1.77Q}{H^2}\frac{m^2 n^2}{(m^2 + n^2)^3} \tag{12.18}$$

For $m \leq 0.4$,

$$\sigma_h' = \frac{0.28Q}{H^2}\frac{n^2}{(0.16 + n^2)^3} \tag{12.19}$$

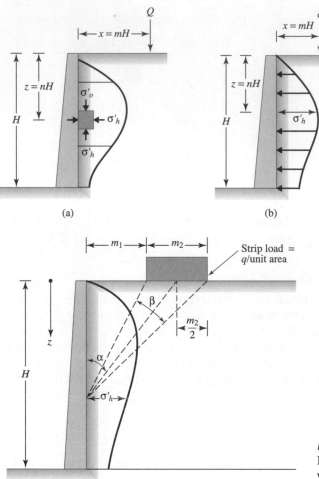

Figure 12.7
Lateral pressure on a retaining wall due to a (a) point load, (b) line load, and (c) strip load

Line-Load Surcharge

Figure 12.7b shows the distribution of lateral pressure against the vertical back face of the wall caused by a line-load surcharge placed parallel to the crest. The modified forms of the equations [similar to Eqs. (12.18) and (12.19) for the case of point-load surcharge] for line-load surcharges are, respectively,

$$\sigma'_h = \frac{4q}{\pi H} \frac{m^2 n}{(m^2 + n^2)^2} \qquad \text{(for } m > 0.4\text{)} \qquad (12.20)$$

and

$$\sigma'_h = \frac{0.203q}{H} \frac{n}{(0.16 + n^2)^2} \qquad \text{(for } m \leq 0.4\text{)} \qquad (12.21)$$

where q = load per unit length of the surcharge.

Strip-Load Surcharge

Figure 12.7e shows a strip-load surcharge with an intensity of q per unit area located at a distance m_1 from a wall of height H. On the basis of the theory of elasticity, the horizontal stress at a depth z on a retaining structure can be given as

$$\sigma'_h = \frac{q}{H}(\beta - \sin\beta\cos 2\alpha) \tag{12.22}$$

The angles α and β are defined in Figure 12.7c. For actual soil behavior (from the wall restraining effect), the preceding equation can be modified to

$$\sigma'_h = \frac{2q}{H}(\beta - \sin\beta\cos 2\alpha) \tag{12.23}$$

The nature of the distribution of σ'_h with depth is shown in Figure 12.7c. The force P per unit length of the wall caused by the strip load alone can be obtained by integration of σ'_h with limits of z from 0 to H.

Example 12.2

Consider the retaining wall shown in Figure 12.8a where $H = 10$ ft. A line load of 800 lb/ft is placed on the ground surface parallel to the crest at a distance of 5 ft from the back face of the wall. Determine the increase in the lateral force per unit length of the wall caused by the line load. Use the modified equation given in Section 12.4.

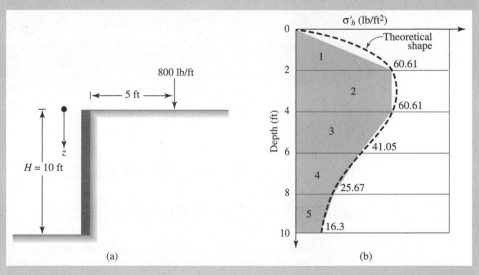

Figure 12.8

Solution

We are given $H = 10$ ft, $q = 800$ lb/ft, and

$$m = \frac{5}{10} = 0.5 > 0.4$$

So Eq. (12.20) will apply:

$$\sigma'_h = \frac{4q}{\pi H} \frac{m^2 n}{(m^2 + n^2)^2}$$

Now the following table can be prepared:

$n = \dfrac{z}{H}$	$\dfrac{4q}{\pi H}$	$\dfrac{m^2 n}{(m^2 + n^2)^2}$	σ'_h (lb/ft²)
0	101.86	0	0
0.2	101.86	0.595	60.61
0.4	101.86	0.595	60.61
0.6	101.86	0.403	41.05
0.8	101.86	0.252	25.67
1.0	101.86	0.16	16.3

Refer to the diagram in Figure 12.8b.

Area no.	Area
1	$\left(\dfrac{1}{2}\right)(2)(60.61) = 60.61$ lb/ft
2	$\left(\dfrac{1}{2}\right)(2)(60.61 + 60.61) = 121.22$ lb/ft
3	$\left(\dfrac{1}{2}\right)(2)(60.61 + 41.05) = 101.66$ lb/ft
4	$\left(\dfrac{1}{2}\right)(2)(41.05 + 25.67) = 66.72$ lb/ft
5	$\left(\dfrac{1}{2}\right)(2)(25.67 + 16.3) = 41.97$ lb/ft

Total = **392.18 lb/ft**

≈ **390 lb/ft**   ∎

RANKINE'S LATERAL EARTH PRESSURE

12.5 *Rankine's Theory of Active Pressure*

The phrase *plastic equilibrium in soil* refers to the condition where every point in a soil mass is on the verge of failure. Rankine (1857) investigated the stress conditions in soil at a state of plastic equilibrium. In this section and in section 12.6, we deal with Rankine's theory of earth pressure.

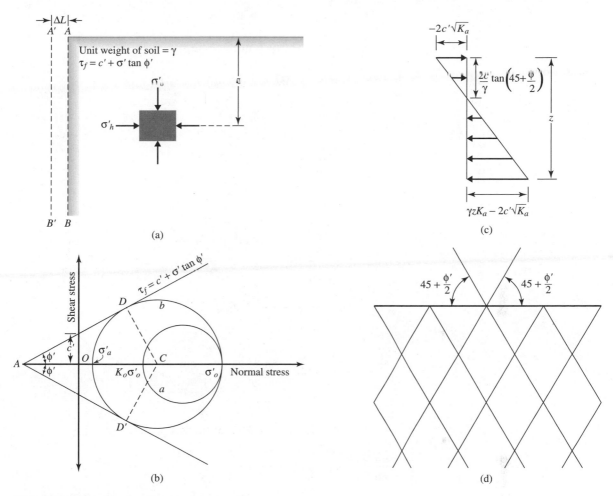

Figure 12.9 Rankine's active earth pressure

Figure 12.9a shows a soil mass that is bounded by a frictionless wall, AB, that extends to an infinite depth. The vertical and horizontal effective principal stresses on a soil element at a depth z are σ'_o and σ'_h, respectively. As we saw in Section 12.2, if the wall AB is not allowed to move, then $\sigma'_h = K_o\sigma'_o$. The stress condition in the soil element can be represented by the Mohr's circle a in Figure 12.9b. However, if the wall AB is allowed to *move away from the soil mass* gradually, the horizontal principal stress will decrease. Ultimately a state will be reached when the stress condition in the soil element can be represented by the Mohr's circle b, the state of plastic equilibrium, and failure of the soil will occur. This situation represents *Rankine's active state,* and the effective pressure σ'_a on the vertical plane (which is a principal plane) is Rankine's *active earth pressure.* We next derive σ'_a in terms of γ, z, c', and ϕ' from Figure 12.8b

$$\sin \phi' = \frac{CD}{AC} = \frac{CD}{AO + OC}$$

But

$$CD = \text{radius of the failure circle} = \frac{\sigma_o' - \sigma_a'}{2}$$

$$AO = c' \cot \phi'$$

and

$$OC = \frac{\sigma_o' + \sigma_a'}{2}$$

So

$$\sin \phi' = \frac{\dfrac{\sigma_o' - \sigma_a'}{2}}{c' \cot \phi' + \dfrac{\sigma_o' + \sigma_a'}{2}}$$

or

$$c' \cos \phi' + \frac{\sigma_o' + \sigma_a'}{2} \sin \phi' = \frac{\sigma_o' - \sigma_a'}{2}$$

or

$$\sigma_a' = \sigma_o' \frac{1 - \sin \phi'}{1 + \sin \phi'} - 2c' \frac{\cos \phi'}{1 + \sin \phi'} \tag{12.24}$$

But

$$\sigma_o' = \text{vertical effective overburden pressure} = \gamma z$$

$$\frac{1 - \sin \phi'}{1 + \sin \phi'} = \tan^2\left(45 - \frac{\phi'}{2}\right)$$

and

$$\frac{\cos \phi'}{1 + \sin \phi'} = \tan\left(45 - \frac{\phi'}{2}\right)$$

Substituting the preceding values into Eq. (12.24), we get

$$\sigma_a' = \gamma z \tan^2\left(45 - \frac{\phi'}{2}\right) - 2c' \tan\left(45 - \frac{\phi'}{2}\right) \tag{12.25}$$

The variation of σ_a' with depth is shown in Figure 12.9c. For cohesionless soils, $c' = 0$ and

$$\sigma_a' = \sigma_o' \tan^2\left(45 - \frac{\phi'}{2}\right) \tag{12.26}$$

The ratio of σ_a' to σ_o' is called the *coefficient of Rankine's active earth pressure* and is given by

$$K_a = \frac{\sigma'_a}{\sigma'_o} = \tan^2\left(45 - \frac{\phi'}{2}\right) \qquad (12.27)$$

Again, from Figure 12.9b we can see that the failure planes in the soil make $\pm(45 + \phi'/2)$-degree angles with the direction of the major principal plane — that is, the horizontal. These are called potential *slip planes* and are shown in Figure 12.9d.

It is important to realize that a similar equation for σ_a could be derived based on the total stress shear strength parameters — that is, $\tau_f = c + \sigma \tan \phi$. For this case,

$$\sigma_a = \gamma z \tan\left(45 - \frac{\phi}{2}\right) - 2c \tan\left(45 - \frac{\phi}{2}\right) \qquad (12.28)$$

12.6 *Theory of Rankine's Passive Pressure*

Rankine's passive state can be explained with the aid of Figure 12.10. AB is a frictionless wall that extends to an infinite depth (Figure 12.10a). The initial stress condition on a soil element is represented by the Mohr's circle a in Figure 12.10b. If the wall is gradually *pushed into the soil mass,* the effective principal stress σ'_h will increase. Ultimately the wall will reach a situation where the stress condition for the soil element can be expressed by the Mohr's circle b. At this time, failure of the soil will occur. This situation is referred to as *Rankine's passive state*. The lateral earth pressure σ'_p, which is the major principal stress, is called *Rankine's passive earth pressure.* From Figure 12.10b, it can be shown that

$$\sigma'_p = \sigma'_o \tan^2\left(45 + \frac{\phi'}{2}\right) + 2c' \tan\left(45 + \frac{\phi'}{2}\right)$$

$$= \gamma z \tan^2\left(45 + \frac{\phi'}{2}\right) + 2c' \tan\left(45 + \frac{\phi'}{2}\right) \qquad (12.29)$$

The derivation is similar to that for Rankine's active state.

Figure 12.10c shows the variation of passive pressure with depth. For cohesionless soils ($c' = 0$),

$$\sigma'_p = \sigma'_o \tan^2\left(45 + \frac{\phi'}{2}\right)$$

or

$$\frac{\sigma'_p}{\sigma'_o} = K_p = \tan^2\left(45 + \frac{\phi'}{2}\right) \qquad (12.30)$$

K_p (the ratio of effective stresses) in the preceding equation is referred to as the *coefficient of Rankine's passive earth pressure.*

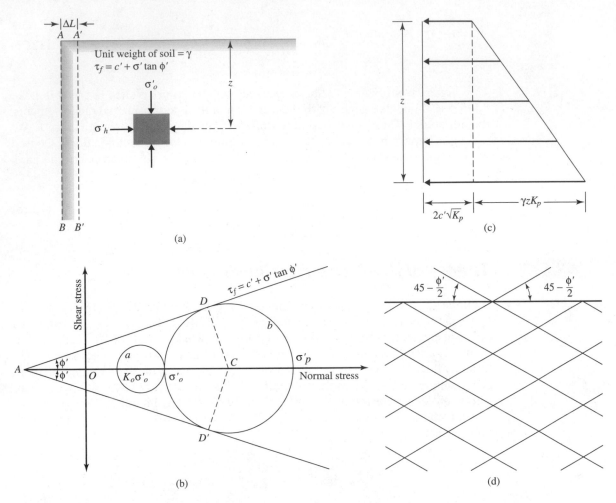

Figure 12.10 Rankine's passive earth pressure

The points D and D' on the failure circle (see Figure 12.10b) correspond to the slip planes in the soil. For Rankine's passive state, the slip planes make $\pm(45 - \phi'/2)$-degree angles with the direction of the minor principal plane — that is, in the horizontal direction. Figure 12.10d shows the distribution of slip planes in the soil mass.

12.7 *Yielding of Wall of Limited Height*

We learned in the preceding discussion that sufficient movement of a frictionless wall extending to an infinite depth is necessary to achieve a state of plastic equilibrium. However, the distribution of lateral pressure against a wall of limited height is very much influenced by the manner in which the wall actually yields. In most retaining walls of limited height, movement may occur by simple translation or, more frequently, by rotation about the bottom.

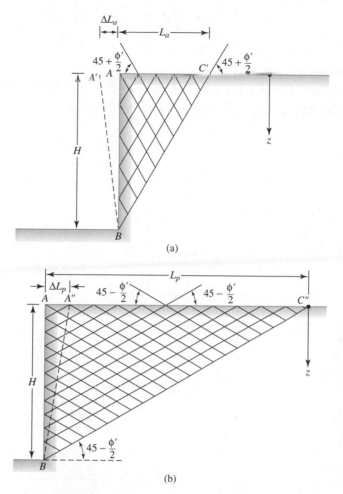

Figure 12.11 Rotation of frictionless wall about the bottom

For preliminary theoretical analysis, let us consider a frictionless retaining wall represented by a plane AB as shown in Figure 12.11a. If the wall AB rotates sufficiently about its bottom to a position $A'B$, then a triangular soil mass ABC' adjacent to the wall will reach Rankine's active state. Because the slip planes in Rankine's active state make angles of $\pm(45 + \phi'/2)$ degrees with the major principal plane, the soil mass in the state of plastic equilibrium is bounded by the plane BC', which makes an angle of $(45 + \phi'/2)$ degrees with the horizontal. The soil inside the zone ABC' undergoes the same unit deformation in the horizontal direction everywhere, which is equal to $\Delta L_a/L_a$. The lateral earth pressure on the wall at any depth z from the ground surface can be calculated by using Eq. (12.25).

In a similar manner, if the frictionless wall AB (Figure 12.11b) rotates sufficiently into the soil mass to a position $A''B$, then the triangular mass of soil ABC'' will reach Rankine's passive state. The slip plane BC'' bounding the soil wedge that is at a state of plastic equilibrium will make an angle of $(45 - \phi'/2)$ degrees with the horizontal. Every point of the soil in the triangular zone ABC'' will undergo the same unit deformation in the horizontal direction, which is equal to $\Delta L_p/L_p$. The passive pressure on the wall at any depth z can be evaluated by using Eq. (12.29).

12.8 Diagrams for Lateral Earth Pressure Distribution against Retaining Walls

Backfill—Cohesionless Soil with Horizontal Ground Surface

Active Case Figure 12.12a shows a retaining wall with cohensionless soil backfill that has a horizontal ground surface. The unit weight and the angle of friction of the soil are γ and ϕ', respectively.

For Rankine's active state, the earth pressure at any depth against the retaining wall can be given by Eq. (12.25):

$$\sigma'_a = K_a\gamma z \qquad (Note:\ c' = 0.)$$

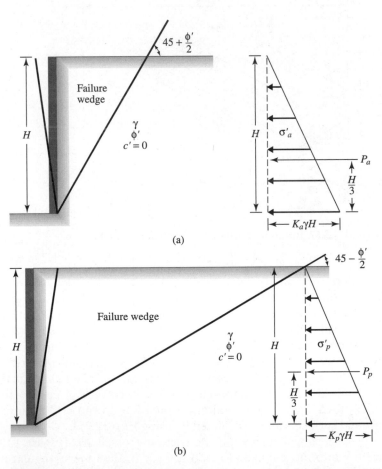

Figure 12.12 Pressure distribution against a retaining wall for cohensionless soil backfill with horizontal ground surface: (a) Rankine's active state; (b) Rankine's passive state

Note that σ_a' increases linearly with depth, and at the bottom of the wall, it is

$$\sigma_a' = K_a \gamma H \tag{12.31}$$

The total force per unit length of the wall is equal to the area of the pressure diagram, so

$$P_a = \tfrac{1}{2} K_a \gamma H^2 \tag{12.32}$$

Passive Case The lateral pressure distribution against a retaining wall of height H for Rankine's passive state is shown in Figure 12.12b. The lateral earth pressure at any depth z [Eq. (12.30), $c' = 0$] is

$$\sigma_p' = K_p \gamma H \tag{12.33}$$

The total force per unit length of the wall is

$$P_p = \tfrac{1}{2} K_p \gamma H^2 \tag{12.34}$$

Backfill—Partially Submerged Cohensionless Soil Supporting a Surcharge

Active Case Figure 12.13a shows a frictionless retaining wall of height H and a backfill of cohensionless soil. The groundwater table is located at a depth of H_1 below the ground surface, and the backfill is supporting a surcharge pressure of q per unit area. From Eq. (12.27), the effective active earth pressure at any depth can be given by

$$\sigma_a' = K_a \sigma_o' \tag{12.35}$$

where σ_o' and σ_a' = the effective vertical pressure and lateral pressure, respectively. At $z = 0$,

$$\sigma_o = \sigma_o' = q \tag{12.36}$$

and

$$\sigma_a' = K_a q \tag{12.37}$$

At depth $z = H_1$,

$$\sigma_o' = (q + \gamma H_1) \tag{12.38}$$

and

$$\sigma_a' = K_a(q + \gamma H_1) \tag{12.39}$$

At depth $z = H$,

$$\sigma_o' = (q + \gamma H_1 + \gamma' H_2) \tag{12.40}$$

and

$$\sigma_a' = K_a(q + \gamma H_1 + \gamma' H_2) \tag{12.41}$$

where $\gamma' = \gamma_{sat} - \gamma_w$. The variation of σ_a' with depth is shown in Figure 12.13b.

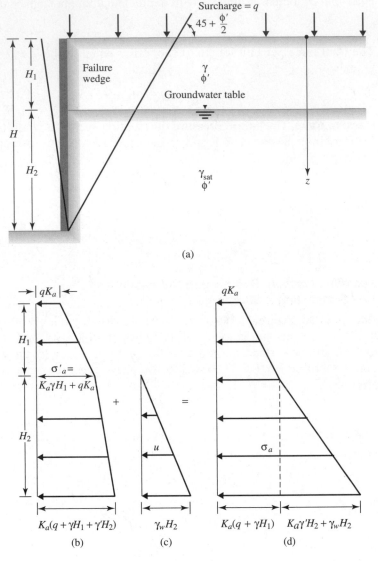

Figure 12.13 Rankine's active earth pressure distribution against a retaining wall with partially submerged cohesionless soil backfill supporting a surcharge

The lateral pressure on the wall from the pore water between $z = 0$ and H_1 is 0, and for $z > H_1$, it increases linearly with depth (Figure 12.13c). At $z = H$,

$$u = \gamma_w H_2$$

The total lateral pressure diagram (Figure 12.13d) is the sum of the pressure diagrams shown in Figures 12.13b and 12.13c. The total active force per unit length of the wall is the area of the total pressure diagram. Thus,

$$P_a = K_a q H + \tfrac{1}{2} K_a \gamma H_1^2 + K_a \gamma H_1 H_2 + \tfrac{1}{2}(K_a \gamma' + \gamma_w) H_2^2 \qquad (12.42)$$

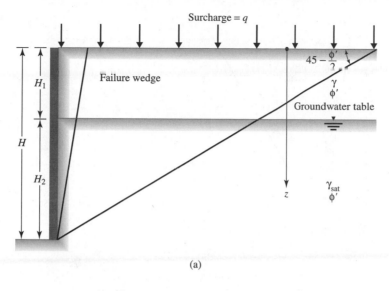

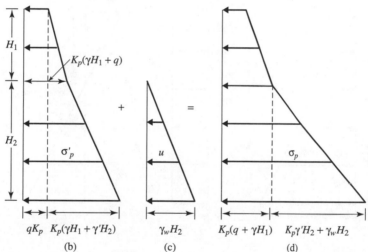

Figure 12.14 Rankine's passive earth pressure distribution against a retaining wall with partially submerged cohesionless soil backfill supporting a surcharge

Passive Case Figure 12.14a shows the same retaining wall as was shown in Figure 12.13a. Rankine's passive pressure at any depth against the wall can be given by Eq. (12.30):

$$\sigma_p' = K_p \sigma_o'$$

Using the preceding equation, we can determine the variation of σ_p' with depth, as shown in Figure 12.14b. The variation of the pressure on the wall from water with depth is shown in Figure 12.14c. Figure 12.14d shows the distribution of the total pressure σ_p with depth. The total lateral passive force per unit length of the wall is the area of the diagram given in Figure 10.11d, or

$$P_p = K_p q H + \tfrac{1}{2} K_p \gamma H_1^2 + K_p \gamma H_1 H_2 + \tfrac{1}{2}(K_p \gamma' + \gamma_w) H_2^2 \qquad (12.43)$$

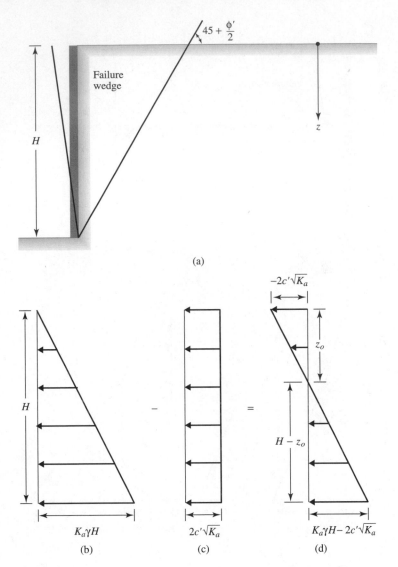

Figure 12.15 Rankine's active earth pressure distribution against a retaining wall with cohesive soil backfill

Backfill—Cohesive Soil with Horizontal Backfill

Active Case Figure 12.15a shows a frictionless retaining wall with a cohesive soil backfill. The active pressure against the wall at any depth below the ground surface can be expressed as [Eq. (12.25)]

$$\sigma'_a = K_a \gamma z - 2\sqrt{K_a}c'$$

The variation of $K_a \gamma z$ with depth is shown in Figure 12.15b, and the variation of $2\sqrt{K_a}c'$ with depth is shown in Figure 12.15c. Note that $2\sqrt{K_a}c'$ is not a function of z; hence, Figure 12.15c is a rectangle. The variation of the net value of σ'_a with depth is plotted in Figure 12.15d. Also note that, because of the effect of cohesion, σ'_a is negative in the upper part of the retaining wall. The depth z_o at which the active pressure becomes equal to 0 can be found from Eq. (12.25) as

$$K_a \gamma z_o - 2\sqrt{K_a}c' = 0$$

or

$$z_o = \frac{2c'}{\gamma\sqrt{K_a}} \tag{12.44}$$

For the undrained condition — that is, $\phi = 0$, $K_a = \tan^2 45 = 1$, and $c = c_u$ (undrained cohesion) — from Eq. (12.28),

$$z_o = \frac{2c_u}{\gamma} \tag{12.45}$$

So, with time, tensile cracks at the soil-wall interface will develop up to a depth z_o.

The total active force per unit length of the wall can be found from the area of the total pressure diagram (Figure 12.15d), or

$$P_a = \tfrac{1}{2}K_a\gamma H^2 - 2\sqrt{K_a}c'H \tag{12.46}$$

For the $\phi = 0$ condition,

$$P_a = \tfrac{1}{2}\gamma H^2 - 2c_uH \tag{12.47}$$

For calculation of the total active force, common practice is to take the tensile cracks into account. Because no contact exists between the soil and the wall up to a depth of z_o after the development of tensile cracks, only the active pressure distribution against the wall between $z = 2c'/(\gamma\sqrt{K_a})$ and H (Figure 12.15d) is considered. In this case,

$$\begin{aligned} P_a &= \tfrac{1}{2}(K_a\gamma H - 2\sqrt{K_a}c')\left(H - \frac{2c'}{\gamma\sqrt{K_a}}\right) \\ &= \tfrac{1}{2}K_a\gamma H^2 - 2\sqrt{K_a}c'H + 2\frac{c'^2}{\gamma} \end{aligned} \tag{12.48}$$

For the $\phi - 0$ condition,

$$P_a = \tfrac{1}{2}\gamma H^2 - 2c_uH + 2\frac{c_u^2}{\gamma} \tag{12.49}$$

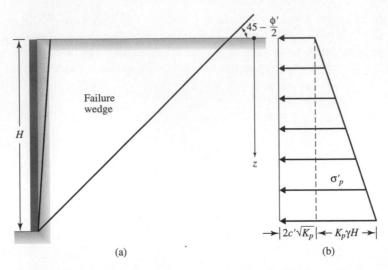

Figure 12.16 Rankine's passive earth pressure distribution against a retaining wall with cohesive soil backfill

Passive Case Figure 12.16a shows the same retaining wall with backfill similar to that considered in Figure 12.15a. Rankine's passive pressure against the wall at depth z can be given by [Eq. (12.29)]

$$\sigma'_p = K_p\gamma z + 2\sqrt{K_p}c'$$

At $z = 0$,

$$\sigma'_p = 2\sqrt{K_p}c' \tag{12.50}$$

and at $z = H$,

$$\sigma'_p = K_p\gamma H + 2\sqrt{K_p}c' \tag{12.51}$$

The variation of σ'_p with depth is shown in Figure 12.16b. The passive force per unit length of the wall can be found from the area of the pressure diagrams as

$$P_p = \tfrac{1}{2}K_p\gamma H^2 + 2\sqrt{K_p}c'H \tag{12.52}$$

For the $\phi = 0$ condition, $K_p = 1$ and

$$P_p = \tfrac{1}{2}\gamma H^2 + 2c_uH \tag{12.53}$$

Example 12.3

An 6 m high retaining wall is shown in Figure 12.17a. Determine

a. The Rankine active force per unit length of the wall and the location of the resultant

b. The Rankine passive force per unit length of the wall and the location of the resultant

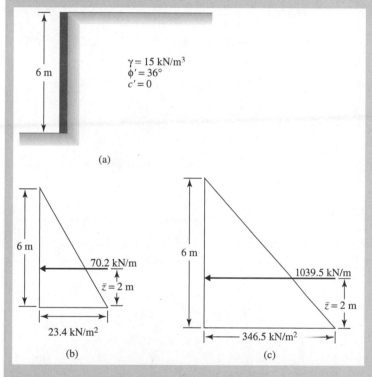

Figure 12.17
Diagrams for determining active, and passive forces

Solution

a. Because $c' = 0$, to determine the active force, we can use Eq. (12.27):

$$\sigma'_a = K_a \sigma'_o = K_a \gamma z$$

$$K_a = \frac{1 - \sin \phi'}{1 + \sin \phi'} = \frac{1 - \sin 36}{1 + \sin 36} = 0.26$$

At $z = 0$, $\sigma'_a = 0$; at $z = 6$ m,

$$\sigma'_a = (0.26)(15)(6) = 23.4 \text{ kN/m}^2$$

The pressure distribution diagram is shown in Figure 12.17b. The active force per unit length of the wall is as follows:

$$P_a = \tfrac{1}{2}(6)(23.4) = \textbf{70.2 kN/m}$$

Also,

$$\bar{z} = \frac{6 \text{ m}}{3} = \textbf{2 m}$$

b. To determine the passive force, we are given that $c' = 0$. So, from Eq. (12.30),

$$\sigma'_p = K_p\sigma'_o = K_p\gamma z$$

$$K_p = \frac{1 + \sin \phi'}{1 - \sin \phi'} = \frac{1 + \sin 36}{1 - \sin 36} = 3.85$$

At $z = 0$, $\sigma'_p = 0$; at $z = 6$ m,

$$\sigma_p = (3.85)(15)(6) = 346.5 \text{ kN/m}^2$$

The pressure distribution diagram is shown in Figure 12.17c. The passive force per unit length of the wall is

$$P_p = \tfrac{1}{2}(6)(346.5) = \mathbf{1039.5 \text{ kN/m}}$$

Also,

$$\bar{z} = \frac{6 \text{ m}}{3} = \mathbf{2 \text{ m}} \qquad \blacksquare$$

Example 12.4

For the retaining wall shown in Figure 12.18a, determine the force per unit width of the wall for Rankine's active state. Also find the location of the resultant.

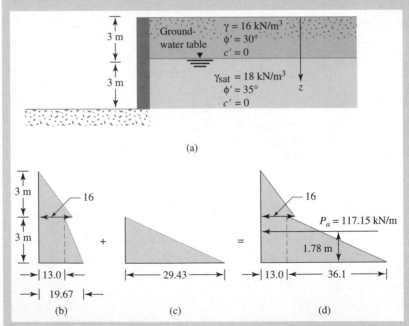

Figure 12.18 Retaining wall and pressure diagrams for determining Rankine's active earth pressure. (*Note:* The units of pressure in (b), (c), and (d) are kN/m²)

Solution

Given that $c' = 0$, we known that $\sigma'_a = K_a\sigma'_o$. For the upper layer of the soil, Rankine's active earth pressure coefficient is

$$K_a = K_{a(1)} = \frac{1 - \sin 30°}{1 + \sin 30°} = \frac{1}{3}$$

For the lower layer,

$$K_a = K_{a(2)} = \frac{1 - \sin 35°}{1 + \sin 35°} = 0.271$$

At $z = 0$, $\sigma'_o = 0$. At $z = 3$ m (just inside the bottom of the upper layer), $\sigma'_o = 3 \times 16 = 48$ kN/m². So

$$\sigma'_a = K_{a(1)}\sigma'_o = \tfrac{1}{3} \times 48 = 16 \text{ kN/m}^2$$

Again, at $z = 3$ m (in the lower layer), $\sigma'_o = 3 \times 16 = 48$ kN/m², and

$$\sigma'_a = K_{a(2)}\sigma'_o = (0.271) \times (48) = 13.0 \text{ kN/m}^2$$

At $z = 6$ m,

$$\sigma'_o = 3 \times 16 + 3(18 - 9.81) = 72.57 \text{ kN/m}^2$$
$$\uparrow$$
$$\gamma_w$$

and

$$\sigma'_a = K_{a(2)}\sigma'_o = (0.271) \times (72.57) = 19.67 \text{ kN/m}^2$$

The variation of σ'_a with depth is shown in Figure 12.18b.

The lateral pressures due to the pore water are as follows:

At $z = 0$: $u = 0$

At $z = 3$ m: $u = 0$

At $z = 6$ m: $u = 3 \times \gamma_w = 3 \times 9.81 = 29.43 \text{ kN/m}^2$

The variation of u with depth is shown in Figure 12.18c, and that for σ_a (total active pressure) is shown in Figure 12.18d. Thus,

$$P_a = (\tfrac{1}{2})(3)(16) + 3(13.0) + (\tfrac{1}{2})(3)(36.1) = 24 + 39.0 + 54.15 = \textbf{117.15 kN/m}$$

The location of the resultant can be found by taking the moment about the bottom of the wall:

$$\bar{z} = \frac{24\left(3 + \dfrac{3}{3}\right) + 39.0\left(\dfrac{3}{2}\right) + 54.15\left(\dfrac{3}{3}\right)}{117.15}$$

$$= \textbf{1.78 m}$$

Example 12.5

A frictionless retaining wall is shown in Figure 12.19a. Determine the active force, P_a, after the tensile crack occurs.

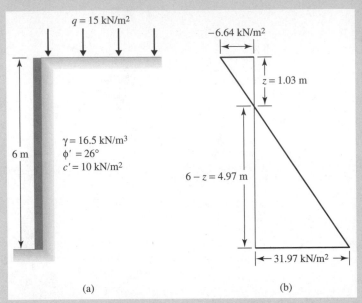

Figure 12.19 (a) Frictionless retaining wall; (b) active pressure distribution diagram

Solution

Given that $\phi' = 26°$, we have

$$K_a = \frac{1 - \sin \phi'}{1 + \sin \phi'} = \frac{1 - \sin 26}{1 + \sin 26} = 0.39$$

From Eq. (12.25),

$$\sigma_a' = K_a \sigma_o' - 2c' \sqrt{K_a}$$

At $z = 0$,

$$\sigma_a' = (0.39)(15) - (2)(10)\sqrt{0.39} = -6.64 \text{ kN/m}^2$$

At $z = 6$ m,

$$\sigma_a' = (0.39)[15 + (6)(16.5)] - (2)(10)\sqrt{0.39} = 31.97 \text{ kN/m}^2$$

The pressure distribution diagram is shown in Figure 12.19b. From this diagram,

$$\frac{6.64}{z} = \frac{31.97}{6 - z}$$

or

$$z = 1.03 \text{ m}$$

After the tensile crack occurs,

$$P_a = \tfrac{1}{2}(6 - z)(31.97) = \tfrac{1}{2}(4.97)(31.97) = \textbf{79.45 kN/m}$$

∎

Example 12.6

A frictionless retaining wall is shown in Figure 12.20a. Find the passive resistance (P_p) on the backfill and the location of the resultant passive force.

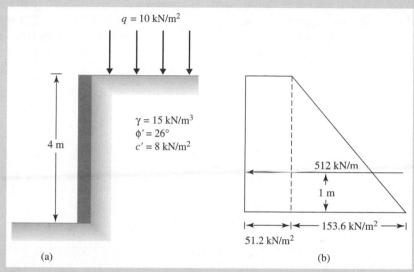

Figure 12.20 (a) Frictionless retaining wall; (b) passive pressure distribution diagram

Solution

Given that $\phi' = 26°$, it follows that

$$K_p = \frac{1 + \sin \phi'}{1 - \sin \phi'} = \frac{1 + \sin 26°}{1 - \sin 26°} = 2.56$$

From Eq. (12.29),

$$\sigma'_p = K_p \sigma'_o + 2\sqrt{K_p}\,c'$$

At $z = 0$, $\sigma'_o = 10$ kN/m²; thus,

$$\sigma'_p = (2.56)(10) + 2\sqrt{2.56}(8) = 25.6 + 25.6 = 51.2 \text{ kN/m}^2$$

Again, at $z = 4$ m, $\sigma'_o = (10 + 4 \times 15) = 70$ kN/m². So

$$\sigma'_p = (2.56)(70) + 2\sqrt{2.56}(8) = 204.8 \text{ kN/m}^2$$

The pressure distribution diagram is shown in Figure 12.20b. The passive resistance per unit width of the wall is as follows:

$$P_p = (51.2)(4) + \tfrac{1}{2}(4)(153.6) = 204.8 + 307.2 = \mathbf{512 \text{ kN/m}}$$

The location of the resultant can be found by taking the moment of the pressure diagram about the bottom of the wall. Thus,

$$\bar{z} = \frac{(51.2)\left(\dfrac{4}{2}\right) + \dfrac{1}{2}(153.6)(4)\left(\dfrac{4}{3}\right)}{512.18} = 1 \text{ m}$$

12.9 Rankine Active and Passive Pressure with Sloping Backfill

In Sections 12.5 through 12.8, we considered retaining walls with vertical backs and horizontal backfills. In some cases, however, the backfill may be continuously sloping at an angle α with the horizontal as shown in Figure 12.21 for active pressure case. In such cases, the direction of Rankine's active or passive pressures are no longer horizontal. Rather, they are inclined at an angle α with the horizontal. If the backfill is a granular soil with a drained friction angle ϕ', and $c' = 0$, then

$$\sigma_a' = \gamma z K_a$$

where

$$K_a = \text{Rankine's active pressure coefficient}$$
$$= \cos \alpha \frac{\cos \alpha - \sqrt{\cos^2 \alpha - \cos^2 \phi'}}{\cos \alpha + \sqrt{\cos^2 \alpha - \cos^2 \phi'}} \tag{12.54}$$

The active force per unit length of the wall can be given as

$$P_a = \frac{1}{2} K_a \gamma H^2 \tag{12.55}$$

The line of action of the resultant acts at a distance of $H/3$ measured from the bottom of the wall. Table 12.2 gives the values of K_a for various combinations of α and ϕ'.

In a similar manner, the *Rankine passive earth pressure* for a wall of height H with a granular sloping backfill can be represented by the equation

$$P_p = \frac{1}{2} \gamma H^2 K_p \tag{12.56}$$

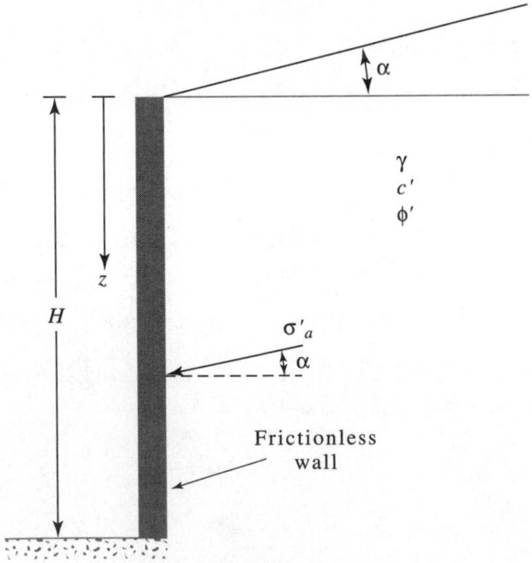

Figure 12.21
Frictionless vertical retaining wall with sloping backfill

Table 12.2 Values of K_a [Eq. (12.54)]

↓ α (deg)	ϕ' (deg) →						
	28	**30**	**32**	**34**	**36**	**38**	**40**
0	0.361	0.333	0.307	0.283	0.260	0.238	0.217
5	0.366	0.337	0.311	0.286	0.262	0.240	0.219
10	0.380	0.350	0.321	0.294	0.270	0.246	0.225
15	0.409	0.373	0.341	0.311	0.283	0.258	0.235
20	0.461	0.414	0.374	0.338	0.306	0.277	0.250
25	0.573	0.494	0.434	0.385	0.343	0.307	0.275

Table 12.3 Passive Earth Pressure Coefficient, K_p [Eq. (12.57)]

↓ α (deg)	ϕ' (deg) →						
	28	**30**	**32**	**34**	**36**	**38**	**40**
0	2.770	3.000	3.255	3.537	3.852	4.204	4.599
5	2.715	2.943	3.196	3.476	3.788	4.136	4.527
10	2.551	2.775	3.022	3.295	3.598	3.937	4.316
15	2.284	2.502	2.740	3.003	3.293	3.615	3.977
20	1.918	2.132	2.362	2.612	2.886	3.189	3.526
25	1.434	1.664	1.894	2.135	2.394	2.676	2.987

where

$$K_p = \cos \alpha \frac{\cos \alpha + \sqrt{\cos^2 \alpha - \cos^2 \phi'}}{\cos \alpha - \sqrt{\cos^2 \alpha - \cos^2 \phi'}} \tag{12.57}$$

is the passive earth pressure coefficient.

As in the case of the active force, the resultant force P_p is inclined at an angle α with the horizontal and intersects the wall at a distance of $H/3$ measured from the bottom of the wall. The values of K_p (passive earth pressure coefficient) for various values of α and ϕ' are given in Table 12.3.

c'-ϕ' Soil

The preceding analysis can be extended to the determination of the active and passive Rankine earth pressure for an inclined backfill with a c'-ϕ' soil. The details of the mathematical derivation are given by Mazindrani and Ganjali (1997). For a c'-ϕ' backfill, the active pressure is given by

$$\sigma'_a = \gamma z K_a = \gamma z K''_a \cos \alpha \tag{12.58}$$

where K_a = Rankine active earth pressure coefficient and

$$K''_a = \frac{K_a}{\cos \alpha} \tag{12.59}$$

The passive pressure is given by

$$\sigma'_p = \gamma z K_p = \gamma z K''_p \cos \alpha \tag{12.60}$$

where K_p = Rankine passive earth pressure coefficient and

$$K''_p = \frac{K_p}{\cos \alpha} \tag{12.61}$$

Also,

$$K_a'', K_p'' = \frac{1}{\cos^2 \phi'}$$

$$\times \left\{ \frac{2\cos^2\alpha + 2\left(\dfrac{c'}{\gamma z}\right)\cos\phi'\sin\phi'}{\pm\sqrt{\left[4\cos^2\alpha(\cos^2\alpha - \cos^2\phi') + 4\left(\dfrac{c'}{\gamma z}\right)^2\cos^2\phi' + 8\left(\dfrac{c'}{\gamma z}\right)\cos^2\alpha\sin\phi'\cos\phi'\right]}} \right\} - 1$$

(12.62)

Table 12.4 Variation of K_a'' and K_p''*

α (deg)	K_a''/K_p''	0.0	0.025	0.05	0.1	0.5	1.0
					$\dfrac{c'}{\gamma z}$		
a	$\phi' = 15°$						
0	K_a''	0.5888	0.5504	0.5121	0.4353	−0.1785	−0.9459
0	K_p''	1.6984	1.7637	1.8287	1.9590	3.0016	4.3048
5	K_a''	0.6069	0.5658	0.5252	0.4449	−0.1804	−0.9518
5	K_p''	1.6477	1.7156	1.7830	1.9169	2.9709	4.2782
10	K_a''	0.6738	0.6206	0.5707	0.4769	−0.1861	−0.9696
10	K_p''	1.4841	1.5641	1.6408	1.7882	2.8799	4.1993
15	K_a''	1.0000	0.7762	0.6834	0.5464	−0.1962	−1.0000
15	K_p''	1.0000	1.2506	1.3702	1.5608	2.7321	4.0718
b	$\phi' = 20°$						
0	K_a''	0.4903	0.4553	0.4203	0.3502	−0.2099	−0.9101
0	K_p''	2.0396	2.1110	2.1824	2.3252	3.4678	4.8959
5	K_a''	0.5015	0.4650	0.4287	0.3565	−0.2119	−0.9155
5	K_p''	1.9940	2.0669	2.1396	2.2846	3.4353	4.8669
10	K_a''	0.5394	0.4974	0.4564	0.3767	−0.2180	−0.9320
10	K_p''	1.8539	1.9323	2.0097	2.1622	3.3392	4.7812
15	K_a''	0.6241	0.5666	0.5137	0.4165	−0.2287	−0.9599
15	K_p''	1.6024	1.6962	1.7856	1.9556	3.1831	4.6422
c	$\phi' = 25°$						
0	K_a''	0.4059	0.3740	0.3422	0.2784	−0.2312	−0.8683
0	K_p''	2.4639	2.5424	2.6209	2.7779	4.0336	5.6033
5	K_a''	0.4133	0.3805	0.3478	0.2826	−0.2332	−0.8733
5	K_p''	2.4195	2.4989	2.5782	2.7367	3.9986	5.5713
10	K_a''	0.4376	0.4015	0.3660	0.2960	−0.2394	−0.8884
10	K_p''	2.2854	2.3680	2.4502	2.6135	3.8950	5.4765
15	K_a''	0.4860	0.4428	0.4011	0.3211	−0.2503	−0.9140
15	K_p''	2.0575	2.1474	2.2357	2.4090	3.7264	5.3228

(*continued*)

Table 12.4 gives the variation of K_a'' and K_p'' with α, $c'/\gamma z$, and ϕ'.
For the *active* case, the depth of the tensile crack can be given as

$$z_o = \frac{2c'}{\gamma} \sqrt{\frac{1 + \sin \phi'}{1 - \sin \phi'}} \tag{12.63}$$

Table 12.4 (*continued*)

α (deg)	K_a''/K_p''	$\dfrac{c'}{\gamma z}$					
		0.0	0.025	0.05	0.1	0.5	1.0
d $\phi' = 30°$							
0	K_a''	0.3333	0.3045	0.2756	0.2179	−0.2440	−0.8214
0	K_p''	3.0000	3.0866	3.1732	3.3464	4.7321	6.4641
5	K_a''	0.3385	0.3090	0.2795	0.2207	−0.2460	−0.8260
5	K_p''	2.9543	3.0416	3.1288	3.3030	4.6935	6.4282
10	K_a''	0.3549	0.3233	0.2919	0.2297	−0.2522	−0.8399
10	K_p''	2.8176	2.9070	2.9961	3.1737	4.5794	6.3218
15	K_a''	0.3861	0.3502	0.3150	0.2462	−0.2628	−0.8635
15	K_p''	2.5900	2.6836	2.7766	2.9608	4.3936	6.1489
e $\phi' = 35°$							
0	K_a''	0.2710	0.2450	0.2189	0.1669	−0.2496	−0.7701
0	K_p''	3.6902	3.7862	3.8823	4.0744	5.6112	7.5321
5	K_a''	0.2746	0.2481	0.2217	0.1688	−0.2515	−0.7744
5	K_p''	3.6413	3.7378	3.8342	4.0271	5.5678	7.4911
10	K_a''	0.2861	0.2581	0.2303	0.1749	−0.2575	−0.7872
10	K_p''	3.4953	3.5933	3.6912	3.8866	5.4393	7.3694
15	K_a''	0.3073	0.2764	0.2459	0.1860	−0.2678	−0.8089
15	K_p''	3.2546	3.3555	3.4559	3.6559	5.2300	7.1715
f $\phi' = 40°$							
0	K_a''	0.2174	0.1941	0.1708	0.1242	−0.2489	−0.7152
0	K_p''	4.5989	4.7061	4.8134	5.0278	6.7434	8.8879
5	K_a''	0.2200	0.1964	0.1727	0.1255	−0.2507	−0.7190
5	K_p''	4.5445	4.6521	4.7597	4.9747	6.6935	8.8400
10	K_a''	0.2282	0.2034	0.1787	0.1296	−0.2564	−0.7308
10	K_p''	4.3826	4.4913	4.5999	4.8168	6.5454	8.6980
15	K_a''	0.2429	0.2161	0.1895	0.1370	−0.2662	−0.7507
15	K_p''	4.1168	4.2275	4.3380	4.5584	6.3041	8.4669

*After Mazindrani and Ganjali (1997)

Example 12.7

Refer to Figure 12.21 on page 392. Given that $H = 6.1$ m, $\alpha = 5°$, $\gamma = 16.5$ kN/m³, $\phi' = 20°$, $c' = 10$ kN/m², determine the Rankine active force P_a on the retaining wall after the tensile crack occurs.

Solution

From Eq. (12.63), the depth of tensile crack is

$$z_o = \frac{2c'}{\gamma}\sqrt{\frac{1 + \sin \phi'}{1 - \sin \phi'}} = \frac{(2)(10)}{16.5}\sqrt{\frac{1 + \sin 20}{1 - \sin 20}} = 1.73 \text{ m}$$

So

$$\text{At } z = 0: \qquad \sigma_a' = 0$$

$$\text{At } z = 6.1 \text{ m:} \quad \sigma_a' = \gamma z K_a'' \cos \alpha$$

$$\frac{c'}{\gamma z} = \frac{10}{(16.5)(6.1)} \approx 0.1$$

From Table 12.4, for $\alpha = 5°$ and $c'/\gamma z = 0.1$, the magnitude of $K_a'' = 0.3565$. So

$$\sigma_a' = (16.5)(6.1)(0.3565)(\cos 5°) = 35.75 \text{ kN/m}^2$$

Hence,

$$P_a = \frac{1}{2}(H - z_o)(35.75) = \frac{1}{2}(6.1 - 1.73)(35.75) = \textbf{78.1 kN/m} \quad \blacksquare$$

COULOMB'S EARTH PRESSURE THEORY

More than 200 years ago, Coulomb (1776) presented a theory for active and passive earth pressures against retaining walls. In this theory, Coulomb assumed that the failure surface is a plane. The *wall friction* was taken into consideration. The following sections discuss the general principles of the derivation of Coulomb's earth pressure theory for a cohesionless backfill (shear strength defined by the equation $\tau_f = \sigma' \tan \phi'$).

12.10   *Coulomb's Active Pressure*

Let AB (Figure 12.22a) be the back face of a retaining wall supporting a granular soil, the surface of which is constantly sloping at an angle α with the horizontal. BC is a trial failure surface. In the stability consideration of the probable failure wedge ABC, the following forces are involved (per unit length of the wall):

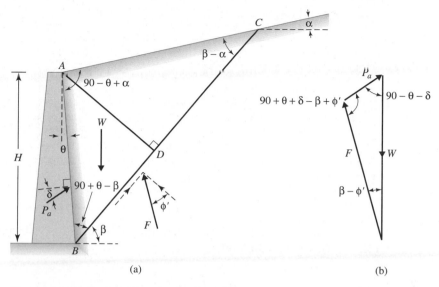

Figure 12.22 Coulomb's active pressure: (a) trial failure wedge; (b) force polygon

1. W, the weight of the soil wedge.
2. F, the resultant of the shear and normal forces on the surface of failure, BC. This is inclined at an angle of ϕ' to the normal drawn to the plane BC.
3. P_a, the active force per unit length of the wall. The direction of P_a is inclined at an angle δ to the normal drawn to the face of the wall that supports the soil. δ is the angle of friction between the soil and the wall.

The force triangle for the wedge is shown in Figure 12.22b. From the law of sines, we have

$$\frac{W}{\sin(90 + \theta + \delta - \beta + \phi')} = \frac{P_a}{\sin(\beta - \phi')} \tag{12.64}$$

or

$$P_a = \frac{\sin(\beta - \phi')}{\sin(90 + \theta + \delta - \beta + \phi')} W \tag{12.65}$$

The preceding equation can be written in the form

$$P_a = \frac{1}{2}\gamma H^2 \left[\frac{\cos(\theta - \beta)\cos(\theta - \alpha)\sin(\beta - \phi')}{\cos^2\theta \sin(\beta - \alpha)\sin(90 + \theta + \delta - \beta + \phi')} \right] \tag{12.66}$$

where $\gamma =$ unit weight of the backfill. The values of γ, H, θ, α, ϕ', and δ are constants, and β is the only variable. To determine the critical value of β for maximum P_a, we have

$$\frac{dP_a}{d\beta} = 0 \tag{12.67}$$

Table 12.5 Values of K_a [Eq. (12.69)] for $\theta = 0°$, $\alpha = 0°$

↓ ϕ' (deg)	δ (deg) →					
	0	5	10	15	20	25
28	0.3610	0.3448	0.3330	0.3251	0.3203	0.3186
30	0.3333	0.3189	0.3085	0.3014	0.2973	0.2956
32	0.3073	0.2945	0.2853	0.2791	0.2755	0.2745
34	0.2827	0.2714	0.2633	0.2579	0.2549	0.2542
36	0.2596	0.2497	0.2426	0.2379	0.2354	0.2350
38	0.2379	0.2292	0.2230	0.2190	0.2169	0.2167
40	0.2174	0.2089	0.2045	0.2011	0.1994	0.1995
42	0.1982	0.1916	0.1870	0.1841	0.1828	0.1831

After solving Eq. (12.67), when the relationship of β is substituted into Eq. (12.66), we obtain Coulomb's active earth pressure as

$$P_a = \tfrac{1}{2}K_a\gamma H^2 \qquad (12.68)$$

where K_a is Coulomb's active earth pressure coefficient and is given by

$$K_a = \frac{\cos^2(\phi' - \theta)}{\cos^2\theta\cos(\delta + \theta)\left[1 + \sqrt{\dfrac{\sin(\delta + \phi')\sin(\phi' - \alpha)}{\cos(\delta + \theta)\cos(\theta - \alpha)}}\,\right]^2} \qquad (12.69)$$

Note that when $\alpha = 0°$, $\theta = 0°$, and $\delta = 0°$, Coulomb's active earth pressure coefficient becomes equal to $(1 - \sin\phi')/(1 + \sin\phi')$, which is the same as Rankine's earth pressure coefficient given earlier in this chapter.

The variation of the values of K_a for retaining walls with a vertical back ($\theta = 0°$) and horizontal backfill ($\alpha = 0°$) is given in Table 12.5. From this table, note that for a given value of ϕ', the effect of wall friction is to reduce somewhat the active earth pressure coefficient.

12.11 *Graphic Solution for Coulomb's Active Earth Pressure*

An expedient method for creating a graphic solution of Coulomb's earth pressure theory was given by Culmann (1875). Culmann's solution can be used for any wall friction, regardless of irregularity of backfill and surcharges. Hence, it provides a powerful technique for estimating lateral earth pressure. The steps in Culmann's solution of active pressure with granular backfill ($c' = 0$) are described next, with reference to Figure 12.23a:

1. Draw the features of the retaining wall and the backfill to a convenient scale.
2. Determine the value of ψ (degrees) $= 90 - \theta - \delta$, where $\theta =$ the inclination of the back face of the retaining wall with the vertical, and $\delta =$ angle of wall friction.

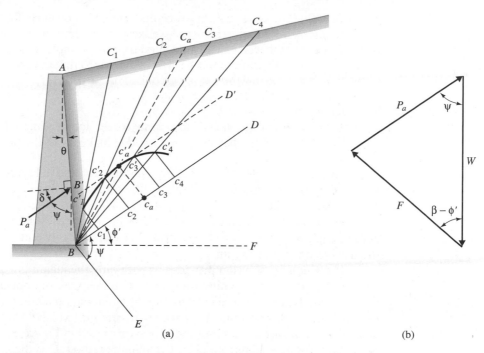

Figure 12.23 Culmann's solution for active earth pressure

3. Draw a line BD that makes an angle ϕ' with the horizontal.
4. Draw a line BE that makes an angle ψ with line BD.
5. To consider some trial failure wedges, draw lines $BC_1, BC_2, BC_3, \ldots, BC_n$.
6. Find the areas of $ABC_1, ABC_2, ABC_3, \ldots, ABC_n$.
7. Determine the weight of soil, W, per unit length of the retaining wall in each of the trial failure wedges as follows:
$$W_1 = (\text{Area of } ABC_1) \times (\gamma) \times (1)$$
$$W_2 = (\text{Area of } ABC_2) \times (\gamma) \times (1)$$
$$W_3 = (\text{Area of } ABC_3) \times (\gamma) \times (1)$$
$$\vdots$$
$$W_n = (\text{Area of } ABC_n) \times (\gamma) \times (1)$$
8. Adopt a convenient load scale and plot the weights $W_1, W_2, W_3, \ldots, W_n$ determined from step 7 on line BD. (*Note:* $Bc_1 = W_1$, $Bc_2 = W_2$, $Bc_3 = W_3, \ldots,$ $Bc_n = W_n$.)
9. Draw $c_1c_1', c_2c_2', c_3c_3', \ldots, c_nc_n'$ parallel to the line BE. (*Note:* $c_1', c_2', c_3', \ldots, c_n'$ are located on lines $BC_1, BC_2, BC_3, \ldots, BC_n$, respectively.)
10. Draw a smooth curve through points $c_1', c_2', c_3', \ldots, c_n'$. This curve is called the *Culmann line*.
11. Draw a tangent $B'D'$ to the smooth curve drawn in step 10. $B'D'$ is parallel to line BD. Let c_a' be the point of tangency.
12. Draw a line c_ac_a' parallel to the line BE.
13. Determine the active force per unit length of wall as
$$P_a = (\text{Length of } c_ac_a') \times (\text{Load scale})$$
14. Draw a line $Bc_a'C_a$. ABC_a is the desired failure wedge.

Note that the construction procedure entails, in essence, drawing a number of force polygons for a number of trial wedges and finding the maximum value of the active force that the wall can be subjected to. For example, Figure 12.23b shows the force polygon for the failure wedge ABC_a (similar to that in Figure 12.22b), in which

W = weight of the failure wedge of soil ABC_a
P_a = active force on the wall
F = the resultant of the shear and normal forces acting along BC_a
$\beta = \angle C_aBF$ (the angle that the failure wedge makes with the horizontal)

The force triangle (Figure 12.23b) is simply rotated in Figure 12.23a and is represented by the triangle Bc_ac_a'. Similarly, the force triangles Bc_1c_1', Bc_2c_2', Bc_3c_3', ..., Bc_nc_n' correspond to the trial wedges ABC_1, ABC_2, ABC_3, ..., ABC_n.

The preceding graphic procedure is given in a step-by-step manner only to facilitate basic understanding. These problems can be easily and effectively solved by the use of computer programs.

The Culmann solution provides us with only the magnitude of the active force per unit length of the retaining wall — not with the point of application of the resultant. The analytic procedure used to find the point of application of the resultant can be tedious. For this reason, an approximate method, which does not sacrifice much accuracy, can be used. This method is demonstrated in Figure 12.24, in which ABC is the failure wedge determined by Culmann's method. O is the center of gravity of the wedge ABC. If a line OO' is drawn parallel to the surface of sliding, BC, the point of intersection of this line with the back face of the wall will give the point of application of P_a. Thus, P_a acts at O' inclined at angle δ with the normal drawn to the back face of the wall.

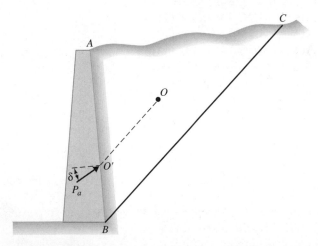

Figure 12.24 Approximate method for finding the point of application of the resultant active force

Example 12.8

A 15-ft-high retaining wall with a granular soil backfill is shown in Figure 12.25. Given that $\gamma = 100 \text{ lb/ft}^3$, $\phi' = 35°$, $\theta = 5°$, and $\delta = 10°$, determine the active thrust per foot length of the wall.

Solution

For this problem, $\psi = 90 - \theta - \delta = 90° - 5° - 10° = 75°$. The graphic construction is shown in Figure 12.25. The weights of the wedges considered are as follows:

Wedge	Weight (lb)
ABC_1	$\frac{1}{2}(4.38)(17.88)(100) = 3{,}916$
ABC_2	$3{,}916 + [\frac{1}{2}(2.36)(18.56)](100) = 6{,}106$
ABC_3	$6{,}106 + [\frac{1}{2}(2.24)(19.54)](100) = 8{,}295$
ABC_4	$8{,}295 + [\frac{1}{2}(2.11)(20.77)](100) = 10{,}486$
ABC_5	$10{,}486 + [\frac{1}{2}(1.97)(22.22)](100) = 12{,}675$

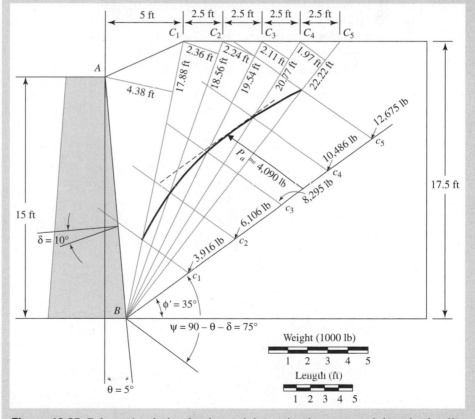

Figure 12.25 Culmann's solution for determining active thrust per unit length of wall

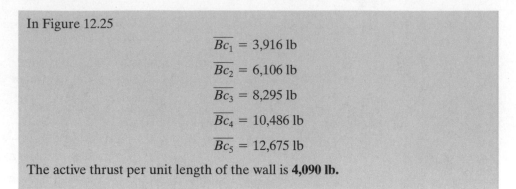

In Figure 12.25

$$\overline{Bc_1} = 3{,}916 \text{ lb}$$

$$\overline{Bc_2} = 6{,}106 \text{ lb}$$

$$\overline{Bc_3} = 8{,}295 \text{ lb}$$

$$\overline{Bc_4} = 10{,}486 \text{ lb}$$

$$\overline{Bc_5} = 12{,}675 \text{ lb}$$

The active thrust per unit length of the wall is **4,090 lb.**

12.12   *Active Force on Retaining Walls with Earthquake Forces*

Coulomb's analysis for active force on retaining walls discussed in Section 12.10 can be conveniently extended to include earthquake forces. To do so, let us consider a retaining wall of height H with a sloping *granular backfill* as shown in Figure 12.26a. Let the unit weight and the friction angle of the granular soil retained by the wall be equal to γ and ϕ', respectively. Also, let δ be the angle of friction between the soil and the wall. ABC is a trial failure wedge. The forces acting on the wedge are as follows:

1. Weight of the soil in the wedge, W
2. Resultant of the shear and normal forces on the failure surface BC, F

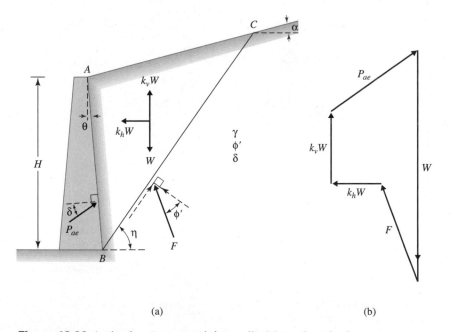

(a)   (b)

Figure 12.26 Active force on a retaining wall with earthquake forces

3. Active force per unit length of the wall, P_{ae}
4. Horizontal inertial force, $k_h W$
5. Vertical inertial force, $k_v W$

Note that

$$k_h = \frac{\text{Horizontal component of earthquake acceleration}}{g} \tag{12.70}$$

$$k_v = \frac{\text{Vertical component of earthquake acceleration}}{g} \tag{12.71}$$

where g = acceleration due to gravity.

The force polygon demonstrating these forces is shown in Figure 12.26b. The dynamic active force on the wall is the maximum value of P_{ae} exerted by any wedge. This value can be expressed as

$$P_{ae} = \tfrac{1}{2}\gamma H^2 (1 - k_v) K_a' \tag{12.72}$$

where

$$K_a' = \frac{\cos^2(\phi' - \theta - \overline{\beta})}{\cos^2\theta \cos\overline{\beta}\cos(\delta + \theta + \overline{\beta})\left\{1 + \left[\dfrac{\sin(\delta + \phi')\sin(\phi' - \alpha - \overline{\beta})}{\cos(\delta + \theta + \overline{\beta})\cos(\theta - \alpha)}\right]^{1/2}\right\}^2} \tag{12.73}$$

and

$$\overline{\beta} = \tan^{-1}\left(\frac{k_h}{1 - k_v}\right) \tag{12.74}$$

Note that with no inertia forces from earthquakes, $\overline{\beta}$ is equal to 0. Hence, $K_a' = K_a$ as given in Eq. (12.69). Equations (12.72) and (12.73) are generally referred to as the *Mononobe-Okabe equations* (Mononobe, 1929; Okabe, 1926). The variation of K_a' with $\theta = 0°$ and $k_v = 0$ is given in Table 12.6.

Considering the active force relation given by Eqs. (12.72) through (12.74), we find that the term $\sin (\phi' - \alpha - \overline{\beta})$ in Eq. (12.73) has two important implications. First, if $\phi' - \alpha - \overline{\beta} < 0$ (i.e., negative), no real solution of K_a' is possible. Physically, this implies that an *equilibrium condition will not exist*. Hence, for stability, the limiting slope of the backfill may be given as

$$\alpha \leq \phi' - \overline{\beta} \tag{12.75}$$

Table 12.6 Values of K_a' [Eq. (12.73)] with $\theta = 0°$ and $k_v = 0$

k_h	δ (deg)	α (deg)	28	30	35	40	45
					ϕ' (deg)		
0.1	0	0	0.427	0.397	0.328	0.268	0.217
0.2			0.508	0.473	0.396	0.382	0.270
0.3			0.611	0.569	0.478	0.400	0.334
0.4			0.753	0.697	0.581	0.488	0.409
0.5			1.005	0.890	0.716	0.596	0.500
0.1	0	5	0.457	0.423	0.347	0.282	0.227
0.2			0.554	0.514	0.424	0.349	0.285
0.3			0.690	0.635	0.522	0.431	0.356
0.4			0.942	0.825	0.653	0.535	0.442
0.5			—	—	0.855	0.673	0.551
0.1	0	10	0.497	0.457	0.371	0.299	0.238
0.2			0.623	0.570	0.461	0.375	0.303
0.3			0.856	0.748	0.585	0.472	0.383
0.4			—	—	0.780	0.604	0.486
0.5			—	—	—	0.809	0.624
0.1	$\phi/2$	0	0.396	0.368	0.306	0.253	0.207
0.2			0.485	0.452	0.380	0.319	0.267
0.3			0.604	0.563	0.474	0.402	0.340
0.4			0.778	0.718	0.599	0.508	0.433
0.5			1.115	0.972	0.774	0.648	0.552
0.1	$\phi/2$	5	0.428	0.396	0.326	0.268	0.218
0.2			0.537	0.497	0.412	0.342	0.283
0.3			0.699	0.640	0.526	0.438	0.367
0.4			1.025	0.881	0.690	0.568	0.475
0.5			—	—	0.962	0.752	0.620
0.1	$\phi/2$	10	0.472	0.433	0.352	0.285	0.230
0.2			0.616	0.562	0.454	0.371	0.303
0.3			0.908	0.780	0.602	0.487	0.400
0.4			—	—	0.857	0.656	0.531
0.5			—	—	—	0.944	0.722
0.1	$\frac{2}{3}\phi$	0	0.393	0.366	0.306	0.256	0.212
0.2			0.486	0.454	0.384	0.326	0.276
0.3			0.612	0.572	0.486	0.416	0.357
0.4			0.801	0.740	0.622	0.533	0.462
0.5			1.177	1.023	0.819	0.693	0.600
0.1	$\frac{2}{3}\phi$	5	0.427	0.395	0.327	0.271	0.224
0.2			0.541	0.501	0.418	0.350	0.294
0.3			0.714	0.655	0.541	0.455	0.386
0.4			1.073	0.921	0.722	0.600	0.509
0.5			—	—	1.034	0.812	0.679
0.1	$\frac{2}{3}\phi$	10	0.472	0.434	0.354	0.290	0.237
0.2			0.625	0.570	0.463	0.381	0.317
0.3			0.942	0.807	0.624	0.509	0.423
0.4			—	—	0.909	0.699	0.573
0.5			—	—	—	1.037	0.800

For no earthquake condition, $\bar{\beta} = 0°$; for stability, Eq. (12.75) gives the familiar relation

$$\alpha \leq \phi' \tag{12.76}$$

Second, for horizontal backfill, $\alpha = 0°$; for stability,

$$\bar{\beta} \leq \phi' \tag{12.77}$$

Because $\bar{\beta} = \tan^{-1}[k_h/(1 - k_v)]$, for stability, combining Eqs. (12.74) and (12.77) results in

$$k_h \leq (1 - k_v)\tan \phi' \tag{12.78a}$$

Hence, the critical value of the horizontal acceleration can be defined as

$$k_{h(\text{cr})} = (1 - k_v)\tan \phi' \tag{12.78b}$$

where $k_{h(\text{cr})}$ = critical of horizontal acceleration (Figure 12.27).

Location of Line of Action of Resultant Force, P_{ae}

Seed and Whitman (1970) proposed a simple procedure to determine the location of the line of action of the resultant, P_{ae}. Their method is as follows:

1. Let

$$P_{ae} = P_a + \Delta P_{ae} \tag{12.79}$$

where P_a = Coulomb's active force as determined from Eq. (12.68)
ΔP_{ae} = additional active force caused by the earthquake effect

2. Calculate P_a [Eq. (12.68)].
3. Calculate P_{ae} [Eq. (12.72)].
4. Calculate $\Delta P_{ae} = P_{ae} - P_a$.

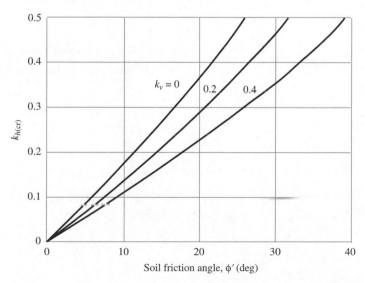

Figure 12.27 Critical values of horizontal acceleration (Eq. 12.78b)

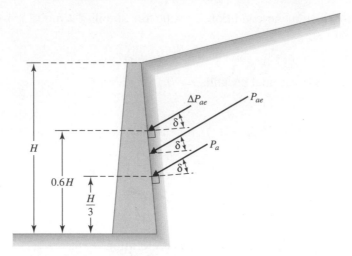

Figure 12.28 Location of the line of action of P_{ae}

5. According to Figure 12.28, P_a will act at a distance of $H/3$ from the base of the wall. Also, ΔP_{ae} will act at a distance of $0.6H$ from the base of the wall.
6. Calculate the location of P_{ae} as

$$\bar{z} = \frac{P_a\left(\dfrac{H}{3}\right) + \Delta P_{ae}(0.6H)}{P_{ae}} \tag{12.80}$$

where $\bar{z}$ = distance of the line of action of P_{ae} from the base of the wall.

Note that the line of action of P_{ae} will be inclined at an angle of δ to the normal drawn to the back face of the retaining wall. It is very important to realize that this method of determining P_{ae} is approximate and does not actually model the soil dynamics.

Example 12.9

For a retaining wall with a cohesionless soil backfill, $\gamma = 15.5$ kN/m³, $\phi' = 30°$, $\delta = 15°$, $\theta = 0°$, $\alpha = 0°$, $H = 4$ m, $k_v = 0$, and $k_h = 0.2$. Determine P_{ae}. Also determine the location of the resultant line of action of P_{ae}—that is, $\bar{z}$.

Solution
To determine P_{ae}, we use Eq. (12.72):

$$P_{ae} = \tfrac{1}{2}\gamma H^2(1 - k_v)K_a'$$

We are given that $\phi' = 30°$ and $\delta = 15°$, so

$$\delta = \frac{\phi}{2}$$

Also, $\theta = 0°$, $\alpha = 0°$, $k_h = 0.2$. From these values and Table 12.6, we find that the magnitude of K'_a is equal to 0.452. Hence,

$$P_{ae} = \tfrac{1}{2}(15.5)(4)^2(1 - 0)(0.452) = \textbf{56.05 kN/m}$$

We now locate the resultant line of action. From Eq. (12.68),

$$P_a = \tfrac{1}{2}K_a\gamma H^2$$

For $\phi' = 30°$ and $\delta = 15°$, $K_a = 0.3014$ (Table 12.5), so

$$P_a = \tfrac{1}{2}(0.3014)(15.5)(4)^2 = 37.37 \text{ kN/m}$$

Hence, $\Delta P_{ae} = 56.05 - 37.37 = 18.68$ kN/m. From Eq. (12.80),

$$\bar{z} = \frac{P_a\left(\dfrac{H}{3}\right) + \Delta P_{ae}(0.6H)}{P_{ae}} = \frac{(37.37)\left(\dfrac{4}{3}\right) + (18.68)(2.4)}{56.05} = \textbf{1.69 m} \quad\blacksquare$$

12.13 *P$_{ae}$ for c'-ϕ' Soil Backfill*

The Mononobe–Okabe equation for estimating P_{ae} for cohesionless backfill described in Section 12.12 can also be extended to c'-ϕ' soil (Prakash and Saran, 1966; Saran and Prakash, 1968). Figure 12.29 shows a retaining wall of height H with a horizontal c'-ϕ' backfill. The depth of tensile crack that may develop in a c'-ϕ' soil was given in Eq. (12.44) as

$$z_o = \frac{2c'}{\gamma\sqrt{K_a}}$$

where $K_a = \tan^2(45 - \phi'/2)$.

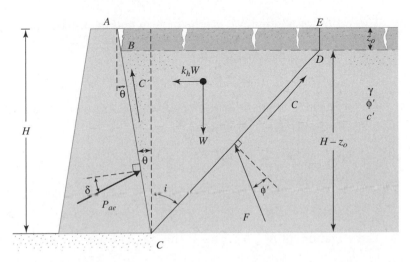

Figure 12.29 Trial failure wedge behind a retaining wall with a c'-ϕ' backfill

Refering to Figure 12.29 the forces acting on the soil wedge (per unit length of the wall) are as follows:

1. The weight of the wedge $ABCDE$, W
2. Resultant of the shear and normal forces on the failure surface CD, F
3. Active force, P_{ae}
4. Horizontal inertia force, $k_h W$
5. Cohesive force along CD, $C = c(\overline{CD})$
6. Adhesive force along BC, $C' = c(\overline{BC})$

It is important to realize that the following two assumptions have been made:

1. The vertical inertia force $(k_v W)$ has been taken to be zero.
2. The *unit adhesion* along the soil-wall interface (BC) has been taken to be equal to the cohesion (c) of the soil.

Considering these forces, we can show that

$$P_{ae} = \gamma(H - z_o)^2 N'_{a\gamma} - c'(H - z_o)N'_{ac} \tag{12.81}$$

where

$$N'_{ac} = \frac{\cos \eta' \sec \theta + \cos \phi' \sec i}{\sin(\eta' + \delta)} \tag{12.82}$$

$$N'_{a\gamma} = \frac{[(n + 0.5)(\tan \theta + \tan i) + n^2 \tan \theta][\cos(i + \phi') + k_h \sin(i + \phi')]}{\sin(\eta' + \delta)} \tag{12.83}$$

in which

$$\eta' = \theta + i + \phi' \tag{12.84}$$

$$n = \frac{z_o}{H - z_o} \tag{12.85}$$

The values of N'_{ac} and $N'_{a\gamma}$ can be determined by optimizing each coefficient separately. Thus, Eq. (12.81) gives the upper bound of P_{ae}.

For the static condition, $k_h = 0$. Thus,

$$P_{ae} = \gamma(H - z_o)^2 N_{a\gamma} - c'(H - z_o)N_{ac} \tag{12.86}$$

The relationships for N_{ac} and $N_{a\gamma}$ can be determined by substituting $k_h = 0$ into Eqs. (12.82) and (12.83). Hence,

$$N_{ac} = N'_{ac} = \frac{\cos \eta' \sec \theta + \cos \phi' \sec i}{\sin(\eta' + \delta)} \tag{12.87}$$

$$N_{a\gamma} = \frac{N'_{a\gamma}}{\lambda} = \frac{[(n + 0.5)(\tan \theta + \tan i) + n^2 \tan \theta]\cos(i + \phi')}{\sin(\eta' + \delta)} \tag{12.88}$$

The variations of N_{ac}, $N_{a\gamma}$, and λ with ϕ' and θ are shown in Figures 12.30 through 12.33.

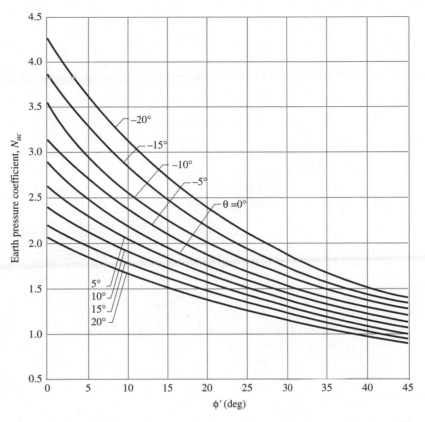

Figure 12.30 Variation of $N_{ac} = N'_{ac}$ with ϕ' and θ (based on Prakash and Saran, 1966, and Saran and Prakash, 1968)

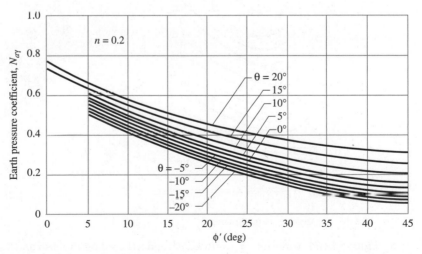

Figure 12.31 Variation of $N_{a\gamma}$ with ϕ' and θ ($n = 0.2$) (based on Prakash and Saran, 1966, and Saran and Prakash, 1968)

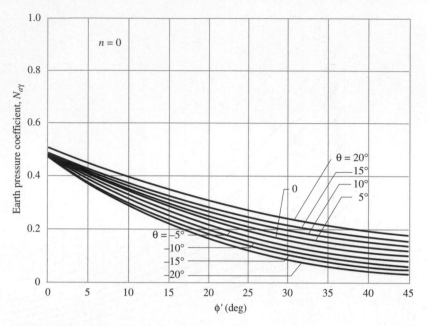

Figure 12.32 Variation of $N_{a\gamma}$ with ϕ' and θ ($n = 0$) (based on Prakash and Saran, 1966, and Saran and Prakash, 1968)

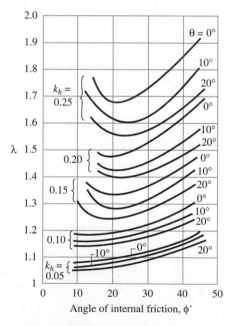

Figure 12.33 Variation of λ with k_h, ϕ', and θ (based on Prakash and Saran, 1966, and Saran and Prakash, 1968)

Example 12.10

For a retaining wall, the following are given:

$$H = 28 \text{ ft} \qquad c' = 210 \text{ lb/ft}^2$$

$$\theta = +10° \qquad \gamma = 118 \text{ lb/ft}^3$$

$$\phi' = 20° \qquad k_h = 0.1$$

Determine the magnitude of the active force, P_{ae}.

Solution
From Eq. (12.44),

$$z_o = \frac{2c'}{\gamma\sqrt{K_a}} = \frac{2c}{\gamma \tan\left(45 - \dfrac{\phi'}{2}\right)} = \frac{(2)(210)}{(118)\tan\left(45 - \dfrac{20}{2}\right)} = 5.08 \text{ ft}$$

From Eq. (12.85),

$$n = \frac{z_o}{H - z_o} = \frac{5.08}{28 - 5.08} = 0.22 \approx 0.2$$

From Eqs. (12.81), (12.87), and (12.88),

$$P_{ae} = \gamma(H - z_o)^2(\lambda N_{a\gamma}) - c'(H - z_o)N_{ac}$$

For $\theta = 10°$, $\phi' = 20°$, $k_h = 0.1$, and $n \approx 0.2$,

$$N_{ac} = 1.67 \quad \text{(Figure 12.30)}$$

$$N_{a\gamma} = 0.375 \quad \text{(Figure 12.31)}$$

$$\lambda = 1.17 \quad \text{(Figure 12.33)}$$

Thus,

$$P_{ae} = (118)(28 - 5.08)^2(1.17 \times 0.375) - (210)(28 - 5.08)(1.67)$$

$$= \textbf{19,160 lb/ft} \qquad \blacksquare$$

12.14 *Coulomb's Passive Pressure*

Figure 12.34a shows a retaining wall with a sloping cohensionless backfill similar to that considered in Figure 12.22a. The force polygon for equilibrium of the wedge ABC for the passive state is shown in Figure 12.34b. P_p is the notation for the passive force. Other notations used are the same as those for the active case (Section 12.10). In a procedure similar to the one that we followed in the active case [Eq. (12.68)], we get

$$P_p = \tfrac{1}{2}K_p\gamma H^2 \tag{12.89}$$

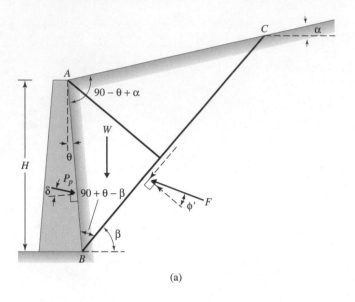

(a)

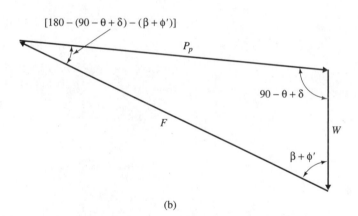

(b)

Figure 12.34 Coulomb's passive pressure: (a) trial failure wedge; (b) force polygon

where K_p = Coulomb's passive earth pressure coefficient, or

$$K_p = \frac{\cos^2(\phi' + \theta)}{\cos^2 \theta \cos(\delta - \theta)\left[1 - \sqrt{\dfrac{\sin(\phi' - \delta)\sin(\phi' + \alpha)}{\cos(\delta - \theta)\cos(\alpha - \theta)}}\right]^2} \qquad (12.90)$$

For a frictionless wall with the vertical back face supporting granular soil backfill with a horizontal surface (that is, $\theta = 0°$, $\alpha = 0°$, and $\delta = 0°$), Eq. (12.90) yields

$$K_p = \frac{1 + \sin \phi'}{1 - \sin \phi'} = \tan^2\left(45 + \frac{\phi'}{2}\right)$$

Table 12.7 Values of K_p [Eq. 12.90] for $\theta = 0°$, $\alpha = 0°$

↓ ϕ' (deg)	δ (deg) →				
	0	**5**	**10**	**15**	**20**
15	1.698	1.900	2.130	2.405	2.735
20	2.040	2.313	2.636	3.030	3.525
25	2.464	2.830	3.286	3.855	4.597
30	3.000	3.506	4.143	4.977	6.105
35	3.690	4.390	5.310	6.854	8.324
40	4.600	5.590	6.946	8.870	11.772

This relationship is the same as that obtained for the passive earth pressure coefficient in Rankine's case, given by Eq. (12.30).

The variation of K_p with ϕ' and δ (for $\theta = 0°$ and $\alpha = 0°$) is given in Table 12.7. We can see from this table that for given value of ϕ', the value of K_p increases with the wall friction.

12.15 Passive Force on Retaining Walls with Earthquake Forces

Figure 12.35 shows the failure wedge analysis for a passive force against a retaining wall of height H with a granular backfill and earthquake forces. As in Figure 12.25, the failure surface is assumed to be a plane. P_{pe} is the passive force. All other notations in Figure 12.35 are the same as those in Figure 12.26. Following a procedure similar to that used in Section 12.12, (after Kapila, 1962) we obtain

$$P_{pe} = \tfrac{1}{2}\gamma H^2(1 - k_v)K'_p \tag{12.91}$$

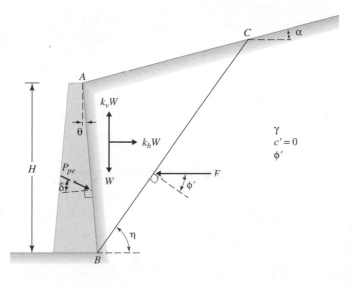

Figure 12.35
Passive force on a retaining wall with earthquake forces

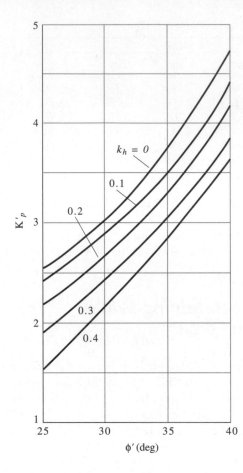

Figure 12.36
Variation of K'_p with k_h for $k_v = \alpha = \theta = \delta = 0$

where

$$K'_p = \frac{\cos^2(\phi' + \theta - \bar{\beta})}{\cos^2 \theta \cos^2 \bar{\beta} \cos(\delta - \theta + \bar{\beta})\left\{1 - \left[\dfrac{\sin(\delta + \phi')\sin(\phi' + \alpha - \bar{\beta})}{\cos(\delta - \theta + \bar{\beta})\cos(\alpha - \theta)}\right]^{1/2}\right\}^2}$$

(12.92)

in which $\bar{\beta} = \tan^{-1}\left(\dfrac{k_h}{1 - k_v}\right)$

Figure 12.36 shows a plot of K'_p with ϕ' for various values of k_h (for $k_v = \alpha = \theta = \delta = 0$).

12.16   *Summary and General Comments*

This chapter covers the general topics of lateral earth pressure, including the following:

1. At-rest earth pressure
2. Active earth pressure — Rankine's and Coulomb's

3. Passive earth pressure — Rankine's and Coulomb's
4. Pressure on retaining wall due to surcharge (based on the theory of elasticity)
5. Active and passive earth pressure, which includes earthquake forces. This is an extension of Coulomb's theory

For design, it is important to realize that the lateral active pressure on a retaining wall can be calculated using Rankine's theory only when the wall moves *sufficiently* outward by rotation about the toe of the footing or by deflection of the wall. If sufficient wall movement cannot occur (or is not allowed to occur) then the lateral earth pressure will be greater than the Rankine active pressure and sometimes may be closer to the at-rest earth pressure. Hence, proper selection of the lateral earth pressure coefficient is crucial for safe and proper design. It is a general practice to assume a value for the soil friction angle (ϕ') of the backfill in order to calculate the Rankine active pressure distribution, ignoring the contribution of the cohesion (c'). The general range of ϕ' used for the design of retaining walls is given in the following table:

Soil type	Soil friction angle, ϕ' (deg)
Soft clay	0–15
Compacted clay	20–30
Dry sand and gravel	30–40
Silty sand	20–30

In Section 12.5, we saw that the lateral earth pressure on a retaining wall is greatly increased in the presence of a water table above the base of the wall. Most retaining walls are not designed to withstand full hydrostatic pressure; hence, it is important that adequate drainage facilities are provided to ensure that the backfill soil does not become fully saturated. This can be achieved by providing weepholes at regular intervals along the length of the wall.

Problems

12.1–12.6 Assuming that the wall shown in Figure 12.37 is restrained from yielding, find the magnitude and location of the resultant lateral force per unit width of the wall.

Problem	H	γ	ϕ'
12.1	10 ft	110 lb/ft^3	32°
12.2	12 ft	98 lb/ft^3	28°
12.3	18 ft	100 lb/ft^3	40°
12.4	3 m	17.6 kN/m^3	36°
12.5	4.5 m	19.95 kN/m^3	42°
12.6	5.5 m	17.8 kN/m^3	37°

12.7 Consider a 5-m-high retaining wall that has a vertical back face with a horizontal backfill. A vertical point load of 10 kN is placed on the ground surface at a distance of 2 m from the wall. Calculate the increase in the lateral force on the wall for the section that contains the point load. Plot the variation of

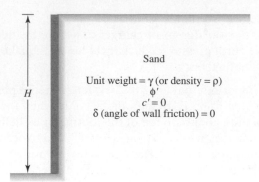

Figure 12.37

the pressure increase with depth. Use the modified equation given in Section 12.4.

12.8–12.11   Assume that the retaining wall shown in Figure 12.37 is frictionless. For each problem, determine the Rankine active force per unit length of the wall, the variation of active earth pressure with depth, and the location of the resultant.

Problem	H	ϕ' (deg)	γ
12.8	15 ft	30	105 lb/ft^3
12.9	18 ft	32	100 lb/ft^3
12.10	4 m	36	18 kN/m^3
12.11	5 m	40	17 kN/m^3

12.12–12.14   A retaining wall is shown in Figure 12.38. For each problem, determine the Rankine active force, P_a, per unit length of the wall and the location of the resultant.

Problem	H	H_1	γ_1	γ_2	ϕ'_1 (deg)	ϕ'_2 (deg)	q
12.12	10 ft	5 ft	105 lb/ft^3	122 lb/ft^3	30	30	0
12.13	20 ft	6 ft	110 lb/ft^3	126 lb/ft^3	34	34	300 lb/ft^2
12.14	6 m	3 m	15.5 kN/m^3	19.0 kN/m^3	30	36	15 kN/m^2

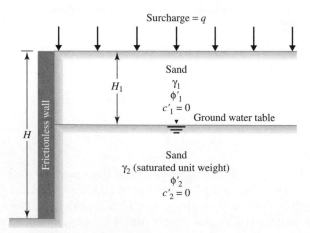

Figure 12.38

12.15 A 15-ft-high retaining wall with a vertical back face retains a homogeneous saturated soft clay. The saturated unit weight of the clay is 122 lb/ft^3. Laboratory tests showed that the undrained shear strength c_u of the clay is equal to 350 lb/ft^2.

    a. Make the necessary calculations and draw the variation of Rankine's active pressure on the wall with depth.

    b. Find the depth up to which a tensile crack can occur.

    c. Determine the total active force per unit length of the wall before the tensile crack occurs.

    d. Determine the total active force per unit length of the wall after the tensile crack occurs. Also find the location of the resultant.

12.16 Redo Problem 12.15 assuming that the backfill is supporting a surcharge of 200 lb/ft^2.

12.17 A 5-m-high retaining wall with a vertical back face has a c'-ϕ' soil for backfill. For the backfill, $\gamma = 19$ kN/m^3, $c' = 26$ kN/m^2, and $\phi' = 16°$. Considering the existence of the tensile crack, determine the active force P_a on the wall for Rankine's active state.

12.18 For the retaining wall shown in Figure 12.39, determine the active force P_a for Rankine's state. Also, find the position of the resultant. Assume that the tensile crack exists.

$$\rho = 2100 \text{ kg/m}^3, \phi = 0°, c = c_u = 30.2 \text{ kN/m}^2$$

12.19 Repeat Problem 12.18 using the following values:

$$\rho = 1950 \text{ kg/m}^3, \phi' = 18°, c' = 19.4 \text{ kN/m}^2$$

12.20–12.23 Assume that the retaining wall shown in Figure 12.37 is frictionless. For each problem, determine the Rankine passive force per unit length of the wall, the variation of lateral pressure with depth, and the location of the resultant.

Problem	H	ϕ' (deg)	γ
12.20	8 ft	34	110 lb/ft^3
12.21	10 ft	36	105 lb/ft^3
12.22	5 m	35	14 kN/m^3
12.23	4 m	30	15 kN/m^3

12.24 For the retaining wall described in Problem 12.12, determine the Rankine passive force per unit length of the wall and the location of the resultant.

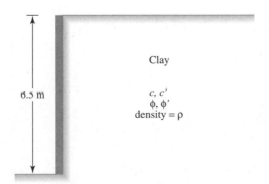

6.5 m

Clay

c, c'
ϕ, ϕ'
density $= \rho$

Figure 12.39

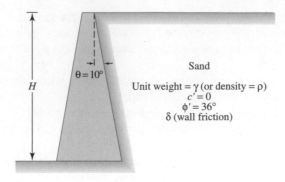

Figure 12.40

12.25 For the retaining wall described in Problem 12.13, determine the Rankine passive force per unit length of the wall and the location of the resultant.

12.26 A retaining wall is shown in Figure 12.40. The height of the wall is 5 m, and the unit weight of the sand backfill is 18 kN/m³. Using Coulomb's equation, calculate the active force P_a on the wall for the following values of the angle of wall friction:

a. $\delta = 18°$

b. $\delta = 24°$

Comment on the direction and location of the resultant.

12.27 Referring to Figure 12.41, determine Coulomb's active force P_a per unit length of the wall for the following cases:

a. $H = 15$ ft, $\beta = 85°$, $n = 1$, $H_1 = 20$ ft, $\gamma = 128$ lb/ft³, $\phi' = 38°$, $\delta = 20°$

b. $H = 18$ ft, $\beta = 90°$, $n = 2$, $H_1 = 22$ ft, $\gamma = 116$ lb/ft³, $\phi' = 34°$, $\delta = 17°$

c. $H = 5.5$ m, $\beta = 80°$, $n = 1$, $H_1 = 6.5$ m, $\gamma = 1680$ kg/m³, $\phi' = 30°$, $\delta = 30°$

Use Culmann's graphic construction procedure.

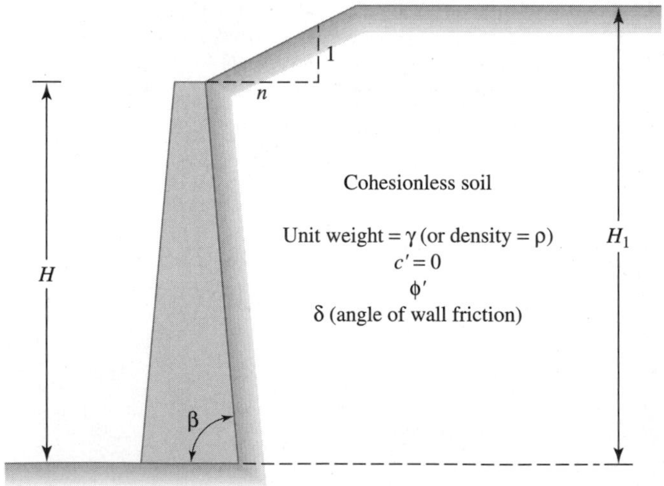

Figure 12.41

12.28 Refer to Figure 12.26. Given that $H = 6$ m, $\theta = 0°$, $\alpha = 0°$, $\gamma = 15$ kN/m^3, $\phi' = 35°$, $\delta = 2/3\ \phi'$, $k_h = 0.3$, and $k_v = 0$, determine the active force P_{ae} per unit length of the retaining wall.

12.29 Refer to Problem 12.28. Determine the location of the point of intersection of the resultant force P_{ae} with the back face of the retaining wall.

12.30 Repeat Problem 12.28 with the following Values: $H = 10$ ft, $\theta = 10°$, $\alpha = 10°$, $\gamma = 110$ lb/ft^3, $\phi' = 30°$, $\delta = 10°$, $k_h = 0.25$, and $k_v = 0$.

12.31 Refer to Figure 12.29. Given that $H = 6$ m, $\theta = 10°$, $\phi' = 15°$, $c' = 20$ kN/m^2, $\gamma = 19$ kN/m^3, and $k_h = 0.15$, using the method described in Section 12.13, determine P_{ae}. Assume that the depth of tensile crack is zero.

12.32 Repeat Problem 12.31 with the following Values: $H = 10$ ft, $\theta = 5°$, $\phi' = 20°$, $c' = 200$ lb/ft^2, $\gamma = 100$ lb/ft^3, and $k_h = 0.25$.

References

COULOMB, C. A. (1776). "Essai sur une Application des Règles de Maximis et Minimis a quelques Problèmes de Statique, relatifs a l'Architecture," *Mem. Roy. des Sciences,* Paris, Vol. 3, 38.

CULMANN, C. (1875). *Die graphische Statik,* Meyer and Zeller, Zurich.

GERBER, E. (1929). *Untersuchungen über die Druckverteilung im Örlich belasteten Sand,* Technische Hochschule, Zurich.

JAKY, J. (1944). "The Coefficient of Earth Pressure at Rest," *Journal of the Society of Hungarian Architects and Engineers,* Vol. 7, 355–358.

KAPILA, J. P. (1962). "Earthquake Resistant Design of Retaining Walls," *Proceedings, 2nd Earthquake Symposium,* University of Roorkee, Roorkee, India.

MASSARSCH, K. R. (1979). "Lateral Earth Pressure in Normally Consolidated Clay," *Proceedings of the Seventh European Conference on Soil Mechanics and Foundation Engineering,* Brighton, England, Vol. 2, 245–250.

MAZINDRANI, Z. H., and GANJALI, M. H. (1997). "Lateral Earth Pressure Problem of Cohesive Backfill with Inclined Surface," *Journal of Geotechnical and Geoenvironmental Engineering,* ASCE, Vol. 123, No. 2, 110–112.

OKABE, S. (1926). "General Theory of Earth Pressure," *Journal of the Japanese Society of Civil Engineers,* Tokyo, Vol. 12, No. 1.

PRAKASH, S., and SARAN, S. (1966). "Static and Dynamic Earth Pressure Behind Retaining Walls," *Proceedings,* 3rd Symposium on Earthquake Engineering, Roorkee, India, Vol. 1, 277–288.

RANKINE, W. M. J. (1857). "On Stability on Loose Earth," *Philosophic Transactions of Royal Society,* London, Part I, 9–27.

SARAN, S., and PRAKASH, S. (1968). "Dimensionless Parameters for Static and Dynamic Earth Pressure for Retaining Walls," *Indian Geotechnical Journal,* Vol. 7, No. 3, 295–310.

SEED, H. B., and WHITMAN, R. V. (1970). "Design of Earth Retaining Structures for Dynamic Loads," *Proceedings,* Specialty Conference on Lateral Stresses in the Ground and Design of Earth Retaining Structures, ASCE, 103–147.

SHERIF, M. A., FANG, Y. S., and SHERIF, R. I. (1984). "K_A and K_O Behind Rotating and Non-Yielding Walls," *Journal of Geotechnical Engineering,* ASCE, Vol. 110, No. GT1, 41–56.

SPANGLER, M. G. (1938). "Horizontal Pressures on Retaining Walls Due to Concentrated Surface Loads," Iowa State University Engineering Experiment Station, *Bulletin,* No. 140.

13

Lateral Earth Pressure—Curved Failure Surface

In Chapter 12, we considered Coulomb's earth pressure theory, in which the retaining wall was considered to be rough. The potential failure surfaces in the backfill were considered to be planes. In reality, most failure surfaces in soil are curved. There are several instances where the assumption of plane failure surfaces in soil may provide unsafe results. Examples of these cases are the estimation of passive pressure and braced cuts. This chapter describes procedures by which passive earth pressure and lateral earth pressure on braced cuts can be estimated using curved failure surfaces in the soil.

13.1 Retaining Walls with Friction

In reality, retaining walls are rough, and shear forces develop between the face of the wall and the backfill. To understand the effect of wall friction on the failure surface, let us consider a rough retaining wall AB with a horizontal granular backfill as shown in Figure 13.1.

In the active case (Figure 13.1a), when the wall AB moves to a position $A'B$, the soil mass in the active zone will be stretched outward. This will cause a downward motion of the soil relative to the wall. This motion causes a downward shear on the wall (Figure 13.1b), and it is called a *positive wall friction in the active case*. If δ is the angle of friction between the wall and the backfill, then the resultant active force P_a will be inclined at an angle δ to the normal drawn to the back face of the retaining wall. Advanced studies show that the failure surface in the backfill can be represented by BCD, as shown in Figure 13.1a. The portion BC is curved, and the portion CD of the failure surface is a straight line. Rankine's active state exists in the zone ACD.

Under certain conditions, if the wall shown in Figure 13.1a is forced downward with reference to the backfill, the direction of the active force P_a will change as shown in Figure 13.1c. This is a situation of negative wall friction ($-\delta$) in the active case. Figure 13.1c also shows the nature of the failure surface in the backfill.

The effect of wall friction for the passive state is shown in Figures 13.1d and e. When the wall AB is pushed to a position $A'B$ (Figure 13.1d), the soil in the passive

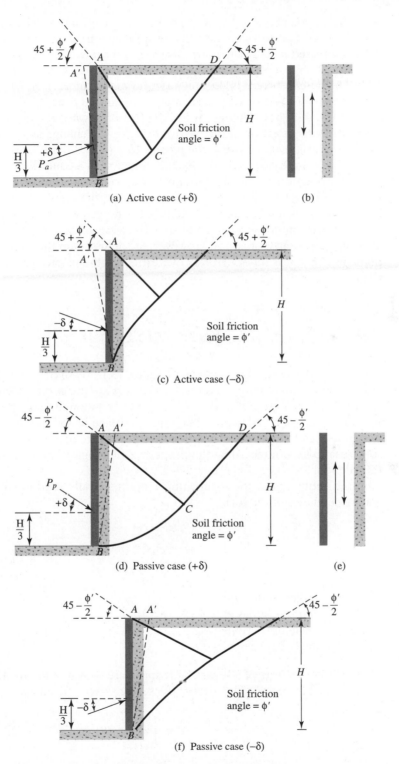

Figure 13.1 Effect of wall friction on failure surface

zone will be compressed. The result is an upward motion relative to the wall. The upward motion of the soil will cause an upward shear on the retaining wall (Figure 13.1e). This is referred to as *positive wall friction in the passive case.* The resultant passive force, P_p, will be inclined at an angle δ to the normal drawn to the back face of the wall. The failure surface in the soil has a curved lower portion *BC* and a straight upper portion *CD*. Rankine's passive state exists in the zone *ACD*.

If the wall shown in Figure 13.1d is forced upward relative to the backfill by a force, then the direction of the passive force P_p will change as shown in Figure 13.1f. This is *negative wall friction in the passive case* ($-\delta$). Figure 13.1f also shows the nature of the failure surface in the backfill under such a condition.

For practical considerations, in the case of loose granular backfill, the angle of wall friction δ is taken to be equal to the angle of friction of soil, ϕ'. For dense granular backfills, δ is smaller than ϕ' and is in the range of $\phi'/2 \leq \delta \leq (2/3)\phi'$.

The assumption of plane failure surface gives reasonably good results while calculating active earth pressure. However, the assumption that the failure surface is a plane in Coulomb's theory grossly overestimates the passive resistance of walls, particularly for $\delta > \phi'/2$.

13.2   *Properties of a Logarithmic Spiral*

The case of passive pressure shown in Figure 13.1d (case of $+\delta$) is the most common one encountered in design and construction. Also, the curved failure surface represented by *BC* in Figure 13.1d is most commonly assumed to be the arc of a logarithmic spiral. In a similar manner, the failure surface in soil in the case of braced cuts (Sections 13.6 to 13.10) is also assumed to be the arc of a logarithmic spiral. Hence, some useful ideas concerning the properties of a logarithmic spiral are described in this section.

The equation of the logarithmic spiral generally used in solving problems in soil mechanics is of the form

$$r = r_o e^{\theta \tan \phi'}$$

(13.1)

where r = radius of the spiral
r_o = starting radius at $\theta = 0$
ϕ' = angle of friction of soil
θ = angle between r and r_o

The basic parameters of a logarithmic spiral are shown in Figure 13.2, in which O is the center of the spiral. The area of the sector *OAB* is given by

$$A = \int_0^\theta \frac{1}{2} r(r\, d\theta)$$

(13.2)

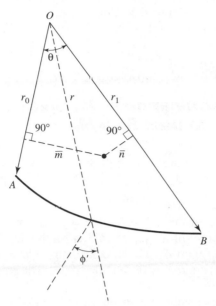

Figure 13.2
General parameters of a logarithmic spiral

Substituting the values of r from Eq. (13.1) into Eq. (13.2), we get

$$A = \int_0^{\theta_1} \frac{1}{2} r_o^2 e^{2\theta \tan \phi'} \, d\theta$$

$$= \frac{r_1^2 - r_o^2}{4 \tan \phi'} \tag{13.3}$$

The location of the centroid can be defined by the distances $\overline{m}$ and $\overline{n}$ (Figure 13.2), measured from OA and OB, respectively, and can be given by the following equations (Hijab, 1956):

$$\overline{m} = \frac{4}{3} r_o \frac{\tan \phi'}{(9 \tan^2 \phi' + 1)} \left[\frac{\left(\frac{r_1}{r_o}\right)^3 (3 \tan \phi' \sin \theta - \cos \theta) + 1}{\left(\frac{r_1}{r_o}\right)^2 - 1} \right] \tag{13.4}$$

$$\overline{n} = \frac{4}{3} r_o \frac{\tan \phi'}{(9 \tan^2 \phi' + 1)} \left[\frac{\left(\frac{r_1}{r_o}\right)^3 - 3 \tan \phi' \sin \theta - \cos \theta}{\left(\frac{r_1}{r_o}\right)^2 \quad 1} \right] \tag{13.5}$$

Another important property of the logarithmic spiral defined by Eq. (13.1) is that any radial line makes an angle ϕ' with the normal to the curve drawn at the point where the radial line and the spiral intersect. This basic property is particularly useful in solving problems related to lateral earth pressure.

PASSIVE EARTH PRESSURE

13.3 Procedure for Determination of Passive Earth Pressure, P_p (Cohesionless Backfill)

Figure 13.1d shows the curved failure surface in the granular backfill of a retaining wall of height H. The shear strength of the granular backfill is expressed as

$$\tau_f = \sigma' \tan \phi' \tag{13.6}$$

The curved lower portion BC of the failure wedge is an arc of a logarithmic spiral defined by Eq. (13.1). The center of the log spiral lies on the line CA (not necessarily within the limits of points C and A). The upper portion CD is a straight line that makes an angle of $(45 - \phi'/2)$ degrees with the horizontal. The soil in the zone ACD is in *Rankine's passive state*.

Figure 13.3 shows the procedure for evaluating the passive resistance by trial wedges (Terzaghi and Peck, 1967). The retaining wall is first drawn to scale as shown in Figure 13.3a. The line C_1A is drawn in such a way that it makes an angle of $(45 - \phi'/2)$ degrees with the surface of the backfill. BC_1D_1 is a trial wedge in which BC_1 is the arc of a logarithmic spiral. According to the equation $r_1 = r_o e^{\theta \tan \phi'}$, O_1 is the center of the spiral. (*Note:* $\overline{O_1B} = r_o$ and $\overline{O_1C_1} = r_1$ and $\angle BO_1C_1 = \theta_1$; refer to Figure 13.2.)

Now let us consider the stability of the soil mass ABC_1C_1' (Figure 13.3b). For equilibrium, the following forces per unit length of the wall are to be considered:

1. Weight of the soil in zone $ABC_1C_1' = W_1 = (\gamma)(\text{Area of } ABC_1C_1')(1)$.
2. The vertical face, C_1C_1', is in the zone of Rankine's passive state; hence, the force acting on this face is

$$P_{d(1)} = \frac{1}{2}\gamma(d_1)^2 \tan^2\left(45 + \frac{\phi'}{2}\right) \tag{13.7}$$

   where $d_1 = \overline{C_1C_1'}$. $P_{d(1)}$ acts horizontally at a distance of $d_1/3$ measured vertically upward from C_1.
3. F_1 is the resultant of the shear and normal forces that act along the surface of sliding, BC_1. At any point on the curve, according to the property of the logarithmic spiral, a radial line makes an angle ϕ' with the normal. Because the resultant, F_1, makes an angle ϕ' with the normal to the spiral at its point of application, its line of application will coincide with a radial line and will pass through the point O_1.
4. P_1 is the passive force per unit length of the wall. It acts at a distance of $H/3$ measured vertically from the bottom of the wall. The direction of the force P_1 is inclined at an angle δ with the normal drawn to the back face of the wall.

Now, taking the moments of $W_1, P_{d(1)}, F_1$, and P_1 about the point O_1, for equilibrium, we have

$$W_1[l_{W(1)}] + P_{d(1)}[l_1] + F_1[0] = P_1[l_{P(1)}] \tag{13.8}$$

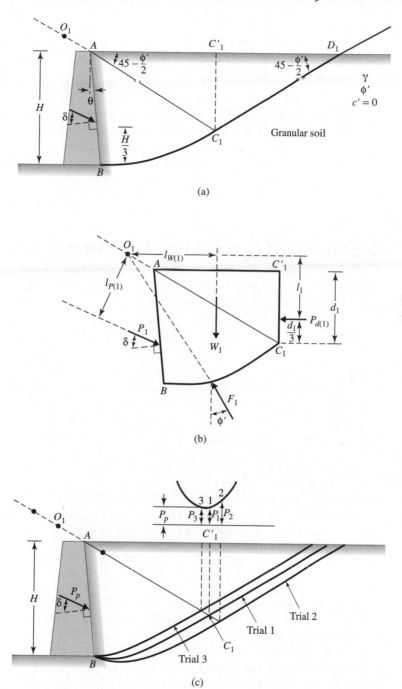

Figure 13.3 Passive earth pressure against retaining wall with curved failure surface

or

$$P_1 = \frac{1}{l_{P(1)}}[W_1 l_{W(1)} + P_{d(1)} l_1] \tag{13.9}$$

where $l_{W(1)}$, l_1, and $l_{P(1)}$ are moment arms for the forces W_1, $P_{d(1)}$, and P_1, respectively.

The preceding procedure for finding the trial passive force per unit length of the wall is repeated for several trial wedges such as those shown in Figure 13.3c. Let $P_1, P_2, P_3, \ldots, P_n$ be the forces that correspond to trial wedges $1, 2, 3, \ldots, n$, respectively. The forces are plotted to some scale as shown in the upper part of the figure. A smooth curve is plotted through the points $1, 2, 3, \ldots, n$. The lowest point of the smooth curve defines the actual passive force, P_p, per unit length of the wall.

13.4 *Coefficient of Passive Earth Pressure (K$_p$)*

The passive force per unit length of a rough retaining wall with a cohesionless horizontal backfill can be calculated as

$$P_p = \frac{1}{2}\gamma H_1^2 K_p \tag{13.10}$$

where K_p = passive pressure coefficient.

For the definition of H_1, see Figure 13.4. The variation of K_p determined by Caquot and Kerisel (1948) is also shown in Figure 13.4.

It is important to note that the K_p values shown in Figure 13.4 are for $\delta/\phi' = 1$. If $\delta/\phi' \neq 1$, the following procedure must be used to determine K_p:

1. Assume δ and ϕ'.
2. Calculate δ/ϕ'.
3. Using the ratio of δ/ϕ' (step 2), determine the reduction factor, R, from Table 13.1.
4. Determine K_p from Figure 13.4 for $\delta/\phi' = 1$.
5. Calculate K_p for the required δ/ϕ' as

$$K_p = (R)[K_{p(\delta/\phi'=1)}] \tag{13.11}$$

Shields and Tolunay (1973) improved the trial wedge solution described in Section 13.3 by using the *method of slices* to consider the stability of the trial soil wedge such as $ABC_1 C_1'$ in Figure 13.3a. The details of the analysis are beyond the scope of this text. However, the values of K_p (passive earth pressure coefficient) obtained by this method are given in Table 13.2, and they seem to be as good as any other set of values available currently. Note that the values of K_p shown in Table 13.1 are for retaining walls with a vertical back (that is, $\theta = 0$ in Figure 13.3) supporting a granular backfill with a horizontal ground surface. The passive pressure for such a case can be given as

$$P_p = \frac{1}{2}\gamma H^2 K_p$$

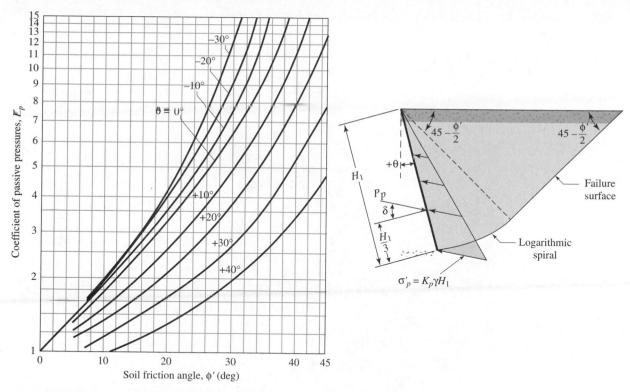

Figure 13.4 Caquot and Kerisel's solution for K_p

Table 13.1 Caquot and Kerisel's Reduction Factor, R, for Passive Pressure Calculation

	δ/ϕ'							
ϕ'	0.7	0.6	0.5	0.4	0.3	0.2	0.1	0.0
10	0.978	0.962	0.946	0.929	0.912	0.898	0.881	0.864
15	0.961	0.934	0.907	0.881	0.854	0.830	0.803	0.775
20	0.939	0.901	0.862	0.824	0.787	0.752	0.716	0.678
25	0.912	0.860	0.808	0.759	0.711	0.666	0.620	0.574
30	0.878	0.811	0.746	0.686	0.627	0.574	0.520	0.467
35	0.836	0.752	0.674	0.603	0.536	0.475	0.417	0.362
40	0.783	0.682	0.592	0.512	0.439	0.375	0.316	0.262
45	0.718	0.600	0.500	0.414	0.339	0.276	0.221	0.174

Table 13.2 Shields and Tolunay's Values of K_p Based on the Method of Slices

	δ *(deg)*									
ϕ' *(deg)*	0	5	10	15	20	25	30	35	40	45
20	2.04	2.26	2.43	2.55	2.70					
25	2.46	2.77	3.03	3.23	3.39	3.63				
30	3.00	3.43	3.80	4.13	4.40	4.64	5.03			
35	3.69	4.29	4.84	5.34	5.80	6.21	6.59	7.25		
40	4.69	5.44	6.26	7.05	7.80	8.51	9.18	9.83	11.03	
45	5.83	7.06	8.30	9.55	10.80	12.04	13.26	14.46	15.60	18.01

Example 13.1

Consider a 3-m-high (H) retaining wall with a vertical back ($\theta = 0°$) and a horizontal granular backfill. Given: $\gamma = 15.7$ kN/m³, $\delta = 15°$, and $\phi' = 30°$. Estimate the passive force, P_p, by using

a. Coulomb's theory
b. curved failure surface assumption (Caquot and Kerisel solution)
c. Shields and Tolunay's solutions

Solution

a. From Eq. (12.89),

$$P_p = \tfrac{1}{2} K_p \gamma H^2$$

From Table 12.7, for $\phi' = 30°$ and $\delta = 15°$, the value of K_p is 4.977. Thus,

$$P = (\tfrac{1}{2})(4.977)(15.7)(3)^2 = \textbf{351.6 kN/m}$$

b. From Eq. (13.10),

$$P_p = \tfrac{1}{2} K_p \gamma H_1^2 = \tfrac{1}{2} K_p \gamma H^2 \qquad \text{(because } \theta = 0°, H = H_1)$$

From Figure 13.4, for $\theta = 0°$, $\delta/\phi' = 1$, and $\phi' = 30°$, the value of K_p is 6.4. From Table 13.1, for $\delta/\phi' = 15/30 = 0.5$, the reduction factor, R, is 0.746. Thus, per Eq. (13.11),

$$K_p = (0.746)(6.4) = 4.77$$

So

$$P_p = \tfrac{1}{2}(4.77)(15.7)(3)^2 = \textbf{337 kN/m}^3$$

c. $P_p = \dfrac{1}{2} K_p \gamma H^2$

From Table 13.2, for $\phi' = 30°$ and $\delta = 15°$, the value of K_p is 4.13. Hence,

$$P_p = \left(\frac{1}{2}\right)(4.13)(15.7)(3)^2 \approx \textbf{292 kN/m}$$

■

13.5   *Passive Force on Walls with Earthquake Forces*

The passive force on retaining walls with earthquake forces was discussed in Section 12.15. In that analysis, the backfill was considered to be a granular soil, and the failure surface in the backfill was assumed to be a plane. It was shown in Sections 13.3

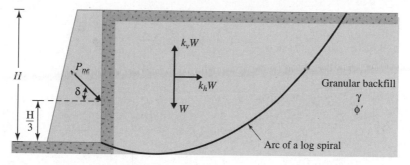

Figure 13.5 Logarithmic spiral failure surface for determination of P_{pe}

and 13.4 that for static conditions and when $\delta > \phi'/2$, the passive force calculated by assuming a plane failure surface in the backfill becomes unsafe. For that reason, Morrison and Ebeling (1995) assumed that the failure surface was an arc of a logarithmic spiral as defined by Eq. (13.1), and they calculated the magnitude of passive force by including earthquake forces (Figure 13.5). In Figure 13.5, the back face of the retaining wall is *vertical* and the backfill is *horizontal.* Also,

$$H = \text{height of retaining wall}$$

$$W = \text{weight of failure wedge}$$

$$P_{pe} = \text{Passive force per unit length of the wall}$$

$$\delta = \text{angle of wall friction}$$

$$k_h = \frac{\text{horizontal component of earthquake acceleration}}{\text{acceleration due to gravity, } g}$$

$$k_v = \frac{\text{vertical component of earthquake acceleration}}{\text{acceleration due to gravity, } g}$$

Based on Morrison and Ebeling's analysis, the passive force can be given as

$$P_{pe} = \frac{1}{2}\gamma H^2 K'_p \tag{13.12}$$

Figure 13.6 shows variation of K'_p with k_h and ϕ' for the Mononobe–Okabe solution [Eq. (12.92)] and for the logarithmic spiral type of failure surface analysis, with $\delta = 2\phi'/3$, $k_v = 0$, $\theta = 0°$, and $\alpha = 0°$. As we can see from the figure, for a given value of k_h, the magnitude of K'_p is always larger when the failure surface is assumed to be a plane (Mononobe–Okabe solution). This is true for all values of ϕ'. Figure 13.7 shows the variation of K'_p with k_h and δ for the Mononobe–Okabe solution and the logarithmic spiral solution, with $k_v = 0$, $\phi' = 30°$, $\theta = 0°$, $\alpha = 0°$, and $\delta = 0$, $\phi'/2$, $2\phi'/3$, and ϕ'.

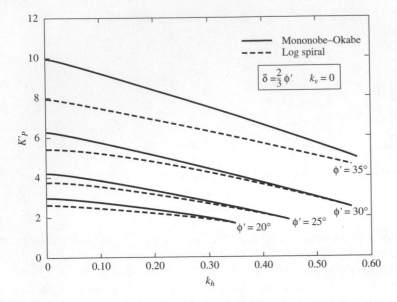

Figure 13.6
Variation of K'_p with k_h and ϕ' ($\delta = \frac{2}{3}\phi'$, $k_v = 0$, $\theta = 0°$, and $\alpha = 0°$) (based on Morrison and Ebeling, 1995)

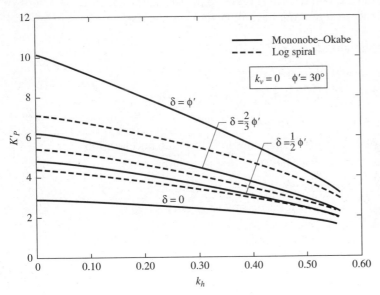

Figure 13.7
Variation of K'_p with k_h and δ ($k_v = 0$, $\phi' = 30°$, $\theta = 0°$, $\alpha = 0°$) (based on Morrison and Ebeling, 1995)

BRACED CUTS

13.6 *Braced Cuts—General*

Frequently during the construction of foundations or utilities (such as sewers), open trenches with vertical soil slopes are excavated. Although most of these trenches are temporary, the sides of the cuts must be supported by proper bracing systems. Figure 13.8 shows one of several bracing systems commonly adopted in construction practice. The bracing consists of sheet piles, wales, and struts.

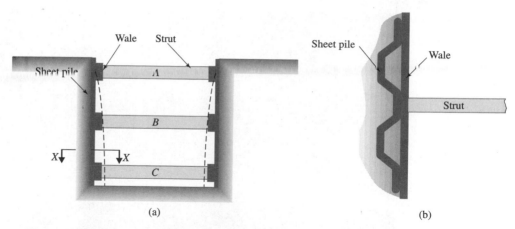

Figure 13.8 Braced cut: (a) cross section; (b) plan (section at X-X).

Proper design of these elements requires a knowledge of the lateral earth pressure exerted on the braced walls. The magnitude of the lateral earth pressure at various depths of the cut is very much influenced by the deformation condition of the sheeting. To understand the nature of the deformation of the braced walls, one needs to follow the sequence of construction. Construction of the unit begins with driving the sheetings. The top row of the wales and struts (marked A in Figure 13.8a) is emplaced immediately after a small cut is made. This emplacement must be done immediately so that the soil mass outside the cut has no time to deform and cause the sheetings to yield. As the sequence of driving the sheetings, excavating the soil, and placing rows of wales and struts (see B and C in Figure 13.8) continues, the sheetings move inward at greater depths. This action is caused by greater earth pressure exerted by the soil outside the cut. The deformation of the braced walls is shown by the broken lines in Figure 13.8a. Essentially, the problem models a condition where the walls are rotating about the level of the top row of struts. A photograph of braced cuts made for subway construction in Chicago is shown in Figure 13.9a. Figures 13.9b and 13.9c are photographs of two braced cuts — one in Seoul, South Korea, and the other in Taiwan.

The deformation of a braced wall differs from the deformation condition of a retaining wall in that, in a braced wall, the rotation is about the top. For this reason, neither Coulomb's nor Rankine's theory will give the actual earth pressure distribution. This fact is illustrated in Figure 13.10 on page 433, in which AB is a frictionless wall with a granular soil backfill. When the wall deforms to position AB', failure surface BC develops. Because the upper portion of the soil mass in the zone ABC does not undergo sufficient deformation, it does not pass into Rankine's active state. The sliding surface DC intersects the ground surface almost at 90°. The corresponding earth pressure will be somewhat parabolic, like acb shown in Figure 13.10b. With this type of pressure distribution, the point of application of the resultant active thrust P_a, will be at a height of $n_a H$ measured from the bottom of the wall, with $n_a > \frac{1}{3}$ (for triangular pressure distribution $n_a = \frac{1}{3}$). Theoretical evaluation and field measurements have shown that n_a could be as high as 0.55.

(a)

(b)

Figure 13.9 Braced cuts: (a) Chicago subway construction (courtesy of Ralph B. Peck); (b) in Seoul, South Korea (courtesy of E. C. Shin, University of Inchon, South Korea); (c) in Taiwan (courtesy of Richard Tsai, C&M Hi-Tech Engineering Co., Ltd., Taipei, Taiwan)

(c)

Figure 13.9 (*continued*)

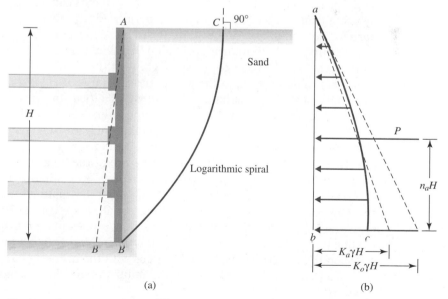

(a)

(b)

Figure 13.10 Earth pressure distribution against a wall with rotation about the top

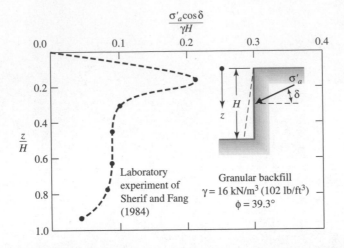

Figure 13.11
Laboratory observation of the distribution of horizontal component of lateral earth pressure on retaining wall rotating about the top.

Figure 13.11 shows the laboratory observations of Sherif and Fang (1984) related to the distribution of the horizontal component of the lateral earth pressure on a model retaining wall with a dry granular backfill rotating about the top. This figure clearly demonstrates the nonhydrostatic distribution of the lateral earth pressure for this type of wall movement.

13.7   *Determination of Active Thrust on Bracing Systems of Open Cuts in Granular Soil*

The active thrust on the bracing system of open cuts can be theoretically estimated by using trial wedges and Terzaghi's general wedge theory (1941). The basic procedure for determination of the active thrust are described in this section.

Figure 13.12a shows a braced wall AB of height H that deforms by rotating about its top. The wall is assumed to be rough, with the angle of wall friction equal to δ. The point of application of the active thrust (that is, n_aH) is assumed to be known. The curve of sliding is assumed to be an arc of a logarithmic spiral. As we discussed in the preceding section, the curve of sliding intersects the horizontal ground surface at 90°. To proceed with the trial wedge solution, let us select a point b_1. From b_1, a line b_1b_1' that makes an angle ϕ' with the ground surface is drawn. (Note that ϕ' = effective angle of friction of the soil.) The arc of the logarithmic spiral, b_1B, which defines the curve of sliding for this trial, can now be drawn, with the center of the spiral (point O_1) located on the line b_1b_1'. Note that the equation for the logarithmic spiral is given by $r_1 = r_oe^{\theta_1 \tan \phi'}$ and, in this case, $\overline{O_1b_1} = r_o$ and $\overline{O_1B} = r_1$. Also, it is interesting to see that the horizontal line that represents the ground surface is the normal to the curve of sliding at the point b_1, and that O_1b_1 is a radial line. The angle between them is equal to ϕ', which agrees with the property of the spiral.

To look at the equilibrium of the failure wedge, let us consider the following forces per unit length of the braced wall:

1. W_1 = the weight of the wedge ABb_1 = (Area of ABb_1) × (γ) × (1).
2. P_1 = the active thrust acting at a point n_aH measured vertically upward from the bottom of the cut and inclined at an angle δ with the horizontal.

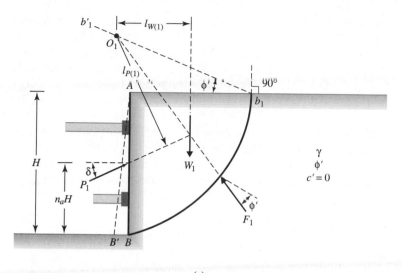

(a)

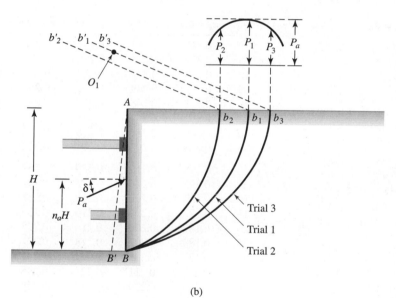

(b)

Figure 13.12
Determination of active force on bracing system of open cut in cohesionless soil

3. F_1 = the resultant of the shear and normal forces that act along the trial failure surface. The line of action of the force F_1 will pass through the point O_1.

Now, taking the moments of these forces about O_1, we have

$$W_1[l_{W(1)}] + F_1(0) - P_1[l_{P(1)}] = 0$$

or

$$P_1 = \frac{W_1 l_{W(1)}}{l_{P(1)}} \tag{13.13}$$

where $l_{W(1)}$ and $l_{P(1)}$ are the moment arms for the forces W_1 and P_1, respectively.

This procedure of finding the active thrust can now be repeated for several wedges such as ABb_2, ABb_3, . . ., ABb_n (Figure 13.12b). Note that the centers of the logarithmic spiral arcs will lie on lines $b_2b'_2$, $b_3b'_3$, . . ., $b_nb'_n$, respectively. The active thrusts P_1, P_2, P_3, . . ., P_n derived from the trial wedges are plotted to some scale in the upper portion of Figure 13.12b. The maximum point of the smooth curve drawn through these points will yield the desired maximum active thrust, P_a, on the braced wall.

Kim and Preber (1969) determined the values of $P_a/0.5\gamma H^2$ for braced excavations for various values of ϕ', δ, and n_a. These values are given in Table 13.3. In general, the average magnitude of P_a is about 10% greater when the wall rotation is about the top as compared with the value obtained by Coulomb's active earth pressure theory.

Table 13.3 $P_a/0.5\gamma H^2$ Against ϕ', δ, and n_a ($c' = 0$) for Braced Cuts*

ϕ' (deg)	δ (deg)	$P_a/0.5\gamma H^2$				ϕ' (deg)	δ (deg)	$P_a/0.5\gamma H^2$			
		$n_a = 0.3$	$n_a = 0.4$	$n_a = 0.5$	$n_a = 0.6$			$n_a = 0.3$	$n_a = 0.4$	$n_a = 0.5$	$n_a = 0.6$
10	0	0.653	0.734	0.840	0.983	35	0	0.247	0.267	0.290	0.318
	5	0.623	0.700	0.799	0.933		5	0.239	0.258	0.280	0.318
	10	0.610	0.685	0.783	0.916		10	0.234	0.252	0.273	0.300
							15	0.231	0.249	0.270	0.296
15	0	0.542	0.602	0.679	0.778		20	0.231	0.248	0.269	0.295
	5	0.518	0.575	0.646	0.739		25	0.232	0.250	0.271	0.297
	10	0.505	0.559	0.629	0.719		30	0.236	0.254	0.276	0.302
	15	0.499	0.554	0.623	0.714		35	0.243	0.262	0.284	0.312
20	0	0.499	0.495	0.551	0.622	40	0	0.198	0.213	0.230	0.252
	5	0.430	0.473	0.526	0.593		5	0.192	0.206	0.223	0.244
	10	0.419	0.460	0.511	0.575		10	0.189	0.202	0.219	0.238
	15	0.413	0.454	0.504	0.568		15	0.187	0.200	0.216	0.236
	20	0.413	0.454	0.504	0.569		20	0.187	0.200	0.216	0.235
25	0	0.371	0.405	0.447	0.499		25	0.188	0.202	0.218	0.237
	5	0.356	0.389	0.428	0.477		30	0.192	0.205	0.222	0.241
	10	0.347	0.378	0.416	0.464		35	0.197	0.211	0.228	0.248
	15	0.342	0.373	0.410	0.457		40	0.205	0.220	0.237	0.259
	20	0.341	0.372	0.409	0.456	45	0	0.156	0.167	0.180	0.196
	25	0.344	0.375	0.413	0.461		5	0.152	0.163	0.175	0.190
30	0	0.304	0.330	0.361	0.400		10	0.150	0.160	0.172	0.187
	5	0.293	0.318	0.347	0.384		15	0.148	0.159	0.171	0.185
	10	0.286	0.310	0.339	0.374		20	0.149	0.159	0.171	0.185
	15	0.282	0.306	0.334	0.368		25	0.150	0.160	0.173	0.187
	20	0.281	0.305	0.332	0.367		30	0.153	0.164	0.176	0.190
	25	0.284	0.307	0.335	0.370		35	0.158	0.168	0.181	0.196
	30	0.289	0.313	0.341	0.377		40	0.164	0.175	0.188	0.204
							45	0.173	0.184	0.198	0.213

*After Kim and Preber (1969)

Table 13.4 Values of $P_a/0.5\gamma H^2$ for Cuts in a c'-ϕ' Soil with the Assumption $c_a = c'(\tan \delta/\tan \phi')$*

δ (deg)	$n_a = 0.3$ and $c'/\gamma H = 0.1$	$n_a = 0.4$ and $c'/\gamma H = 0.1$	$n_a = 0.5$ and $c'/\gamma H = 0.1$
$\phi' = 15°$			
0	0.254	0.285	0.322
5	0.214	0.240	0.270
10	0.187	0.210	0.238
15	0.169	0.191	0.218
$\phi' = 20°$			
0	0.191	0.210	0.236
5	0.160	0.179	0.200
10	0.140	0.156	0.173
15	0.122	0.127	0.154
20	0.113	0.124	0.140
$\phi' = 25°$			
0	0.138	0.150	0.167
5	0.116	0.128	0.141
10	0.099	0.110	0.122
15	0.085	0.095	0.106
20	0.074	0.083	0.093
25	0.065	0.074	0.083
$\phi' = 30°$			
0	0.093	0.103	0.113
5	0.078	0.086	0.094
10	0.066	0.073	0.080
15	0.056	0.060	0.067
20	0.047	0.051	0.056
25	0.036	0.042	0.047
30	0.029	0.033	0.038

*After Kim and Preber (1969)

13.8 Determination of Active Thrust on Bracing Systems for Cuts in Cohesive Soil

Using the principles of the general wedge theory, we can also determine the active thrust on bracing systems for cuts made in c'-ϕ' soil. Table 13.4 gives the variation of P_a in a nondimensional form for various values of ϕ', δ, n_a, and $c'/\gamma H$.

For the $\phi = 0$ condition, $c = c_u$. For this condition, it can be shown (Das and Seeley, 1975) that

$$P_a = \frac{1}{2(1 - n_a)}(0.677 - KN_c)\gamma H^2 \qquad (13.14)$$

where

$$N_c = \left(\frac{c_u}{\gamma H}\right) \tag{13.15}$$

$$K = f\left(\frac{c_a}{c_u}\right) \tag{13.16}$$

where c_a = adhesion along the face of sheet piles.

The values of K are

$\left(\dfrac{c_a}{c_u}\right)$	K
0	2.762
0.5	3.056
1.0	3.143

13.9 Pressure Variation for Design of Sheetings, Struts, and Wales

The active thrust against sheeting in a braced cut, calculated by using the general wedge theory, does not explain the variation of the earth pressure with depth that is necessary for design work. An important difference between bracings in open cuts and retaining walls is that retaining walls fail as single units, whereas bracings in an open cut undergo progressive failure where one or more struts fail at one time.

Empirical lateral pressure diagrams against sheetings for the design of bracing systems have been given by Peck (1969). These pressure diagrams for cuts in sand, soft to medium clay, and stiff clay are given in Figure 13.13. Strut loads may be determined by assuming that the vertical members are hinged at each strut level except

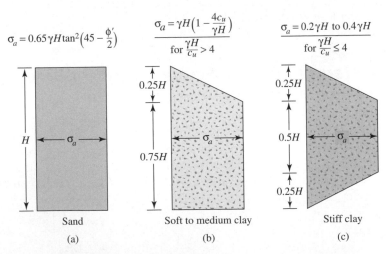

Figure 13.13 Peck's pressure diagrams for design of bracing systems

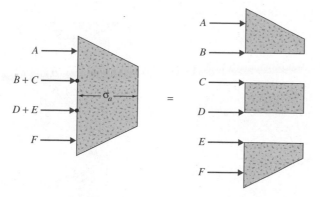

Figure 13.14 Determination of strut loads from empirical lateral pressure diagrams

the topmost and bottommost ones (Figure 13.14). Example 13.2 illustrates the procedure for the calculation of strut loads.

Example 13.2

A 7-m-deep braced cut in sand is shown in Figure 13.15. In the plan, the struts are placed at $s = 2$ m center to center. Using Peck's empirical pressure diagram, calculate the design strut loads.

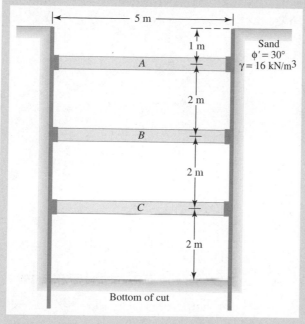

Figure 13.15 Braced cut in sand

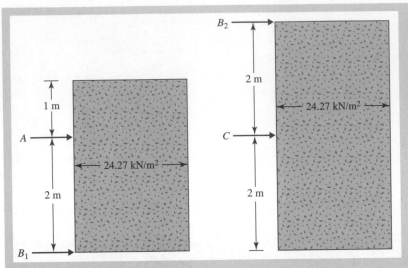

Figure 13.16 Calculation of strut loads from pressure envelope

Solution

Refer to Figure 13.13a. For the lateral earth pressure diagram,

$$\sigma_a = 0.65\gamma H \tan^2\left(45 - \frac{\phi'}{2}\right) = (0.65)(16)(7)\tan^2\left(45 - \frac{30}{2}\right) = 24.27 \text{ kN/m}^2$$

Assume that the sheeting is hinged at strut level B. Now refer to the diagram in Figure 13.16 We need to find reactions at A, B_1, B_2, and C. Taking the moment about B_1, we have

$$2A = (24.27)(3)(\tfrac{3}{2}); \qquad A = 54.61 \text{ kN/m}$$

Hence,

$$B_1 = (24.27)(3) - 54.61 = 18.2 \text{ kN/m}$$

Again, taking the moment about B_2, we have

$$2C = (24.27)(4)(\tfrac{4}{2})$$

$$C = 97.08 \text{ kN/m}$$

So

$$B_2 = (24.27)(4) - 97.08 = 0$$

The strut loads are as follows:

At level A: $\qquad (A)(s) = (54.61)(2) = \textbf{109.22 kN}$

At level B: $\quad (B_1 + B_2)(s) = (18.2 + 0)(2) = \textbf{36.4 kN}$

At level C: $\qquad (C)(s) = (97.08)(2) = \textbf{194.16 kN}$

13.10 *Dynamic Earth Pressure Distribution Behind a Wall Rotating about the Top*

Lateral earth pressure on braced cuts is essentially a problem in which the wall rotates about the top. On the basis of laboratory model test results, Sherif and Fang (1984) reported the dynamic earth pressure distribution behind a rigid retaining wall ($H = 1$ m) with dense sand as backfill material and rotation about its top. Figure 13.17 shows a plot of $\sigma'_a \cos \delta$ versus depth for various values of k_h (for $k_v = 0$). The magnitude of active thrust, P_{ae}, can be obtained from the equation

$$P_{ae} \cos \delta = \int_0^H (\sigma'_a \cos \delta) \, dz$$

or

$$P_{ae} = \frac{1}{\cos \delta} \int_0^H (\sigma'_a \cos \delta) \, dz \tag{13.17}$$

For a given value of k_h, the magnitude of P_{ae} is 15 to 20% higher than that obtained by using Eq. (12.72) (i.e., the case of wall rotation about the bottom).

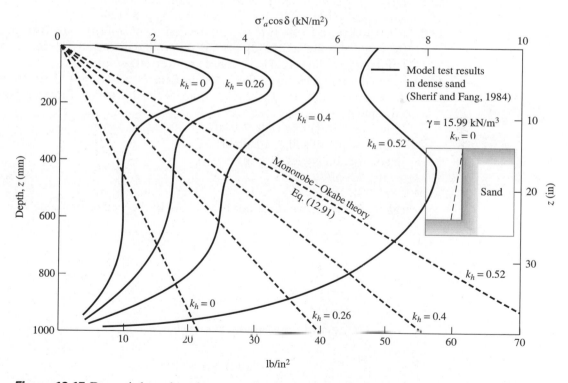

Figure 13.17 Dynamic lateral earth presure distribution behind a rigid model retaining wall rotating about the top

With the model test results just described, Sherif and Fang (1984) suggested that the location of the point of application of P_{ae} with the wall rotating about the top may be assumed to be $0.55H$ measured from the bottom of the wall.

13.11 Summary

This chapter covers two major topics: (a) estimation of passive pressure using curved failure surface in soil; and (b) lateral earth pressure on braced cuts using the general wedge theory and pressure envelopes for design of struts, wales, and sheet piles.

Passive pressure calculations using curved failure surface is essential for the case in which $\delta > \phi'/2$, since plane failure surface assumption provides results on the unsafe side for design.

In the case of braced cuts, although the general wedge theory provides the force per unit length of the cut, it does not provide the nature of distribution of earth pressure with depth. For that reason, pressure envelopes are necessary for practical design.

Problems

13.1 Draw a logarithmic spiral according to the equation $r = r_o e^{\theta \tan \phi'}$, with θ varying from 0° to 180°. Use $\phi' = 40°$ and $r_o = 30$ mm.

13.2 Refer to Figure 13.18. If $H = 6$ m, the density of soil $(\rho) = 1850$ kg/m³, and the angle of wall friction $(\delta) = 17.5°$, determine the passive force, P_p, per unit length of the wall. Use Caquot and Kerisel's solution.

13.3 Repeat Problem 13.2 with the following data: $H = 10$ ft, $\gamma = 110$ lb/ft³, $\delta = 14°$.

13.4 A retaining wall has a vertical back face with a horizontal granular backfill. Given that

    height of retaining wall = 15 ft
    unit weight of soil = 100 lb/ft³
    soil friction angle, $\phi' = 30°$
    $\delta = 2/3 \, \phi'$,
    $c' = 0$

calculate the passive force per foot length of the wall using Table 13.2.

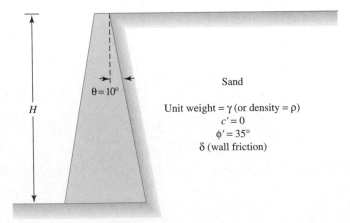

Figure 13.18

13.5 Refer to Figure 13.5. Given that $H = 5$ m, $\gamma = 16$ kN/m³, $\phi' = 30°$, $\delta = 15°$, $k_v = 0$, and $k_h = 0.3$,

    a. Calculate P_{pe} for the retaining wall using the Mononobe–Okabe solution (Section 12.15);

    b. Calculate P_{pe} for the retaining wall using the logarithmic spiral solution (Section 13.5).

13.6 Using the theory described in the section on general wedge theory, determine the active thrust, P_a, for the braced wall shown in Figure 13.19.

13.7 The elevation and plan of a bracing system for an open cut in sand are shown in Figure 13.20. Assuming $\gamma_{sand} = 105$ lb/ft³ and $\phi' = 38°$, determine the strut loads.

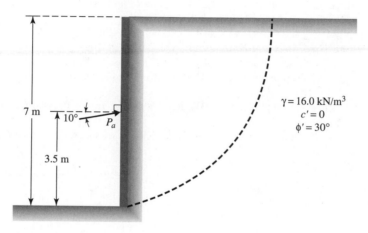

Figure 13.19

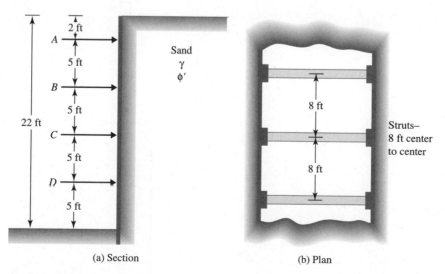

(a) Section   (b) Plan

Figure 13.20

References

CAQUOT, A., and KERISEL, J. (1948). *Tables for the Calculation of Passive Pressure, Active Pressure, and Bearing Capacity of Foundations,* Gauthier-Villars, Paris.

DAS, B. M., and SEELEY, G. R. (1975). "Active Thrust on Braced Cut in Clay," *Journal of the Construction Division,* ASCE, Vol. 101, No. CO4, 945–949.

HIJAB, W. (1956). "A Note on the Centroid of a Logarithmic Spiral Sector," *Geotechnique,* Vol. 4, No. 2, 96–99.

KIM, J. S., and PREBER, T. (1969). "Earth Pressure Against Braced Excavations," *Journal of the Soil Mechanics and Foundations Division,* ASCE, Vol. 95, No. SM6, 1581–1584.

MORRISON, E. E., JR., and EBELING, R. M. (1995). "Limit Equilibrium Computation of Dynamic Passive Earth Pressure," *Canadian Geotechnical Journal,* Vol. 32, No. 3, 481–487.

PECK, R. B. (1969). "Deep Excavation and Tunneling in Soft Ground," *Proceedings,* 7th International Conference on Soil Mechanics and Foundation Engineering, Mexico City, State-of-the-Art Vol., 225–290.

SHERIF, M. A., and FANG, Y. S. (1984). "Dynamic Earth Pressure on Walls Rotating About the Top," *Soils and Foundations,* Vol. 24, No. 4, 109–117.

SHIELDS, D. H., and TOLUNAY, A. Z. (1973). "Passive Pressure Coefficients by Method of Slices," *Journal of the Soil Mechanics and Foundations Division,* ASCE, Vol. 99, No. SM12, 1043–1053.

TERZAGHI, K. (1941). "General Wedge Theory of Earth Pressure," *Transactions,* ASCE, Vol. 106, 68–97.

TERZAGHI, K., and PECK, R. B. (1967). *Soil Mechanics in Engineering Practice,* 2nd ed., Wiley, New York.

14

Slope Stability

An exposed ground surface that stands at an angle with the horizontal is called an *unrestrained slope*. The slope can be natural or man-made. If the ground surface is not horizontal, a component of gravity will tend to move the soil downward as schematically shown in Figure 14.1a. If the component of gravity is large enough, slope failure can occur — that is, the soil mass in zone *abcdea* can slide downward. The driving force overcomes the resistance from the shear strength of the soil along the rupture surface. Figures 14.1b and 14.1c show photographs of two slope failures in the field.

Civil engineers often are expected to make calculations to check the safety of natural slopes, slopes of excavations, and compacted embankments. This check involves determining the shear stress developed along the most likely rupture surface and comparing it with the shear strength of the soil. This process is called *slope stability analysis*. The most likely rupture surface is the critical surface that has the minimum factor of safety.

The stability analysis of a slope is difficult to perform. Evaluation of variables such as the soil stratification and its in-place shear strength parameters may prove to

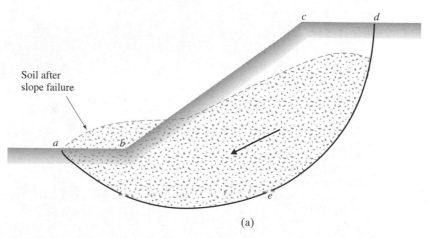

(a)

Figure 14.1 Slope failure: (a) schematic diagram; (b) a failure in Taiwan (courtesy of Richard Tsai, C&M Hi-Tech Engineering Co., Ltd., Taipei, Taiwan); (c) a failure in Alabama (courtesy of E. C. Shin, University of Inchon, South Korea)

(b)

(c)

Figure 14.1 (*continued*)

be a formidable task. Seepage through the slope and the choice of a potential slip surface add to the complexity of the problem.

This chapter explains the basic principles involved in slope stability analysis.

14.1   *Factor of Safety*

The task of the engineer charged with analyzing slope stability is to determine the factor of safety. Generally, the factor of safety is defined as

$$F_s = \frac{\tau_f}{\tau_d} \tag{14.1}$$

where F_s = factor of safety with respect to strength
τ_f = average shear strength of the soil
τ_d = average shear stress developed along the potential failure surface

The shear strength of a soil consists of two components, cohesion and friction, and may be written as

$$\tau_f = c' + \sigma' \tan \phi' \tag{14.2}$$

where c' = cohesion
ϕ' = angle of friction
σ' = normal stress on the potential failure surface

In a similar manner, we can write

$$\tau_d = c'_d + \sigma' \tan \phi'_d \tag{14.3}$$

where c'_d and ϕ'_d are, respectively, the cohesion and the angle of friction that develop along the potential failure surface. Substituting Eqs. (14.2) and (14.3) into Eq. (14.1), we get

$$F_s = \frac{c' + \sigma' \tan \phi'}{c'_d + \sigma' \tan \phi'_d} \tag{14.4}$$

Now we can introduce some other aspects of the factor of safety — that is, the factor of safety with respect to cohesion, F'_c, and the factor of safety with respect to friction, F'_ϕ. They are defined as

$$F_{c'} = \frac{c'}{c'_d} \tag{14.5}$$

and

$$F_{\phi'} = \frac{\tan \phi'}{\tan \phi'_d} \tag{14.6}$$

When we compare Eqs. (14.4) through (14.6), we can see that when $F_{c'}$ becomes equal to $F_{\phi'}$, it gives the factor of safety with respect to strength. Or, if

$$\frac{c'}{c'_d} = \frac{\tan \phi'}{\tan \phi'_d}$$

then we can write

$$F_s = F_{c'} = F_{\phi'} \qquad (14.7)$$

When F_s is equal to 1, the slope is in a state of impending failure. Generally, a value of 1.5 for the factor of safety with respect to strength is acceptable for the design of a stable slope.

14.2 Stability of Infinite Slopes

In considering the problem of slope stability, let us start with the case of an infinite slope as shown in Figure 14.2. The shear strength of the soil may be given by Eq. (14.2):

$$\tau_f = c' + \sigma' \tan \phi'$$

Assuming that the pore water pressure is zero, we will evaluate the factor of safety against a possible slope failure along a plane AB located at a depth H below the ground surface. The slope failure can occur by the movement of soil above the plane AB from right to left.

Let us consider a slope element $abcd$ that has a unit length perpendicular to the plane of the section shown. The forces, F, that act on the faces ab and cd are equal and opposite and may be ignored. The weight of the soil element is

$$W = (\text{Volume of soil element}) \times (\text{Unit weight of soil}) = \gamma LH \qquad (14.8)$$

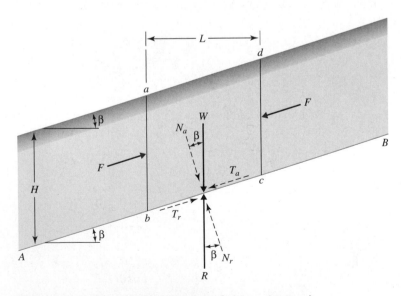

Figure 14.2 Analysis of infinite slope (without seepage)

The weight W can be resolved into two components:

1. Force perpendicular to the plane $AB = N_a = W \cos \beta = \gamma LH \cos \beta$.
2. Force parallel to the plane $AB = T_a = W \sin \beta = \gamma LH \sin \beta$. Note that this is the force that tends to cause the slip along the plane.

Thus, the effective normal stress and the shear stress at the base of the slope element can be given, respectively, as

$$\sigma' = \frac{N_a}{\text{Area of base}} = \frac{\gamma LH \cos \beta}{\left(\dfrac{L}{\cos \beta} \right)} = \gamma H \cos^2 \beta \qquad (14.9)$$

and

$$\tau = \frac{T_a}{\text{Area of base}} = \frac{\gamma LH \sin \beta}{\left(\dfrac{L}{\cos \beta} \right)} = \gamma H \cos \beta \sin \beta \qquad (14.10)$$

The reaction to the weight W is an equal and opposite force R. The normal and tangential components of R with respect to the plane AB are

$$N_r = R \cos \beta = W \cos \beta \qquad (14.11)$$

and

$$T_r = R \sin \beta = W \sin \beta \qquad (14.12)$$

For equilibrium, the resistive shear stress that develops at the base of the element is equal to $(T_r)/(\text{Area of base}) = \gamma H \sin \beta \cos \beta$. The resistive shear stress may also be written in the same form as Eq. (14.3):

$$\tau_d = c'_d + \sigma' \tan \phi'_d$$

The value of the normal stress is given by Eq. (14.9). Substitution of Eq. (14.9) into Eq. (14.3) yields

$$\tau_d = c'_d + \gamma H \cos^2 \beta \tan \phi'_d \qquad (14.13)$$

Thus,

$$\gamma H \sin \beta \cos \beta = c'_d + \gamma H \cos^2 \beta \tan \phi'_d$$

or

$$\frac{c'_d}{\gamma H} = \sin \beta \cos \beta - \cos^2 \beta \tan \phi'_d$$

$$= \cos^2 \beta (\tan \beta - \tan \phi'_d) \qquad (14.14)$$

The factor of safety with respect to strength has been defined in Eq. (14.7), from which we get

$$\tan \phi'_d = \frac{\tan \phi'}{F_s} \quad \text{and} \quad c'_d = \frac{c'}{F_s}$$

Substituting the preceding relationships into Eq. (14.14), we obtain

$$F_s = \frac{c'}{\gamma H \cos^2 \beta \tan \beta} + \frac{\tan \phi'}{\tan \beta} \qquad (14.15)$$

For granular soils, $c' = 0$, and the factor of safety, F_s, becomes equal to $(\tan \phi')/(\tan \beta)$. This indicates that in an infinite slope in sand, the value of F_s is independent of the height H and the slope is stable as long as $\beta < \phi'$.

If a soil possesses cohesion and friction, the depth of the plane along which critical equilibrium occurs may be determined by substituting $F_s = 1$ and $H = H_{cr}$ into Eq. (14.15). Thus,

$$H_{cr} = \frac{c'}{\gamma} \frac{1}{\cos^2 \beta(\tan \beta - \tan \phi')} \qquad (14.16)$$

Stability With Seepage

Figure 14.3a shows an infinite slope. We will assume that there is seepage through the soil and that the groundwater level coincides with the ground surface. The shear strength of the soil is given by

$$\tau_f = c' + \sigma' \tan \phi' \qquad (14.17)$$

To determine the factor of safety against failure along the plane *AB*, consider the slope element *abcd*. The forces that act on the vertical faces *ab* and *cd* are equal and opposite. The total weight of the slope element of unit length is

$$W = \gamma_{sat}LH \qquad (14.18)$$

where γ_{sat} = saturated unit weight of soil.

The components of W in the directions normal and parallel to plane *AB* are

$$N_a = W \cos \beta = \gamma_{sat}LH \cos \beta \qquad (14.19)$$

and

$$T_a = W \sin \beta = \gamma_{sat}LH \sin \beta \qquad (14.20)$$

The reaction to the weight W is equal to R. Thus,

$$N_r = R \cos \beta = W \cos \beta = \gamma_{sat}LH \cos \beta \qquad (14.21)$$

and

$$T_r = R \sin \beta = W \sin \beta = \gamma_{sat}LH \sin \beta \qquad (14.22)$$

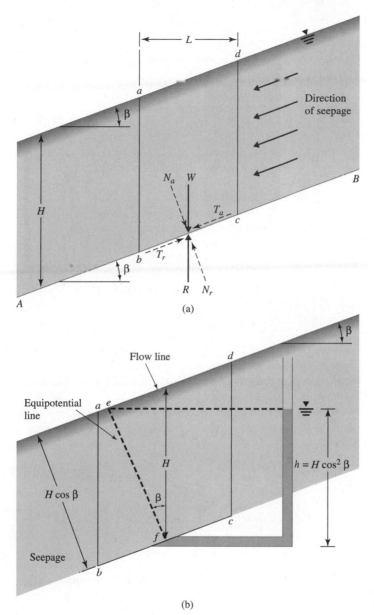

Figure 14.3
Analysis of infinite
slope (with seepage)

The total normal stress and the shear stress at the base of the element are, respectively,

$$\sigma = \frac{N_r}{\left(\dfrac{L}{\cos \beta}\right)} = \gamma_{sat}H \cos^2 \beta \tag{14.23}$$

and

$$\tau = \frac{T_r}{\left(\dfrac{L}{\cos \beta}\right)} = \gamma_{sat}H \cos \beta \sin \beta \qquad (14.24)$$

The resistive shear stress developed at the base of the element can also be given by

$$\tau_d = c'_d + \sigma' \tan \phi_d = c'_d + (\sigma - u)\tan \phi'_d \qquad (14.25)$$

where u = pore water pressure. Referring to Figure 14.3b, we see that

$$u = (\text{Height of water in piezometer placed at } f)(\gamma_w) = h\gamma_w$$

and

$$h = \overline{ef} \cos \beta = (H \cos \beta)(\cos \beta) = H \cos^2 \beta$$

so

$$u = \gamma_w H \cos^2 \beta$$

Substituting the values of σ [Eq. (14.23)] and u into Eq. (14.25), we get

$$\tau_d = c'_d + (\gamma_{sat}H \cos^2 \beta - \gamma_w H \cos^2 \beta)\tan \phi'_d$$
$$= c'_d + \gamma'H \cos^2 \beta \tan \phi'_d \qquad (14.26)$$

Now, setting the right-hand sides of Eqs. (14.24) and (14.26) equal to each other gives

$$\gamma_{sat}H \cos \beta \sin \beta = c'_d + \gamma'H \cos^2 \beta \tan \phi'_d$$

or

$$\frac{c'_d}{\gamma_{sat}H} = \cos^2 \beta \left(\tan \beta - \frac{\gamma'}{\gamma_{sat}} \tan \phi'_d \right) \qquad (14.27)$$

where $\gamma' = \gamma_{sat} - \gamma_w$ = effective unit weight of soil.

The factor of safety with respect to strength can be found by substituting $\tan \phi'_d = (\tan \phi')/F_s$ and $c'_d = c'/F_s$ into Eq. (14.27), or

$$F_s = \frac{c'}{\gamma_{sat}H \cos^2 \beta \tan \beta} + \frac{\gamma'}{\gamma_{sat}} \frac{\tan \phi'}{\tan \beta} \qquad (14.28)$$

Example 14.1

An infinite slope is shown in Figure 14.4. There is ground water seepage, and the ground water table coincides with the ground surface. Determine the factor of safety, F_s.

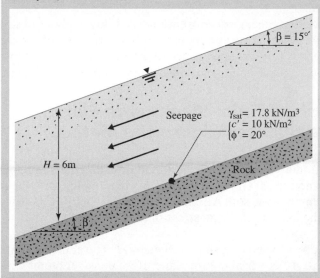

Figure 14.4

Solution

$$\gamma_{sat} = 17.8 \text{ kN/m}^3 \quad \text{and} \quad \gamma_w = 9.81 \text{ kN/m}^3$$

So,

$$\gamma' = \gamma_{sat} - \gamma_w = 17.8 - 9.81 = 7.99 \text{ kN/m}^3$$

From Eq. (14.28),

$$F_s = \frac{c'}{\gamma_{sat} H \cos^2 \beta \tan \beta} + \frac{\gamma'}{\gamma_{sat}} \frac{\tan \phi'}{\tan \beta}$$

$$= \frac{10}{(17.8)(6)(\cos 15)^2 \tan 15} + \frac{7.99}{17.8} \frac{\tan 20}{\tan 15}$$

$$= 0.375 + 0.61 = \textbf{0.985}$$

The value of F_s would be less than 1; hence, the slope would be unstable. ∎

14.3 Finite Slopes—General

When the value of H_{cr} approaches the height of the slope, the slope may generally be considered finite. For simplicity, when analyzing the stability of a finite slope in a homogeneous soil, we need to make an assumption about the general shape of the surface of potential failure. Although considerable evidence suggests that slope failures usually occur on curved failure surfaces, Culmann (1875) approximated the surface of potential failure as a plane. The factor of safety, F_s, calculated by using Culmann's approximation, gives fairly good results for near-vertical slopes only. After extensive investigation of slope failures in the 1920s, a Swedish geotechnical commission recommended that the actual surface of sliding may be approximated to be circularly cylindrical.

Since that time, most conventional stability analyses of slopes have been made by assuming that the curve of potential sliding is an arc of a circle. However, in many circumstances (for example, zoned dams and foundations on weak strata), stability analysis using plane failure of sliding is more appropriate and yields excellent results.

14.4 Analysis of Finite Slopes with Plane Failure Surfaces (Culmann's Method)

Culmann's analysis is based on the assumption that the failure of a slope occurs along a plane when the average shearing stress tending to cause the slip is more than the shear strength of the soil. Also, the most critical plane is the one that has a minimum ratio of the average shearing stress that tends to cause failure to the shear strength of soil.

Figure 14.5 shows a slope of height H. The slope rises at an angle β with the horizontal. AC is a trial failure plane. If we consider a unit length perpendicular to the section of the slope, we find that the weight of the wedge ABC is equal to

$$W = \tfrac{1}{2}(H)(\overline{BC})(1)(\gamma) = \tfrac{1}{2}H(H\cot\theta - H\cot\beta)\gamma$$

$$= \frac{1}{2}\gamma H^2\left[\frac{\sin(\beta - \theta)}{\sin\beta\sin\theta}\right] \tag{14.29}$$

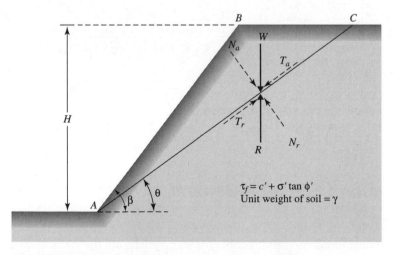

Figure 14.5 Finite slope analysis — Culmann's method

The normal and tangential components of W with respect to the plane AC are as follows:

$$N_a = \text{normal component} = W \cos \theta = \frac{1}{2}\gamma H^2\left[\frac{\sin(\beta - \theta)}{\sin \beta \sin \theta}\right]\cos \theta \qquad (14.30)$$

$$T_a = \text{tangential component} = W \sin \theta = \frac{1}{2}\gamma H^2\left[\frac{\sin(\beta - \theta)}{\sin \beta \sin \theta}\right]\sin \theta \quad (14.31)$$

The average effective normal stress and the average shear stress on the plane AC are, respectively,

$$\sigma' = \frac{N_a}{(AC)(1)} = \frac{N_a}{\left(\dfrac{H}{\sin \theta}\right)}$$

$$= \frac{1}{2}\gamma H\left[\frac{\sin(\beta - \theta)}{\sin \beta \sin \theta}\right]\cos \theta \sin \theta \qquad (14.32)$$

and

$$\tau = \frac{T_a}{(AC)(1)} = \frac{T_a}{\left(\dfrac{H}{\sin \theta}\right)}$$

$$= \frac{1}{2}\gamma H\left[\frac{\sin(\beta - \theta)}{\sin \beta \sin \theta}\right]\sin^2 \theta \qquad (14.33)$$

The average resistive shearing stress developed along the plane AC may also be expressed as

$$\tau_d = c'_d + \sigma' \tan \phi'_d$$

$$= c'_d + \frac{1}{2}\gamma H\left[\frac{\sin(\beta - \theta)}{\sin \beta \sin \theta}\right]\cos \theta \sin \theta \tan \phi'_d \qquad (14.34)$$

Now, from Eqs. (14.33) and (14.34),

$$\frac{1}{2}\gamma H\left[\frac{\sin(\beta - \theta)}{\sin \beta \sin \theta}\right]\sin^2 \theta = c'_d + \frac{1}{2}\gamma H\left[\frac{\sin(\beta - \theta)}{\sin \beta \sin \theta}\right]\cos \theta \sin \theta \tan \phi'_d \quad (14.35)$$

or

$$c_d = \frac{1}{2}\gamma H\left[\frac{\sin(\beta - \theta)(\sin \theta - \cos \theta \tan \phi'_d)}{\sin \beta}\right] \qquad (14.36)$$

The expression in Eq. (14.36) is derived for the trial failure plane AC. In an effort to determine the critical failure plane, we must use the principle of maxima and minima (for a given value of ϕ'_d) to find the angle θ where the developed cohesion would be maximum. Thus, the first derivative of c_d with respect to θ is set equal to zero, or

$$\frac{\partial c'_d}{\partial \theta} = 0 \qquad (14.37)$$

Because γ, H, and β are constants in Eq. (14.36), we have

$$\frac{\partial}{\partial \theta}\left[\sin(\beta - \theta)(\sin \theta - \cos \theta \tan \phi'_d)\right] = 0 \tag{14.38}$$

Solving Eq. (14.38) gives the critical value of θ, or

$$\theta_{cr} = \frac{\beta + \phi'_d}{2} \tag{14.39}$$

Substitution of the value of $\theta = \theta_{cr}$ into Eq. (14.36) yields

$$c'_d = \frac{\gamma H}{4}\left[\frac{1 - \cos(\beta - \phi'_d)}{\sin \beta \cos \phi'_d}\right] \tag{14.40}$$

The preceding equation can also be written as

$$\frac{c'_d}{\gamma H} = m = \frac{1 - \cos(\beta - \phi'_d)}{4 \sin \beta \cos \phi'_d} \tag{14.41}$$

where m = stability number.

The maximum height of the slope for which critical equilibrium occurs can be obtained by substituting $c'_d = c'$ and $\phi'_d = \phi'$ into Eq. (14.40). Thus,

$$H_{cr} = \frac{4c'}{\gamma}\left[\frac{\sin \beta \cos \phi'}{1 - \cos(\beta - \phi')}\right] \tag{14.42}$$

Example 14.2

A cut is to be made in a soil that has $\gamma = 16$ kN/m^3, $c' = 28$ kN/m^2, and $\phi' = 20°$. The side of the cut slope will make an angle of $45°$ with the horizontal. What should be the depth of the cut slope that will have a factor of safety, F_s, of 3.5?

Solution
We are given that $\phi' = 20°$ and $c' = 28$ kN/m^2. If $F_s = 3.5$, then, from Eq. (14.7), $F_{c'}$ and $F_{\phi'}$ should both be equal to 3.5. From Eq. (14.5),

$$F_{c'} = \frac{c'}{c'_d}$$

or

$$c'_d = \frac{c'}{F_{c'}} = \frac{c'}{F_s} = \frac{28}{3.5} = 8 \text{ kN/m}^2$$

Similarly, from Eq. (14.6),

$$F_{\phi'} = \frac{\tan \phi'}{\tan \phi'_d}$$

$$\tan \phi'_d = \frac{\tan \phi'}{F_{\phi'}} = \frac{\tan \phi'}{F_s} = \frac{\tan 20}{3.5}$$

or

$$\phi'_d = \tan^{-1}\left[\frac{\tan 20}{3.5}\right] = 5.9°$$

Substituting the preceding values of c'_d and ϕ'_d into Eq. (14.40) gives

$$H = \frac{4c'_d}{\gamma}\left[\frac{\sin \beta \cos \phi'_d}{1 - \cos(\beta - \phi'_d)}\right] = \frac{4 \times 8}{16}\left[\frac{\sin 45 \cos 5.9}{1 - \cos(45 - 5.9)}\right]$$

$$= \textbf{6.28 m} \qquad ■$$

14.5   *Analysis of Finite Slopes with Circular Failure Surfaces—General*

Modes of Failure

In general, finite slope failure occurs in one of the following modes (Figure 14.6):

1. When the failure occurs in such a way that the surface of sliding intersects the slope at or above its toe, it is called a *slope failure* (Figure 14.6a). The failure circle is referred to as a *toe circle* if it passes through the toe of the slope and as a *slope circle* if it passes above the toe of the slope. Under certain circumstances, a *shallow slope failure* can occur, as shown in Figure 14.6b.
2. When the failure occurs in such a way that the surface of sliding passes at some distance below the toe of the slope, it is called a *base failure* (Figure 14.6c). The failure circle in the case of base failure is called a *midpoint circle*.

Types of Stability Analysis Procedures

Various procedures of stability analysis may, in general, be divided into two major classes:

1. *Mass procedure:* In this case, the mass of the soil above the surface of sliding is taken as a unit. This procedure is useful when the soil that forms the slope is assumed to be homogeneous, although this is not the case in most natural slopes.
2. *Method of slices:* In this procedure, the soil above the surface of sliding is divided into a number of vertical parallel slices. The stability of each slice is calculated separately. This is a versatile technique in which the nonhomogeneity

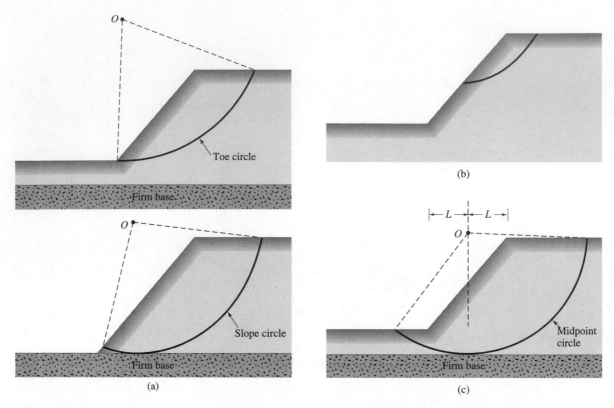

Figure 14.6 Modes of failure of finite slope: (a) slope failure; (b) shallow slope failure; (c) base failure

of the soils and pore water pressure can be taken into consideration. It also accounts for the variation of the normal stress along the potential failure surface.

The fundamentals of the analysis of slope stability by mass procedure and method of slices are given in the following sections.

14.6 Mass Procedure—Slopes in Homogeneous Clay Soil with $\phi = 0$

Figure 14.7 shows a slope in a homogeneous soil. The undrained shear strength of the soil is assumed to be constant with depth and may be given by $\tau_f = c_u$. To perform the stability analysis, we choose a trial potential curve of sliding, AED, which is an arc of a circle that has a radius r. The center of the circle is located at O. Considering a unit length perpendicular to the section of the slope, we can give the weight of the soil above the curve AED as $W = W_1 + W_2$, where

$$W_1 = (\text{Area of } FCDEF)(\gamma)$$

and

$$W_2 = (\text{Area of } ABFEA)(\gamma)$$

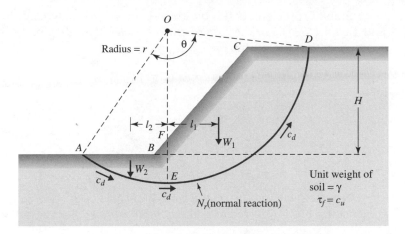

Figure 14.7 Stability analysis of slope in homogeneous saturated clay soil ($\phi = 0$)

Failure of the slope may occur by sliding of the soil mass. The moment of the driving force about O to cause slope instability is

$$M_d = W_1 l_1 - W_2 l_2 \tag{14.43}$$

where l_1 and l_2 are the moment arms.

The resistance to sliding is derived from the cohesion that acts along the potential surface of sliding. If c_d is the cohesion that needs to be developed, the moment of the resisting forces about O is

$$M_R = c_d(\widehat{AED})(1)(r) = c_d r^2 \theta \tag{14.44}$$

For equilibrium, $M_R = M_d$; thus,

$$c_d r^2 \theta = W_1 l_1 - W_2 l_2$$

or

$$c_d = \frac{W_1 l_1 - W_2 l_2}{r^2 \theta} \tag{14.45}$$

The factor of safety against sliding may now be found:

$$F_s = \frac{\tau_f}{c_d} = \frac{c_u}{c_d} \tag{14.46}$$

Note that the potential curve of sliding, AED, was chosen arbitrarily. The critical surface is that for which the ratio of c_u to c_d is a minimum. In other words, c_d is maximum. To find the critical surface for sliding, one must make a number of trials for different trial circles. The minimum value of the factor of safety thus obtained is the factor of safety against sliding for the slope, and the corresponding circle is the critical circle.

Stability problems of this type have been solved analytically by Fellenius (1927) and Taylor (1937). For the case of *critical circles,* the developed cohesion can be expressed by the relationship

$$c_d = \gamma H m$$

or

$$\frac{c_d}{\gamma H} = m \qquad (14.47)$$

Note that the term m on the right-hand side of the preceding equation is non-dimensional and is referred to as the *stability number.* The critical height (i.e., $F_s = 1$)

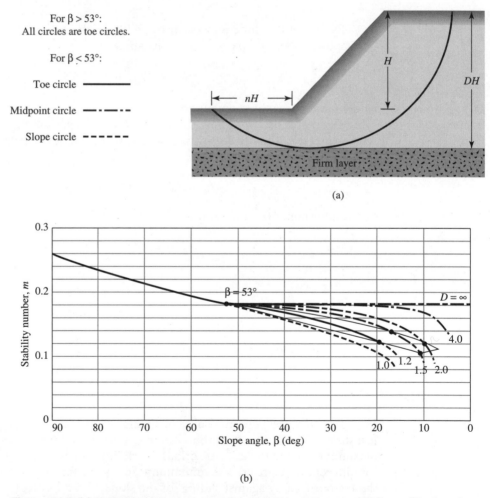

For $\beta > 53°$:
All circles are toe circles.

For $\beta < 53°$:

Toe circle ———

Midpoint circle —·—·—··

Slope circle ------

(a)

(b)

Figure 14.8 (a) Definition of parameters for midpoint circle type of failure; (b) plot of stability number against slope angle (redrawn from Terzaghi and Peck, 1967)

of the slope can be evaluated by substituting $H = H_{cr}$ and $c_d = c_u$ (full mobilization of the undrained shear strength) into the preceding equation. Thus,

$$H_{cr} = \frac{c_u}{\gamma m} \qquad (14.48)$$

Values of the stability number, m, for various slope angles, β, are given in Figure 14.8. Terzaghi used the term $\gamma H/c_d$, the reciprocal of m and called it the *stability factor*. Readers should be careful in using Figure 14.8 and note that it is valid for slopes of saturated clay and is applicable to only undrained conditions ($\phi = 0$).

In reference to Figure 14.8, the following must be pointed out:

1. For a slope angle β greater than 53°, the critical circle is always a toe circle.
2. For $\beta < 53°$, the critical circle may be a toe, slope, or midpoint circle, depending on the location of the firm base under the slope. This is called the *depth function,* which is defined as

$$D = \frac{\text{Vertical distance from top of slope to firm base}}{\text{Height of slope}} \qquad (14.49)$$

3. When the critical circle is a midpoint circle (i.e., the failure surface is tangent to the firm base), its position can be determined with the aid of Figure 14.9.
4. The maximum possible value of the stability number for failure as a midpoint circle is 0.181.

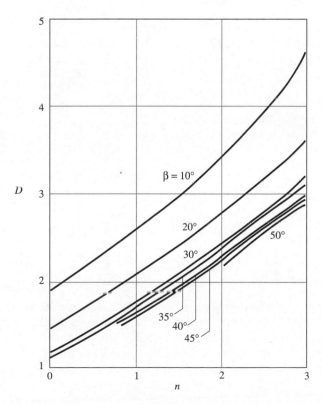

Figure 14.9
Location of midpoint circle

Example 14.3

A cut slope in saturated clay (Figure 14.10) makes an angle of 56° with the horizontal.

    a. Determine the maximum depth up to which the cut could be made. Assume that the critical surface for sliding is circularly cylindrical. What will be the nature of the critical circle (i.e., toe, slope, or midpoint)?

    b. How deep should the cut be made if a factor of safety of 2 against sliding is required?

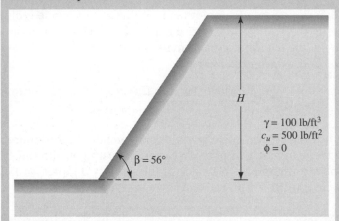

$\gamma = 100\ \text{lb/ft}^3$
$c_u = 500\ \text{lb/ft}^2$
$\phi = 0$

$\beta = 56°$

Figure 14.10

Solution

    a. Since the slope angle $\beta = 56° > 53°$, the critical circle is a **toe circle.** From Figure 14.8, for $\beta = 56°$, $m = 0.185$. Using Eq. (14.48), we have

$$H_{cr} = \frac{c_u}{\gamma m} = \frac{500}{(110)(0.185)} = \textbf{24.57 ft}$$

    b. The developed cohesion is

$$c_d = \frac{c_u}{F_s} = \frac{500}{2} = 250\ \text{lb/ft}^2$$

From Figure 14.8, for $\beta = 56°$, $m = 0.185$. Thus, we have

$$H = \frac{c_d}{\gamma m} = \frac{250}{(110)(0.185)} = \textbf{12.29 ft}$$ ∎

Example 14.4

A cut slope was excavated in a saturated clay. The slope made an angle of 40° with the horizontal. Slope failure occurred when the cut reached a depth of 6.1 m. Previous soil explorations showed that a rock layer was located at a depth of 9.15 m below the ground surface. Assume an undrained condition and $\gamma_{sat} = 17.29\ \text{kN/m}^3$.

a. Determine the undrained cohesion of the clay (use Figure 14.8).
b. What was the nature of the critical circle?
c. With reference to the toe of the slope, at what distance did the surface of sliding intersect the bottom of the excavation?

Solution

a. Referring to Figure 14.8, we find that

$$D = \frac{9.15}{6.1} = 1.5$$

$$\gamma_{sat} = 17.29 \text{ kN/m}^3$$

and

$$H_{cr} = \frac{c_u}{\gamma m}$$

a. From Figure 14.8, for $\beta = 40°$ and $D = 1.5$, $m = 0.175$, so

$$c_u = (H_{cr})(\gamma)(m) = (6.15)(17.29)(0.175) = \textbf{18.6 kN/m}^2$$

b. **Midpoint circle**
c. Again, from Figure 14.9, for $D = 1.5$, $\beta = 40°$; $n = 0.9$, so distance = $(n)(H_{cr}) = (0.9)(6.1) = \textbf{5.49 m}$ ∎

14.7   *Mass Procedure for Stability of Saturated Clay Slope ($\phi = 0$ condition) with Earthquake Forces*

The stability of saturated clay slopes ($\phi = 0$ condition) with earthquake forces has been analyzed by Koppula (1984). Figure 14.11 shows a clay slope with a potential

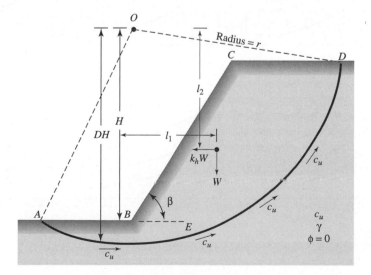

Figure 14.11
Stability analysis of slope in homogeneous saturated clay with earthquake forces ($\phi = 0$ condition)

curve of sliding *AED,* which is an arc of a circle that has radius *r*. The center of the circle is located at *O*. Taking a unit length perpendicular to the slope, we consider these forces for stability analysis:

1. Weight of the soil wedge:

$$W = (\text{area of } ABCDEA)(\gamma)$$

2. Horizontal inertia force, $k_h W$:

$$k_h = \frac{\text{horizontal component of earthquake acceleration}}{g}$$

(g = acceleration from gravity)

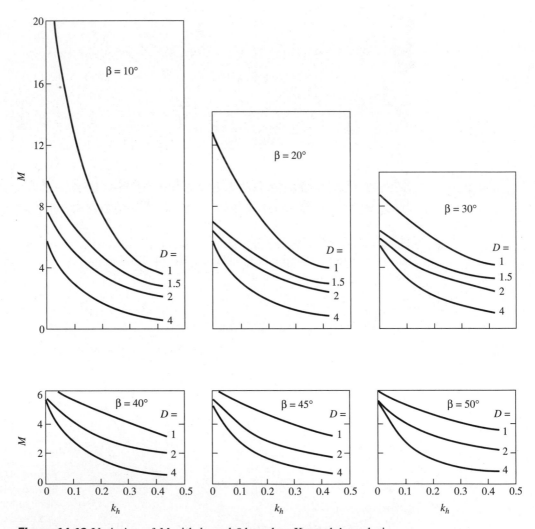

Figure 14.12 Variation of M with k_h and β based on Koppula's analysis

3. Cohesive force along the surface of sliding, which will have a magnitude of $(\overparen{AED})c_u$.

The moment of the driving forces about O can now be given as

$$M_d = Wl_1 + k_h Wl_2 \qquad (14.50)$$

Similarly, the moment of the resisting forces about O is

$$M_r = (\overparen{AED})(c_u)r \qquad (14.51)$$

Thus, the factor of safety against sliding is

$$F_s = \frac{M_r}{M_d} = \frac{(\overparen{AED})(c_u)(r)}{Wl_1 + k_h Wl_2} = \frac{c_u}{\gamma H} M \qquad (14.52)$$

where M = stability factor.

The variations of the stability factor M with slope angle β and k_h based on Koppula's (1984) analysis are given in Figures 14.12 and 14.13.

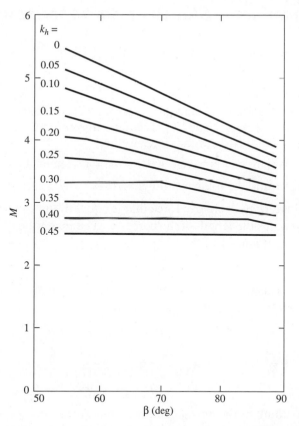

Figure 14.13 Variation of M with k_h based on Koppula's analysis (for $\beta \geq 55°$)

Example 14.5

Solve parts (a) and (b) of Example 14.3 assuming $k_h = 0.25$.

Solution

a. For the critical height of the slopes, $F_s = 1$. So, from Eq. (14.52),

$$H_{cr} = \frac{c_u M}{\gamma}$$

From Figure 14.13, for $\beta = 56°$ and $k_h = 0.25$, $M = 3.66$. Thus,

$$H_{cr} = \frac{(500)(3.66)}{110} = \mathbf{16.64 \ ft}$$

b. From Eq. (14.52),

$$H = \frac{c_u M}{\gamma F_s} = \frac{(500)(3.66)}{(110)(2)} = \mathbf{8.32 \ ft} \qquad \blacksquare$$

14.8 *Mass Procedure—Slopes in Homogeneous c'-ϕ' Soil*

A slope in a homogeneous soil is shown in Figure 14.14a. The shear strength of the soil is given by

$$\tau_f = c' + \sigma' \tan \phi'$$

The pore water pressure is assumed to be zero. $\overset{\frown}{AC}$ is a trial circular arc that passes through the toe of the slope, and O is the center of the circle. Considering a unit length perpendicular to the section of the slope, we find

$$\text{Weight of soil wedge } ABC = W = (\text{Area of } ABC)(\gamma)$$

For equilibrium, the following other forces are acting on the wedge:

1. C_d—resultant of the cohesive force that is equal to the cohesion per unit area developed times the length of the cord $\overline{AC}$. The magnitude of C_d is given by the following (Figure 13.17b):

$$C_d = c'_d(\overline{AC}) \qquad (14.53)$$

C_d acts in a direction parallel to the cord $\overline{AC}$ (see Figure 14.14b) and at a distance a from the center of the circle O such that

$$C_d(a) = c'_d(\overset{\frown}{AC})r$$

or

$$a = \frac{c'_d(\overset{\frown}{AC})r}{C_d} = \frac{\overset{\frown}{AC}}{\overline{AC}}r \qquad (14.54)$$

2. F—the resultant of the normal and frictional forces along the surface of sliding. For equilibrium, the line of action of F will pass through the point of intersection of the line of action of W and C_d.

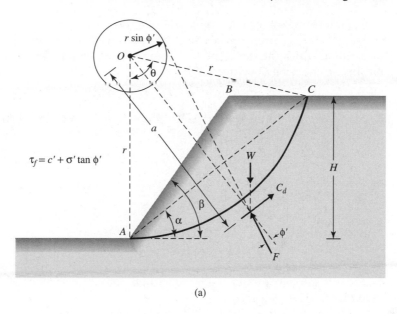

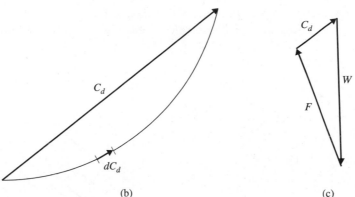

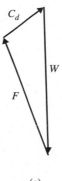

(b)   (c)

Figure 14.14 Stability analysis of slope in homogeneous c'-ϕ' soil

Now, if we assume that full friction is mobilized ($\phi'_d = \phi'$ or $F_{\phi'} = 1$), the line of action of F will make an angle of ϕ' with a normal to the arc and will thus be a tangent to a circle with its center at O and having a radius of $r \sin \phi'$. This circle is called the *friction circle*. Actually, the radius of the friction circle is a little larger than $r \sin \phi'$.

Because the directions of W, C_d, and F are known and the magnitude of W is known, a force polygon, as shown in Figure 14.14c, can be plotted. The magnitude of C_d can be determined from the force polygon. So the cohesion per unit area developed can be found:

$$c'_d = \frac{C_d}{AC}$$

Determination of the magnitude of c_d' described previously is based on a trial surface of sliding. Several trials must be made to obtain the most critical sliding surface, along which the developed cohesion is a maximum. Thus, we can express the maximum cohesion developed along the critical surface as

$$c_d' = \gamma H[f(\alpha, \beta, \theta, \phi')] \tag{14.55}$$

For critical equilibrium — that is, $F_{c'} = F_{\phi'} = F_s = 1$ — we can substitute $H = H_{cr}$ and $c_d' = c'$ into Eq. (14.55) and write

$$c' = \gamma H_{cr}[f(\alpha, \beta, \theta, \phi')]$$

or

$$\frac{c'}{\gamma H_{cr}} = f(\alpha, \beta, \theta, \phi') = m \tag{14.56}$$

where m = stability number. The values of m for various values of ϕ' and β are given in Figure 14.15.

Calculations have shown that for $\phi > \sim 3°$, the critical circles are all *toe circles*. Using Taylor's method of slope stability, Singh (1970) provided graphs of equal factors of safety, F_s, for various slopes. These graphs are given in Figure 14.16. In these charts, the pore water pressure is assumed to be zero.

The technique used to develop Figure 14.16 is illustrated in Example 14.7.

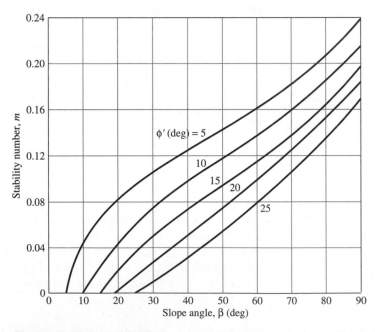

Figure 14.15 Plot of stability number with slope angle (after Taylor, 1937)

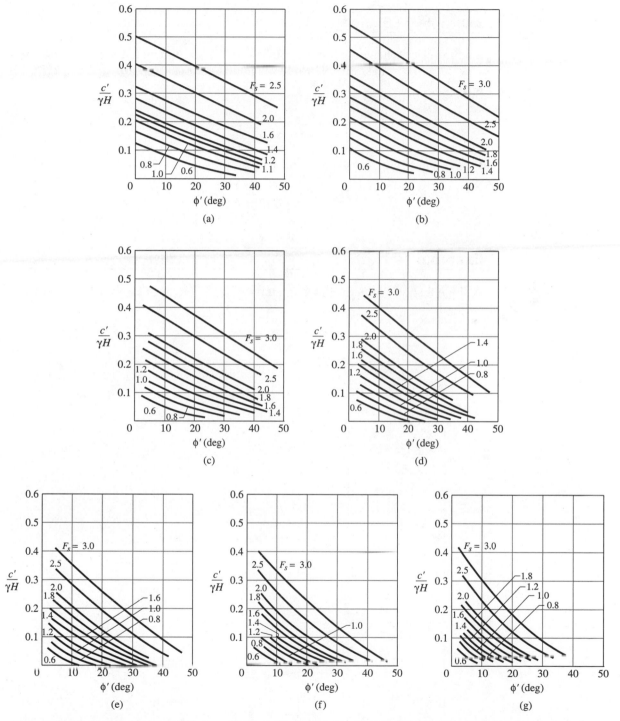

Figure 14.16 Contours of equal factors of safety: (a) slope −1 vertical to 0.5 horizontal; (b) slope −1 vertical to 0.75 horizontal; (c) slope −1 vertical to 1 horizontal; (d) slope −1 vertical to 1.5 horizontal; (e) slope −1 vertical to 2 horizontal; (f) slope −1 vertical to 2.5 horizontal; (g) slope −1 vertical to 3 horizontal (after Singh, 1970)

Example 14.6

Find the critical height of a slope with $\beta = 45°$ to be constructed with a soil having $\phi' = 20°$ and $c' = 15\ kN/m^2$. The unit weight of the compacted soil will be $17\ kN/m^3$.

Solution
We have

$$m = \frac{c'}{\gamma H_{cr}}$$

From Figure 14.15, for $\beta = 45°$ and $\phi' = 20°$, $m = 0.062$. So

$$H_{cr} = \frac{c'}{\gamma m} = \frac{15}{17 \times 0.062} = \textbf{14.2 m}$$

∎

Example 14.7

A slope is shown in Figure 14.17a. Determine the factor of safety with respect to strength.

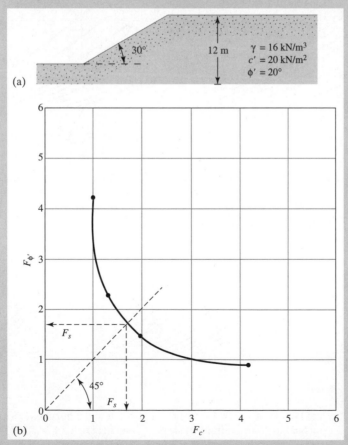

(a)

(b)

$F_{c'}$

Figure 14.17

Solution

If we assume that full friction is mobilized, then, referring to Figure 14.15 (for $\beta = 30°$ and $\phi'_d = \phi' = 20°$), we obtain

$$m = 0.025 = \frac{c'_d}{\gamma H}$$

or

$$c'_d = (0.025)(16)(12) = 4.8 \text{ kN/m}^2$$

Thus,

$$F_{\phi'} = \frac{\tan \phi'}{\tan \phi'_d} = \frac{\tan 20}{\tan 20} = 1$$

and

$$F_{c'} = \frac{c'}{c'_d} = \frac{20}{4.8} = 4.17$$

Since $F_{c'} \neq F_{\phi'}$, this is not the factor of safety with respect to strength.

Now we can make another trial. Let the developed angle of friction, ϕ'_d, be equal to 15°. For $\beta = 30°$ and the friction angle equal to 15°,

$$m = 0.046 = \frac{c'_d}{\gamma H} \qquad \text{(Figure 14.15)}$$

or

$$c'_d = 0.046 \times 16 \times 12 = 8.83 \text{ kN/m}^2$$

For this trial,

$$F_{\phi'} = \frac{\tan \phi'}{\tan \phi'_d} = \frac{\tan 20}{\tan 15} = 1.36$$

and

$$F_{c'} = \frac{c'}{c'_d} = \frac{20}{8.83} = 2.26$$

Similar calculations of $F_{\phi'}$ and $F_{c'}$ for various assumed values of ϕ'_d can be made and appear in the following table

ϕ'_d	$\tan \phi'_d$	$F_{\phi'}$	m	c'_d (kN/m²)	$F_{c'}$
20	0.364	1	0.025	4.8	4.17
15	0.268	1.36	0.046	8.83	2.26
10	0.176	2.07	0.075	14.4	1.39
5	0.0875	4.16	0.11	21.12	0.95

The values of $F_{\phi'}$ have been plotted against their corresponding values of $F_{c'}$ in Figure 14.17b, from which we get

$$F_{c'} = F_{\phi'} = F_s = \mathbf{1.73} \qquad \blacksquare$$

14.9   *Ordinary Method of Slices*

Stability analysis by using the method of slices can be explained with the use of Figure 14.18a, in which AC is an arc of a circle representing the trial failure surface. The soil above the trial failure surface is divided into several vertical slices. The width of each slice need not be the same. Considering a unit length perpendicular to the cross section shown, the forces that act on a typical slice (nth slice) are shown in Figure 14.18b. W_n is the weight of the slice. The forces N_r and T_r, respectively, are the normal and tangential components of the reaction R. P_n and P_{n+1} are the normal forces that act on the sides of the slice. Similarly, the shearing forces that act on the sides of the slice are T_n and T_{n+1}. For simplicity, the pore water pressure is assumed to be zero. The forces P_n, P_{n+1}, T_n, and T_{n+1} are difficult to determine. However, we can make an approximate assumption that the resultants of P_n and T_n are equal in magnitude to the resultants of P_{n+1} and T_{n+1} and that their lines of action coincide.

For equilibrium consideration,

$$N_r = W_n \cos \alpha_n$$

The resisting shear force can be expressed as

$$T_r = \tau_d(\Delta L_n) = \frac{\tau_f(\Delta L_n)}{F_s} = \frac{1}{F_s}[c' + \sigma' \tan \phi']\Delta L_n \tag{14.57}$$

The normal stress, σ', in Eq. (14.57) is equal to

$$\frac{N_r}{\Delta L_n} = \frac{W_n \cos \alpha_n}{\Delta L_n}$$

For equilibrium of the trial wedge ABC, the moment of the driving force about O equals the moment of the resisting force about O, or

$$\sum_{n=1}^{n=p} W_n r \sin \alpha_n = \sum_{n=1}^{n=p} \frac{1}{F_s}\left(c' + \frac{W_n \cos \alpha_n}{\Delta L_n} \tan \phi' \right)(\Delta L_n)(r)$$

or

$$F_s = \frac{\displaystyle\sum_{n=1}^{n=p}(c'\Delta L_n + W_n \cos \alpha_n \tan \phi')}{\displaystyle\sum_{n=1}^{n=p} W_n \sin \alpha_n} \tag{14.58}$$

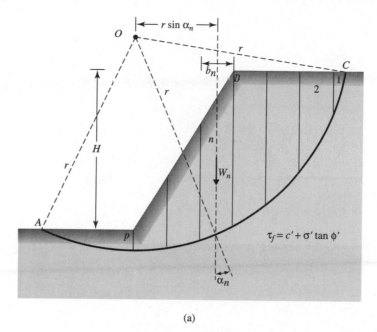

(a)

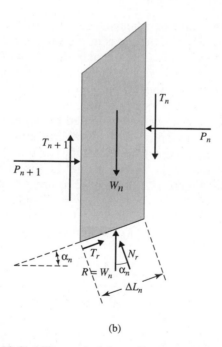

(b)

Figure 14.18 Stability analysis by ordinary method of slices: (a) trial failure surface; (b) forces acting on nth slice

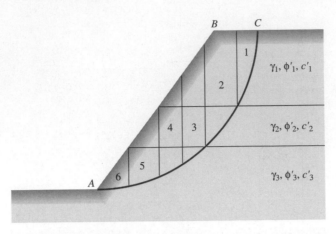

Figure 14.19 Stability analysis, by ordinary method of slices, for slope in layered soils

[*Note:* ΔL_n in Eq. (14.58) is approximately equal to $(b_n)/(\cos \alpha_n)$, where $b_n =$ the width of the nth slice.]

Note that the value of α_n may be either positive or negative. The value of α_n is positive when the slope of the arc is in the same quadrant as the ground slope. To find the minimum factor of safety — that is, the factor of safety for the critical circle — one must make several trials by changing the center of the trial circle. This method is generally referred to as the *ordinary method of slices.*

For convenience, a slope in a homogeneous soil is shown in Figure 14.18. However, the method of slices can be extended to slopes with layered soil, as shown in Figure 14.19. The general procedure of stability analysis is the same. However, some minor points should be kept in mind. When Eq. (14.58) is used for the factor of safety calculation, the values of ϕ' and c' will not be the same for all slices. For example, for slice no. 3 (see Figure 14.19), we have to use a friction angle of $\phi' = \phi'_3$ and cohesion $c' = c'_3$; similarly, for slice no. 2, $\phi' = \phi'_2$ and $c' = c'_2$.

It is of interest to note that if total shear strength parameters (that is, $\tau_f = c + \tan \phi$) were used, Eq. (14.58) would take the form

$$F_s = \frac{\sum\limits_{n=1}^{n=p} (c\Delta L_n + W_n \cos \alpha_n \tan \phi)}{\sum\limits_{n=1}^{n=p} W_n \sin \alpha_n} \tag{14.59}$$

Example 14.8

For the slope shown in Figure 14.20, find the factor of safety against sliding for the trial slip surface AC. Use the ordinary method of slices.

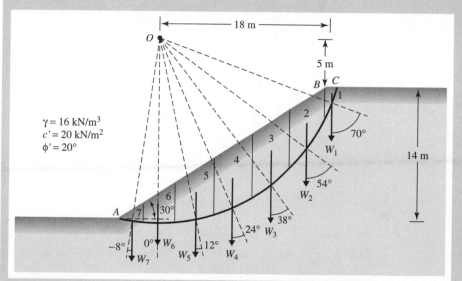

Figure 14.20 Stability analysis of a slope by ordinary method of slices

Solution

The sliding wedge is divided into seven slices. Now the following table can be prepared:

Slice no. (1)	W (kN/m) (2)	α_n (deg) (3)	$\sin \alpha_n$ (4)	$\cos \alpha_n$ (5)	ΔL_n (m) (6)	$W_n \sin \alpha_n$ (kN/m) (7)	$W_n \cos \alpha_n$ (kN/m) (8)
1	22.4	70	0.94	0.342	2.924	21.1	7.66
2	294.4	54	0.81	0.588	6.803	238.5	173.1
3	435.2	38	0.616	0.788	5.076	268.1	342.94
4	435.2	24	0.407	0.914	4.376	177.1	397.8
5	390.4	12	0.208	0.978	4.09	81.2	381.8
6	268.8	0	0	1	4	0	268.8
7	66.58	−8	−0.139	0.990	3.232	−9.25	65.9
					Σ Col. 6 = 30.501 m	Σ Col. 7 = 776.75 kN/m	Σ Col. 8 = 1638 kN/m

$$F_s = \frac{(\Sigma \text{ Col. 6})(c') + (\Sigma \text{ Col. 8})\tan \phi'}{\Sigma \text{ Col. 7}}$$

$$= \frac{(30.501)(20) + (1638)(\tan 20)}{776.75} = \mathbf{1.55}$$

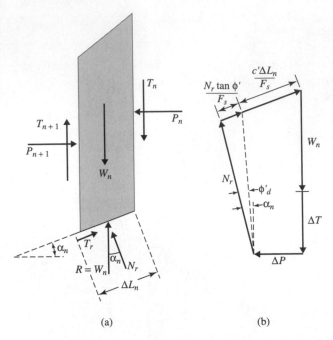

Figure 14.21 Bishop's simplified method of slices: (a) forces acting on the nth slice; (b) force polygon for equilibrium

Bishop's Simplified Method of Slices

In 1955, Bishop proposed a more refined solution to the ordinary method of slices. In this method, the effect of forces on the sides of each slice are accounted for to some degree. We can study this method by referring to the slope analysis presented in Figure 14.18. The forces that act on the nth slice shown in Figure 14.18b have been redrawn in Figure 14.21a. Now, let $P_n - P_{n+1} = \Delta P$ and $T_n - T_{n+1} = \Delta T$. Also, we can write

$$T_r = N_r(\tan \phi'_d) + c'_d \Delta L_n = N_r\left(\frac{\tan \phi'}{F_s}\right) + \frac{c' \Delta L_n}{F_s} \tag{14.60}$$

Figure 14.21b shows the force polygon for equilibrium of the nth slice. Summing the forces in the vertical direction gives

$$W_n + \Delta T = N_r \cos \alpha_n + \left[\frac{N_r \tan \phi'}{F_s} + \frac{c' \Delta L_n}{F_s}\right]\sin \alpha_n$$

or

$$N_r = \frac{W_n + \Delta T - \dfrac{c' \Delta L_n}{F_s} \sin \alpha_n}{\cos \alpha_n + \dfrac{\tan \phi' \sin \alpha_n}{F_s}} \tag{14.61}$$

For equilibrium of the wedge ABC (Figure 14.18a), taking the moment about O gives

$$\sum_{n=1}^{n=p} W_n r \sin \alpha_n = \sum_{n=1}^{n=p} T_r r \qquad (14.62)$$

where

$$T_r = \frac{1}{F_s}(c' + \sigma' \tan \phi')\Delta L_n$$

$$= \frac{1}{F_s}(c'\Delta L_n + N_r \tan \phi') \qquad (14.63)$$

Substitution of Eqs. (14.61) and (14.63) into Eq. (14.62) gives

$$F_s = \frac{\displaystyle\sum_{n=1}^{n=p} (c'b_n + W_n \tan \phi' + \Delta T \tan \phi')\frac{1}{m_{\alpha(n)}}}{\displaystyle\sum_{n=1}^{n=p} W_n \sin \alpha_n} \qquad (14.64)$$

where

$$m_{\alpha(n)} = \cos \alpha_n + \frac{\tan \phi' \sin \alpha_n}{F_s} \qquad (14.65)$$

For simplicity, if we let $\Delta T = 0$, Eq. (14.64) becomes

$$F_s = \frac{\displaystyle\sum_{n=1}^{n=p} (c'b_n + W_n \tan \phi')\frac{1}{m_{\alpha(n)}}}{\displaystyle\sum_{n=1}^{n=p} W_n \sin \alpha_n} \qquad (14.66)$$

Note that the term F_s is present on both sides of Eq. (14.66). Hence, we must adopt a trial-and-error procedure to find the value of F_s. As in the method of ordinary slices, a number of failure surfaces must be investigated so that we can find the critical surface that provides the minimum factor of safety.

Bishop's simplified method is probably the most widely used. When incorporated into computer programs, it yields satisfactory results in most cases. The ordinary method of slices is presented in this chapter as a learning tool only. It is rarely used now because it is too conservative.

14.11 Stability Analysis by Method of Slices for Steady State Seepage

The fundamentals of the ordinary method of slices and Bishop's simplified method of slices were presented in Sections 14.9 and 14.10, respectively, and we assumed the

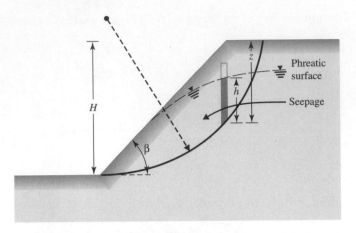

Figure 14.22 Stability analysis of slope with steady state seepage

pore water pressure to be zero. However, for steady state seepage through slopes, as is the situation in many practical cases, the pore water pressure must be considered when effective shear strength parameters are used. So we need to modify Eqs. (14.58) and (14.66) slightly.

Figure 14.22 shows a slope through which there is steady state seepage. For the nth slice, the average pore water pressure at the bottom of the slice is equal to $u_n = h_n \gamma_w$. The total force caused by the pore water pressure at the bottom of the nth slice is equal to $u_n \Delta L_n$.

Thus, Eq. (14.58) for the ordinary method of slices will be modified to read as follows:

$$F_s = \frac{\sum\limits_{n=1}^{n=p} (c' \Delta L_n + (W_n \cos \alpha_n - u_n \Delta L_n)] \tan \phi'}{\sum\limits_{n=1}^{n=p} W_n \sin \alpha_n} \tag{14.67}$$

Similarly, Eq. (14.66) for Bishop's simplified method of slices will be modified to the form

$$F_s = \frac{\sum\limits_{n=1}^{n=p} [c' b_n + (W_n - u_n b_n) \tan \phi'] \dfrac{1}{m_{(\alpha)n}}}{\sum\limits_{n=1}^{n=p} W_n \sin \alpha_n} \tag{14.68}$$

Note that W_n in Eqs. (14.67) and (14.68) is the *total weight* of the slice.

Using Eq. (14.68), Bishop and Morgenstern (1960) developed tables for the calculation of F_s for simple slopes. The principles of these developments can be explained as follows. In Eq. (14.68),

$$W_n = \text{total weight of the } n\text{th slice} = \gamma b_n z_n \qquad (14.69)$$

where z_n = average height of the nth slice. Also in Eq. (14.68)

$$u_n = h_n \gamma_w$$

So we can let

$$r_{u(n)} = \frac{u_n}{\gamma z_n} = \frac{h_n \gamma_w}{\gamma z_n} \qquad (14.70)$$

Note that $r_{u(n)}$ is a nondimensional quantity. Substituting Eqs. (14.69) and (14.70) into Eq. (14.68) and simplifying, we obtain

$$F_s = \left[\frac{1}{\sum\limits_{n=1}^{n=p} \dfrac{b_n}{H} \dfrac{z_n}{H} \sin \alpha_n} \right] \times \sum_{n=1}^{n=p} \left\{ \frac{\dfrac{c'}{\gamma H} \dfrac{b_n}{H} + \dfrac{b_n}{H} \dfrac{z_n}{H}[1 - r_{u(n)}]\tan \phi'}{m_{\alpha(n)}} \right\} \qquad (14.71)$$

For a steady-state seepage condition, a weighted average value of $r_{u(n)}$ can be taken, which is a constant. Let the weighted averaged value of $r_{u(n)}$ be r_u. For most practical cases, the value of r_u may range up to 0.5. Thus,

$$F_s = \left[\frac{1}{\sum\limits_{n=1}^{n=p} \dfrac{b_n}{H} \dfrac{z_n}{H} \sin \alpha_n} \right] \times \sum_{n=1}^{n=p} \left\{ \frac{\left[\dfrac{c'}{\gamma H} \dfrac{b_n}{H} + \dfrac{b_n}{H} \dfrac{z_n}{H}(1 - r_u)\tan \phi' \right]}{m_{\alpha(n)}} \right\}$$

$$(14.72)$$

The factor of safety based on the preceding equation can be solved and expressed in the form

$$F_s = m' - n' r_u \qquad (14.73)$$

where m' and n' = stability coefficients. Table 14.1 gives the values of m' and n' for various combinations of $c'/\gamma H$, D, ϕ', and β.

To determine F_s from Table 14.1, we must use the following step-by-step procedure:

1. Obtain ϕ', β, and $c'/\gamma H$.
2. Obtain r_u (weighted average value).
3. From Table 14.1, obtain the values of m' and n' for $D = 1, 1.25$, and 1.5 (for the required parameters ϕ', β, r_u, and $c'/\gamma H$).
4. Determine F_s, using the values of m' and n' for each value of D.
5. The required value of F_s is the smallest one obtained in step 4.

Table 14.1 Values of m' and n' [Eq. (14.73)]

a. Stability coefficients m' and n' for $c'/\gamma H = 0$

	Stability coefficients for earth slopes							
	Slope 2:1		Slope 3:1		Slope 4:1		Slope 5:1	
ϕ'	m'	n'	m'	n'	m'	n'	m'	n'
10.0	0.353	0.441	0.529	0.588	0.705	0.749	0.882	0.917
12.5	0.443	0.554	0.665	0.739	0.887	0.943	1.109	1.153
15.0	0.536	0.670	0.804	0.893	1.072	1.139	1.340	1.393
17.5	0.631	0.789	0.946	1.051	1.261	1.340	1.577	1.639
20.0	0.728	0.910	1.092	1.213	1.456	1.547	1.820	1.892
22.5	0.828	1.035	1.243	1.381	1.657	1.761	2.071	2.153
25.0	0.933	1.166	1.399	1.554	1.865	1.982	2.332	2.424
27.5	1.041	1.301	1.562	1.736	2.082	2.213	2.603	2.706
30.0	1.155	1.444	1.732	1.924	2.309	2.454	2.887	3.001
32.5	1.274	1.593	1.911	2.123	2.548	2.708	3.185	3.311
35.0	1.400	1.750	2.101	2.334	2.801	2.977	3.501	3.639
37.5	1.535	1.919	2.302	2.558	3.069	3.261	3.837	3.989
40.0	1.678	2.098	2.517	2.797	3.356	3.566	4.196	4.362

b. Stability coefficients m' and n' for $c'/\gamma H = 0.025$ and $D = 1.00$

	Stability coefficients for earth slopes							
	Slope 2:1		Slope 3:1		Slope 4:1		Slope 5:1	
ϕ'	m'	n'	m'	n'	m'	n'	m'	n'
10.0	0.678	0.534	0.906	0.683	1.130	0.846	1.367	1.031
12.5	0.790	0.655	1.066	0.849	1.337	1.061	1.620	1.282
15.0	0.901	0.776	1.224	1.014	1.544	1.273	1.868	1.534
17.5	1.012	0.898	1.380	1.179	1.751	1.485	2.121	1.789
20.0	1.124	1.022	1.542	1.347	1.962	1.698	2.380	2.050
22.5	1.239	1.150	1.705	1.518	2.177	1.916	2.646	2.317
25.0	1.356	1.282	1.875	1.696	2.400	2.141	2.921	2.596
27.5	1.478	1.421	2.050	1.882	2.631	2.375	3.207	2.886
30.0	1.606	1.567	2.235	2.078	2.873	2.622	3.508	3.191
32.5	1.739	1.721	2.431	2.285	3.127	2.883	3.823	3.511
35.0	1.880	1.885	2.635	2.505	3.396	3.160	4.156	3.849
37.5	2.030	2.060	2.855	2.741	3.681	3.458	4.510	4.209
40.0	2.190	2.247	3.090	2.993	3.984	3.778	4.885	4.592

(continued)

Table 14.1 (*continued*)

c. *Stability coefficients m' and n' for c'/γH = 0.025 and D = 1.25*

	Stability coefficients for earth slopes							
	Slope 2:1		Slope 3:1		Slope 4:1		Slope 5:1	
ϕ'	m'	n'	m'	n'	m'	n'	m'	n'
10.0	0.737	0.614	0.901	0.726	1.085	0.867	1.285	1.014
12.5	0.878	0.759	1.076	0.908	1.299	1.098	1.543	1.278
15.0	1.019	0.907	1.253	1.093	1.515	1.311	1.803	1.545
17.5	1.162	1.059	1.433	1.282	1.736	1.541	2.065	1.814
20.0	1.309	1.216	1.618	1.478	1.961	1.775	2.334	2.090
22.5	1.461	1.379	1.808	1.680	2.194	2.017	2.610	2.373
25.0	1.619	1.547	2.007	1.891	2.437	2.269	2.879	2.669
27.5	1.783	1.728	2.213	2.111	2.689	2.531	3.196	2.976
30.0	1.956	1.915	2.431	2.342	2.953	2.806	3.511	3.299
32.5	2.139	2.112	2.659	2.686	3.231	3.095	3.841	3.638
35.0	2.331	2.321	2.901	2.841	3.524	3.400	4.191	3.998
37.5	2.536	2.541	3.158	3.112	3.835	3.723	4.563	4.379
40.0	2.753	2.775	3.431	3.399	4.164	4.064	4.958	4.784

d. *Stability coefficients m' and n' for c'/γH = 0.05 and D = 1.00*

	Stability coefficients for earth slopes							
	Slope 2:1		Slope 3:1		Slope 4:1		Slope 5:1	
ϕ'	m'	n'	m'	n'	m'	n'	m'	n'
10.0	0.913	0.563	1.181	0.717	1.469	0.910	1.733	1.069
12.5	1.030	0.690	1.343	0.878	1.688	1.136	1.995	1.316
15.0	1.145	0.816	1.506	1.043	1.904	1.353	2.256	1.567
17.5	1.262	0.942	1.671	1.212	2.117	1.565	2.517	1.825
20.0	1.380	1.071	1.840	1.387	2.333	1.776	2.783	2.091
22.5	1.500	1.202	2.014	1.568	2.551	1.989	3.055	2.365
25.0	1.624	1.338	2.193	1.757	2.778	2.211	3.336	2.651
27.5	1.753	1.480	1.380	1.952	3.013	2.444	3.628	2.948
30.0	1.888	1.630	2.574	2.157	3.261	2.693	3.934	3.259
32.5	2.029	1.789	2.777	2.370	3.523	2.961	4.256	3.585
35.0	2.178	1.958	2.990	2.592	3.803	3.253	4.597	3.927
37.5	2.336	2.138	3.215	2.826	4.103	3.574	4.959	4.288
40.0	2.505	2.332	3.451	3.071	4.425	3.926	5.344	4.668

(*continued*)

Table 14.1 (*continued*)

e. Stability coefficients m' and n' for c'/γH = 0.05 and D = 1.25

| | Stability coefficients for earth slopes | | | | | | | |
| | Slope 2:1 | | Slope 3:1 | | Slope 4:1 | | Slope 5:1 | |
ϕ'	m'	n'	m'	n'	m'	n'	m'	n'
10.0	0.919	0.633	1.119	0.766	1.344	0.886	1.594	1.042
12.5	1.065	0.792	1.294	0.941	1.563	1.112	1.850	1.300
15.0	1.211	0.950	1.471	1.119	1.782	1.338	2.109	1.562
17.5	1.359	1.108	1.650	1.303	2.004	1.567	2.373	1.831
20.0	1.509	1.266	1.834	1.493	2.230	1.799	2.643	2.107
22.5	1.663	1.428	2.024	1.690	2.463	2.038	2.921	2.392
25.0	1.822	1.595	2.222	1.897	2.705	2.287	3.211	2.690
27.5	1.988	1.769	2.428	2.113	2.957	2.546	3.513	2.999
30.0	2.161	1.950	2.645	2.342	3.221	2.819	3.829	3.324
32.5	2.343	2.141	2.873	2.583	3.500	3.107	4.161	3.665
35.0	2.535	2.344	3.114	2.839	3.795	3.413	4.511	4.025
37.5	2.738	2.560	3.370	3.111	4.109	3.740	4.881	4.405
40.0	2.953	2.791	3.642	3.400	4.442	4.090	5.273	4.806

f. Stability coefficients m' and n' for c'/γH = 0.05 and D = 1.50

| | Stability coefficients for earth slopes | | | | | | | |
| | Slope 2:1 | | Slope 3:1 | | Slope 4:1 | | Slope 5:1 | |
ϕ'	m'	n'	m'	n'	m'	n'	m'	n'
10.0	1.022	0.751	1.170	0.828	1.343	0.974	1.547	1.108
12.5	1.202	0.936	1.376	1.043	1.589	1.227	1.829	1.399
15.0	1.383	1.122	1.583	1.260	1.835	1.480	2.112	1.690
17.5	1.565	1.309	1.795	1.480	2.084	1.734	2.398	1.983
20.0	1.752	1.501	2.011	1.705	2.337	1.993	2.690	2.280
22.5	1.943	1.698	2.234	1.937	2.597	2.258	2.990	2.585
25.0	2.143	1.903	2.467	2.179	2.867	2.534	3.302	2.902
27.5	2.350	2.117	2.709	2.431	3.148	2.820	3.626	3.231
30.0	2.568	2.342	2.964	2.696	3.443	3.120	3.967	3.577
32.5	2.798	2.580	3.232	2.975	3.753	3.436	4.326	3.940
35.0	3.041	2.832	3.515	3.269	4.082	3.771	4.707	4.325
37.5	3.299	3.102	3.817	3.583	4.431	4.128	5.112	4.735
40.0	3.574	3.389	4.136	3.915	4.803	4.507	5.543	5.171

(*continued*)

Table 14.1 (*continued*)

g. *Stability coefficients m' and n' for c'/γH = 0.075 and toe circles*

	Stability coefficients for earth slopes							
	Slope 2:1		Slope 3:1		Slope 4:1		Slope 5:1	
ϕ'	m'	n'	m'	n'	m'	n'	m'	n'
20	1.593	1.158	2.055	1.516	2.498	1.903	2.934	2.301
25	1.853	1.430	2.426	1.888	2.980	2.361	3.520	2.861
30	2.133	1.730	2.826	2.288	3.496	2.888	4.150	3.461
35	2.433	2.058	3.253	2.730	4.055	3.445	4.846	4.159
40	2.773	2.430	3.737	3.231	4.680	4.061	5.609	4.918

h. *Stability coefficients m' and n' for c'/γH = 0.075 and D = 1.00*

	Stability coefficients for earth slopes							
	Slope 2:1		Slope 3:1		Slope 4:1		Slope 5:1	
ϕ'	m'	n'	m'	n'	m'	n'	m'	n'
20	1.610	1.100	2.141	1.443	2.664	1.801	3.173	2.130
25	1.872	1.386	2.502	1.815	3.126	2.259	3.742	2.715
30	2.142	1.686	2.884	2.201	3.623	2.758	4.357	3.331
35	2.443	2.030	3.306	2.659	4.177	3.331	5.024	4.001
40	2.772	2.386	3.775	3.145	4.785	3.945	5.776	4.759

i. *Stability coefficients m' and n' for c'/γH = 0.075 and D = 1.25*

	Stability coefficients for earth slopes							
	Slope 2:1		Slope 3:1		Slope 4:1		Slope 5:1	
ϕ'	m'	n'	m'	n'	m'	n'	m'	n'
20	1.688	1.285	2.071	1.543	2.492	1.815	2.954	2.173
25	2.004	1.641	2.469	1.957	2.972	2.315	3.523	2.730
30	2.352	2.015	2.888	2.385	3.499	2.857	4.149	3.357
35	2.728	2.385	3.357	2.870	4.079	3.457	4.831	4.043
40	3.154	2.841	3.889	3.428	4.729	4.128	5.603	4.830

j. *Stability coefficients m' and n' for c'/γH = 0.075 and D = 1.50*

	Stability coefficients for earth slopes							
	Slope 2:1		Slope 3:1		Slope 4:1		Slope 5:1	
ϕ'	m'	n'	m'	n'	m'	n'	m'	n'
20	1.918	1.514	2.199	1.728	2.548	1.985	2.931	2.272
25	2.308	1.914	2.660	2.200	3.083	2.530	3.552	2.915
30	2.735	2.355	3.158	2.714	3.659	3.128	4.128	3.585
35	3.211	2.854	3.708	3.285	4.302	3.786	4.961	4.343
40	3.742	3.397	4.332	3.926	5.026	4.527	5.788	5.185

(*continued*)

Table 14.1 (*continued*)

k. *Stability coefficients m′ and n′ for c′/γH = 0.100 and toe circles*

	Stability coefficients for earth slopes							
	Slope 2:1		Slope 3:1		Slope 4:1		Slope 5:1	
ϕ'	m′	n′	m′	n′	m′	n′	m′	n′
20	1.804	2.101	2.286	1.588	2.748	1.974	3.190	2.361
25	2.076	1.488	2.665	1.945	3.246	2.459	3.796	2.959
30	2.362	1.786	3.076	2.359	3.770	2.961	4.442	3.576
35	2.673	2.130	3.518	2.803	4.339	3.518	5.146	4.249
40	3.012	2.486	4.008	3.303	4.984	4.173	5.923	5.019

l. *Stability coefficients m′ and n′ for c′/γH = 0.100 and D = 1.00*

	Stability coefficients for earth slopes							
	Slope 2:1		Slope 3:1		Slope 4:1		Slope 5:1	
ϕ'	m′	n′	m′	n′	m′	n′	m′	n′
20	1.841	1.143	2.421	1.472	2.982	1.815	3.549	2.157
25	2.102	1.430	2.785	1.845	3.458	2.303	4.131	2.743
30	2.378	1.714	3.183	2.258	3.973	2.830	4.751	3.372
35	2.692	2.086	3.612	2.715	4.516	3.359	5.426	4.059
40	3.025	2.445	4.103	3.230	5.144	4.001	6.187	4.831

m. *Stability coefficients m′ and n′ for c′/γH = 0.100 and D = 1.25*

	Stability coefficients for earth slopes							
	Slope 2:1		Slope 3:1		Slope 4:1		Slope 5:1	
ϕ'	m′	n′	m′	n′	m′	n′	m′	n′
20	1.874	1.301	2.283	1.558	2.751	1.843	3.253	2.158
25	2.197	1.642	2.681	1.972	3.233	2.330	3.833	2.758
30	2.540	2.000	3.112	2.415	3.753	2.858	4.451	3.372
35	2.922	2.415	3.588	2.914	4.333	3.458	5.141	4.072
40	3.345	2.855	4.119	3.457	4.987	4.142	5.921	4.872

n. *Stability coefficients m′ and n′ for c′/γH = 0.100 and D = 1.50*

	Stability coefficients for earth slopes							
	Slope 2:1		Slope 3:1		Slope 4:1		Slope 5:1	
ϕ'	m′	n′	m′	n′	m′	n′	m′	n′
20	2.079	1.528	2.387	1.742	2.768	2.014	3.158	2.285
25	2.477	1.942	2.852	2.215	3.297	2.542	3.796	2.927
30	2.908	2.385	3.349	2.728	3.881	3.143	4.468	3.614
35	3.385	2.884	3.900	3.300	4.520	3.800	5.211	4.372
40	3.924	3.441	4.524	3.941	5.247	4.542	6.040	5.200

Example 14.9

A slope is 15 m high. Given that slopes = 2 horizontal and 1 vertical, $\phi' = 20°$, $c' = 20$ kN/m², $\gamma = 18$ kN/m³, and $r_u = 0.2$, determine the factor of safety, F_s.

Solution

$$\frac{c'}{\gamma H} = \frac{20}{(18)(15)} = 0.074$$

We are given that $\phi' = 20°$, slopes = 2H · 1V, and $r_u = 0.2$. The following table can now be prepared:

Toe circle or D	m'	n'	$F_s = m' - n' r_u$
Toe circle	1.593	1.158	1.361
1.0	1.610	1.100	1.39
1.25	1.688	1.285	1.431
1.5	1.918	1.514	1.615

So

$$F_s \approx 1.361 \approx \mathbf{1.36}$$

■

14.12 *Morgenstern's Method of Slices for Rapid Drawdown Condition*

Morgenstern (1963) used Bishop's method of slices (Section 14.10) to determine the factor of safety, F_s, during rapid drawdown. In preparing the solution, Morgenstern used the following notation (Figure 14.23):

1. L = height of drawdown
2. H = height of embankment
3. β = angle that the slope makes with the horizontal

Morgenstern also assumed that

1. the embankment is made of homogeneous material and rests on an impervious base;
2. initially, the water level coincides with the top of the embankment;
3. during drawdown, pore water pressure does not dissipate; and
4. the unit weight of saturated soil (γ_{sat}) = $2\gamma_w$ (γ_w = unit weight of water).

Figures 14.24 through 14.26 provide the drawdown stability charts developed by Morgenstern.

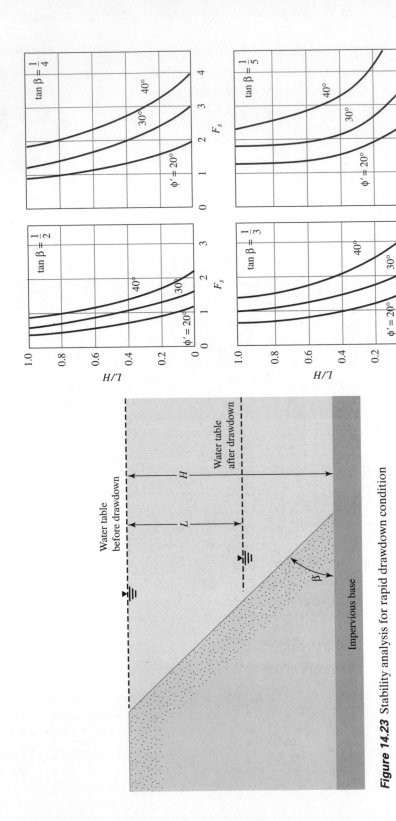

Figure 14.24 Morgenstern's drawdown stability chart for $c'/\gamma H = 0.0125$

Figure 14.23 Stability analysis for rapid drawdown condition

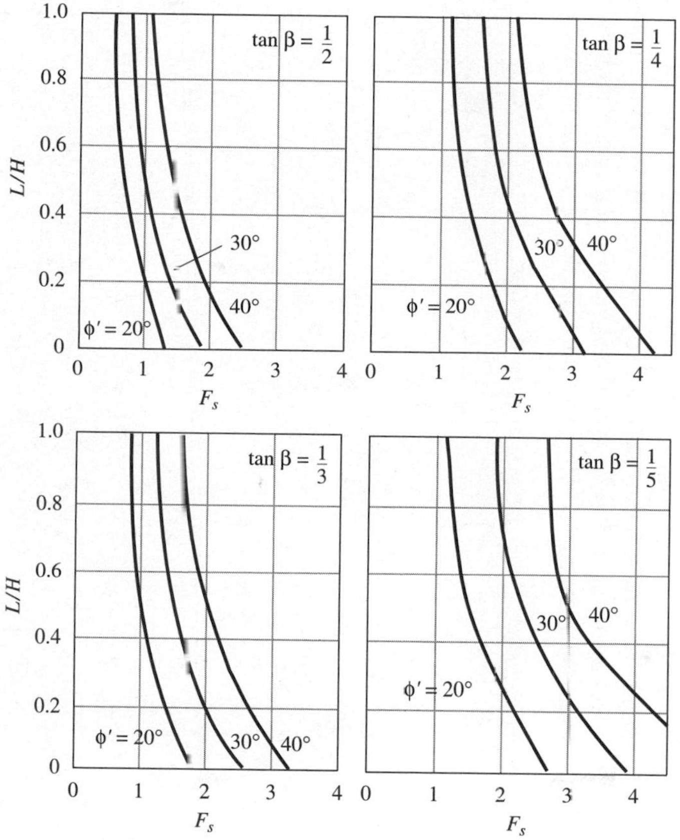

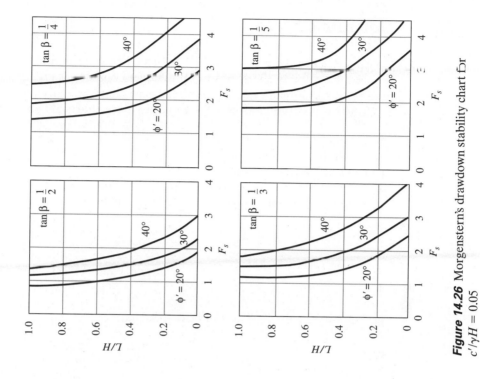

Figure 14.26 Morgenstern's drawdown stability chart for $c'/\gamma H = 0.05$

Figure 14.25 Morgenstern's drawdown stability chart for $c'/\gamma H = 0.025$

14.13   *Cousins Charts*

Cousins (1978) used a variation of Taylor's friction circle method (Section 14.8) to prepare the stability charts of simple homogeneous slopes that considers the effects of pore water pressure due to seepage. These charts are given in Figures 14.27 and 14.28. Various parameters used in preparing these charts are as follows:

a. Height of the slope, H
b. Depth function [as defined in Eq. (14.49)], D
c. Bulk unit weight of soil, γ
d. Effective shear strength parameters for the soil, c' and ϕ'
e. Pore water pressure ratio r_u. According to Figure 14.22, it is defined as

$$r_u = \frac{h\gamma_w}{\gamma z} = \frac{u}{\gamma z} \tag{14.74}$$

f.
$$\lambda_{c'\phi'} = \frac{\gamma H \tan \phi'}{c'} \tag{14.75}$$

g. Stability factor, N_s. This can be defined as

$$N_s = \frac{\gamma H F_s}{c'} \tag{14.76}$$

Note that Figure 14.27 is for critical toe circles, and Figure 14.28 is for critical circles with various depth functions, D.

The pore water pressure ratio r_u to be used in these figures is the average value along the critical failure surface. To determine the minimum factor of safety with respect to strength, checks should be made for toe circles and also for circles with various depth functions. The procedure for doing so is illustrated in Example 14.10.

Example 14.10

Determine the minimum factor of safety (F_s) of a slope with the following parameters: $H = 10.5$ m; $\beta =$ slope angle $= 25°$; $c' = 10.5$ kN/m^2; $\phi' = 25°$; $\gamma = 18.5$ kN/m^3; $r_u = 0.35$ (average value).

Solution
Step 1
From Eq. (14.75),

$$\lambda_{c'\phi'} = \frac{\gamma H \tan \phi}{c'} = \frac{(18.5)(10.5)(\tan 25°)}{10.5} = 8.63$$

Step 2
Using Figure 14.27, check the possibility of toe failure. From Figure 14.27b and c, for $r_u = 0.25$ and $\beta = 25°$, $N_s \simeq 26$ with $D \simeq 1.05$. Also, for $r_u = 0.5$ and $\beta = 25°$, $N_s \simeq 19$ with $D \simeq 1.05$.

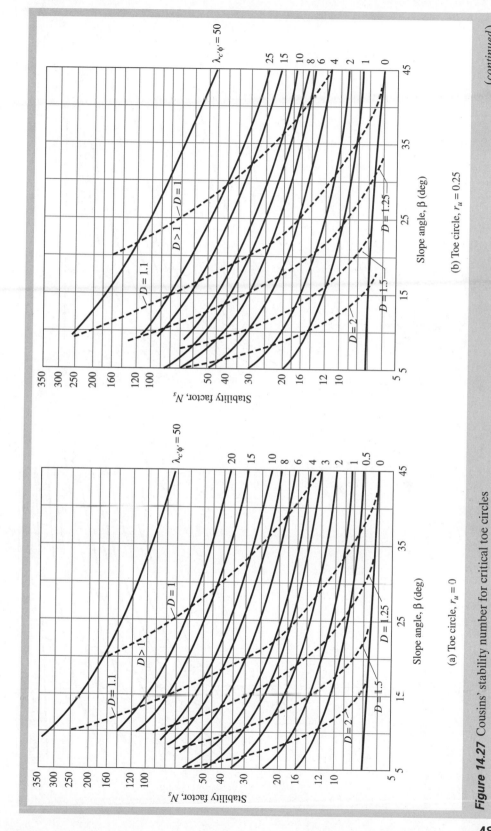

Figure 14.27 Cousins' stability number for critical toe circles

(continued)

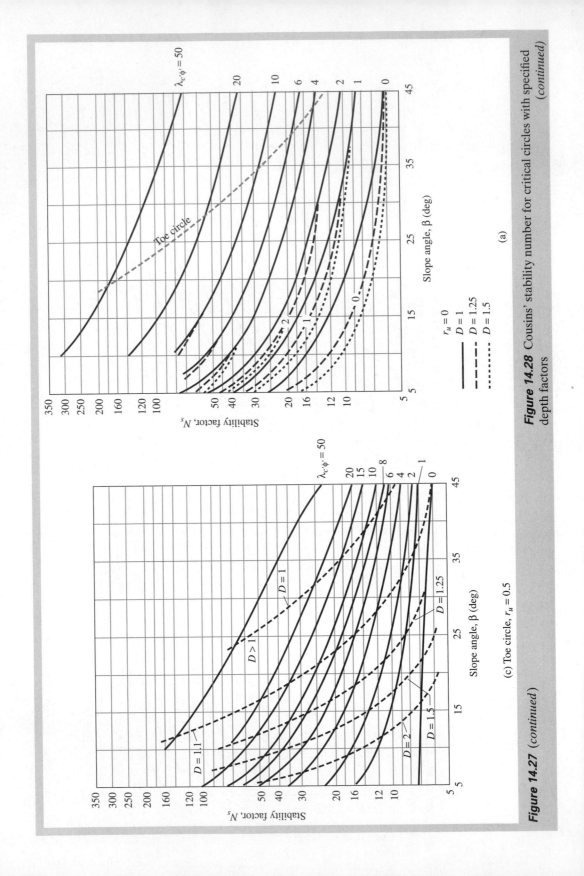

Figure 14.28 Cousins' stability number for critical circles with specified depth factors *(continued)*

(a)

$r_u = 0$
$D = 1$
$D = 1.25$
$D = 1.5$

(c) Toe circle, $r_u = 0.5$

Figure 14.27 *(continued)*

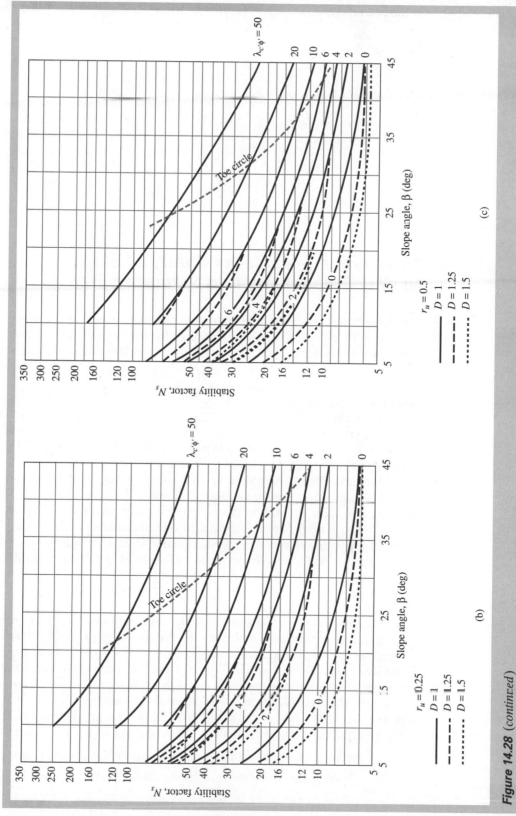

Figure 14.28 (*continued*)

Step 3
Using Figure 14.28b and c, for $r_u = 0.25$ and $\beta = 25°$, $N_s \simeq 25$ with $D \simeq 1.0$. Also, for $r_u = 0.5$ and $\beta = 25°$, $N_s \simeq 19$ with $D \simeq 1.0$.

Step 4
Based on the results in Steps 2 and 3, by interpolation it appears that the failure circle passes close to $D \simeq 1.0$. For $r_u = 0.35$, the minimum value of N_s is about 21.5. Applying Eq. (14.76) we get

$$F_s = \frac{N_s c'}{\gamma H} = \frac{(21.5)(10.5)}{(18.5)(10.5)} = \mathbf{1.16}$$ ∎

14.14 Fluctuation of Factor of Safety of Slopes in Clay Embankment on Saturated Clay

Figure 14.29a shows a clay embankment constructed on a *saturated soft clay*. Let P be a point on a potential failure surface APB that is an arc of a circle. Before construction of the embankment, the pore water pressure at P can be expressed as

$$u = h\gamma_w \tag{14.77}$$

Under ideal conditions, let us assume that the height of the fill needed for the construction of the embankment is placed uniformly, as shown in Figure 14.29b. At time $t = t_1$, the embankment height is equal to H, and it remains constant thereafter (that is, $t > t_1$). The average shear stress increase, τ, on the potential failure surface caused by the construction of the embankment is also shown in Figure 14.29b. The value of τ will increase linearly with time up to time $t = t_1$ and remain constant thereafter.

The pore water pressure at point P (Figure 14.29a) will continue to increase as construction of the embankment progresses, as shown in Figure 14.29c. At time $t = t_1$, $u = u_1 > h\gamma_w$. This is because of the slow rate of drainage from the clay layer. However, after construction of the embankment is completed (that is, $t > t_1$), the pore water pressure will gradually decrease with time as the drainage, and thus consolidation, progresses. At time $t \simeq t_2$,

$$u = h\gamma_w$$

For simplicity, if we assume that the embankment construction is rapid and that practically no drainage occurs during the construction period, the average *shear strength* of the clay will remain constant from $t = 0$ to $t = t_1$, or $\tau_f = c_u$ (undrained shear strength). This is shown in Figure 14.29d. For time $t > t_1$, as consolidation progresses, the magnitude of the shear strength, τ_f, will gradually increase. At time $t \geq t_2$ — that is, after consolidation is completed — the average shear strength of the clay will be equal to $\tau_f = c' + \sigma' \tan \phi'$ (drained shear strength) (Figure 14.29d). The factor of safety of the embankment along the potential surface of sliding can be given as

$$F_s = \frac{\text{Average shear strength of clay, } \tau_f, \text{ along sliding surface (Figure 14.29d)}}{\text{Average shear stress, } \tau, \text{ along sliding surface (Figure 14.29b)}}$$

$$\tag{14.78}$$

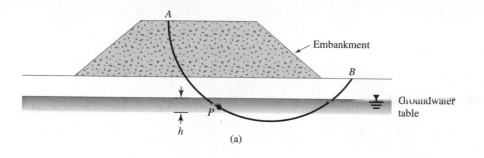

(a)

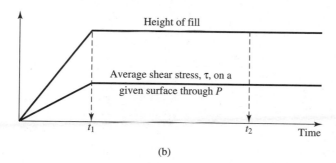

(b)

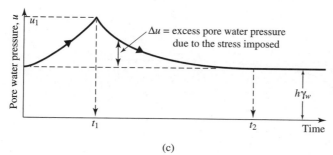

(c)

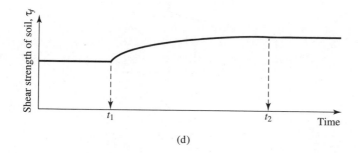

(d)

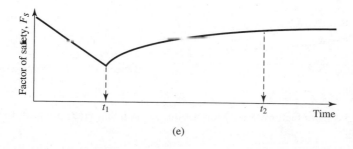

(e)

Figure 14.29
Factor of safety variation with time for embankment on soft clay (redrawn after Bishop and Bjerrum, 1960)

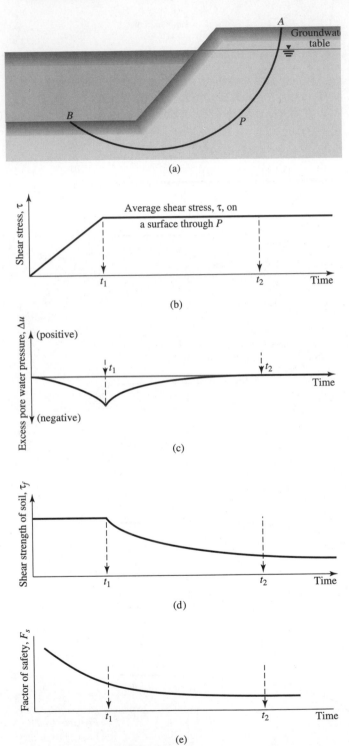

Figure 14.30 Variation of factor of safety for cut slope in soft clay (redrawn after Bishop and Bjerrum, 1960)

The general nature of the variation of the factor of safety, F_s, with time is shown in Figure 14.29e. As we can see from this figure, the magnitude of F_s initially decreases with time. At the end of construction (time $t = t_1$), the value of the factor of safety is a minimum. Beyond this point, the value of F_s continues to increase with drainage up to time $t = t_2$.

Cuts in Saturated Clay

Figure 14.30a shows a cut slope in a saturated soft clay in which APB is a circular potential failure surface. During advancement of the cut, the average shear stress, τ, on the potential failure surface passing through P will increase. The maximum value of the average shear stress, τ, will be attained at the end of construction — that is, at time $t = t_1$. This property is shown in Figure 14.30b.

Because of excavation of the soil, the effective overburden pressure at point P will decrease, which will induce a reduction in the pore water pressure. The variation of the net change of pore water pressure, Δu, is shown in Figure 14.30c. After excavation is complete (time $t > t_1$), the net negative excess pore water pressure will gradually dissipate. At time $t \geq t_2$, the magnitude of Δu will be equal to 0.

The variation of the average shear strength, τ_f, of the clay with time is shown in Figure 14.30d. Note that the shear strength of the soil after excavation gradually decreases. This decrease occurs because of dissipation of the negative excess pore water pressure.

If the factor of safety of the cut slope, F_s, along the potential failure surface is defined by Eq. (14.78), its variation will be as shown in Figure 14.30e. Note that the magnitude of F_s decreases with time, and its minimum value is obtained at time $t \geq t_2$.

14.15 *Summary and General Comments*

In this chapter, the general principles of analyzing the stability of slopes were presented. Several general observations can be made regarding this subject:

1. Culmann's method of slope analysis using a planar failure surface passing through the toe of the slope was presented in Section 14.4. The accuracy of this method is good for steep slopes or vertical banks; however, it will underestimate the factor of safety for flatter slopes. A planar failure surface can be used to analyze slope stability when a sloping mass of soil rests on an inclined layer of stiff soil or rock (Figure 14.31). When water seeps from the top and

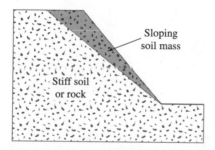

Figure 14.31 Planar failure surface in slope stability analysis

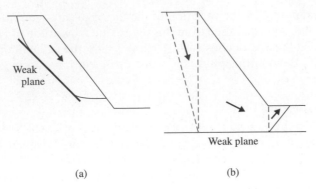

Figure 14.32 (a) Noncircular failure surface in a slope; (b) sliding block failure

then along the plane of discontinuity, it will weaken the soil along the plane of contact and thus may make the upper sloping soil mass unstable.

2. Several other mechanisms by which slopes may fail were not covered in this chapter. Figure 14.32a shows a *noncircular failure surface* in a slope. Another type of slope failure is shown in Figure 14.32b, which is called *sliding block failure.*

   In many situations, the potential failure surface can be approximated by several straight lines. An example of this configuration is shown in Figure 14.33. The soil mass in the potential failure zone is broken up into several wedges and the stability is then analyzed.

3. For a short-term stability analysis of nonfissured clay, the total stress method assuming the condition in which $\phi = 0$ will be satisfactory. However, for long-term slope stability analysis, the effective stress method of analysis should be used. For overconsolidated clay slope analysis the residual strength of the soil should be used. It was shown in Chapter 11 that for overconsolidated clays, the stress-strain plots will be of the nature shown in Figure 14.34a. Figure 14.34b shows the peak and residual effective strength envelopes for an overconsolidated clay. Note that the residual strength can be expressed approximately as

$$\tau_r \approx \sigma' \tan \phi'_r \qquad (14.79)$$

For normally consolidated clays, the difference between peak and residual strengths is small.

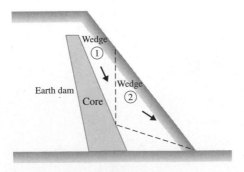

Figure 14.33 Wedge analysis

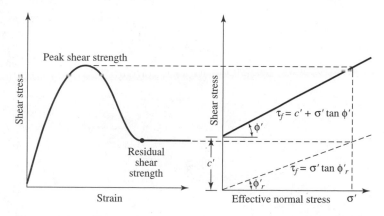

Figure 14.34 Peak and residual strengths of overconsolidated clay

4. Sections 14.9 and 14.10 provided the procedure for stability analysis by the method of slices. Several computer programs are now available for quick routine analysis using Bishop's simplified method of slices.

Problems

14.1 For the slope shown in Figure 14.35, find the height H for critical equilibrium. It is given that $\beta = 22°$, $\gamma = 100$ lb/ft^3, $\phi' = 15°$, and $c' = 200$ lb/ft^2.

14.2 Refer to Figure 14.35. It is given that $\gamma = 15.5$ kN/m^3, $\phi' = 20°$, and $c' = 15$ kN/m^2.

   a. If $\beta = 20°$ and $H = 4$ m, what is the factor of safety of the slope against sliding along the soil-rock interface?

   b. For $\beta = 30°$, find the height H that will have a factor of safety of 2 against sliding along the soil-rock interface.

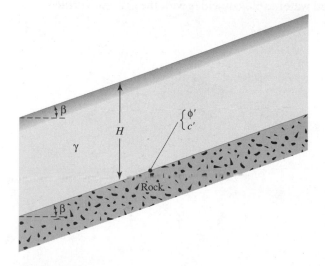

Figure 14.35

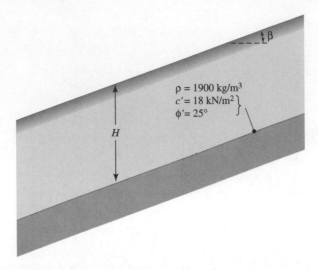

$$\rho = 1900 \text{ kg/m}^3$$
$$c' = 18 \text{ kN/m}^2$$
$$\phi' = 25°$$

H

Figure 14.36

14.3   An infinite slope is shown in Figure 14.36. The shear strength parameters at the interface of soil and rock are $c' = 18$ kN/m² and $\phi' = 25°$.

   a. If $H = 6$ m and $\beta = 18°$, find the factor of safety against sliding on the rock surface.

   b. If $\beta = 30°$, find the height H for which $F_s = 1.5$. (Assume the pore water pressure to be 0.)

14.4   Refer to Figure 14.36. If there were seepage through the soil and the ground water table coincided with the ground surface, what would be the value of F_s? Use $H = 8$ m, $\rho_{sat} = 1900$ kg/m³, and $\beta = 20°$.

14.5   For the infinite slope shown in Figure 14.37, find the factor of safety against sliding along the place *AB*, given that $H = 20$ ft, $G_s = 2.7$, $e = 0.6$, $\phi' = 20°$, and $c' = 500$ lb/ft². Note that there is seepage through the soil and that the ground water table coincides with the ground surface.

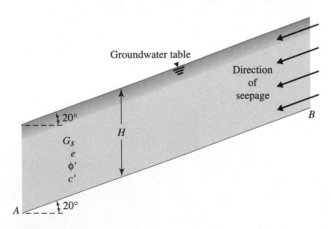

Groundwater table

Direction of seepage

B

20°

G_s
e
ϕ'
c'

H

A   20°

Figure 14.37

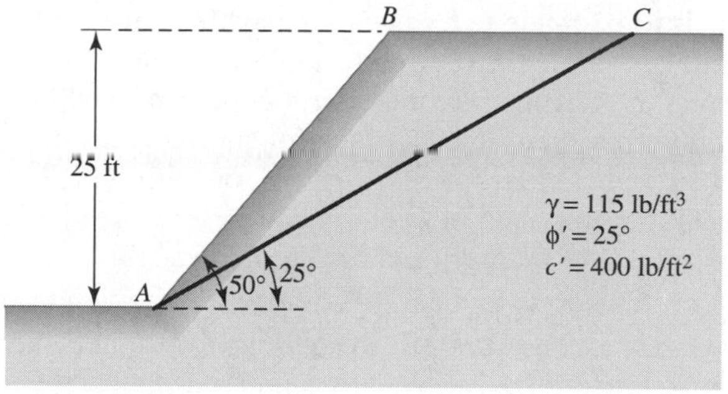

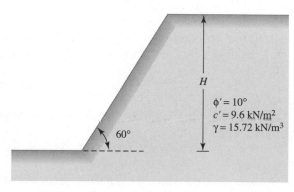

B   C

25 ft

$\gamma = 115$ lb/ft^3
$\phi' = 25°$
$c' = 400$ lb/ft^2

A   50° 25°

Figure 14.38

14.6 Repeat Problem 14.5 with $H = 4$ m, $G_s = 2.65$, $e = 0.7$, $\phi' = 20°$, and $c' = 15$ kN/m^2.

14.7 A slope is shown in Figure 14.38. AC represents a trial failure plane. For the wedge ABC, find the factor of safety against sliding.

14.8 A finite slope is shown in Figure 14.39. Assuming that the slope failure would occur along a plane (Culman's assumption), find the height of the slope for critical equilibrium.

14.9 Referring to Figure 14.39, find the height of the slope H that will have a factor of safety of 2 against sliding. Assume that the critical surface for sliding is a plane.

14.10 Refer to Figure 14.39. If $H = 3.96$ m, what will be the factor of safety with respect to sliding? Assume that the critical surface is a plane.

14.11 Determine the height of a finite slope, 1 vertical to 2 horizontal, that should have a factor of safety of 2 against sliding. For the soil, the following values are given:

  $c' = 18$ kN/m^2
  $\phi' = 20°$
  $\rho = 1700$ kg/m^3

Assume the critical surface for sliding to be a plane.

H

$\phi' = 10°$
$c' = 9.6$ kN/m^2
$\gamma = 15.72$ kN/m^3

60°

Figure 14.39

14.12 A cut slope is to be made in a saturated clay. It is given that $c_u = 25$ kN/m^2 ($\phi = 0$ condition) and $\gamma = 18$ kN/m^3. The slope makes an angle β of 65° with the horizontal. Determine the maximum depth up to which the cut could be made. Assume that the critical surface for sliding is circular. What is the nature of the critical circle (that is, toe, slope, or midpoint)?

14.13 For the cut slope described in Problem 14.12, if we need a factor of safety of 2 against sliding, how deep should the cut be made?

14.14 Using the graph given in Figure 14.8, determine the height of a slope, 1 vertical to $\frac{1}{2}$ horizontal, in saturated clay with an undrained shear strength of 750 lb/ft^2. The desired factor of safety against sliding is 2.5. It is given that $\gamma = 120$ lb/ft^3.

14.15 Refer to Problem 14.14. What should be the critical height of the slope? What is the nature of the critical circle?

14.16 A cut slope was excavated in a saturated clay. The slope angle β is equal to 40° with the horizontal. Slope failure occurred when the cut reached a depth of 8.5 m. Previous soil explorations showed that a rock layer was located at a depth of 12 m below the ground surface. Assuming an undrained condition and that $\gamma_{sat} = 18.5$ kN/m^3,
   a. Determine the undrained cohesion of the clay (Figure 14.8).
   b. What was the nature of the critical circle?
   c. With reference to the top of the slope, at what distance did the surface of sliding intersect the bottom of the excavation?

14.17 If the cut slope described in Problem 14.16 is excavated in a manner such that $H_{cr} = 10$ m, what angle would the slope make with the horizontal? (Use Figure 14.8 and the results of Problem 14.16.)

14.18 A clay slope is built over a layer of rock. For the slope with values
   Height, $H = 16$ m
   Slope angle, $\beta = 30°$
   Saturated unit weight of soil, $\gamma_{sat} = 17$ kN/m^3
   Undrained shear strength, $c_u = 50$ kN/m^2
determine the factor of safety if $k_h = 0.4$. (Use the procedure outlined in Section 14.7.)

14.19 For a slope in clay, $H = 50$ ft, $\gamma = 115$ lb/ft^3, $\beta = 60°$, and $c_u = 1000$ lb/ft^2, determine the factor of safety for $k_h = 0.3$. (Use the procedure outlined in Section 14.7.)

14.20 Refer to Figure 14.40. Use Figure 14.16 ($\phi' > 0$) to solve the following:
   a. If $n' = 2$, $\phi' = 10°$, $c' = 700$ lb/ft^2, and $\gamma = 110$ lb/ft^3, find the critical height of the slope.
   b. If $n' = 1$, $\phi' = 20°$, $c' = 400$ lb/ft^2, and $\gamma = 115$ lb/ft^3, find the critical height of the slope.
   c. If $n' = 2.5$, $\phi' = 12°$, $c' = 23.94$ kN/m^2, and $\gamma = 16.51$ kN/m^3, find the critical height of the slope.
   d. If $n' = 1$, $\phi' = 15°$, $c' = 18$ kN/m^2, and $\gamma = 17.1$ kN/m^3, find the critical height of the slope.

14.21 Refer to Figure 14.40. Using Figure 14.15, find the factor of safety, F_s, with respect to strength for a slope with the following:

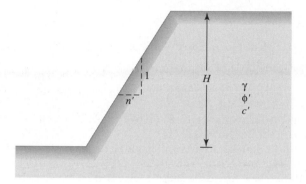

Figure 14.40

a. $n' = 2$, $\phi' = 10°$, $c' = 700$ lb/ft^2, $\gamma = 110$ lb/ft^3, and $H = 50$ ft
b. $n' = 1$, $\phi' = 20°$, $c' = 400$ lb/ft^2, $\gamma = 115$ lb/ft^3, and $H = 30$ ft
c. $n' = 2.5$, $\phi' = 12°$, $c' = 23.94$ kN/m^2, $\gamma = 16.51$ kN/m^3, and $H = 12$ m
d. $n' = 1$, $\phi' = 15°$, $c' = 18$ kN/m^2, $\gamma = 17.1$ kN/m^3, and $H = 5$ m

14.22 Referring to Figure 14.41 and using the ordinary method of slices, find the factor of safety with respect to sliding for the following trial cases:
  a. $\beta = 45°$, $\phi' = 20°$, $c' = 400$ lb/ft^2, $\gamma = 115$ lb/ft^3, $H = 30$ ft, $\alpha = 30°$, and $\theta = 70°$
  b. $\beta = 45°$, $\phi' = 15°$, $c' = 18$ kN/m^2, $\gamma = 17.1$ kN/m^3, $H = 5$ m, $\alpha = 30°$, and $\theta = 80°$

14.23 Determine the minimum factor of safety of a slope with the following parameters: $H = 6$ m, $\beta = 26.57°$, $\phi' = 25°$, $c' = 6$ kN/m^2, $\gamma = 20$ kN/m^3, and $r_u = 0.5$. Use Bishop and Morgenstern's method.

14.24 Determine the minimum factor of safety, F_s, of a slope with the following parameters: $H = 25$ ft, $\beta = 30°$, $\phi' = 20°$, $c' = 100$ lb/ft^2, $\gamma = 115$ lb/ft^3, and $r_u = 0.25$. Use Cousins' charts.

14.25 Repeat Problem 14.24 using the following values: $H = 15$ m, $\beta = 20°$, $\phi' = 15°$, $c' = 20$ kN/m^2, $\gamma = 17.5$ kN/m^3, and $r_u = 0.5$.

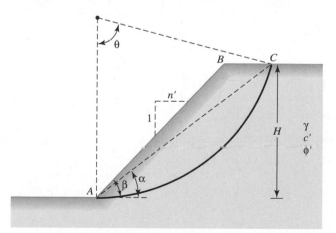

Figure 14.41

References

BISHOP, A. W. (1955). "The Use of Slip Circle in the Stability Analysis of Earth Slopes," *Geotechnique,* Vol. 5, No. 1, 7–17.

BISHOP, A. W., and BJERRUM, L. (1960). "The Relevance of the Triaxial Test to the Solution of Stability Problems," *Proceedings,* Research Conference on Shear Strength of Cohesive Soils, ASCE, 437–501.

BISHOP, A. W., and MORGENSTERN, N. R. (1960). "Stability Coefficients for Earth Slopes," *Geotechnique,* Vol. 10, No. 4, 129–147.

COUSINS, B. F. (1978). "Stability Charts for Simple Earth Slopes," *Journal of the Geotechnical Engineering Division,* ASCE, Vol. 104, No. GT2, 267–279.

CULMANN, C. (1875). *Die Graphische Statik,* Meyer and Zeller, Zurich.

FELLENIUS, W. (1927). *Erdstatische Berechnungen,* rev. ed., W. Ernst u. Sons, Berlin.

MORGENSTERN, N. R. (1963). "Stability Charts for Earth Slopes During Rapid Drawdown," *Geotechnique,* Vol. 13, No. 2, 121–133.

PILOT, G., TRAK, B., and LaROCHELLE, P. (1982). "Effective Stress Analysis of the Stability of Embankments on Soft Soils," *Canadian Geotechnical Journal,* Vol. 19, No. 44, 433–450.

SINGH, A. (1970). "Shear Strength and Stability of Man-Made Slopes," *Journal of the Soil Mechanics and Foundations Division,* ASCE, Vol. 96, No. SM6, 1879–1892.

TAYLOR, D. W. (1937). "Stability of Earth Slopes," *Journal of the Boston Society of Civil Engineers,* Vol. 24, 197–246.

TERZAGHI, K., and PECK, R. B. (1967). *Soil Mechanics in Engineering Practice,* 2nd ed., Wiley, New York.

15

Soil-Bearing Capacity for Shallow Foundations

The lowest part of a structure is generally referred to as the *foundation*. Its function is to transfer the load of the structure to the soil on which it is resting. A properly designed foundation transfers the load throughout the soil without overstressing the soil. Overstressing the soil can result in either excessive settlement or shear failure of the soil, both of which cause damage to the structure. Thus, geotechnical and structural engineers who design foundations must evaluate the bearing capacity of soils.

Depending on the structure and soil encountered, various types of foundations are used. Figure 15.1 shows the most common types of foundations. A *spread footing* is simply an enlargement of a load-bearing wall or column that makes it possible to spread the load of the structure over a larger area of the soil. In soil with low load-bearing capacity, the size of the spread footings required is impracticably large. In that case, it is more economical to construct the entire structure over a concrete pad. This is called a *mat foundation*.

Pile and *drilled shaft foundations* are used for heavier structures when great depth is required for supporting the load. Piles are structural members made of timber, concrete, or steel that transmit the load of the superstructure to the lower layers of the soil. According to how they transmit their load into the subsoil, piles can be divided into two categories: friction piles and end-bearing piles. In the case of friction piles, the superstructure load is resisted by the shear stresses generated along the surface of the pile. In the end-bearing pile, the load carried by the pile is transmitted at its tip to a firm stratum.

In the case of drilled shafts, a shaft is drilled into the subsoil and is then filled with concrete. A metal casing may be used while the shaft is being drilled. The casing may be left in place or may be withdrawn during the placing of concrete. Generally, the diameter of a drilled shaft is much larger than that of a pile. The distinction between piles and drilled shafts becomes hazy at an approximate diameter of 1 m (3 ft), and the definitions and nomenclature are inaccurate.

Spread footings and mat foundations are generally referred to as *shallow foundations,* whereas pile and drilled shaft foundations are classified as *deep foundations*. In a more general sense, shallow foundations are foundations that have a depth-of-embedment-to-width ratio of approximately less than four. When the depth-of-embedment-to-width ratio of a foundation is greater than four, it may be classified as a deep foundation.

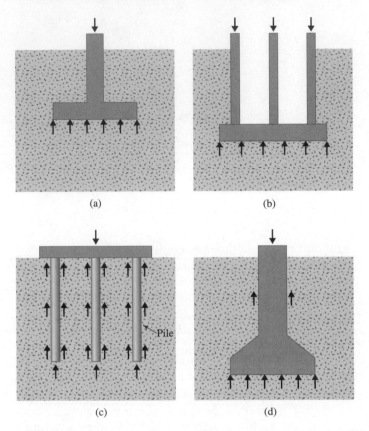

Figure 15.1 Common types of foundations: (a) spread footing; (b) mat foundation; (c) pile foundation; (d) drilled shaft foundation

In this chapter, we discuss the soil-bearing capacity for shallow foundations. As mentioned before, for a foundation to function properly, (1) the settlement of soil caused by the load must be within the tolerable limit, and (2) shear failure of the soil supporting the foundation must not occur. Compressibility of soil — consolidation and elasticity theory — was introduced in Chapter 10. This chapter introduces the load-carrying capacity of shallow foundations based on the criteria of shear failure in soil.

15.1 Ultimate Soil-Bearing Capacity for Shallow Foundations

To understand the concept of the ultimate soil-bearing capacity and the mode of shear failure in soil, let us consider the case of a long rectangular footing of width B located at the surface of a dense sand layer (or stiff soil) shown in Figure 15.2a. When a uniformly distributed load of q per unit area is applied to the footing, it settles. If the uniformly distributed load (q) is increased, the settlement of the footing gradually increases. When the value of $q = q_u$ is reached (Figure 15.2b), bearing capacity

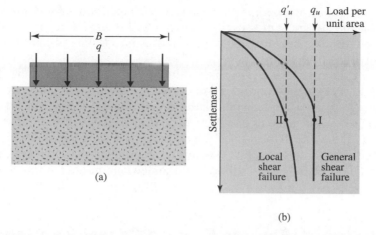

(a)

(b)

Figure 15.2 Ultimate soil-bearing capacity for shallow foundation: (a) model footing; (b) load settlement relationship

failure occurs; the footing undergoes a very large settlement without any further increase of q. The soil on one or both sides of the foundation bulges, and the slip surface extends to the ground surface. The load-settlement relationship is like curve I shown in Figure 15.2b. In this case, q_u is defined as the ultimate bearing capacity of soil.

The bearing capacity failure just described is called a *general shear failure* and can be explained with reference to Figure 15.3a. When the foundation settles under

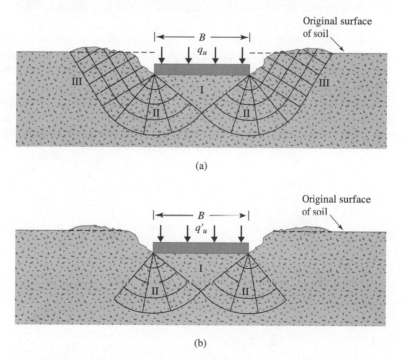

(a)

(b)

Figure 15.3 Modes of bearing capacity failure in soil: (a) general shear failure of soil; (b) local shear failure of soil

the application of a load, a triangular wedge-shaped zone of soil (marked I) is pushed down, and, in turn, it presses the zones marked II and III sideways and then upward. At the ultimate pressure, q_u, the soil passes into a state of plastic equilibrium and failure occurs by sliding.

If the footing test is conducted instead in a loose to medium dense sand, the load-settlement relationship is like curve II in Figure 15.2b. Beyond a certain value of $q = q'_u$, the load-settlement relationship becomes a steep inclined straight line. In this case, q'_u is defined as the ultimate bearing capacity of soil. This type of soil failure is referred to as *local shear failure* and is shown in Figure 15.3b. The triangular wedge-shaped zone (marked I) below the footing moves downward, but unlike general shear failure, the slip surfaces end somewhere inside the soil. Some signs of soil bulging are seen, however.

15.2 Terzaghi's Ultimate Bearing Capacity Equation

In 1921, Prandtl published the results of his study on the penetration of hard bodies, such as metal punches, into a softer material. Terzaghi (1943) extended the plastic failure theory of Prandtl to evaluate the bearing capacity of soils for shallow strip footings. For practical considerations, a long wall footing (length-to-width ratio more than about five) may be called a *strip footing*. According to Terzaghi, a foundation may be defined as a shallow foundation if the depth D_f is less than or equal to its width B (Figure 15.4). He also assumed that for ultimate soil-bearing capacity calculations, the weight of soil above the base of the footing may be replaced by a uniform surcharge, $q = \gamma D_f$.

The failure mechanism assumed by Terzaghi for determining the ultimate soil-bearing capacity (general shear failure) for a rough strip footing located at a depth D_f measured from the ground surface is shown in Figure 15.5a. The soil wedge ABJ (zone I) is an elastic zone. Both AJ and BJ make an angle ϕ' with the horizontal. Zones marked II (AJE and BJD) are the radial shear zones, and zones marked III are the Rankine passive zones. The rupture lines JD and JE are arcs of a logarithmic spiral, and DF and EG are straight lines. AE, BD, EG, and DF make angles of

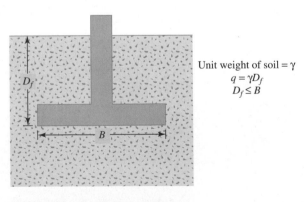

Figure 15.4 Shallow strip footing

Unit weight of soil = γ
$$q = \gamma D_f$$
$$D_f \le B$$

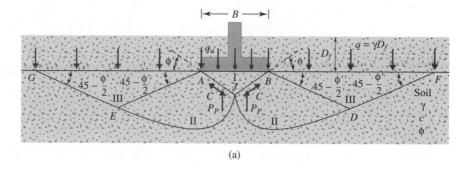

(a)

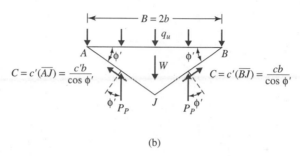

(b)

Figure 15.5 Terzaghi's bearing capacity analysis

$45 - \phi'/2$ degrees with the horizontal. The equation of the arcs of the logarithmic spirals *JD* and *JE* may be given as

$$r = r_o e^{\theta \tan \phi'}$$

If the load per unit area, q_u, is applied to the footing and general shear failure occurs, the passive force P_p is acting on each of the faces of the soil wedge *ABJ*. This concept is easy to conceive of if we imagine that *AJ* and *BJ* are two walls that are pushing the soil wedges *AJEG* and *BJDF*, respectively, to cause passive failure. P_p should be inclined at an angle δ (which is the angle of wall friction) to the perpendicular drawn to the wedge faces (that is, *AJ* and *BJ*). In this case, δ should be equal to the angle of friction of soil, ϕ'. Because *AJ* and *BJ* are inclined at an angle ϕ' to the horizontal, the direction of P_p should be vertical.

Now let us consider the free body diagram of the wedge *ABJ* as shown in Figure 15.5b. Considering the unit length of the footing, we have, for equilibrium,

$$(q_u)(2b)(1) = -W + 2C \sin \phi' + 2P_p \tag{15.1}$$

where $b = B/2$

W = weight of soil wedge $ABJ = \gamma b^2 \tan \phi'$

C = cohesive force acting along each face , *AJ* and *BJ*, that is equal to the unit cohesion times the length of each face = $c'b/(\cos \phi')$

Thus,

$$2bq_u = 2P_p + 2bc' \tan \phi' - \gamma b^2 \tan \phi' \tag{15.2}$$

or

$$q_u = \frac{P_p}{b} + c' \tan \phi' - \frac{\gamma b}{2} \tan \phi' \tag{15.3}$$

The passive pressure in Eq. (15.2) is the sum of the contribution of the weight of soil γ, cohesion c', and surcharge q. Figure 15.6 shows the distribution of passive pressure from each of these components on the wedge face BJ. Thus, we can write

$$P_p = \tfrac{1}{2}\gamma(b \tan \phi')^2 K_\gamma + c'(b \tan \phi')K_c + q(b \tan \phi')K_q \tag{15.4}$$

where K_γ, K_c, and K_q are earth pressure coefficients that are functions of the soil friction angle, ϕ'.

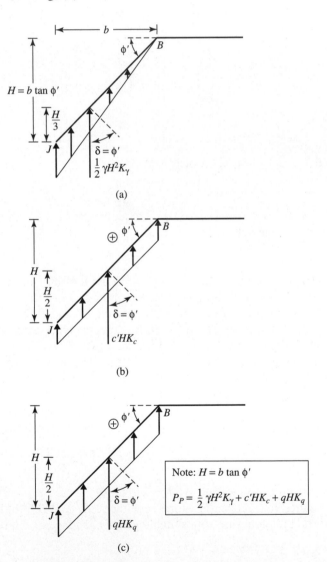

Figure 15.6 Passive force distribution on the wedge face BJ shown in Figure 15.5: (a) contribution of soil weight γ; (b) contribution of cohesion c'; (c) contribution of surcharge q.

Combining Eqs. (15.3) and (15.4), we obtain

$$q_u = c'N_c + qN_q + \frac{1}{2}\gamma BN_\gamma$$

where

$$N_c = \tan \phi'(K_c + 1) \qquad (15.5)$$

$$N_q = K_q \tan \phi' \qquad (15.6)$$

$$N_\gamma = \tfrac{1}{2} \tan \phi'(K_\gamma \tan \phi' - 1) \qquad (15.7)$$

The terms N_c, N_q, and N_γ are, respectively, the contributions of cohesion, surcharge, and unit weight of soil to the ultimate load-bearing capacity. It is extremely tedious to evaluate K_c, K_q, and K_γ. For this reason, Terzaghi used an approximate method to determine the ultimate bearing capacity, q_u. The principles of this approximation follow:

1. If $c' = 0$ and surcharge $(q) = 0$ (that is, $D_f = 0$), then

$$q_u = q_\gamma = \tfrac{1}{2}\gamma BN_\gamma \qquad (15.8)$$

2. If $\gamma = 0$ (that is, weightless soil) and $q = 0$, then

$$q_u = q_c = c'N_c \qquad (15.9)$$

3. If $\gamma = 0$ (weightless soil) and $c' = 0$, then

$$q_u = q_q = qN_q \qquad (15.10)$$

By the method of superimposition, when the effects of the unit weight of soil, cohesion, and surcharge are considered, we have

$$q_u = q_c + q_q + q_\gamma = c'N_c + qN_q + \tfrac{1}{2}\gamma BN_\gamma \qquad (15.11)$$

Equation (15.11) is referred to as *Terzaghi's bearing capacity equation*. The terms N_c, N_q, and N_γ are called the *bearing capacity factors*. The values of these factors are given in Table 15.1.

For square and circular footings, Terzaghi suggested the following equations for ultimate soil-bearing capacity:
The square footing is

$$q_u = 1.3c'N_c + qN_q + 0.4\gamma BN_\gamma \qquad (15.12)$$

The circular footing is

$$q_u = 1.3c'N_c + qN_q + 0.3\gamma BN_\gamma \qquad (15.13)$$

where B = diameter of the footing.

Table 15.1 Terzaghi's Bearing Capacity Factors — N_c, N_q and N_γ — Eqs. (15.11), (15.12), and (15.13)

ϕ' (deg)	N_c	N_q	$N_\gamma{}^a$	ϕ' (deg)	N_c	N_q	$N_\gamma{}^a$
0	5.70	1.00	0.00	26	27.09	14.21	9.84
1	6.00	1.10	0.01	27	29.24	15.90	11.60
2	6.30	1.22	0.04	28	31.61	17.81	13.70
3	6.62	1.35	0.06	29	34.24	19.98	16.18
4	6.97	1.49	0.10	30	37.16	22.46	19.13
5	7.34	1.64	0.14	31	40.41	25.28	22.65
6	7.73	1.81	0.20	32	44.04	28.52	26.87
7	8.15	2.00	0.27	33	48.09	32.23	31.94
8	8.60	2.21	0.35	34	52.64	36.50	38.04
9	9.09	2.44	0.44	35	57.75	41.44	45.41
10	9.61	2.69	0.56	36	63.53	47.16	54.36
11	10.16	2.98	0.69	37	70.01	53.80	65.27
12	10.76	3.29	0.85	38	77.50	61.55	78.61
13	11.41	3.63	1.04	39	85.97	70.61	95.03
14	12.11	4.02	1.26	40	95.66	81.27	115.31
15	12.86	4.45	1.52	41	106.81	93.85	140.51
16	13.68	4.92	1.82	42	119.67	108.75	171.99
17	14.60	5.45	2.18	43	134.58	126.50	211.56
18	15.12	6.04	2.59	44	151.95	147.74	261.60
19	16.56	6.70	3.07	45	172.28	173.28	325.34
20	17.69	7.44	3.64	46	196.22	204.19	407.11
21	18.92	8.26	4.31	47	224.55	241.80	512.84
22	20.27	9.19	5.09	48	258.28	287.85	650.67
23	21.75	10.23	6.00	49	298.71	344.63	831.99
24	23.36	11.40	7.08	50	347.50	415.14	1072.80
25	25.13	12.72	8.34				

^{a}From Kumbhojkar (1993)

Equation (15.11) was derived on the assumption that the bearing capacity failure of soil takes place by general shear failure. In the case of local shear failure, we may assume that

$$\bar{c}' = \tfrac{2}{3}c' \tag{15.14}$$

and

$$\tan \overline{\phi}' = \tfrac{2}{3}\tan \phi' \tag{15.15}$$

The ultimate bearing capacity of soil for a strip footing may be given by

$$q'_u = \bar{c}'N'_c + qN'_q + \tfrac{1}{2}\gamma BN'_\gamma \tag{15.16}$$

Table 15.2 Terzaghi's Modified Bearing Capacity Factors $-N'_c$, N'_q, and N'_γ—Eqs. (15.16), (15.17), and (15.18)

ϕ' (deg)	N_c	N_q	N_γ	ϕ' (dog)	N_c	N_q	N_γ
0	5.70	1.00	0.00	26	15.53	6.05	2.59
1	5.90	1.07	0.005	27	16.30	6.54	2.88
2	6.10	1.14	0.02	28	17.13	7.07	3.29
3	6.30	1.22	0.04	29	18.03	7.66	3.76
4	6.51	1.30	0.055	30	18.99	8.31	4.39
5	6.74	1.39	0.074	31	20.03	9.03	4.83
6	6.97	1.49	0.10	32	21.16	9.82	5.51
7	7.22	1.59	0.128	33	22.39	10.69	6.32
8	7.47	1.70	0.16	34	23.72	11.67	7.22
9	7.74	1.82	0.20	35	25.18	12.75	8.35
10	8.02	1.94	0.24	36	26.77	13.97	9.41
11	8.32	2.08	0.30	37	28.51	15.32	10.90
12	8.63	2.22	0.35	38	30.43	16.85	12.75
13	8.96	2.38	0.42	39	32.53	18.56	14.71
14	9.31	2.55	0.48	40	34.87	20.50	17.22
15	9.67	2.73	0.57	41	37.45	22.70	19.75
16	10.06	2.92	0.67	42	40.33	25.21	22.50
17	10.47	3.13	0.76	43	43.54	28.06	26.25
18	10.90	3.36	0.88	44	47.13	31.34	30.40
19	11.36	3.61	1.03	45	51.17	35.11	36.00
20	11.85	3.88	1.12	46	55.73	39.48	41.70
21	12.37	4.17	1.35	47	60.91	44.54	49.30
22	12.92	4.48	1.55	48	66.80	50.46	59.25
23	13.51	4.82	1.74	49	73.55	57.41	71.45
24	14.14	5.20	1.97	50	81.31	65.60	85.75
25	14.80	5.60	2.25				

The modified bearing capacity factors N'_c, N'_q, and N'_γ are calculated by using the same general equation as that for N_c, N_q, and N_γ, but by substituting $\overline{\phi}' = \tan^{-1}(\frac{2}{3}\tan\phi')$ for ϕ'. The values of the bearing capacity factors for a local shear failure are given in Table 15.2. The ultimate soil-bearing capacity for square and circular footings for the local shear failure case may now be given as follows [similar to Eqs. (15.12) and (15.13)]:

The square footing is

$$q'_u = 1.3\overline{c}'N'_c + qN'_q + 0.4\gamma BN'_\gamma \qquad (15.17)$$

The circular footing is

$$q'_u = 1.3\overline{c}'N'_c + qN'_q + 0.3\gamma BN'_\gamma \qquad (15.18)$$

For undrained condition with $\phi = 0$ and $\tau_f = c_u$, the bearing capacity factors are $N_\gamma = N_\gamma' = 0$ and $N_q = N_q' = 1$. Also, $N_c = N_c' = 5.7$. In that case, Eqs. (15.11), (15.12), and (15.13) (which are the cases for general shear failure) take the forms

$$q_u = 5.7c_u + q \qquad \text{(strip footing)} \qquad (15.18a)$$

and

$$q_u = (1.3)(5.7)c_u + q = 7.41c_u + q \qquad \text{(square and circular footing)}$$
$$(15.18b)$$

In a similar manner, Eqs. (15.16), (15.17), and (15.18), which are for the case of local shear failure, will take the forms

$$q_u' = \left(\frac{2}{3}c_u\right)(5.7) + q = 3.8c_u + q \qquad \text{(strip footing)} \qquad (15.19a)$$

and

$$q_u' = (1.3)\left(\frac{2}{3}c_u\right)(5.7) + q = 4.94c_u + q \qquad \text{(square and circular footing)}$$
$$(15.19b)$$

15.3 General Bearing Capacity Equation

After the development of Terzaghi's bearing capacity equation, several investigators worked in this area and refined the solution (that is, Meyerhof, 1951, 1963; Lundgren and Mortensen, 1953; Balla, 1962). Different solutions show that the bearing capacity factors N_c and N_q do not change much. However, for a given value of ϕ', the values of N_γ obtained by different investigators vary widely. This difference is because of the variation of the assumption of the wedge shape of soil located directly below the footing, as explained in the following paragraph.

While deriving the bearing capacity equation for a strip footing, Terzaghi used the case of a rough footing and assumed that the sides AJ and BJ of the soil wedge ABJ (see Figure 15.5a) make an angle ϕ' with the horizontal. Later model tests (for example, DeBeer and Vesic, 1958) showed that Terzaghi's assumption of the general nature of the rupture surface in soil for bearing capacity failure is correct. However, tests have shown that the sides AJ and BJ of the soil wedge ABJ make angles of about $45 + \phi'/2$ degrees, instead of ϕ', with the horizontal. This type of failure mechanism is shown in Figure 15.7. It consists of a Rankine active zone ABJ (zone I), two radial shear zones (zones II), and two Rankine passive zones (zones III). The curves JD and JE are arcs of a logarithmic spiral.

On the basis of this type of failure mechanism, the ultimate bearing capacity of a strip footing may be evaluated by the approximate method of superimposition described in Section 15.2 as

$$q_u = q_c + q_q + q_\gamma \qquad (15.20)$$

where q_c, q_q, and q_γ are the contributions of cohesion, surcharge, and unit weight of soil, respectively.

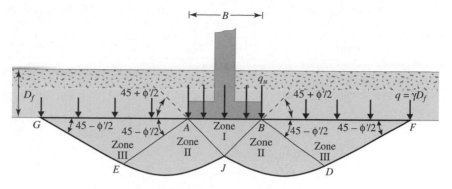

Figure 15.7 Soil-bearing capacity calculation — general shear failure

Reissner (1924) expressed q_q as

$$q_q = qN_q \tag{15.21}$$

where

$$N_q = e^{\pi \tan \phi'} \tan^2\left(45 + \frac{\phi'}{2}\right) \tag{15.22}$$

Prandtl (1921) showed that

$$q_c = c'N_c \tag{15.23}$$

where

$$N_c = (N_q - 1)\cot \phi' \\ \uparrow \\ \text{Eq. (15.22)} \tag{15.24}$$

Meyerhof (1963) expressed q_γ as

$$q_\gamma = \tfrac{1}{2}B\gamma N_\gamma \tag{15.25}$$

where

$$N_\gamma = (N_q - 1)\tan(1.4\phi') \\ \uparrow \\ \text{Eq. (15.22)} \tag{15.26}$$

Combining Eqs. (15.20), (15.21), (15.23), and (15.26), we obtain

$$q_u = c'N_c + qN_q + \tfrac{1}{2}\gamma BN_\gamma \tag{15.27}$$

This equation is in the same general form as that given by Terzaghi [Eq. (15.11)]; however, the values of the bearing capacity factors are not the same. The values of N_q, N_c, and N_γ, defined by Eqs. (15.22), (15.24), and (15.26), are given in Tables 15.3 and

Table 15.3 Bearing Capacity Factors N_c, and N_q [Eqs. (15.22) and (15.24)]

ϕ' (deg)	N_c	N_q	ϕ' (deg)	N_c	N_q
0	5.14	1.00	26	22.25	11.85
1	5.38	1.09	27	23.94	13.20
2	5.63	1.20	28	25.80	14.72
3	5.90	1.31	29	27.86	16.44
4	6.19	1.43	30	30.14	18.40
5	6.49	1.57	31	32.67	20.63
6	6.81	1.72	32	35.49	23.18
7	7.16	1.88	33	38.64	26.09
8	7.53	2.06	34	42.16	29.44
9	7.92	2.25	35	46.12	33.30
10	8.35	2.47	36	50.59	37.75
11	8.80	2.71	37	55.63	42.92
12	9.28	2.97	38	61.35	48.93
13	9.81	3.26	39	67.87	55.96
14	10.37	3.59	40	75.31	64.20
15	10.98	3.94	41	83.86	73.90
16	11.63	4.34	42	93.71	85.38
17	12.34	4.77	43	105.11	99.02
18	13.10	5.26	44	118.37	115.31
19	13.93	5.80	45	133.88	134.88
20	14.83	6.40	46	152.10	158.51
21	15.82	7.07	47	173.64	187.21
22	16.88	7.82	48	199.26	222.31
23	18.05	8.66	49	229.93	265.51
24	19.32	9.60	50	266.89	319.07
25	20.72	10.66			

15.4, but for all practical purposes, Terzaghi's bearing capacity factors will yield good results. Differences in bearing capacity factors are usually minor compared with the unknown soil parameters.

The soil-bearing capacity equation for a strip footing given by Eq. (15.27) can be modified for general use by incorporating the following factors:

1. *depth factor:* to account for the shearing resistance developed along the failure surface in soil above the base of the footing;
2. *shape factor:* to determine the bearing capacity of rectangular and circular footings; and
3. *inclination factor:* to determine the bearing capacity of a footing on which the direction of load application is inclined at a certain angle to the vertical.

Thus, the modified general ultimate bearing capacity equation can be written as

$$q_u = c'\lambda_{cs}\lambda_{cd}\lambda_{ci}N_c + q\lambda_{qs}\lambda_{qd}\lambda_{qi}N_q + \tfrac{1}{2}\lambda_{\gamma s}\lambda_{\gamma d}\lambda_{\gamma i}\gamma B N_\gamma \qquad (15.28)$$

where λ_{cs}, λ_{qs}, and $\lambda_{\gamma s}$ = shape factors
λ_{cd}, λ_{qd}, and $\lambda_{\gamma d}$ = depth factors
λ_{ci}, λ_{qi}, and $\lambda_{\gamma i}$ = inclination factors

Table 15.4 Bearing Capacity Factor N_γ [Eq. (15.26)]

ϕ' (deg)	N_γ	ϕ' (deg)	N_γ
0	0.000	27	9.463
1	0.002	28	11.190
2	0.010	29	13.236
3	0.023	30	15.668
4	0.042	31	18.564
5	0.070	32	22.022
6	0.106	33	26.166
7	0.152	34	31.145
8	0.209	35	37.152
9	0.280	36	44.426
10	0.367	37	53.270
11	0.471	38	64.073
12	0.596	39	77.332
13	0.744	40	93.690
14	0.921	41	113.985
15	1.129	42	139.316
16	1.375	43	171.141
17	1.664	44	211.406
18	2.003	45	262.739
19	2.403	46	328.728
20	2.871	47	414.322
21	3.421	48	526.444
22	4.066	49	674.908
23	4.824	50	873.843
24	5.716	51	1143.934
25	6.765	52	1516.051
26	8.002	53	2037.258

The approximate values of these shape, depth, and inclination factors recommended by Meyerhof are given in Table 15.5.

For undrained condition, if the footing is subjected to vertical loading (that is, $\alpha = 0°$), then

$$\phi = 0$$

$$c = c_u$$

$$N_\gamma = 0$$

$$N_q = 1$$

$$N_c = 1$$

$$\lambda_{ci} = \lambda_{qi} = \lambda_{\gamma i} = 1$$

So Eq. (15.28) transforms to

$$q_u = 5.14c_u\left[1 + 0.2\left(\frac{B}{L}\right)\right]\left[1 + 0.2\left(\frac{D_f}{B}\right)\right] + q \qquad (15.29)$$

Table 15.5 Meyerhof's Shape, Depth, and Inclination
Factors for a Rectangular Footing[a]

Shape factors

For $\phi = 0°$:

$$\lambda_{cs} = 1 + 0.2\left(\frac{B}{L}\right)$$

$$\lambda_{qs} = 1$$
$$\lambda_{\gamma s} = 1$$

For $\phi' \geq 10°$:

$$\lambda_{cs} = 1 + 0.2\left(\frac{B}{L}\right)\tan^2\left(45 + \frac{\phi'}{2}\right)$$

$$\lambda_{qs} = \lambda_{\gamma s} = 1 + 0.1\left(\frac{B}{L}\right)\tan^2\left(45 + \frac{\phi'}{2}\right)$$

Depth factors

For $\phi = 0°$:

$$\lambda_{cd} = 1 + 0.2\left(\frac{D_f}{B}\right)$$

$$\lambda_{qd} = \lambda_{\gamma d} = 1$$

For $\phi' \geq 10°$:

$$\lambda_{cd} = 1 + 0.2\left(\frac{D_f}{B}\right)\tan\left(45 + \frac{\phi'}{2}\right)$$

$$\lambda_{qd} = \lambda_{\gamma d} = 1 + 0.1\left(\frac{D_f}{B}\right)\tan\left(45 + \frac{\phi'}{2}\right)$$

Inclination factors

$$\lambda_{ci} = \left(1 - \frac{\alpha°}{90°}\right)^2$$

$$\lambda_{qi} = \left(1 - \frac{\alpha°}{90°}\right)^2$$

$$\lambda_{\gamma i} = \left(1 - \frac{\alpha°}{\phi'°}\right)^2$$

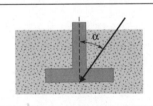

[a]B = width of footing; L = length of footing

15.4   *Effect of Groundwater Table*

In developing the bearing capacity equations given in the preceding sections we assumed that the groundwater table is located at a depth much greater than the width, B of the footing. However, if the groundwater table is close to the footing, some changes are required in the second and third terms of Eqs. (15.11) to (15.13), Eqs. (15.16) to (15.18), and Eq. 15.28. Three different conditions can arise regarding the location of the groundwater table with respect to the bottom of the foundation. They are shown in Figure 15.8. Each of these conditions is briefly described next.

- *Case I (Figure 15.8a):* If the groundwater table is located at a distance D above the bottom of the foundation, the magnitude of q in the second term of the bearing capacity equation should be calculated as

$$q = \gamma(D_f - D) + \gamma'D \tag{15.30}$$

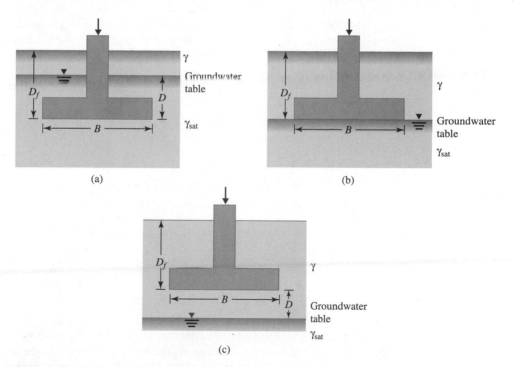

Figure 15.8 Effect of the location of groundwater table on the bearing capacity of shallow foundations: (a) Case I; (b) Case II; (c) Case III

where $\gamma' = \gamma_{sat} - \gamma_w$ = effective unit weight of soil. Also, the unit weight of soil, γ, that appears in the third term of the bearing capacity equations should be replaced by γ'.

- *Case II (Figure 15.8b):* If the groundwater table coincides with the bottom of the foundation, the magnitude of q is equal to γD_f. However, the unit weight, γ, in the third term of the bearing capacity equations should be replaced by γ'.

- *Case III (Figure 15.8c):* When the groundwater table is at a depth D below the bottom of the foundation, $q = \gamma D_f$. The magnitude of γ in the third term of the bearing capacity equations should be replaced by γ_{av}

$$\gamma_{av} = \frac{1}{B}[\gamma D + \gamma'(B - D)] \quad \text{(for } D \leq B) \tag{15.31}$$

$$\gamma_{av} = \gamma \quad \text{(for } D > B) \tag{15.32}$$

15.5   *Factor of Safety*

Generally, a factor of safety, F_s, of about 3 or more is applied to the ultimate soil-bearing capacity to arrive at the value of the allowable bearing capacity. An F_s of 3 or more is not considered too conservative. In nature, soils are neither homogeneous nor isotropic. Much uncertainty is involved in evaluating the basic shear strength parameters of soil.

There are two basic definitions of the allowable bearing capacity of shallow foundations. They are gross allowable bearing capacity, and net allowable bearing capacity.

The *gross allowable bearing capacity* can be calculated as

$$q_{all} = \frac{q_u}{F_s} \tag{15.33}$$

As defined by Eq. (15.33) q_{all} is the allowable load per unit area to which the soil under the foundation should be subjected to avoid any chance of bearing capacity failure. It includes the contribution (Figure 15.9) of (a) the dead and live loads above the ground surface, $W_{(D+L)}$; (b) the self-weight of the foundation, W_F; and (c) the weight of the soil located immediately above foundation, W_S. Thus,

$$q_{all} = \frac{q_u}{F_s} = \left[\frac{W_{(D+L)} + W_F + W_S}{A}\right]\frac{1}{F_S} \tag{15.34}$$

where A = area of the foundation.

The *net allowable bearing capacity* is the allowable load per unit area of the foundation in excess of the existing vertical effective stress at the level of the foundation. The vertical effective stress at the foundation level is equal to $q = \gamma D_f$. So the net ultimate load is

$$q_{u(net)} = q_u - q \tag{15.35a}$$

Hence,

$$q_{all(net)} = \frac{q_{u(net)}}{F_s} = \frac{q_u - q}{F_s} \tag{15.35b}$$

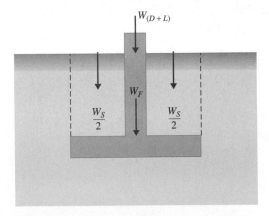

Figure 15.9 Contributions to q_{all}

If we assume that the weight of the soil and the weight of the concrete from which the foundation is made are approximately the same, then

$$q = \gamma D_f \approx \frac{W_S + W_F}{A}$$

Hence,

$$q_{all(net)} = \frac{W_{(D+L)}}{A} = \frac{q_u - q}{F_S} \tag{15.36}$$

Example 15.1

The plan of a 4-ft-square footing is shown in Figure 15.10. Determine the gross allowable load, Q_{all} ($Q_{all} = q_{all} \times$ area of the footing) that the footing can carry. A factor of safety of 3 is needed. Use Terzaghi's equation and assume general shear failure in soil.

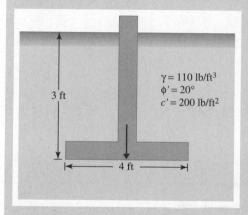

$\gamma = 110$ lb/ft^3
$\phi' = 20°$
$c' = 200$ lb/ft^2

3 ft

4 ft

Figure 15.10

Solution
Assuming general shear failure of soil, we have

$$q_u = 1.3c'N_c + qN_q + 0.4\gamma BN_\gamma \qquad \text{[Eq. (15.12)]}$$

From Table 15.1, for $\phi' = 20°$, $N_c = 17.69$, $N_q = 7.44$, and $N_\gamma = 3.64$,

$$q = \gamma D_f = 110 \times 3 = 330 \text{ lb/ft}^2$$

So

$$q_u = (1.3)(200)(17.69) + (330)(7.44) + (0.4)(110)(4)(3.64)$$
$$= 4599 + 2455 + 641 = 7695 \text{ lb/ft}^2$$

$$q_{all} = \frac{q_u}{F_s} = \frac{7695}{3} = 2565 \text{ lb/ft}^2$$

Hence,

$$Q_{all} = 2565 \times B^2 = 2565 \times 16 = \textbf{41,040 lb}$$

■

Example 15.2

Redo Example Problem 15.1 assuming local shear failure in soil. Use Eq. 15.17.

Solution
From Eq. (15.17),

$$q'_u = 1.3\bar{c}'N'_c + qN'_q + 0.4\gamma BN'_\gamma$$

and

$$\bar{c}' = \frac{2}{3}(200) = 133.3 \text{ lb/ft}^2$$

From Table 15.2, $N'_c = 11.85$, $N'_q = 3.88$, and $N'_\gamma = 1.12$. So,

$$q'_u = (1.3)(133.3)(11.85) + (110 \times 3)3.88 + 0.4(110)(4)(1.12)$$

$$= 2054 + 1280 + 197 = 3531 \text{ lb/ft}^2$$

and

$$q_{\text{all}} = \frac{q'_u}{3} = \frac{3531}{3} = 1177 \text{ lb/ft}^2$$

Hence,

$$Q_{\text{all}} = 1177 \times B^2 = 1177 \times 16 = \mathbf{18{,}832 \text{ lb}} \qquad \blacksquare$$

Example 15.3

A square footing is shown in Figure 15.11. The footing will carry a gross mass of 30,000 kg. Using a factor of safety of 3, determine the size of the footing — that is, the size of B. Use Eq. (15.12).

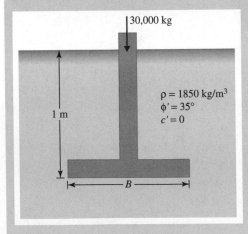

30,000 kg

1 m

$\rho = 1850 \text{ kg/m}^3$
$\phi' = 35°$
$c' = 0$

B

Figure 15.11

Solution

It is given that soil density = 1850 kg/m³. So

$$\gamma = \frac{1850 \times 9.81}{1000} = 18.15 \text{ kN/m}^3$$

Total gross load to be supported by the footing is

$$\frac{(30,000)9.81}{1000} = 294.3 \text{ kN} = Q_{all}$$

From Eq. (15.12)

$$q_u = 1.3c'N_c + qN_q + 0.4\gamma BN_\gamma$$

With a factor of safety of 3

$$q_{all} = \frac{q_u}{3} = \frac{1}{3}(1.3c'N_c + qN_q + 0.4\gamma BN_\gamma) \tag{a}$$

Also,

$$q_{all} = \frac{Q_{all}}{B^2} = \frac{294.3}{B^2} \tag{b}$$

From Eqs. (a) and (b),

$$\frac{294.3}{B^2} = \frac{1}{3}(1.3c'N_c + qN_q + 0.4\gamma BN_\gamma) \tag{c}$$

From Table 15.1, for $\phi' = 35°$, $N_c = 57.75$, $N_q = 41.44$, and $N_\gamma = 45.41$. Substituting these values into Eq. (c) yields

$$\frac{294.3}{B^2} = \frac{1}{3}[(1.3)(0)(57.75) + (18.15 \times 1)(41.44) + 0.4(18.15)(B)(45.41)]$$

or

$$\frac{294.3}{B^2} = 250.7 + 109.9$$

The preceding equation may now be solved by trial and error, and from that we get

$$B \simeq \textbf{0.95 m}$$

Example 15.4

Refer to Example 15.1. Determine the net allowable load $Q_{\text{all(net)}}$ with an $F_s = 3$ against the net ultimate bearing capacity.

Solution
From Example 15.1

$$q_u = 7695 \text{ lb/ft}^2$$

$$q_{u(\text{net})} = q_u - q = 7695 - 330 = 7365 \text{ lb/ft}^2$$

$$q_{\text{all(net)}} = \frac{q_{u(\text{net})}}{F_s} = \frac{7365}{3} = 2455 \text{ lb/ft}^2$$

So

$$Q_{\text{all(net)}} = (q_{\text{all(net)}})(B^2) = (2455)(4^2) = \mathbf{39{,}280 \text{ lb}} \qquad \blacksquare$$

Example 15.5

A square footing is shown in Figure 15.12. Determine the safe gross load (factor of safety of 3) that the footing can carry. Use Eq. (15.28).

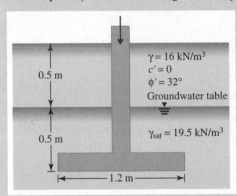

$\gamma = 16 \text{ kN/m}^3$
$c' = 0$
$\phi' = 32°$
Groundwater table

$\gamma_{\text{sat}} = 19.5 \text{ kN/m}^3$

0.5 m

0.5 m

1.2 m

Figure 15.12

Solution
From Eq. (15.28),

$$q_u = c'\lambda_{cs}\lambda_{cd}N_c + q\lambda_{qs}\lambda_{qd}N_q + \tfrac{1}{2}\gamma'\lambda_{\gamma s}\lambda_{\gamma d}BN_\gamma$$

(*Note:* λ_{ci}, λ_{qi}, and $\lambda_{\gamma i}$ are all equal to 1 because the load is vertical.)
    Because $c' = 0$,

$$q_u = q\lambda_{qs}\lambda_{qd}N_q + \tfrac{1}{2}\gamma'\lambda_{\gamma s}\lambda_{\gamma d}BN_\gamma$$

From Tables 15.3 and 15.4, for $\phi' = 32°$, $N_q = 23.18$ and $N_\gamma = 22.02$. From Table 15.5,

$$\lambda_{qs} = \lambda_{\gamma s} = 1 + 0.1\left(\frac{B}{L}\right)\tan^2\left(45 + \frac{\phi'}{2}\right)$$

$$= 1 + 0.1\left(\frac{1.2}{1.2}\right)\tan^2\left(45 + \frac{32}{2}\right) = 1.325$$

$$\lambda_{qd} = \lambda_{\gamma d} = 1 + 0.1\left(\frac{D_f}{B}\right)\tan\left(45 + \frac{\phi'}{2}\right)$$

$$= 1 + 0.1\left(\frac{1}{1.2}\right)\tan\left(45 + \frac{32}{2}\right) = 1.15$$

The groundwater table is located above the bottom of the foundation, so, from Eq. (15.30),

$$q = (0.5)(16) + (0.5)(19.5 - 9.81) = 12.845 \text{ kN/m}^2$$

Thus,

$$q_u = (12.845)(1.325)(1.15)(23.18) + (\tfrac{1}{2})(19.5 - 9.81)(1.325)(1.15)(1.2)(22.02)$$

$$= 453.7 + 195.1 = 648.8 \text{ kN/m}^2$$

$$q_{\text{all}} = \frac{q_u}{3} = \frac{648.8}{3} = 216.3 \text{ kN/m}^2$$

Hence, the gross load is as follows:

$$Q = q_{\text{all}}(B^2) = 216.3(1.2)^2 = \mathbf{311.5 \text{ kN}} \qquad \blacksquare$$

15.6   Ultimate Load for Shallow Foundations under Eccentric Load

One-Way Eccentricity

To calculate the bearing capacity of shallow foundations with eccentric loading, Meyerhof (1953) introduced the concept of *effective area*. This concept can be explained with reference to Figure 15.13, in which a footing of length L and width B is subjected to an eccentric load, Q_u. If Q_u is the ultimate load on the footing, it may be approximated as follows:

1. Referring to Figures 15.13b and 15.13c, calculate the effective dimensions of the foundation. If the eccentricity (e) is in the x direction (Figure 15.13b), the *effective dimensions* are

$$X = B - 2e$$

and

$$Y = L$$

However, if the eccentricity is in the y direction (Figure 15.13c), the effective dimensions are

$$Y = L - 2e$$

and

$$X = B$$

2. The lower of the two effective dimensions calculated in step 1 is the *effective width* (B') and the other is the *effective length* (L'). Thus,

$$B' = X \text{ or } Y, \text{ whichever is smaller}$$

$$L' = X \text{ or } Y, \text{ whichever is larger}$$

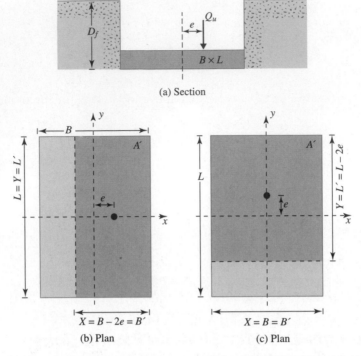

Figure 15.13 Ultimate load for shallow foundation under eccentric load

3. So the effective area is equal to B' times L'. Now, using the effective width, we can rewrite Eq. (15.28) as

$$q_u = c'\lambda_{cs}\lambda_{cd}N_c + q\lambda_{qs}\lambda_{qd}N_q + \tfrac{1}{2}\lambda_{\gamma s}\lambda_{\gamma d}\gamma B'N_\gamma \qquad (15.37)$$

Note that the preceding equation is obtained by substituting B' for B in Eq. (15.28). While computing the shape and depth factors, one should use B' for B and L' for L.

4. Once the value of q_u is calculated from Eq. (15.37), we can obtain the total gross ultimate load as follows:

$$Q_u = q_u(B'L') = q_u A' \qquad (15.38)$$

where A' = effective area.

Two-Way Eccentricity

When foundations are subjected to loads with two-way eccentricity, as shown in Figure 15.14, the effective area is determined such that the centroid coincides with the load. The procedure for finding the effective dimensions, B' and L', are beyond the scope of this text and readers may refer to Das (1999). Once B' and L' are determined, Eqs. (15.37) and (15.38) may be used to determine the ultimate load.

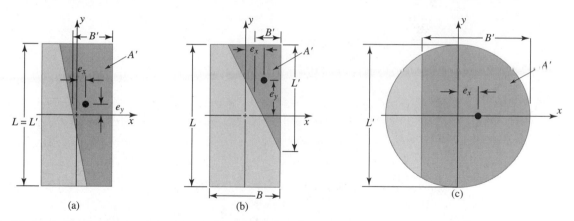

Figure 15.14 Foundation subjected to two-way eccentricity

Example 15.6

A rectangular footing 1.5 m × 1 m is shown in Figure 15.15. Determine the magnitude of the gross ultimate load applied eccentrically for bearing capacity failure in soil.

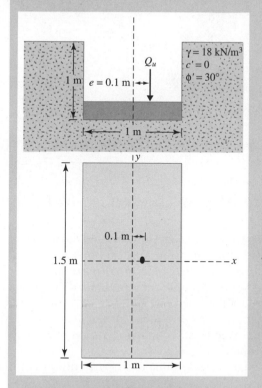

Figure 15.15

Solution

From Figure 15.13b and 15.15,

$$X = B - 2e = 1 - 2e = 1 - (2)(0.1) = 0.8 \text{ m}$$

$$Y = L = 1.5 \text{ m}$$

So the effective width $(B') = 0.8$ m and the effective length $(L') = 1.5$ m. From Eq. (15.37),

$$q_u = q\lambda_{qs}\lambda_{qd}N_q + \tfrac{1}{2}\lambda_{\gamma s}\lambda_{\gamma d}\gamma B' N_\gamma$$

From Tables 15.3 and 15.4, for $\phi' = 30°$, $N_q = 18.4$ and $N_\gamma = 15.668$. From Table 15.5,

$$\lambda_{qs} = \lambda_{\gamma s} = 1 + 0.1\left(\frac{B'}{L'}\right)\tan^2\left(45 + \frac{\phi'}{2}\right)$$

$$= 1 + 0.1\left(\frac{0.8}{1.5}\right)\tan^2\left(45 + \frac{30}{2}\right) = 1.16$$

$$\lambda_{qd} = \lambda_{\gamma d} = 1 + 0.1\left(\frac{D_f}{B'}\right)\tan\left(45 + \frac{\phi'}{2}\right)$$

$$= 1 + 0.1\left(\frac{1}{0.8}\right)\tan\left(45 + \frac{30}{2}\right) = 1.217$$

So

$$q_u = (1 \times 18)(1.16)(1.217)(18.4)$$
$$+ (\tfrac{1}{2})(1.16)(1.217)(18)(0.8)(15.668) \approx 627 \text{ kN/m}^2$$

Hence, from Eq. (15.38),

$$Q_u = q_u(B'L') = (627)(0.8)(1.5) \approx \textbf{752 kN} \qquad \blacksquare$$

15.7 *Bearing Capacity of Sand Based on Settlement*

Obtaining undisturbed specimens of cohesionless sand during a soil exploration program is usually difficult. For this reason, the results of standard penetration tests (SPTs) performed during subsurface exploration are commonly used to predict the allowable soil-bearing capacity of foundations on sand. (The procedure for conducting SPTs is discussed in detail in Chapter 17.)

Meyerhof (1956) proposed a correlation for the *net allowable bearing pressure* for foundations with the corrected standard penetration resistance, N_{cor}. The net allowable pressure was defined in Eq. (15.36).

(For definition of the corrected standard penetration resistance, please see Section 17.5.)

According to Meyerhof's theory, for 25 mm (1 in.) of estimated maximum settlement,

$$q_{all(net)}(kN/m^2) = 11.98N_{cor} \qquad \text{(for } B \le 1.22 \text{ m)} \qquad (15.39)$$

$$q_{all(net)}(kN/m^2) = 7.99N_{cor}\left(\frac{3.28B + 1}{3.28B}\right)^2 \qquad \text{(for } B > 1.22 \text{ m)} \qquad (15.40)$$

where N_{cor} = corrected standard penetration number
Note that in Eqs. (15.39) and (15.40) B is in meters.
In English units,

$$q_{all(net)}(kip/ft^2) = \frac{N_{cor}}{4} \qquad \text{(for } B \le 4 \text{ ft)} \qquad (15.41)$$

and

$$q_{all(net)}(kip/ft^2) = \frac{N_{cor}}{6}\left(\frac{B + 1}{B}\right)^2 \qquad \text{(for } B > 4 \text{ ft)} \qquad (15.42)$$

Since Meyerhof proposed his original correlation, researchers have observed that its results are rather conservative. Later, Meyerhof (1965) suggested that the net allowable bearing pressure should be increased by about 50%. Bowles (1977) proposed that the modified form of the bearing pressure equations be expressed as

$$q_{all(net)}(kN/m^2) = 19.16N_{cor}F_d\left(\frac{S_e}{25}\right) \qquad \text{(for } B \le 1.22 \text{ m)} \qquad (15.43)$$

and

$$q_{all(net)}(kN/m^2) = 11.98N_{cor}\left(\frac{3.28B + 1}{3.28B}\right)^2 F_d\left(\frac{S_e}{25}\right) \qquad \text{(for } B > 1.22 \text{ m)}$$

$$(15.44)$$

where F_d = depth factor = $1 + 0.33(D_f/B) \le 1.33$ (15.45)
    S_e = tolerable elastic settlement, in mm

Again, the unit of B is meters.
In English units,

$$q_{all(net)}(kip/ft^2) = \frac{N_{cor}}{2.5}F_dS_e \qquad \text{(for } B \le 4 \text{ ft)} \qquad (15.46)$$

and

$$q_{all(net)}(kip/ft^2) = \frac{N_{cor}}{4}\left(\frac{B + 1}{B}\right)^2 F_dS_e \qquad \text{(for } B > 4 \text{ ft)} \qquad (15.47)$$

where F_d is given by Eq. (15.45)
    S_e = tolerable elastic settlement, in in.

The empirical relations just presented may raise some questions. For example, which value of the standard penetration number should be used, and what is the effect of the water table on the net allowable bearing capacity? The design value of N_{cor} should be determined by taking into account the N_{cor} values for a depth of 2B to 3B, measured from the bottom of the foundation. Many engineers are also of the opinion that the N_{cor} value should be reduced somewhat if the water table is close to the foundation. However, the author believes that this reduction is not required because the penetration resistance reflects the location of the water table.

15.8   *Plate Load Test*

In some cases, conducting field load tests to determine the soil-bearing capacity of foundations is desirable. The standard method for a field load test is given by the American Society for Testing and Materials (ASTM) under Designation D-1194 (ASTM, 1997). Circular steel bearing plates 152 to 760 mm (6 to 30 in.) in diameter and 305 mm × 305 mm (1 ft × 1 ft) square plates are used for this type of test.

A diagram of the load test is shown in Figure 15.16. To conduct the test, one must have a pit of depth D_f excavated. The width of the test pit should be at least four times the width of the bearing plate to be used for the test. The bearing plate is placed on the soil at the bottom of the pit, and an incremental load on the bearing plate is applied. After the application of an incremental load, enough time is allowed for settlement to occur. When the settlement of the bearing plate becomes negligible, another incremental load is applied. In this manner, a load-settlement plot can be obtained, as shown in Figure 15.17.

From the results of field load tests, the ultimate soil-bearing capacity of actual footings can be approximated as follows:

For clays,

$$q_{u(footing)} = q_{u(plate)} \tag{15.48}$$

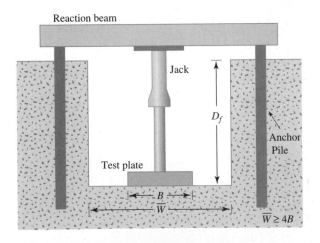

Figure 15.16 Diagram of plate load test

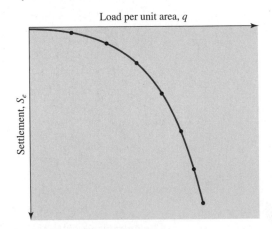

Figure 15.17 Typical load-settlement curve obtained from plate load test

For sandy soils,

$$q_{u(\text{footing})} = q_{u(\text{plate})} \frac{B_{(\text{footing})}}{B_{(\text{plate})}} \tag{15.49}$$

For a given intensity of load q, the settlement of the actual footing can also be approximated from the following equations:

In clay,

$$S_{e(\text{footing})} = S_{e(\text{plate})} \frac{B_{(\text{footing})}}{B_{(\text{plate})}} \tag{15.50}$$

In sandy soil,

$$S_{e(\text{footing})} = S_{e(\text{plate})} \left[\frac{2B_{(\text{footing})}}{B_{(\text{footing})} + B_{(\text{plate})}} \right]^2 \tag{15.51}$$

Housel (1929) also proposed a method for obtaining the soil-bearing capacity of a footing that rests on a cohesive soil for a given settlement S_e. According to this procedure, the total load carried by a footing of area A and perimeter P can be given by

$$Q = Aq + Ps \tag{15.52}$$

where q = compression stress below the footing
s = unit shear stress at the perimeter

Note that q and s are the two unknowns that must be determined from the results of the field load tests conducted on two different-size plates. If Q_1 and Q_2 are the loads required to produce a settlement S_e in plates 1 and 2, respectively, then

$$Q_1 = A_1 q + P_1 s \tag{15.53}$$

and

$$Q_2 = A_2 q + P_2 s \tag{15.54}$$

Solution of Eqs. (15.53) and (15.54) yields the values of q and s. Housel's method is not widely used in practice.

Example 15.7

The ultimate bearing capacity of a 700 mm diameter plate as determined from field load tests is 280 kN/m². Estimate the ultimate bearing capacity of a circular footing with a diameter of 1.5 m. The soil is sandy.

Solution
From Eq. (15.49),

$$\dot{q}_{u(\text{footing})} = q_{u(\text{plate})} \frac{B_{(\text{footing})}}{B_{(\text{plate})}} = 280 \left(\frac{1.5}{0.7} \right)$$

$$= \textbf{680 kN/m}^2 \qquad \blacksquare$$

Example 15.8

Following are the results of two plate load tests in a cohesive soil:

Plate size (ft)	Settlement (in.)	Total load, Q (lb)
1.5 × 1.5	0.5	15,750
2.5 × 2.5	0.5	33,750

If a square footing 5.75 ft × 5.75 ft is to be constructed and the allowable settlement is 0.5 in., what is the magnitude of the total load that it can carry?

Solution
From Eq. (15.52),

$$Q = Aq + Ps$$

So

$$15{,}750 = (1.5)^2 q + (4 \times 1.5)s \qquad \text{(a)}$$

$$33{,}750 = (2.5)^2 q + (4 \times 2.5)s \qquad \text{(b)}$$

From Eqs. (a) and (b),

$$q = 3{,}000 \text{ lb/ft}^2 \quad \text{and} \quad s = 1{,}500 \text{ lb/ft}$$

$$Q = Aq + Ps = (5.75)^2(3{,}000) + (4 \times 5.75)(1{,}500)$$

$$= 133{,}687.5 \text{ lb} \approx \textbf{133.69 kip}$$

∎

15.9 *Ultimate Bearing Capacity on Layered Soil*

The ultimate and allowable bearing capacities of shallow foundations on weaker (loose) sands and soft clays can be increased by placing a layer of compact (dense) sand over it. This is essentially a bearing capacity problem on a layered soil, which is the subject of discussion in this section. It is divided into two parts — the first discusses the bearing capacity on layered sand (dense over loose) followed by an evaluation of the bearing capacity on a stronger sand layer underlain by a weaker saturated clay layer.

Foundations on Layered Sand—Dense over Loose

A simple theory for determining the ultimate bearing capacity of a foundation that rests on a layer of dense sand underlain by loose sand has been proposed by Meyerhof and Hanna (1978). The basic principle of this theory can be explained with the aid of Figure 15.18, which is for a strip foundation. When the top dense sand layer is

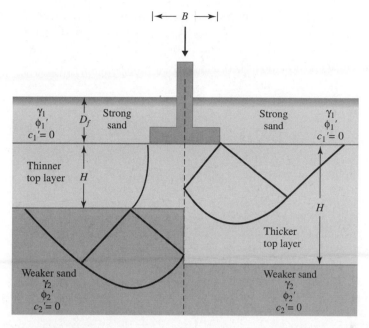

Figure 15.18 Bearing capacity in layered sand — strong sand underlain by weak sand

relatively thick, as shown by the right-hand side of Figure 15.18, the failure surface in soil under the foundation will be fully located inside the dense sand. For this case,

$$q_u = q_{u(t)} = \gamma_1 D_f N_{q(1)} + \frac{1}{2}\gamma_1 B N_{\gamma(1)}$$

(for strip foundations) (15.55)

$$q_u = q_{u(t)} = \gamma_1 D_f N_{q(1)} + 0.3\gamma_1 B N_{\gamma(1)}$$

(for circular or square foundations) (15.56)

and

$$q_u = q_{u(t)} = \gamma_1 D_f N_{q(1)} + \frac{1}{2}\left[1 - 0.4\left(\frac{B}{L}\right)\right]\gamma_1 B N_{\gamma(1)} \qquad (15.57)$$

(for rectangular foundations)

where γ_1 = unit weight of top layer (dense sand in this case)

$N_{q(1)}$ and $N_{\gamma(1)}$ = bearing capacity factors with reference to the soil friction angle, ϕ_1' (Tables 15.3 and 15.4)

Note that Eqs. (15.55), (15.56), and (15.57) are similar to Eq. (15.28). However, the depth factors have not been incorporated; they can be assumed to be somewhat conservative.

If the thickness of the dense sand layer under the foundation H is relatively thin, the failure in soil would take place by *punching* in the dense sand layer followed by a general shear failure in the bottom (or weaker) sand layer, as shown in the left-hand side of Figure 15.18. For such a case, the ultimate bearing capacity for the foundation can be given as

$$q_u = q_{u(b)} + \gamma_1 H^2 \left(1 + \frac{2D_f}{H}\right) K_s \frac{\tan \phi_1'}{B} - \gamma_1 H \leq q_{u(t)}$$

$$\uparrow$$
$$[\text{Eq. (15.55)}]$$

(for strip foundations)   (15.58)

$$q_u = q_{u(b)} + 2\gamma_1 H^2 \left(1 + \frac{2D_f}{H}\right)\left(\frac{K_s \tan \phi_1'}{B}\right)\lambda_s' - \gamma_1 H \leq q_{u(t)}$$

$$\uparrow$$
$$[\text{Eq. (15.56)}]$$

(for square or circular foundation)   (15.59)

and

$$q_u = q_{u(b)} + \left(1 + \frac{B}{L}\right)\gamma_1 H^2 \left(1 + \frac{2D_f}{H}\right)\left(\frac{K_s \tan \phi_1'}{B}\right)\lambda_s' - \gamma_1 H \leq q_{u(t)}$$

$$\uparrow$$
$$[\text{Eq. (15.57)}]$$

(for rectangular foundations)

(15.60)

where K_s = punching shear coefficient
$\quad\lambda_s'$ = shape factor
$\quad q_{u(b)}$ = ultimate bearing capacity of the bottom soil layer

The value of the shape factor λ_s' can be taken to be approximately 1. The punching shear coefficient is

$$K_s = f(\gamma_1, \gamma_2, N_{\gamma(1)}, N_{\gamma(2)}) \tag{15.61}$$

where γ_2 = unit weight of the lower layer of sand
$\quad N_{\gamma(2)}$ = bearing capacity factor for the soil friction angle, ϕ_2'

The variation of K_s is shown in Figure 15.19. The term $q_{u(b)}$ in Eqs. (15.58), (15.59), and (15.60) is given by the relationships

$$q_{u(b)} = \gamma_1(D_f + H)N_{q(2)} + \frac{1}{2}\gamma_2 B N_{\gamma(2)}$$

(for strip foundations)   (15.62)

$$q_{u(b)} = \gamma_1(D_f + H)N_{q(2)} + 0.3\gamma_2 B N_{\gamma(2)}$$

(for circular or square foundations)   (15.63)

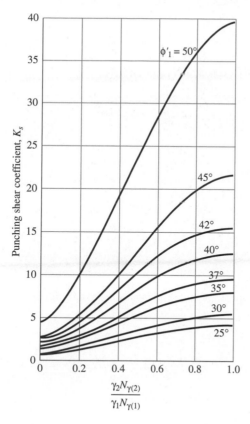

Figure 15.19
Variation of K_s with $(\gamma_2 N_{\gamma(2)})/(\gamma_1 N_{\gamma(1)})$

and

$$q_{u(b)} = \gamma_1(D_f + H)N_{q(2)} + \frac{1}{2}\left[1 - 0.4\left(\frac{B}{L}\right)\right]\gamma_2 B N_{\gamma(2)}$$

(for rectangular foundations)   (15.64)

Foundations on Dense or Compacted Sand Overlying Soft Clay

If the thickness of the sand layer under the foundation is relatively small, the failure surface may extend into the soft clay layer. This is shown in the left half of Figure 15.20. However, if the sand layer under the foundation is large, the failure surface will lie entirely in the sand layer, as shown in the right half of Figure 15.20. According to Meyerhof and Hanna (1978), in this case the ultimate bearing capacity of a strip foundation may be given by

$$q_u = c_u N_c + \gamma H^2\left(1 + \frac{2D_f}{H}\right)K_s \frac{\tan \phi'}{B} + \gamma D_f$$

(15.65)

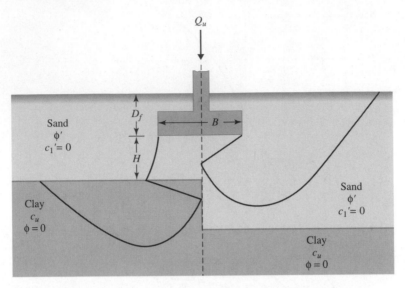

Figure 15.20 Foundation on compacted sand layer overlying soft clay

with a maximum of

$$q_u = \frac{1}{2}\gamma B N_\gamma + \gamma D_f N_q \tag{15.66}$$

where ϕ' = angle of friction of top sand layer
$\quad \gamma$ = unit weight of sand
$\quad K_s$ = punching shear resistance coefficient

N_γ and N_q correspond to the angle of friction, ϕ', for sand (Tables 15.3 and 15.4). *Note*: for $\phi = 0$, $N_c = 5.14$, as determined from Table 15.3.

For rectangular foundations,

$$q_u = \left(1 + 0.2\frac{B}{L}\right)c_u N_c + \left(1 + \frac{B}{L}\right)\gamma H^2\left(1 + \frac{2D_f}{H}\right)K_s\frac{\tan\phi'}{B} + \gamma D_f$$

$$\tag{15.67}$$

with a maximum of

$$q_u = \frac{1}{2}\left(1 - 0.4\frac{B}{L}\right)\gamma B N_\gamma + \gamma D_f N_q \tag{15.68}$$

The variation of the punching shear resistance factor, K_s, is given in Figure 15.21. Equations (15.66) and (15.68) are estimates of the values of q_u for strip and rectangular foundations, respectively, in the upper sand layer. This condition corresponds to that shown in the right half of Figure 15.20.

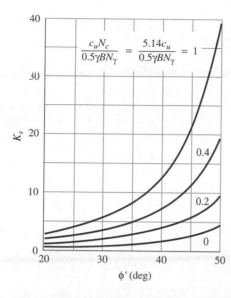

$$\frac{c_u N_c}{0.5\gamma B N_\gamma} = \frac{5.14 c_u}{0.5\gamma B N_\gamma} = 1$$

Figure 15.21
Variation of K_s with ϕ' (according to Meyerhof and Hanna)

Example 15.9

Figure 15.22 shows a rectangular foundation with $B = 4$ ft and $L = 6$ ft. Using a factor of safety of 3, determine the net allowable load the foundation can carry.

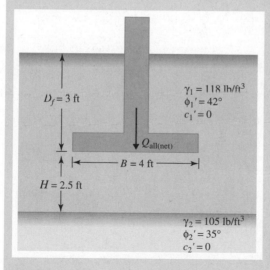

$\gamma_1 = 118$ lb/ft^3
$\phi_1' = 42°$
$c_1' = 0$

$D_f = 3$ ft

$Q_{all(net)}$

$B = 4$ ft

$H = 2.5$ ft

$\gamma_2 = 105$ lb/ft^3
$\phi_2' = 35°$
$c_2' = 0$

Figure 15.22

Solution

The top layer of sand is dense since it has $\phi_1' = 42°$, which is greater than $\phi_2' = 35°$. Also, $\gamma_1 > \gamma_2$. So Eq. (15.60) should be used to calculate q_u:

$$q_u = q_{u(b)} + \left(1 + \frac{B}{L}\right)\gamma_1 H^2\left(1 + \frac{2D_f}{H}\right)\left(\frac{K_s \tan\phi_1'}{B}\right)\lambda_s' - \gamma_1 H$$

It is given that $\gamma_1 = 118 \text{ lb/ft}^3$, $\gamma_2 = 105 \text{ lb/ft}^3$, $\phi_1' = 42°$, and $\phi_2' = 35°$. Also, from Table 15.4, $N_{\gamma(1)} = 139.32$ and $N_{\gamma(2)} = 37.12$. So

$$\frac{\gamma_2 N_{\gamma(2)}}{\gamma_1 N_{\gamma(1)}} = \frac{(105)(37.12)}{(118)(139.32)} = 0.237$$

From Figure 15.19, for $\phi_1' = 42°$, the value of K_s is 6. Thus, from Eq. (15.60),

$$q_u = q_{u(b)} + \left(1 + \frac{4}{6}\right)(118)(2.5)^2 \left[1 + \frac{(2)(3)}{2.5}\right]\left[\frac{(6)(\tan 42°)}{4}\right](1) - (118)(2.5)$$

$$= q_{u(b)} + 5349 \tag{a}$$

From Eq. (15.64),

$$q_{u(b)} = \gamma_1(D_f + H)N_{q(2)} + \frac{1}{2}\left[1 - 0.4\left(\frac{B}{L}\right)\right]\gamma_2 B N_{\gamma(2)}$$

From Tables 15.3 and 15.4, for $\phi_2' = 35°$, the values of $N_{q(2)} = 33.3$ and $N_{\gamma(2)} = 37.15$. Hence,

$$q_{u(b)} = (118)(3 + 2.5)(33.3) + \frac{1}{2}\left[1 - 0.4\left(\frac{4}{6}\right)\right](105)(4)(37.15) = 27{,}333 \text{ lb/ft}^2 \tag{b}$$

From Eqs. (a) and (b),

$$q_u = 27{,}333 + 5349 = 32{,}682 \text{ lb/ft}^2 \tag{c}$$

We also need to check Eq. (15.57):

$$q_u = \gamma_1 D_f N_{q(1)} + \frac{1}{2}\left[1 - 0.4\left(\frac{B}{L}\right)\right]\gamma_1 B N_{\gamma(1)}$$

From Tables 15.3 and 15.4, for $\phi_1' = 42°$, $N_{\gamma(1)} = 139.32$ and $N_{q(1)} = 85.38$, so

$$q_u = (118)(3)(85.38) + \left(\frac{1}{2}\right)\left[1 - 0.4\left(\frac{4}{6}\right)\right](118)(4)(139.32)$$

$$= 54{,}336 \text{ lb/ft}^2 \tag{d}$$

Comparing Eqs. (c) and (d), $q_u = 32{,}682 \text{ lb/ft}^2$, we have

$$q_{u(\text{net})} = q_u - \gamma_1 D_f = 32{,}682 - (3)(118) = 32{,}328 \text{ lb/ft}^2$$

$$Q_{\text{all}} = \frac{q_{u(\text{net})}BL}{F_s} = \frac{(32{,}328)(4)(6)}{(1000)(3)} \approx \textbf{258.6 kips} \qquad \blacksquare$$

Example 15.10

Refer to Figure 15.20. For sand

$$\gamma = 117 \text{ lb/ft}^3$$
$$\phi' = 40°$$

and for clay

$$c_u = 400 \text{ lb/ft}^2$$

For the foundation

$$B = 3 \text{ ft}$$
$$L = 4.5 \text{ ft}$$
$$D_f = 3 \text{ ft}$$
$$H = 4 \text{ ft}$$

Determine the gross ultimate bearing capacity of the foundation.

Solution

The foundation is rectangular, so Eqs. (15.67) and (15.68) will apply. For $\phi' = 40°$, from Table 15.4, $N_\gamma = 93.69$ and

$$\frac{c_u N_c}{0.5\gamma B N_\gamma} = \frac{(400)(5.14)}{(0.5)(117)(3)(93.69)} = 0.125$$

From Figure 15.21, for $c_u N_c/0.5\gamma B N_\gamma = 0.125$ and $\phi' = 40°$, the value of $K_s \approx 2.5$. Equation (15.67) gives

$$q_u = \left[1 + (0.2)\left(\frac{B}{L}\right)\right]c_u N_c + \left(1 + \frac{B}{L}\right)\gamma H^2\left(1 + \frac{2D_f}{H}\right)K_s \frac{\tan \phi'}{B} + \gamma D_f$$

$$= \left[1 + (0.2)\left(\frac{3}{4.5}\right)\right](400)(5.14) + \left(1 + \frac{3}{4.5}\right)(117)(4)^2$$

$$\times \left[1 + \frac{(2)(3)}{4}\right](2.5)\frac{\tan 40}{3} + (117)(3)$$

$$= 2330 + 5454 + 351 = 8135 \text{ lb/ft}^2$$

Again, from Eq. (15.68),

$$q_u = \frac{1}{2}\left[1 - (0.4)\left(\frac{B}{L}\right)\right]\gamma B N_\gamma + \gamma D_f N_q$$

For $\phi = 40°$, $N_q = 64.20$ (Table 15.3) and

$$q_u = (0.5)\left[1 - (0.4)\left(\frac{3}{4.5}\right)\right](117)(3)(93.69) + (117)(3)(64.20)$$

$$= 12,058 + 22,534 = 34,592 \text{ lb/ft}^2$$

Hence,

$$q_u = \mathbf{8135 \text{ lb/ft}^2} \qquad \blacksquare$$

15.10   *Summary and General Comments*

In this chapter, theories for estimating the ultimate and allowable bearing capacities of shallow foundations were presented. Procedures for field load tests and estimation of the allowable bearing capacity of granular soil based on limited settlement criteria were briefly discussed.

Several building codes now used in the United States and elsewhere provide presumptive bearing capacities for various types of soil. It is extremely important to realize that they are *approximate values only*. The bearing capacity of foundations depends on several factors:

1. Subsoil stratification
2. Shear strength parameters of the subsoil
3. Location of the ground water table
4. Environmental factors
5. Building size and weight
6. Depth of excavation
7. Type of structure.

Hence, it is important that the allowable bearing capacity at a given site be determined based on the findings of soil exploration at that site, past experience of foundation construction, and fundamentals of the geotechnical engineering theories for bearing capacity.

The allowable bearing capacity relationships based on settlement considerations such as those given in Section 15.7 do not take into account the settlement caused by consolidation of the clay layers. Excessive settlement usually causes the building to crack, which may ultimately lead to structural failure. Uniform settlement of a structure does not produce cracking; on the other hand, differential settlement may produce cracks and damage to a building.

Problems

15.1 For the continuous footing shown in Figure 15.23, determine the gross allowable bearing capacity. Use Terzaghi's bearing capacity factors and a factor of safety of 4. Assume general bearing capacity failure.

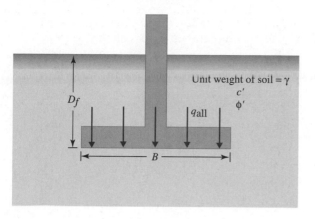

Figure 15.23

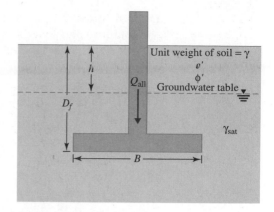

Figure 15.24

a. $\gamma = 120$ lb/ft³, $c' = 0$, $\phi' = 40°$, $D_f = 3$ ft, $B = 3.5$ ft
b. $\gamma = 115$ lb/ft³, $c' = 600$ lb/ft³, $\phi' = 25°$, $D_f = 3.5$ ft, $B = 4$ ft
c. $\gamma = 17.5$ kN/m³, $c' = 14$ kN/m², $\phi' = 20°$, $D_f = 1.0$ m, $B = 1.2$ m
d. $\gamma = 118$ lb/ft³, $c' = 450$ lb/ft², $\phi' = 28°$, $D_f = 4$ ft, $B = 4$ ft
e. $\gamma = 17.7$ kN/m³, $c = 48$ kN/m², $\phi = 0°$, $D_f = 0.6$ m, $B = 0.8$ m

15.2 Repeat Problem 15.1 assuming local shear failure.

15.3 Redo Problem 15.1 using the Prandtl, Reissner, and Meyerhof bearing capacity factors given in Tables 15.3 and 15.4 and Eq. (15.28).

15.4 A square footing has the following values:
Gross allowable load = 42,260 lb
Factor of safety = 3
$D_f = 3$ ft
Soil properties: $\gamma = 110$ lb/ft³
$\phi' = 20°$
$c' = 200$ lb/ft³
Use Eq. (15.12) to determine the size of the footing.

15.5 Repeat Problem 15.4 for the following data:
Gross allowable load = 1870 kN
Factor of safety = 3
$D_f = 1$ m
$\gamma = 17$ kN/m³
$\phi' = 35°$
$c' = 0$

15.6 A square footing is shown in Figure 15.24. For the following cases, determine the gross allowable load, Q_{all}, that the footing can carry. Use Terzaghi's equation for general shear failure ($F_s = 3$).
a. $\gamma = 105$ lb/ft³, $\gamma_{sat} = 118$ lb/ft³, $c' = 0$, $\phi' = 35°$, $B = 5$ ft, $D_f = 4$ ft, $h = 2$ ft
b. $\rho = 1800$ kg/m³, $\rho_{sat} = 1980$ kg/m³, $c' = 23.94$ kN/m², $\phi' = 25°$, $B = 1.8$ m, $D_f = 1.2$ m, $h = 2$ m

15.7   A square footing is shown in Figure 15.24. Use Eq. (15.28) for general shear failure and a factor of safety of 3. Determine the safe gross allowable load. Use the following values:

$$\gamma = 100 \text{ lb/ft}^3$$
$$\gamma_{sat} = 115 \text{ lb/ft}^3$$
$$c' = 0$$
$$\phi' = 30°$$
$$B = 4 \text{ ft}$$
$$D_f = 3.5 \text{ ft}$$
$$h = 2 \text{ ft}$$

15.8   Solve Problem 15.7 with the following values:

$$\gamma = 15.72 \text{ kN/m}^3$$
$$\gamma_{sat} = 18.55 \text{ kN/m}^3$$
$$c' = 0$$
$$\phi' = 35°$$
$$B = 1.53 \text{ m}$$
$$D_f = 1.22 \text{ m}$$
$$h = 0.61 \text{ m}$$

15.9   Solve Problem 15.7 with the following data:

$$\gamma = 115 \text{ lb/ft}^3$$
$$\gamma_{sat} = 122.4 \text{ lb/ft}^3$$
$$c' = 100 \text{ lb/ft}^2$$
$$\phi' = 30°$$
$$B = 4 \text{ ft}$$
$$D_f = 3 \text{ ft}$$
$$h = 4 \text{ ft}$$

15.10   The square footing shown in Figure 15.25 is subjected to an eccentric load. For the following cases, determine the gross allowable load that the footing could carry (use $F_s = 3$):

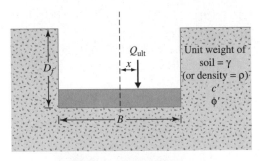

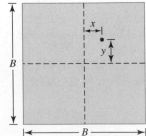

Figure 15.25

    a. $\gamma = 16$ kN/m³, $c' = 0$, $\phi' = 30°$, $B = 1.5$ m, $D_f = 1$ m, $x = 0.15$ m, $y = 0$
    b. $\rho = 2000$ kg/m³, $c' = 0$, $\phi' = 42°$, $B = 2.5$ m, $D_f = 1.5$ m, $x = 0.2$ m, $y = 0$
    c. $\rho = 1950$ kg/m³, $c' = 0$, $\phi' = 40°$, $B = 3$ m, $D_f = 1.4$ m, $x = 0.3$ m, $y = 0$

15.11 For a square footing supported by a sand, given that $B = 2$ m, $D_f = 1.5$ m, corrected standard penetration number $N_{cor} = 9$, allowable settlement $S_e = 20$ mm, estimate the net allowable bearing capacity.

15.12 A plate load test was conducted in a sandy soil in which the size of the bearing plate was 1 ft × 1 ft. The ultimate load per unit area (q_u) for the test was found to be 4200 lb/ft². Estimate the total allowable load (Q_{all}) for a footing of size 5.5 ft × 5.5 ft. Use a factor of safety of 4.

15.13 A plate load test (bearing plate of 762 mm diameter) was conducted in clay. The ultimate load per unit area, q_u, for the test was found to be 200 kN/m². What should be the total allowable load, Q_{all}, for a column footing 1.75 m in diameter? Use a factor of safety of 3.

15.14 The results of two field load tests in a clay soil are given in the following table:

Plate diameter (mm)	Settlement (mm)	Total load (kN)
204.8	15	49.5
457.2	15	133.1

Based on these results, determine the size of a square footing that will carry a total load of 300 kN with a maximum settlement of 15 mm.

15.15 Figure 15.26 shows a footing on layered sand. Determine the net allowable load it can carry, given the following conditions:
    Square footing: $B = 5$ ft
    Factor of safety required = 4
    $D_f = 3.5$ ft
    $H = 2$ ft
    $\gamma_1 = 118$ lb/ft³
    $\gamma_2 = 105$ lb/ft³
    $\phi_1' = 40°$
    $\phi_2' = 30°$

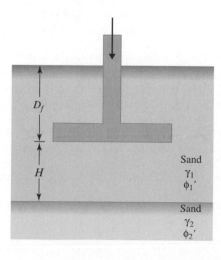

Figure 15.26

15.16 Redo Problem 15.15 with the following values:

Rectangular footing: $B = 1$ m; $L = 1.5$ m

Factor of safety required = 3

$D_f = 1$ m

$H = 0.6$ m

$\gamma_1 = 17.5$ kN/m^3

$\gamma_2 = 15$ kN/m^3

$\phi'_1 = 42°$

$\phi'_2 = 34°$

15.17 Refer to Figure 15.20. The foundation is 1 m × 2 m in plan. $D_f = 1$ m and $H = 1.5$ m. For the sand layer, $\phi' = 35°$, $c' = 0$, $\gamma = 17.8$ kN/m^3; and, for the clay layer, $\phi = 0°$, $c_u = 60$ kN/m^2, $\gamma = 18.2$ kN/m^3. Determine the gross allowable load that the foundation could carry. Use $F_s = 4$.

15.18 Redo Problem 15.17 with the following data:

Foundation: $B \times L = 3$ ft × 6 ft

$D_f = 2.5$ ft

$H = 3$ ft

Sand: $\phi' = 40°$

$c' = 0$

$\gamma = 115$ lb/ft^3

Clay: $\phi = 0°$

$c_u = 750$ lb/ft^2

$\gamma = 118$ lb/ft^3

References

AMERICAN SOCIETY FOR TESTING AND MATERIALS (1997). *Annual Book of Standards,* ASTM, Vol. 04.08, West Conshohocken, Pa.

BOWLES, J. E. (1977). *Foundation Analysis and Design,* 2nd ed., McGraw-Hill, New York.

DAS, B. M. (1999), *Principles of Foundation Engineering,* 4th ed., Plus Publishing Co., Boston.

DEBEER, E. E., and VESIC, A. S. (1958). "Etude Experimentale de la Capacité Portante du Sable Sous des Fondations Directes Etablies en Surface," *Ann. Trav. Publics Belg.,* Vol. 59, No. 3.

HOUSEL, W. S. (1929). "A Practical Method for the Selection of Foundations Based on Fundamental Research in Soil Mechanics," University of Michigan Research Station, *Bulletin No. 13,* Ann Arbor.

KUMBHOJKAR, A. S. (1993). "Numerical Evaluation of Terzaghi's N_γ," *Journal of Geotechnical Engineering,* ASCE, Vol. 119, No. GT3, 598–607.

MEYERHOF, G. G. (1953). "The Bearing Capacity of Foundations Under Eccentric and Inclined Loads," *Proceedings,* 3rd International Conference on Soil Mechanics and Foundation Engineering, Vol. 1, 440–445.

MEYERHOF, G. G. (1956). "Penetration Tests and Bearing Capacity of Cohesionless Soils, *Journal of the Soil Mechanics and Foundations Division,* American Society of Civil Engineers, Vol. 82, No. SM1, pp. 1–19.

MEYERHOF, G. G. (1963). "Some Recent Research on the Bearing Capacity of Foundations," *Canadian Geotechnical Journal,* Vol. 1, 16–26.

MEYERHOF, G. G. (1965). "Shallow Foundations," *Journal of the Soil Mechanics and Foundations Division,* ASCE, Vol. 91, No. SM2, pp. 21–31.

MEYERHOF, G. G., and HANNA, A. M. (1978). "Ultimate Bearing Capacity of Foundations on Layered Soil under Inclined Load," *Canadian Geotechnical Journal,* Vol. 15, No. 4, 565–572.

PRANDTL, L. (1921). "Über die Eindringungsfestigkeit (Harte) plasticher Baustoffe und die Festigkeit von Schneiden," *Zeitschrift für Angewandte Mathematik und Mechanik,* Basel, Switzerland, Vol. 1, No. 1, 15–20.

REISSNER, H. (1924). "Zum Erddruckproblem," *Proceedings,* 1st International Congress of Applied Mechanics, 295–311.

TERZAGHI, K. (1943). *Theoretical Soil Mechanics,* Wiley, New York.

16

Landfill Liners and Geosynthetics

Enormous amounts of solid waste are generated every year in the United States and other industrialized countries. These waste materials can, in general, be classified into four major categories: (1) municipal waste, (2) industrial waste, (3) hazardous waste, and (4) low-level radioactive waste. Table 16.1 lists the waste material generated in 1984 in the United States in these four categories (Koerner, 1994).

The waste materials are generally placed in landfills. The landfill materials interact with moisture received from rainfall and snow to form a liquid called *leachate*. The chemical composition of leachates varies widely depending on the waste material involved. Leachates are a main source of groundwater pollution; therefore, they must be properly contained in all landfills, surface impoundments, and waste piles, within some type of liner system. In the following sections of this chapter, various types of liner systems and the materials used in them are discussed.

16.1 Landfill Liners—Overview

Until about 1982, the predominant liner material used in landfills was clay. Proper clay liners have a hydraulic conductivity of about 10^{-7} cm/sec or less. In 1984, the U.S. Environmental Protection Agency's minimum technological requirements for hazardous waste landfill design and construction were introduced by the U.S. Congress in Hazardous and Solid Waste amendments. In these amendments, Congress stipulated that all new landfills should have double liners and systems for leachate collection and removal.

To understand the construction and functioning of the double-liner system, we must review the general properties of the component materials involved in the system — that is, clay soil and geosynthetics (such as geotextiles, geomembranes, and geonets).

In Sections 6.9 and 6.10, discussion on compaction of a clay soil for a liner system was discussed in more detail. A brief review of the essential properties of geosynthetics is given in the following sections.

Table 16.1 Waste Material Generation in the United States

Waste type	Approximate quantity in 1984 (millions of metric tons)
Municipal	300
Industrial (building debris, degradable waste, nondegradable waste, and near hazardous)	600
Hazardous	150
Low-level radioactive	15

16.2 Geosynthetics

In general, geosynthetics are fabriclike material made from polymers such as polyester, polyethylene, polypropylene, polyvinyl chloride (PVC), nylon, chlorinated polyethylene, and others. The term *geosynthetics* includes the following:

1. Geotextiles
2. Geomembranes
3. Geogrids
4. Geonets
5. Geocomposites

Each type of geosynthetic performs one or more of the following five major functions:

1. Separation
2. Reinforcement
3. Filtration
4. Drainage
5. Moisture barrier

Geosynthetics have been used in civil engineering construction since the late 1970s, and their use is currently growing rapidly. In this chapter, it is not possible to provide detailed descriptions of manufacturing procedures, properties, and uses of all types of geosynthetics. However, an overview of geotextiles, geomembranes, and geonets is given. For further information, refer to a geosynthetics text, such as that by Koerner (1994).

16.3 Geotextiles

Geotextiles are textiles in the traditional sense; however, the fabrics are usually made from petroleum products such as polyester, polyethylene, and polypropylene. They may also be made from fiberglass. Geotextiles are not prepared from natural fabrics, which decay too quickly. They may be *woven, knitted, or nonwoven.*

Woven geotextiles are made of two sets of parallel filaments or strands of yarn systematically interlaced to form a planar structure. *Knitted geotextiles* are formed by interlocking a series of loops of one or more filaments or strands of yarn to form a planar structure. *Nonwoven geotextiles* are formed from filaments or short fibers arranged in an oriented or a random pattern in a planar structure. These filaments, or short fibers, are first arranged into a loose web. They are then bonded by using one or a combination of the following processes:

1. *Chemical bonding* — by glue, rubber, latex, cellulose derivative, and so forth
2. *Thermal bonding* — by heat for partial melting of filaments
3. *Mechanical bonding* — by needle punching

The *needle-punched nonwoven* geotextiles are thick and have high in-plane hydraulic conductivity.

Geotextiles have four major uses:

1. *Drainage:* The fabrics can rapidly channel water from soil to various outlets.
2. *Filtration:* When placed between two soil layers, one coarse grained and the other fine grained, the fabric allows free seepage of water from one layer to the other. At the same time, it protects the fine-grained soil from being washed into the coarse-grained soil.
3. *Separation:* Geotextiles help keep various soil layers separate after construction. For example, in the construction of highways, a clayey subgrade can be kept separate from a granular base course.
4. *Reinforcement:* The tensile strength of geotextiles increases the load-bearing capacity of the soil.

Geotextiles currently available commercially have thicknesses that vary from about 0.25 to 7.6 mm (0.01 to 0.3 in.). The mass per unit area of these geotextiles ranges from about 150 to 700 g/cm^2.

One of the major functions of geotextiles is filtration. For this purpose, water must be able to flow freely through the fabric of the geotextile (Figure 16.1). Hence, the *cross-plane hydraulic conductivity* is an important parameter for design purposes. It should be realized that geotextile fabrics are compressible, however, and

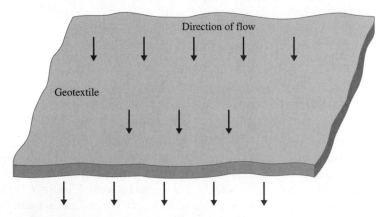

Figure 16.1 Cross-plane flow through geotextile

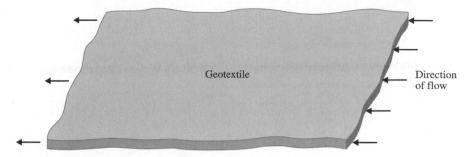

Figure 16.2 In-plane flow in geotextile

their thickness may change depending on the effective normal stress to which they are being subjected. The change in thickness under normal stress also changes the cross-plane hydraulic conductivity of a geotextile. Thus, the cross-plane capability is generally expressed in terms of a quantity called *permittivity,* or

$$P = \frac{k_n}{t} \tag{16.1}$$

where P = permittivity
k_n = hydraulic conductivity for cross-plane flow
t = thickness of the geotextile

In a similar manner, to perform the function of drainage satisfactorily, geotextiles must possess excellent in-plane permeability. For reasons stated previously, the in-plane hydraulic conductivity also depends on the compressibility, and, hence, the thickness of the geotextile. The in-plane drainage capability can thus be expressed in terms of a quantity called *transmissivity,* or

$$T = k_p t \tag{16.2}$$

where T = transmissivity
k_p = hydraulic conductivity for in-plane flow (Figure 16.2)

The units of k_n and k_p are cm/sec or ft/min; however, the unit of permittivity P is sec^{-1} or min^{-1}. In a similar manner, the unit of transmissivity T is $\text{m}^3/\text{sec} \cdot \text{m}$ or $\text{ft}^3/\text{min} \cdot \text{ft}$. Depending on the type of geotextile, k_n and P, and k_p and T can vary widely. Following are some typical values for k_n, P, k_p, and T:

- *Hydraulic conductivity k_n:* 1×10^{-3} to 2.5×10^{-1} cm/sec
- *Permittivity P:* 2×10^{-2} to 2.0 sec^{-1}
- *Hydraulic conductivity k_p:*
  Nonwoven: 1×10^{-3} to 5×10^{-2} cm/sec
  Woven: 2×10^{-3} to 4×10^{-3} cm/sec
- *Transmissivity T:*
  Nonwoven: 2×10^{-6} to 2×10^{-9} $\text{m}^3/\text{sec} \cdot \text{m}$
  Woven: 1.5×10^{-8} to 2×10^{-8} $\text{m}^3/\text{sec} \cdot \text{m}$

When a geotextile is being considered for use in the design and construction of landfill liners, certain properties must be measured by tests on the geotextile to determine its applicability. A partial list of these tests follows:

1. Mass per unit area
2. Percentage of open area
3. Equivalent opening size
4. Thickness
5. Ultraviolet resistivity
6. Permittivity
7. Transmissivity
8. Puncture resistance
9. Resistance to abrasion
10. Compressibility
11. Tensile strength and elongation properties
12. Chemical resistance

16.4   Geomembranes

Geomembranes are impermeable liquid or vapor barriers made primarily from continuous polymeric sheets that are flexible. The type of polymeric material used for geomembranes may be *thermoplastic* or *thermoset.* The thermoplastic polymers include PVC, polyethylene, chlorinated polyethylene, and polyamide. The thermoset polymers include ethylene vinyl acetate, polychloroprene, and isoprene-isobutylene. Although geomembranes are thought to be impermeable, they are not. Water vapor transmission tests show that the hydraulic conductivity of geomembranes is in the range of 10^{-10} to 10^{-13} cm/sec; hence, they are only "essentially impermeable."

Many scrim-reinforced geomembranes manufactured in single piles have thicknesses that range from 0.25 to about 0.4 mm (0.01 to 0.016 in.). These single piles of geomembranes can be laminated together to make thicker geomembranes. Some geomembranes made from PVC and polyethylene may be as thick as 4.5 to 5 mm (0.18 to 0.2 in.).

Following is a partial list of tests that should be conducted on geomembranes when they are to be used as landfill liners:

1. Density
2. Mass per unit area
3. Water vapor transmission capacity
4. Tensile behavior
5. Tear resistance
6. Resistance to impact
7. Puncture resistance
8. Stress cracking
9. Chemical resistance
10. Ultraviolet light resistance
11. Thermal properties
12. Behavior of seams

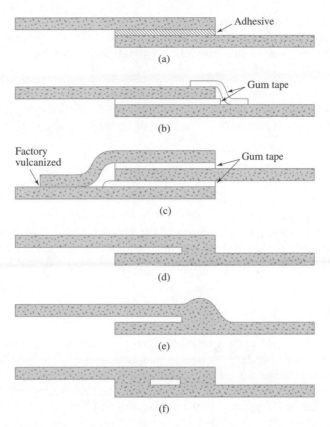

Figure 16.3 Configurations of field geomembrane seams: (a) lap seam; (b) lap seam with gum tape; (c) tongue-and-groove splice; (d) extrusion weld lap seam; (e) fillet weld lap seam; (f) double hot air or wedge seam (after U.S. Environmental Protection Agency, 1989)

The most important aspect of construction with geomembranes is the preparation of seams. Otherwise, the basic reason for using geomembranes as a liquid or vapor barrier will be defeated. Geomembrane sheets are generally seamed together in the factory to prepare larger sheets. These larger sheets are field seamed into their final position. There are several types of seams, some of which are briefly described next.

Lap Seam with Adhesive
• A solvent adhesive is used for this type of seam (Figure 16.3a). After application of the solvent, the two sheets of geomembrane are overlapped, then roller pressure is applied.

Lap Seam with Gum Tape
• This type of seam (Figure 16.3b) is used mostly in dense thermoset material such as isoprene-isobutylene.

Tongue-and-Groove Splice
• A schematic diagram of the tongue-and-groove splice is shown in Figure 16.3c. The tapes used for the splice are double sided.

Extrusion Weld Lap Seam

- Extrusion or fusion welding is done on geomembranes made from polyethylene. A ribbon of molten polymer is extruded between the two surfaces to be joined (Figure 16.3d).

Fillet Weld Lap Seam

- This seam is similar to an extrusion weld lap seam; however, for fillet welding, the extrudate is placed over the edge of the seam (Figure 16.3e).

Double Hot Air or Wedge Seam

- In the hot air seam, hot air is blown to melt the two opposing surfaces. For melting, the temperatures should rise to about 500°F or more. After the opposite surfaces are melted, pressure is applied to form the seam (Figure 16.3f). For hot wedge seams, an electrically heated element like a blade is passed between the two opposing surfaces of the geomembrane. The heated element helps to melt the geomembrane, after which pressure is applied by a roller to form the seam.

16.5   Geonets

Geonets are formed by the continuous extrusion of polymeric ribs at acute angles to each other. They have large openings in a netlike configuration. The primary function of geonets is drainage. Figure 16.4 is a photograph of a typical piece of geonet. Most geonets currently available are made of medium-density and high-density polyethylene. They are available in rolls with widths of 1.8 to 2.1 m (≈6 to 7 ft) and lengths of 30 to 90 m (≈100 to 300 ft). The approximate aperture sizes vary from 30 mm × 30 mm (≈1.2 in. × 1.2 in.) to about 6 mm × 6 mm (≈0.25 in. × 2.5 in.). The thickness of geonets available commercially can vary from 3.8 to 7.6 mm (≈0.15 to 0.3 in.).

Seaming of geonets is somewhat more difficult. For this purpose, staples, threaded loops, and wire are sometimes used.

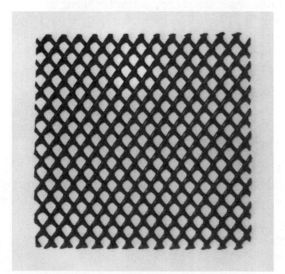

Figure 16.4 Geonet

Single Clay Liner and Single Geomembrane Liner Systems

Until about 1982 — that is, before the guidelines for the minimum technological requirements for hazardous waste landfill design and construction were mandated by the U.S. Environmental Protection Agency — most landfill liners were *single clay liners*. Figure 16.5 shows the cross section of a single clay liner system for a landfill. It consists primarily of a compacted clay liner over the native foundation soil. The thickness of the compacted clay liner varies between 0.9 and 1.8 m (3 and 6 ft). The *maximum* required hydraulic conductivity k is 10^{-7} cm/sec. Over the clay liner is a layer of gravel with perforated pipes for leachate collection and removal. Over the gravel layer is a layer of filter soil. The filter is used to protect the holes in the perforated pipes against the movement of fine soil particles. In most cases, the filter is medium coarse to fine sandy soil. It is important to note that this system does not have any leak-detection capability.

Around 1982, single layers of geomembranes were also used as a liner material for landfill sites. As shown in Figure 16.6, the geomembrane is laid over native foundation soil. Over the geomembrane is a layer of gravel with perforated pipes for leachate collection and removal. A layer of filter soil is placed between the solid waste material and the gravel. As in the single clay liner system, no provision is made for leak detection.

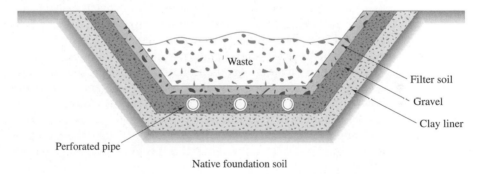

Figure 16.5 Cross section of single clay liner system for a landfill

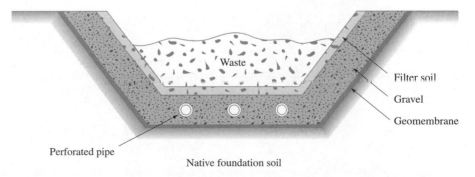

Figure 16.6 Cross section of single geomembrane liner system for a landfill

16.7	**Recent Advances in the Liner Systems for Landfills**

Since 1984, most landfills developed for solid and hazardous wastes have double liners. The two liners are an upper primary liner and a lower secondary liner. Above the top liner is a *primary* leachate collection and removal system. In general, the primary leachate collection system must be able to maintain a leachate head of 0.3 m ($\approx$12 in.) or less. Between the primary and secondary liners is a system for leak detection, collection, and removal (LDCR) of leachates. The general guidelines for the primary leachate collection system and the LDCR system are as follows:

1. It can be a granular drainage layer or a geosynthetic drainage material such as a geonet.
2. If a granular drainage layer is used, it should have a minimum thickness of 0.3 m ($\approx$12 in.)
3. The granular drainage layer (or the geosynthetic) should have a hydraulic conductivity k greater than 10^{-2} cm/sec.
4. If a granular drainage layer is used, it should have a granular filter or a layer of geotextile over it to prevent clogging. A layer of geotextile is also required over the geonet when it is used as the drainage layer.
5. The granular drainage layer, when used, must be chemically resistant to the waste material and the leachate that are produced. It should also have a network of perforated pipes to collect the leachate effectively and efficiently.

In the design of the liner systems, the compacted clay layers should be at least 1 m ($\approx$3 ft) thick, with $k \leq 10^{-7}$ cm/sec. Figures 16.7 and 16.8 show schematic diagrams of two double-liner systems. In Figure 16.7, the primary leachate collection system is made of a granular material with perforated pipes and a filter system above it. The primary liner is a geomembrane. The LDCR system is made of a geonet. The secondary liner is a *composite* liner made of a geomembrane with a compacted clay layer below it. In Figure 16.8, the primary leachate collection system is similar to that shown in Figure 16.7; however, the primary and secondary liners are both composite liners (geomembrane-clay). The LDCR system is a geonet with a layer of geotextile over it. The layer of geotextile acts as a filter and separator.

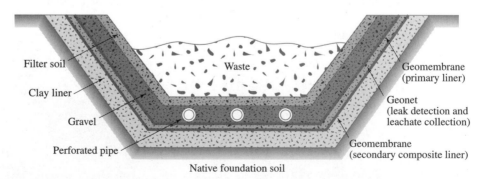

Figure 16.7 Cross section of double-liner system (note the secondary composite liner)

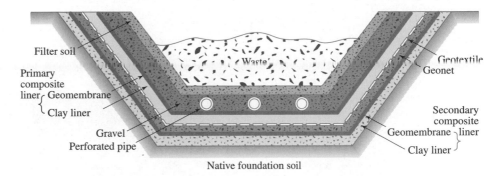

Figure 16.8 Cross section of double-liner system (note the primary and secondary composite liners)

The geomembranes used for landfill lining must have a minimum thickness of 0.76 mm (0.03 in.); however, all geomembranes that have a thickness of 0.76 mm (0.03 in.) may not be suitable in all situations. In practice, most geomembranes used as liners have thicknesses ranging from 1.8 to 2.54 mm (0.7 to 0.1 in.).

16.8 Leachate Removal Systems

The bottom of a landfill must be properly graded so that the leachate collected from the primary collection system and the LDCR system will flow to a low point by gravity. Usually a grade of 2% or more is provided for large landfill sites. The low point of the leachate collection system ends at a sump. For primary leachate collection, a manhole is located at the sump, which rises through the waste material. Figure 16.9 shows a schematic diagram of the leachate removal system with a low-volume sump.

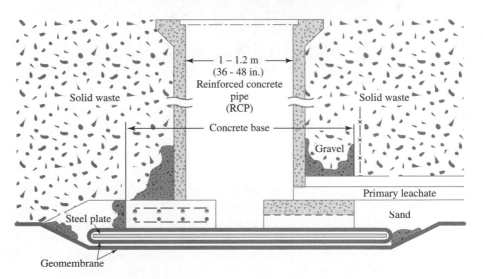

Figure 16.9 Primary leachate removal system with a low-volume sump (after U.S. Environmental Protection Agency, 1989).

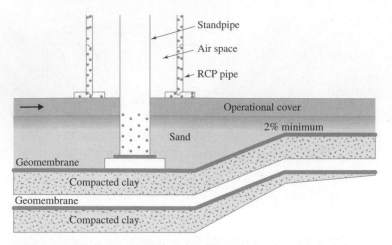

Figure 16.10 Primary leachate removal system with a high-volume sump (after U.S. Environmental Protection Agency, 1989)

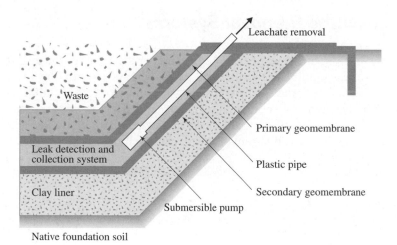

Figure 16.11 Secondary leak detection, collection, and removal (LDCR) system — by means of pumping. *Note:* The plastic pipe penetrates the primary geomembrane

A typical leachate removal system for high-volume sumps (for primary collection) is shown in Figure 16.10.

Leachate can be removed from the LDCR system by means of pumping, as shown in Figure 16.11, or by gravity monitoring, as shown in Figure 16.12. When leachate is removed by pumping, the plastic pipe used for removal must penetrate the primary liner. On the other land, if gravity monitoring is used, the pipe will penetrate the secondary liner.

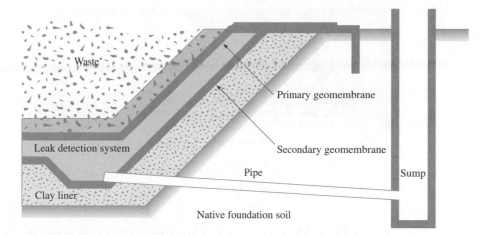

Figure 16.12 Secondary LDCR system, by means of gravity monitoring. *Note:* The plastic pipe penetrates the secondary geomembrane

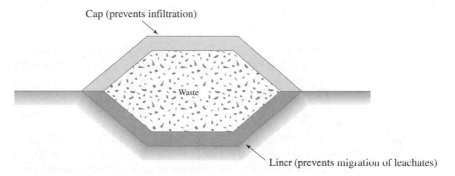

Figure 16.13 Landfill with liner and cap

16.9   *Closure of Landfills*

When the landfill is complete and no more waste can be placed into it, a cap must be put on it (Figure 16.13). This cap will reduce and ultimately eliminate leachate generation. A schematic diagram of the layering system recommended by the U.S. Environmental Protection Agency (1979, 1986) and Koerner (1994) for hazardous waste landfills is shown in Figure 16.14. Essentially, it consists of a compacted clay cap over the solid waste, a geomembrane liner, a drainage layer, and a cover of topsoil. The manhole used for leachate collection penetrates the landfill cover. Leachate removal continues until its generation is stopped. For hazardous waste landfill sites, the EPA (1989) recommends this period to be about 30 years.

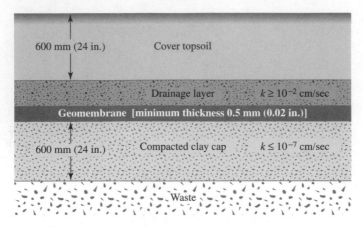

Figure 16.14 Schematic diagram of the layering system for landfill cap

16.10   *Summary and General Comments*

This chapter provided a brief overview of the problems associated with solid and hazardous waste landfills. The general concepts for the construction of landfill liners using compacted clayey soil and geosynthetics (that is, geotextiles, geomembranes, and geonets) were discussed. Several areas were not addressed, however, because they are beyond the scope of the text. Areas not discussed include the following:

1. *Selection of material.* The chemicals contained in leachates generated from hazardous and nonhazardous waste may interact with the liner materials. For this reason, it is essential that representative leachates are used to test the chemical compatibility so that the liner material remains intact during the periods of landfill operation and closure, and possibly longer. Selection of the proper leacheates becomes difficult because of the extreme variations encountered in the field. The mechanical properties of geomembranes are also important. Properties such as workability, creep, stress cracking, and the thermal coefficient of expansion should be investigated thoroughly.

2. *Stability of side slope liner.* The stability and slippage checks of the side slope liners of a landfill site are important and complicated because of the variation of the frictional characters of the composite materials involved in liner construction. For a detailed treatment of this topic, refer to any book on geosynthetics (e.g., Koerner, 1994).

3. *Leak response action plan.* It is extremely important that any leaks or clogging of the drainage layer(s) in a given waste disposal site be detected as quickly as possible. Leaks or cloggings are likelihoods at a site even with good construction quality control. Each waste disposal facility should have a *leak response action plan.*

References

KOERNER, R. M. (1994). *Designing with Geosynthetics,* 3rd ed., Prentice-Hall, Englewood Cliffs, N.J.

U.S. ENVIRONMENTAL PROTECTION AGENCY (1979). *Design and Construction for Solid Waste Landfills,* Publication No. EPA-600/2-79-165, Cincinnati, Ohio.

U.S. ENVIRONMENTAL PROTECTION AGENCY (1986). *Cover for Uncontrolled Hazardous Waste Sites,* Publication No. EPA-540/2-85-002, Cincinnati, Ohio.

U.S. ENVIRONMENTAL PROTECTION AGENCY (1989). *Requirements for Hazardous Waste Landfill Design, Construction, and Closure,* Publication No. EPA-625/4-89-022, Cincinnati, Ohio.

17

Subsoil Exploration

The preceding chapters reviewed the fundamental properties of soils and their behavior under stress and strain in idealized conditions. In practice, natural soil deposits are not homogeneous, elastic, or isotropic. In some places, the stratification of soil deposits may even change greatly within a horizontal distance of 15 to 30 m ($\simeq$50 to 100 ft). For foundation design and construction work, one must know the actual soil stratification at a given site, the laboratory test results of the soil samples obtained from various depths, and the observations made during the construction of other structures built under similar conditions. For most major structures, adequate subsoil exploration at the construction site must be conducted. The purposes of subsoil exploration include the following:

1. Determining the nature of soil at the site and its stratification
2. Obtaining disturbed and undisturbed soil samples for visual identification and appropriate laboratory tests
3. Determining the depth and nature of bedrock, if and when encountered
4. Performing some *in situ* field tests, such as permeability tests (Chapter 6), vane shear tests (Chapter 11), and standard penetration tests
5. Observing drainage conditions from and into the site
6. Assessing any special construction problems with respect to the existing structure(s) nearby
7. Determining the position of the water table

This chapter briefly summarizes subsoil exploration techniques. For additional information, refer to the *Manual of Foundation Investigations* of the American Association of State Highway and Transportation Officials (1967).

17.1   *Planning for Soil Exploration*

A soil exploration program for a given structure can be broadly divided into four phases:

1. *Compilation of the existing information regarding the structure*: This phase includes gathering information such as the type of structure to be constructed

and its future use, the requirements of local building codes, and the column and load-bearing wall loads. If the exploration is for the construction of a bridge foundation, one must have an idea of the length of the span and the anticipated loads to which the piers and abutments will be subjected.

2. *Collection of existing information for the subsoil condition*: Considerable savings in the exploration program can sometimes be realized if the geotechnical engineer in charge of the project thoroughly reviews the existing information regarding the subsoil conditions at the site under consideration. Useful information can be obtained from the following sources:

a. Geologic survey maps

b. County soil survey maps prepared by the U.S. Department of Agriculture and the Soil Conservation Service

c. Soil manuals published by the state highway departments

d. Existing soil exploration reports prepared for the construction of nearby structures

Information gathered from the preceding sources provides insight into the type of soil and problems that might be encountered during actual drilling operations.

3. *Reconnaissance of the proposed construction site*: The engineer should visually inspect the site and the surrounding area. In many cases, the information gathered from such a trip is invaluable for future planning. The type of vegetation at a site may, in some instances, indicate the type of subsoil that will be encountered. The accessibility of a site and the nature of drainage into and from it can also be determined. Open cuts near the site provide an indication about the subsoil stratification. Cracks in the walls of nearby structure(s) may indicate settlement from the possible existence of soft clay layers or the presence of expansive clay soils.

4. *Detailed site investigation*: This phase consists of making several test borings at the site and collecting disturbed and undisturbed soil samples from various depths for visual observation and for laboratory tests. No hard-and-fast rule exists for determining the number of borings or the depth to which the test borings are to be advanced. For most buildings, at least one boring at each corner and one at the center should provide a start. Depending on the uniformity of the subsoil, additional test borings may be made. Table 17.1 gives guidelines for initial planning of borehole spacing.

Table 17.1 Spacing of Borings

Project	Boring spacings	
	m	**ft**
One-story buildings	25–30	75–100
Multistory buildings	15–25	50–75
Highways	250–300	750–1000
Earth dams	25–50	75–150
Residential subdivision planning	60–100	200–300

The test borings should extend through unsuitable foundation materials to firm soil layers. Sowers and Sowers (1970) provided a rough estimate of the minimum depth of borings (unless bedrock is encountered) for multistory buildings. They can be given by the following equations, applicable to light steel or narrow concrete buildings:

$$z_b \, (\text{m}) = 3S^{0.7} \tag{17.1a}$$

$$z_b \, (\text{ft}) = 10S^{0.7} \tag{17.1b}$$

Heavy steel or wide concrete buildings:

$$z_b \, (\text{m}) = 6S^{0.7} \tag{17.2a}$$

$$z_b \, (\text{ft}) = 20S^{0.7} \tag{17.2b}$$

In Eqs. (17.1) and (17.2), z_b is the approximate depth of boring and S is the number of stories.

The American Society of Civil Engineers (1972) recommended the following rules of thumb for estimating the boring depths for buildings:

1. Estimate the variation of the net effective stress increase, $\Delta\sigma'$, that will result from the construction of the proposed structure with depth. This variation can be estimated by using the principles outlined in Chapter 9. Determine the depth D_1 at which the value of $\Delta\sigma'$ is equal to 10% of the average load per unit area of the structure.
2. Plot the variation of the effective vertical stress, σ_o', in the soil layer with depth. Compare this with the net stress increase variation, $\Delta\sigma'$, with depth as determined in step 1. Determine the depth D_2 at which $\Delta\sigma' = 0.05\sigma_o'$.
3. The smaller of the two depths, D_1 and D_2, is the approximate minimum depth of the boring.

When the soil exploration is for the construction of dams and embankments, the depth of boring may range from one-half to two times the embankment height.

The general techniques used for advancing test borings in the field and the procedure for the collection of soil samples are described in the following sections.

17.2   *Boring Methods*

The test boring can be advanced in the field by several methods. The simplest is the *use of augers*. Figure 17.1 shows two types of hand augers that can be used for making boreholes up to a depth of about 3 to 5 m ($\approx$10 to 15 ft). They can be used for soil exploration work for highways and small structures. Information regarding the types of soil present at various depths is obtained by noting the soil that holds to the auger. The soil samples collected in this manner are disturbed, but they can be used to conduct laboratory tests such as grain-size determination and Atterberg limits.

When the boreholes are to be advanced to greater depths, the most common method is to use continuous flight augers, which are power operated. The power for

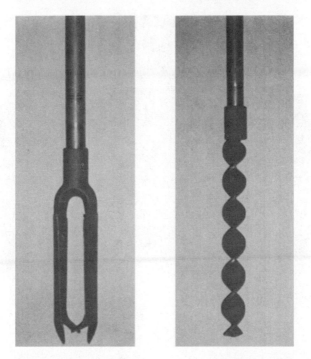

Figure 17.1 Hand augers: (a) Iwan auger; (b) slip auger

drilling is delivered by truck- or tractor-mounted drilling rigs. Continuous flight augers are commercially available in 1 to 1.5 m (3 to 5 ft) sections. During the drilling operation, section after section of auger can be added, and the hole extended downward. Continuous flight augers can be solid stem or hollow stem. Some of the commonly used solid-stem augers have outside diameters of 67 mm ($2\frac{5}{8}$ in.), 83 mm ($3\frac{1}{4}$ in.), 102 mm (4 in.), and 114 mm ($4\frac{1}{2}$ in.). The inside and outside diameters of some hollow-stem augers are given in Table 17.2.

Flight augers bring the loose soil from the bottom of the hole to the surface. The driller can detect the change in soil type encountered by the change of speed

Table 17.2 Dimensions of Commonly Used Hollow-Stem Augers

Inside diameter		Outside diameter	
mm	in.	mm	in.
63.5	2.5	158.75	6.25
69.85	2.75	177.8	7.0
76.2	3.0	203.2	8.0
88.9	3.5	228.6	9.0
101.6	4.0	254.0	10.0

Figure 17.2 Drilling with flight augers (courtesy of Danny R. Anderson, El Paso, Texas)

and the sound of drilling. Figure 17.2 shows a drilling operation with flight augers. When solid-stem augers are used, the auger must be withdrawn at regular intervals to obtain soil samples and to conduct other operations such as standard penetration tests. Hollow-stem augers have a distinct advantage in this respect — they do not have to be removed at frequent intervals for sampling or other tests. As shown in Figure 17.3, the outside of the auger acts like a casing. A removable plug is attached to the bottom of the auger by means of a center rod. During the drilling, the plug can be pulled out with the auger in place, and soil sampling and standard penetration tests can be performed. When hollow-stem augers are used in sandy soils below the groundwater table, the sand might be pushed several feet into the stem of the auger by excess hydrostatic pressure immediately after the removal of the plug. In such conditions, the plug should not be used, and, instead, water inside the hollow stem should be maintained at a higher level than the groundwater table.

Rotary drilling is a procedure by which rapidly rotating drilling bits attached to the bottom of drilling rods cut and grind the soil and advance the borehole down. Several types of drilling bits are available for such work. Rotary drilling can be used in sand, clay, and rock (unless badly fissured). Water or drilling mud is forced down the drilling rods to the bits, and the return flow forces the cuttings to the surface. Drilling mud is a slurry prepared by mixing bentonite and water (bentonite is a montmorillonite clay formed by the weathering of volcanic ash). Boreholes with diameters ranging from 50 to 200 mm (2 to 8 in.) can easily be made by using this technique.

Wash boring is another method of advancing boreholes. In this method, a casing about 2 to 3 m (6 to 10 ft) long is driven into the ground. The soil inside the cas-

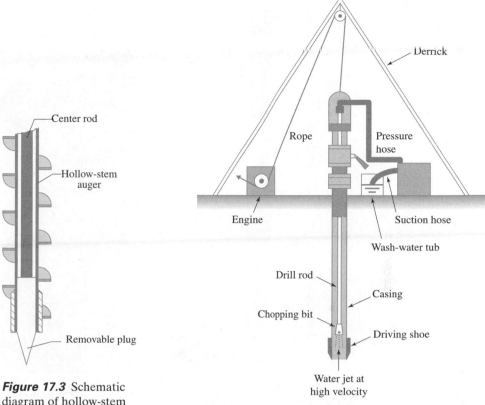

Figure 17.3 Schematic diagram of hollow-stem auger with removable plug

Figure 17.4 Wash boring

ing is then removed by means of a *chopping bit* that is attached to a drilling rod. Water is forced through the drilling rod, and it goes out at a very high velocity through the holes at the bottom of the chopping bit (Figure 17.4). The water and the chopped soil particles rise upward in the drill hole and overflow at the top of the casing through a T-connection. The wash water is then collected in a container. The casing can be extended with additional pieces as the borehole progresses; however, such extension is not necessary if the borehole will stay open without caving in.

Percussion drilling is an alternative method of advancing a borehole, particularly through hard soil and rock. In this technique, a heavy drilling bit is raised and lowered to chop the hard soil. Casing for this type of drilling may be required. The chopped soil particles are brought up by the circulation of water.

17.3 Common Sampling Methods

Various methods for advancing boreholes were discussed in the preceding section. During the advancement of the boreholes, soil samples are collected at various depths for further analysis. This section briefly discusses some of the methods of sample collection.

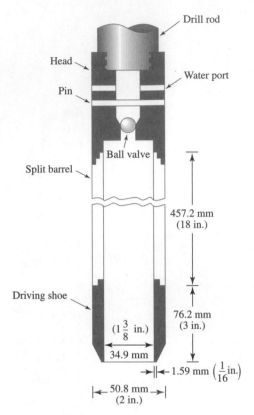

Figure 17.5 Diagram of standard split-spoon sampler

Figure 17.6 Split-spoon sampler, unassembled (courtesy of Soiltest, Inc., Lake Bluff, Illinois)

Sampling by Standard Split Spoon

Figure 17.5 shows a diagram of a split-spoon sampler. It consists of a tool-steel driving shoe at the bottom, a steel tube (that is split longitudinally into halves) in the middle, and a coupling at the top. The steel tube in the middle has inside and outside diameters of 34.9 mm ($1\frac{3}{8}$ in.) and 50.8 mm (2 in.), respectively. Figure 17.6 shows a photograph of an unassembled split-spoon sampler.

When the borehole is advanced to a desired depth, the drilling tools are removed. The split-spoon sampler is attached to the drilling rod and then lowered to the bottom of the borehole. The sampler is driven into the soil at the bottom of the borehole by means of hammer blows. The hammer blows occur at the top of the drilling rod. The hammer weighs 623 N (140 lb). For each blow, the hammer drops a distance of 0.762 m (30 in.). The number of blows required for driving the sampler through three 152.4 mm (6 in.) intervals is recorded. The sum of the number of blows required for driving the last two 152.4 mm (6 in.) intervals is referred to as the *standard penetration number, N.* It is also commonly called the *blow count.* The interpretation of the standard penetration number is given in Section 17.5. After driving is completed, the sampler is withdrawn and the shoe and coupling are removed. The soil sample collected inside the split tube is then removed and transported to the lab-

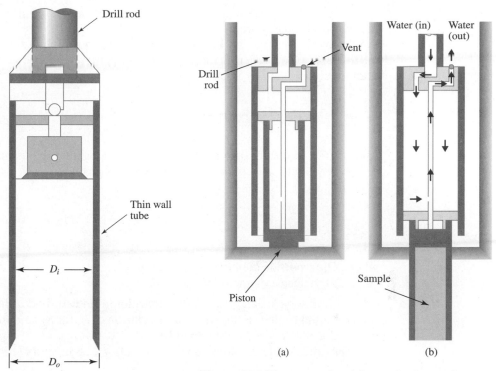

Figure 17.7 Thin wall tube sampler

Figure 17.8 Piston sampler: (a) sampler lowered to bottom of borehole; (b) pressure released through hole in piston rod

oratory in small glass jars. Determination of the standard penetration number and collection of split-spoon samples are usually done at 1.5 m (≈5 ft) intervals.

Sampling by Thin Wall Tube

Sampling by thin wall tube is used for obtaining fairly undisturbed soil samples. The thin wall tubes are made of seamless thin tubes and are commonly referred to as *Shelby tubes* (Figure 17.7). To collect samples at a given depth in a borehole, one must first remove the drilling tools. The sampler is atached to a drilling rod and lowered to the bottom of the borehole. After this, it is hydraulically pushed into the soil. It is then spun to shear off the base and is pulled out. The sampler with the soil inside is sealed and taken to the laboratory for testing. Most commonly used thin wall tube samplers have outside diameters of 76.2 mm (3 in.).

Sampling by Piston Sampler

Piston samplers are particularly useful when highly undisturbed samples are required. The cost of recovering such samples is, of course, higher. Several types of piston samplers can be used; however, the sampler proposed by Osterberg (1952) is the most advantageous (Figure 17.8). It consists of a thin wall tube with a piston. Initially, the piston closes the end of the thin wall tube. The sampler is first lowered to the bottom of the borehole (Figure 17.8a), then the thin wall tube is pushed into the soil hydraulically, past the piston. After this, the pressure is released through a hole in the piston rod (Figure 17.8b). The presence of the piston prevents distortion in the sample by neither letting the soil squeeze into the sampling tube very fast nor admitting excess soil. Samples obtained in this manner are consequently less disturbed than those obtained by Shelby tubes.

17.4   Sample Disturbance

The degree of disturbance of the sample collected by various methods can be expressed by a term called the *area ratio,* which is given by

$$A_r(\%) = \frac{D_o^2 - D_i^2}{D_i^2} \times 100 \tag{17.3}$$

where D_o = outside diameter of the sampler
D_i = inside diameter of the sampler

A soil sample can generally be considered undisturbed if the area ratio is less than or equal to 10%. Following is a calculation of A_r for a standard split-spoon sampler and a 50.8 mm (2 in.) Shelby tube:

For the standard spit-spoon sampler, $D_i = 1.38$ in. and $D_o = 2$ in. Hence,

$$A_r(\%) = \frac{(2)^2 - (1.38)^2}{(1.38)^2} \times 100 = 110\%$$

For the Shelby tube sampler (2-in. diameter), $D_i = 1.875$ in. and $D_o = 2$ in. Hence,

$$A_r(\%) = \frac{(2)^2 - (1.875)^2}{(1.875)^2} \times 100 = 13.8\%$$

The preceding calculation indicates that the sample collected by split spoons is highly disturbed. The area ratio (A_r) of the 50.8 mm (2 in.) diameter Shelby tube samples is slightly higher than the 10% limit stated previously. For practical purpose, however, it can be treated as an undisturbed sample.

The disturbed but representative soil samples recovered by split-spoon samplers can be used for laboratory tests such as grain-size distribution, liquid limit, plastic limit, and shrinkage limit. However, undisturbed soil samples are necessary for performing tests such as consolidation, triaxial compression, and unconfined compression.

17.5   Correlations for Standard Penetration Test

The procedure for conducting the standard penetration test was outlined in Section 17.3. The standard penetration number N is commonly used to correlate several physical parameters of soils. In Chapter 11, a qualitative description of the consistency for clay soils on the basis of their unconfined compression strengths (q_u) was

Table 17.3 Approximate Correlation of Standard Penetration Number and Consistency of Clay

Standard penetration number, N	Consistency	Unconfined compression strength	
		kN/m²	lb/ft²
0 -		0	0
	Very soft		
2 -		25	500
	Soft		
4 -		50	1000
	Medium stiff		
8 -		100	2000
	Stiff		
16 -		200	4000
	Very stiff		
32		400	8000
>32	Hard	>400	>8000

presented. The unconfined compression strength of clay soils can also be approximately correlated to the standard penetration number N. Table 17.3 gives the approximate relationship among the standard penetration number at a given depth, the consistency, and the unconfined compression strength of clayey soils.

It is important to point out that the correlation between N and q_u given in Table 15.3 is only approximate. The sensitivity S_t of clay soil also plays an important role in the actual N value obtained in the field. In any case, for clays of a given geology a reasonable correlation between N and q_u can be obtained, as shown in Figure 17.9. In this figure, the notation p_a is the atmospheric pressure (in the same unit as q_u). For

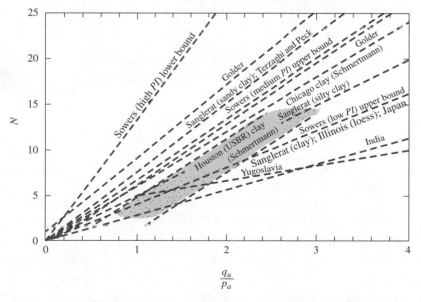

Figure 17.9 Correlation between N and q_u/p_a (based on Djoenaidi, 1985)

the data shown in Figure 17.9, the reported regression is given by (Kulhawy and Mayne, 1990)

$$\frac{q_u}{p_a} = 0.58 N^{0.72} \tag{17.4}$$

In granular soils, the standard penetration number is highly dependent on the effective overburden pressure, σ'_o.

A number of empirical relationships have been proposed to convert the field standard penetration number N_F to a *standard effective overburden pressure, σ'_o,* of 96 kN/m² (2000 lb/ft²). The general form is

$$N_{cor} = C_N N_F \tag{17.5}$$

Several correlations have been developed over the years for the correction factor, C_N. In this text, two of these are recommended for use.

Correlation of Liao and Whitman (1986)

In SI units,

$$C_N = 9.78 \sqrt{\frac{1}{\sigma'_o}} \tag{17.6}$$

where σ'_o effective overburden pressure in kN/m².

In English units,

$$C_N = \sqrt{\frac{1}{\sigma'_o}} \tag{17.7}$$

where σ'_o = effective overburden pressure in U.S. ton/ft² (1 U.S. ton = 2000 lb).

Correlation of Skempton (1986)

In SI units,

$$C_N = \frac{2}{1 + 0.01\sigma'_o} \tag{17.8}$$

where σ'_o = effective overburden pressure in kN/m².

In English units,

$$C_N = \frac{2}{1 + \sigma'_o} \tag{17.9}$$

where σ'_o = effective overburden pressure in U.S. ton/ft².

Table 17.4 shows approximate correlations for the standard penetration number N_{cor}, and relative density D_r.

Cubrinovski and Ishihara (1999) proposed a correlation between N_F and the relative density of granular soils (D_r) in the form

$$D_r (\%) = \left[\frac{N_F \left(0.23 + \dfrac{0.06}{D_{50}} \right)^{1.7}}{9} \left(\frac{98}{\sigma'_o} \right) \right]^{0.5} (100) \tag{17.10}$$

where σ'_o = effective overburden pressure in kN/m²

D_{50} = sieve size through which 50% of soil will pass (mm)

Table 17.4 Approximate Relationship Between Corrected Standard Penetration Number and Relative Density of Sand

Corrected standard penetration number, N_{cor}	Relative density, D_r (%)
0–5	0–5
5–10	5–30
10–30	30–60
30–50	60–95

The drained angle of friction of *granular soils*, ϕ', has also been correlated to the standard penetration number. Peck, Hanson, and Thornburn (1974) gave a correlation between N_{cor} and ϕ' in a graphic form, which can be approximated as (Wolff, 1989)

$$\phi' \ (\text{deg}) = 27.1 + 0.3N_{cor} - 0.00054N_{cor}^2 \tag{17.11}$$

Schmertmann (1975) also provided a correlation for N_F versus σ_o'. After Kulhawy and Mayne (1990), this correlation can be approximated as

$$\phi' = \tan^{-1}\left[\frac{N_F}{12.2 + 20.3\left(\dfrac{\sigma_o'}{p_a}\right)}\right]^{0.34} \tag{17.12}$$

where p_a = atmospheric pressure (same unit as σ_o').

The standard penetration number is a useful guideline in soil exploration and assessment of subsoil conditions, provided that the results are interpreted correctly. Note that all equations and correlations relating to the standard penetration numbers are approximate. Because soil is not homogeneous, a wide variation in the N_F value may be obtained in the field. For soil deposits that contain large boulders and gravel, the standard penetration numbers may be erratic.

Example 17.1

Following are the results of a standard penetration test in sand. Determine the corrected standard penetration numbers, N_{cor}, at various depths. Note that the water table was not observed within a depth of 35 ft below the ground surface. Assume that the average unit weight of sand is 110 lb/ft³. Use Skempton's relationship given in Eq. (17.9).

Depth, z (ft)	N_F
5	8
10	7
15	12
20	14
25	13

Solution

From Eq. (17.9), $C_N = 2/(1 + \sigma')$. Now the following table can be prepared:

Depth, z (ft)	$\sigma' = z\gamma$ (ton /ft²)	C_N	N_F	N_1
5	0.275	1.57	8	≈13
10	0.55	1.29	7	≈9
15	0.825	1.096	12	≈13
20	1.1	0.952	14	≈13
25	1.375	0.842	13	≈11

■

Example 17.2

Refer to Example 17.1. Estimate the drained soil friction angle at a depth $z = 15$ to 20 ft. Use Eq. (17.10).

Solution

From Example 17.1, at $z = 15$ to 20 ft, $N_{cor} \approx 13$. From Eq. (17.11),

$$\phi' = 27.1 + 0.3N_{cor} - 0.00054N_{cor}^2$$

$$= 27.1 + 0.3(13) - 0.00054(13^2) \approx \mathbf{30.9°}$$

■

17.6 *Other* In Situ *Tests*

Depending on the type of project and the complexity of the subsoil, several types of *in situ* tests can be conducted during the exploration period. In many cases, the soil properties evaluated from the *in situ* tests yield more representative values. This better accuracy results primarily because the sample disturbance during soil exploration is eliminated. Following are some of the common tests that can be conducted in the field.

Vane Shear Test

The principles and the application of the vane shear test were discussed in Chapter 11. When soft clay is encountered during the advancement of a borehole, the undrained shear strength of clay c_u can be determined by conducting a vane shear test in the borehole. This test provides valuable information about the strength in undisturbed clay.

Borehole Pressuremeter Test

The pressuremeter is a device that was originally developed by Menard in 1965 for *in situ* measurement of the stress-strain modulus. This device basically consists of a pressure cell and two guard cells (Figure 17.10). The test involves expanding the pressure cell inside a borehole and measuring the expansion of its volume. The test

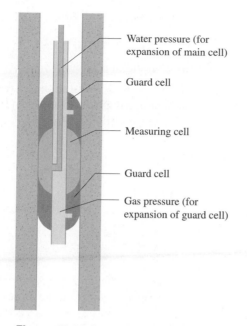

Water pressure (for expansion of main cell)

Guard cell

Measuring cell

Guard cell

Gas pressure (for expansion of guard cell)

Figure 17.10 Schematic diagram for pressuremeter test

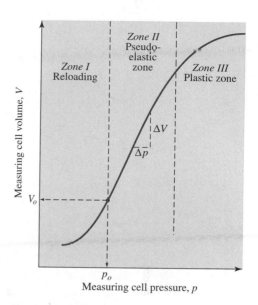

Figure 17.11 Relationship between measuring pressure and measuring volume for Menard pressuremeter

data are interpreted on the basis of the theory of expansion of an infinitely thick cylinder of soil. Figure 17.11. shows the variation of the pressure cell volume with changes in the cell pressure. In this figure, zone I represents the reloading portion, during which the soil around the borehole is pushed back to its initial state — that is, the state it was in before drilling. Zone II represents a pseudoelastic zone, in which the cell volume versus cell pressure is practically linear. The zone marked III is the plastic zone. For the pseudoelastic zone,

$$E_s = 2(1 + \mu_s)V_o\frac{\Delta p}{\Delta V} \tag{17.13}$$

where E = modulus of elasticity of soil
    μ_s = Poisson's ratio of soil
    V_o = cell volume corresponding to pressure p_o (that is, the cell pressure corresponding to the beginning of zone II)
$\Delta p/\Delta V$ = Slope of straight-line plot of zone II

Menard recommended a value of $\mu_s = 0.33$ for use in Eq. 17.13, but other values can be used. With $\mu_s = 0.33$,

$$E_s = 2.66\, V_o\frac{\Delta p}{\Delta V} \tag{17.14}$$

From the theory of elasticity, the relationship between the modulus of elasticity and the shear modulus can be given as

$$E = 2(1 + \mu_s)G \tag{17.15}$$

where G = shear modulus of soil. Hence, combining Eqs. (17.13) and (17.15) gives

$$G = V_o \frac{\Delta p}{\Delta V} \tag{17.16}$$

Pressuremeter test results can be used to determine the at-rest earth pressure coefficient, K_o (Chapter 12). This coefficient can be obtained from the ratio of p_o and σ'_o (σ'_o = effective vertical stress at the depth of the test), or

$$K_o = \frac{p_o}{\sigma'} \tag{17.17}$$

Note that p_o (see Figure 17.11) represents the *in situ* lateral pressure.

The pressuremeter tests are very sensitive to the conditions of a borehole before the test.

Cone Penetration Test

The Dutch cone penetrometer is a device by which a 60° cone with a base area of 10 cm² (1.54 in²) (Figure 17.12) is pushed into the soil, and the cone end resistance, q_c, to penetration is measured. Most cone penetrometers that are commonly used have friction sleeves that follow the point. This allows independent determination of the cone resistance and the frictional resistance of the soil above it. The friction sleeves have an exposed surface area of about 150 cm² ($\approx$23 in²).

In the past, the cone penetration test was used in Europe more commonly than in the United States. Recently, however, this test has attracted considerable interest in the United States. One of the major advantages of the cone penetration test is that boreholes are not necessary to conduct the test. Unlike the standard penetration test, however, soil samples cannot be recovered for visual observation and laboratory tests.

Robertson and Campanella (1983) provided correlations between the vertical effective stress (σ'_o), drained soil friction angle (ϕ'), and q_c for sand. The relationship between σ'_o, ϕ', and q'_c can be approximated (after Kulhawy and Mayne, 1990) as

$$\phi' = \tan^{-1}\left[0.1 + 0.38 \log\left(\frac{q_c}{\sigma'_o}\right)\right] \tag{17.18}$$

The cone penetration, resistance has also been correlated with the equivalent modulus of elasticity E_s of soils by various investigators. Schmertmann (1970) gave a simple correlation for sand as

$$E_s = 2q_c \tag{17.19}$$

Trofimenkov (1974) also gave the following correlations for the modulus of elasticity in sand and clay:

$$E_s = 3q_c \quad \text{(for sands)} \tag{17.20}$$

$$E_s = 7q_c \quad \text{(for clays)} \tag{17.21}$$

Correlations such as Eqs. (17.19) through (17.21) can be used in the calculation of the elastic settlement of foundations (Chapter 10).

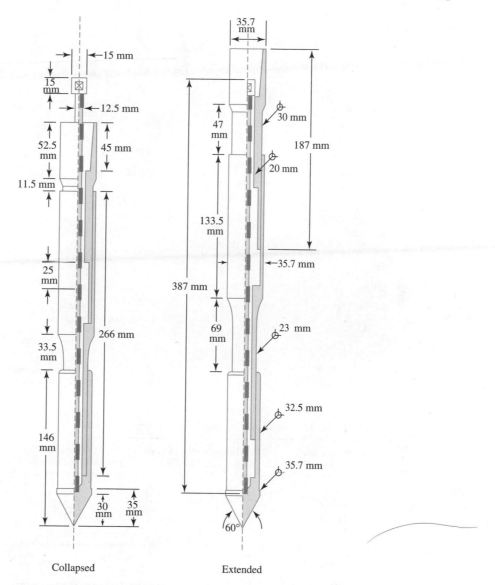

Figure 17.12 Dutch cone penetrometer with friction sleeve [From *Annual Book of ASTM Standards*, 04.08. Copyright © 1991 American Society for Testing and Materials. Reprinted with permission.]

17.7   *Rock Coring*

It may be necessary to core rock if bedrock is encountered at a certain depth during drilling. It is always desirable that coring be done for at least 3 m (≈10 ft). If the bedrock is weathered or irregular, the coring may need to be extended to a greater depth. For coring, a core barrel is attached to the drilling rod. A coring bit is attached to the bottom of the core barrel. The cutting element in the bit may be diamond, tungsten,

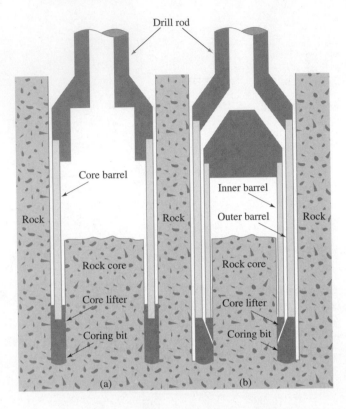

Figure 17.13
Rock coring: (a) single-tube core barrel; (b) double-tube core barrel

Table 17.5 Details of Core Barrel Designations, Bits, and Core Samples

Casing and core barrel designation	Outside diameter of core barrel bit, mm (in.)	Diameter of core sample, mm (in.)
EX	36.5 ($1\frac{7}{16}$)	22.2 ($\frac{7}{8}$)
AX	47.6 ($1\frac{7}{8}$)	28.6 ($1\frac{1}{8}$)
BX	58.7 ($2\frac{5}{16}$)	41.3 ($1\frac{5}{8}$)
NX	74.6 ($2\frac{15}{16}$)	54.0 ($2\frac{1}{8}$)

or carbide. The coring is advanced by rotary drilling. Water is circulated through the drilling rod during coring, and the cuttings are washed out. Figure 17.13a shows a diagram of rock coring by the use of a single-tube core barrel. Rock cores obtained by such barrels can be fractured because of torsion. To avoid this problem, one can use double-tube core barrels (Figure 17.13b). Table 17.5 gives the details of various types of casings and core barrels, diameters of core barrel bits, and diameters of core samples obtained. The core samples smaller than the BX size tend to break away during coring.

On the basis of the length of the rock core obtained from each run, the following quantities can be obtained for evaluation of the quality of rock:

1. Recovery ratio = $\dfrac{\text{Length of rock core recovered}}{\text{Length of coring}}$ (17.22)

Table 17.6 Qualitative Description of Rocks Based on *RQD*

RQD	Rock quality
1–0.9	Excellent
0.9–0.75	Good
0.75–0.5	Fair
0.5–0.25	Poor
0.25–0	Very poor

2. Rock quality designation $(RQD) = \dfrac{\Sigma \text{ Length of rock pieces recovered having lengths of 101.6 mm (4 in.) or more}}{\text{Length of coring}}$ (17.23)

A recovery ratio equal to 1 indicates intact rock. However, highly fractured rocks have a recovery ratio of 0.5 or less. Deere (1963) proposed the classification system in Table 17.6 for *in situ* rocks on the basis of their *RQD*.

17.8 Soil Exploration Report

At the end of the soil exploration program, the soil and rock samples collected from the field are subjected to visual observation and laboratory tests. Then, a soil exploration report is prepared for use by the planning and design office. Any soil exploration report should contain the following information:

1. Scope of investigation
2. General description of the proposed structure for which the exploration has been conducted
3. Geologic conditions of the site
4. Drainage facilities at the site
5. Details of boring
6. Description of subsoil conditions as determined from the soil and rock samples collected
7. Groundwater table as observed from the boreholes
8. Details of foundation recommendations and alternatives
9. Any anticipated construction problems
10. Limitations of the investigation

The following graphic presentations also need to be attached to the soil exploration report:

1. Site location map
2. Location of borings with respect to the proposed structure
3. Boring logs
4. Laboratory test results
5. Other special presentations

BORING LOG

PROJECT TITLE ___Shopping center___

LOCATION ___Intersection Hill Street and Miner Street___   DATE ___June 7, 1997___

BORING NUMBER ___4___   TYPE OF BORING ___Hollow-stem auger___   GROUND ELEVATION ___132.2 ft___

DESCRIPTION OF SOIL	DEPTH (ft) AND SAMPLE NUMBER	STANDARD PENETRATION NUMBER, N_F	MOISTURE CONTENT, w (%)	COMMENTS
Tan sandy silt	1			
	2			
	3			
	4			
Light brown silty clay (CL)	SS-1 5 6	13	11	Liquid limit = 32 $PI = 9$
	7			
	8			
	9			
Groundwater table June 14, 1997	SS-2 10 11	5	24	
	12			
	13			
	14			Liquid limit = 44 $PI = 26$ q_u = unconfined compression strength = 850 lb/ft^2
Soft clay (CL)	ST-1 15 16	6	28	
	17			
	18			
	19			
Compact sand and gravel End of boring @ 22 ft	SS-3 20 21 22	32		

Figure 17.14 Typical boring log. *Note:* SS = split-spoon sample; ST = Shelby tube sample

The boring log is the graphic presentation of the details gathered from each borehole. Figure 17.14 shows a typical boring log.

Problems

17.1 Determine the area ratio of a Shelby tube sampler that has outside and inside diameters of 114 mm and 111 mm, respectively.

17.2 Repeat Problem 17.1 with outside diameter = 3 in. and inside diameter = 2.875 in.

17.3 Following are the results of a standard penetration test in sand. Determine the corrected standard penetration numbers, N_{cor}, at the various depths given. Note that the water table was not observed within a depth of 12 m below the ground surface. Assume that the average unit weight of sand is 15.5 kN/m³. Use Eq. (17.6).

Depth (m)	N_F
1.5	8
3	7
4.5	12
6	14
7.5	13

17.4 Using the values of N_F given in Problem 17.3 and Eq. (17.12), estimate the average soil friction angle, ϕ'.

17.5 A boring log in a sandy soil is shown in Figure 17.15. Determine the corrected standard penetration numbers from Eq. (17.7).

17.6 The standard penetration number N_F of a clay at a certain depth is 12. Estimate its undrained shear strength.

17.7 Repeat Problem 17.6 for $N_F = 8$.

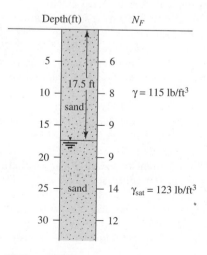

Figure 17.15

17.8 **a.** From the results of Problem 17.3, estimate a design value of N_{cor} (corrected standard penetration number) for the construction of a shallow foundation.

  b. Refer to Eq. (15.44). For a 2-m-square column foundation in plan, what allowable load could the column carry? The bottom of the foundation is to be located at a depth of 1 m below the ground surface. The maximum tolerable settlement is 25 mm.

17.9 The average cone penetration resistance at a certain depth in a sandy soil is 205 kN/m². Estimate the modulus of elasticity of the soil at that depth.

17.10 During a field exploration program, rock was cored for a length of 4 ft. The length of the rock core recovered was 3 ft. Determine the recovery ratio.

References

AMERICAN ASSOCIATION OF STATE HIGHWAY AND TRANSPORTATION OFFICIALS (1967). *Manual of Foundation Investigations,* National Press Building, Washington, D.C.

AMERICAN SOCIETY OF CIVIL ENGINEERS (1972). "Subsurface Investigation for Design and Construction of Foundations of Buildings, Part I," *Journal of the Soil Mechanics and Foundations Division,* ASCE, Vol. 98, No. SM5, 481–490.

AMERICAN SOCIETY FOR TESTING AND MATERIALS (1991). *Annual Book of ASTM Standards,* Vol. 04.08. Philadelphia, Pa.

CUBRINOVSKI, M., and ISHIHARA, K. (1999). "Empirical Correlation Between SPT *N*–Value and Relative Density for Sandy Soils," *Soils and Foundations,* Vol. 39, No. 5, 61–92.

DEERE, D. U. (1963). "Technical Description of Rock Cores for Engineering Purposes," *Felsmechanik und Ingenieurgeologie,* Vol. 1, No. 1, 16–22.

DJOENAIDI, W. J. (1985). "A Compendium of Soil Properties and Correlations." Master's thesis, University of Sydney, Australia.

KULHAWY, F. H., and MAYNE, P. W. (1990). *Manual on Estimating Soil Properties for Foundation Design,* Final Report (EL-6800) submitted to Electric Power Research Institute (EPRI), Palo Alto, Calif.

LIAO, S., and WHITMAN, R. V. (1986). "Overburden Correction Factor for SPT in Sand," *Journal of Geotechnical Engineering,* ASCE, Vol. 112, No. 3, 373–377.

MENARD, L. (1965). "Rules for Calculation of Bearing Capacity and Foundation Settlement Based on Pressuremeter Tests," *Proceedings,* 6th International Conference on Soil Mechanics and Foundation Engineering, Montreal, Canada, Vol. 2, 295–299.

OSTERBERG, J. O. (1952). "New Piston-Type Sampler," *Engineering News Solutions,* April 24.

PECK, R. B., HANSON, W. E., and Thornburn, T. H. (1974). *Foundation Engineering,* 2nd ed., Wiley, New York.

ROBERTSON, P. K., and CAMPANELLA, R. G. (1983). "Interpretation of Cone Penetration Tests. Part I: Sand," *Canadian Geotechnical Journal,* Vol. 20, No. 4, 718–733.

SCHMERTMANN, J. H. (1970). "Static Cone to Compute Static Settlement Over Sand," *Journal of the Soil Mechanics and Foundations Division,* ASCE, Vol. 96, No. SM3, 1011–1043.

SCHMERTMANN, J. H. (1975). "Measurement of *in situ* Shear Strength," *Proceedings,* Specialty Conference on *in situ* Measurement of Soil Properties, ASCE, Vol. 2, 57–138.

SKEMPTON, A. W. (1986). "Standard Penetration Test Procedures and the Effect in Sands of Overburden Pressure, Relative Density, Particle Size, Aging and Overconsolidation," *Geotechnique,* Vol. 36, No. 3, 425–447.

SOWERS, G. B., and SOWERS, G. F. (1970). *Introductory Soil Mechanics and Foundations,* Macmillan, New York.

TROFIMENKOV, J. G. (1974). "General Reports: Eastern Europe," *Proceedings,* European Symposium of Penetration Testing, Stockholm, Sweden, Vol. 2.1, 24–39.

WOLFF, T. F. (1989). "Pile Capacity Prediction Using Parameter Functions," in *Predicted and Observed Axial Behavior of Piles, Results of a Pile Prediction Symposium,* sponsored by Geotechnical Engineering Division, ASCE, Evanston, Ill., June 1989, ASCE Geotechnical Special Publication No. 23, 96–106.

Answers to Selected Problems

Chapter 2

2.1 $C_u = 2.63$; $C_z = 0.66$

2.3 **b.** $D_{10} = 0.12$ mm,
$D_{30} = 0.22$ mm, $D_{60} = 0.4$ mm
c. 3.33 **d.** 1.01

2.5 **b.** $D_{10} = 0.17$ mm, $D_{30} =$
0.39 mm, $D_{60} = 0.46$ mm
c. $C_u = 2.71$ **d.** $C_z = 1.95$

2.7 **a.** Gravel – 0% Sand – 17%
Silt – 57% Clay – 26%
b. Gravel – 0% Sand – 11%
Silt – 63% Clay – 26%

2.9 **a.** Gravel – 0% Sand – 17%
Silt – 72% Clay – 11%
b. Gravel – 0% Sand – 8%
Silt – 81% Clay – 11%

2.11 0.0046 mm

Chapter 3

3.5 **a.** 14.94% **b.** 120 lb/ft^3
c. 104.4 lb/ft^3 **d.** 0.572
e. 0.364 **f.** 68.7%

3.7 **a.** 106.6 lb/ft^3 **b.** 0.45
c. 2.48

3.9 **a.** 15.88 kN/m^3 **b.** 0.649
c. 0.39 **d.** 44.4%

3.11 **a.** 91.3 lb/ft^3 **b.** 0.458
c. 47.9% **d.** 14.9 lb/ft^3

3.13 **a.** 20.59 kN/m^3 **b.** 2.58%

3.15 **a.** 117.4 lb/ft^3 **b.** 96.6 lb/ft^3
c. 77.7%

3.17 **a.** 2.75 **b.** 0.89

3.19 107.3 lb/ft^3

3.21 109.1 lb/ft^3

3.23 12%

3.25 56.9%

3.27 −19%

Chapter 4

4.1

Soil	Classification	GI
1	A-6	5
2	A-7-6	32
3	A-6	7
4	A-4	1
5	A-6	7
6	A-2-4	0
7	A-3	0
8	A-6	9
9	A-2-4	0
10	A-7-5	33

4.3

Soil	Classification	GI
A	A-1-b	0
B	A-1-b	0
C	A-7-5	23
D	A-7-6	27
E	A-6	5

4.5

Soil	Group symbol	Group name
A	SP	Poorly graded sand
B	SW-SM	Poorly graded sand with silt
C	MH	Elastic silt with sand
D	CH	Fat clay
E	SC	Clayey sand

4.7 **a.** A-1-b (0) **b.** SM-SC, silty clayey sand

Chapter 5

5.1

w (%)	γ_{zav} (lb/ft^3)
5	149.4
8	139.4
10	133.5
12	128.0
15	120.6

5.3

w (%)	S (%)	γ_d (lb/ft^3)
10	80	19.5
	90	20.1
	100	20.6
20	80	15.6
	90	16.4
	100	17.0

5.5 $e = 0.358$ $S = 94\%$
5.7 90%
5.9 Borrow pit B
5.11 $R = 96\%$ $\gamma_{d(\text{field})} = 15.84 \, \text{kN/m}^3$
5.13 **a.** 15.94 kN/m^3 **b.** 83.8%
5.15 $S_N = 20.65$; Rating = Fair

Chapter 6

6.1 45×10^{-3} m^3/hr/m
6.3 0.0279 in./sec
6.5 **a.** 2.34×10^{-2} cm/sec
b. 0.075 cm/sec
6.7 0.31 cm^3
6.9 0.024 cm/sec
6.11 7.73×10^{-2} cm/sec
6.13 9.7×10^{-3} cm/sec
6.15 0.309×10^{-6} cm/sec
6.17 6×10^{-9} cm/sec
6.19 23.79

Chapter 7

7.1 77.76×10^{-2} m^3/m/day
7.3 2.1 m^3/m/day
7.5 0.378 m^3/day/m
7.7 0.361 m^3/day/m

Chapter 8

8.1

Point	σ (lb/ft^2)	u (lb/ft^2)	σ' (lb/ft^2)
A	0	0	0
B	560	0	560
C	1280	374.4	905.6
D	2280	873.6	1406.4

8.3

Point	σ (kN/m^2)	u (kN/m^2)	σ' (kN/m^2)
A	0	0	0
B	45	0	45
C	109	39.24	63.76
D	199	88.29	110.71

8.5

Point	σ (kN/m^2)	u (kN/m^2)	σ' (kN/m^2)
A	0	0	0
B	65	0	65
C	126.95	29.43	97.52
D	153.94	44.15	109.79

8.7

Point	σ (kN/m^2)	u (kN/m^2)	σ' (kN/m^2)
A	0	0	0
B	64.84	0	64.84
C	169.24	49.05	120.19

8.9 13.62 ft
8.11 **a.** 360 cm^3/sec **b.** 1.12; no boiling **c.** 2.8 m

8.13

Depth (ft)	σ (lb/ft^2)	u (lb/ft^2)	σ' (lb/ft^2)
0	0	0	0
6	663.96	0	663.96
		−124.8	788.76
10	1103.96	0	1103.96
19	2160.92	561.6	1599.32

8.15 3.52

Chapter 9

9.1 $\sigma_1 = 129.24$ kN/m^2 $\sigma_3 = 30.76$ kN/m^2 $\sigma_n = 51.03$ kN/m^2 $\tau_n = 39.82$ kN/m^2

9.3 $\sigma_1 = 161.1$ kN/m^2 $\sigma_3 = 68.9$ kN/m^2 $\sigma_n = 138.5$ kN/m^2 $\tau_n = 39.7$ kN/m^2

9.5 $\sigma_1 = 36.54$ kN/m^2 $\sigma_3 = -1.54$ kN/m^2 $\sigma_n = 36.5$ kN/m^2 $\tau_n = 1.13$ kN/m^2

9.7 1.27 kN/m^2

9.9 16.53 kN/m^2

9.11 117.3 kN/m

9.13 1435 kN/m^2

9.15 At A 3229 lb/ft^2; at B 3088 lb/ft^2; at C 293 lb/ft^2

9.17

z (ft)	$\Delta\sigma_z$ (lb/ft^2)
1.5	4140
3	3824
6	2715
9	1781
12	1195

9.19 104 kN/m^2

9.21 **a.** 18 kN/m^2 **b.** 42.8 kN/m^2 **c.** 2.7 kN/m^2

Chapter 10

10.1 0.47 in.

10.3 **b.** 3.1 ton/ft^2 **c.** 0.53

10.5 3.39 in.

10.7 95 mm

10.9 159.6 days

10.11 269 kN/m^2

10.13 3.18 in.

10.15 1.71×10^{-4} cm^2/sec

10.17 **a.** 240.8 days **b.** 7.33 in.

10.19 **a.** 48.9 days **b.** 99.95 days

10.21 20.8 mm

Chapter 11

11.1 $\phi' = 41.2°$, $S = 0.739$ kN

11.3 37.5°

11.5 **a.** 20 lb/in.^{2} **b.** 60° **c.** $\sigma' = 25$ lb/in.^{2}; $\tau = 8.66$ lb/in.^{2}

11.7 32.1°

11.9 16.95 lb/in.^{2}

11.11 $\phi' = 18°$; $c' = 55.4$ kN/m^2

11.13 1 ton/ft^2

11.15 $\phi' = 32.3°$; $\phi = 18.3°$

11.17 **a.** 18° **b.** 64.9 kN/m^2

11.19 $\sigma_1 = 2.64$ ton/ft^2 $\Delta u_{d(f)} = 0.56$ ton/ft^2

11.21 $\phi = 9°$; $c = 50$ kN/m^2

11.23 $\Delta u_{d(f)} = -68.3$ kN/m^2

11.25 **a.** $m = 10.5$ lb/in.^{2}; $\alpha = 14°$ **b.** $\phi' = 14.4°$; $c' = 10.82$ lb/in.^{2}

Chapter 12

12.1 $P_o = 2585$ lb/ft; $\bar{z} = 3.33$ ft

12.3 $P_o = 5783$ lb/ft; $\bar{z} = 6$ ft

12.5 $P_o = 66.9$ kN/m; $\bar{z} = 1.5$ m

12.7 $P_o = 1.57$ kN/m

12.9 $P_a = 4973.4$ lb/ft; $\bar{z} = 6$ ft

12.11 $P_a = 46.11$ kN/m; $\bar{z} = 1.67$ m

12.13 $P_a = 12,751.9$ lb/ft; $\bar{z} = 6.35$ ft

12.15 **b.** 5.74 ft **c.** 3225 lb/ft **d.** $P_a = 5233$ lb/ft; $\bar{z} = 3.09$ ft

12.17 10.02 kN/m

12.19 $P_a = 69.38$ kN/m; $\bar{z} = 1.24$ m

12.21 $P_p = 20,223$ lb/ft; $\bar{z} = 3.33$ ft

12.23 $P_p = 360$ kN/m; $\bar{z} = 1.33$ m

12.25 $P_p = 89,067$ lb/ft; $\bar{z} = 7.68$ ft

12.27 **a.** 5.2 kip/ft **b.** 6.3 kip/ft **c.** 125 kN/m

12.29 $\bar{z} = 2.8$ m

12.31 56.76 kN/m

Chapter 13

13.1

θ (deg)	r (mm)
0	30
30	46.6
60	72.2
90	112.0
120	173.9
150	269.9
180	418.8

13.3 21,105 lb/ft

13.5 **a.** 740 kN/m **b.** 700 kN/m

13.7 At A: 14,002 lb At B: 13,145 lb At C: 7,144 lb At D: 28,576 lb

Chapter 14

14.1 17.1 ft
14.3 **a.** 1.98 **b.** 3.2 m
14.5 1.12
14.7 2.19
14.9 2.48 m
14.11 23.74 m
14.13 3.47 m
14.15 31.9 m; toe circle
14.17 33°
14.19 0.57
14.21 **a.** 1.5 **b.** 1.36 **c.** 1.75
　　　d. 1.8
14.23 0.955
14.25 1.1

Chapter 15

15.1 **a.** 13,368 lb/ft^2 **b.** 5529 lb/ft^2
　　　c. 104 kN/m^2 **d.** 6466 lb/ft^2
　　　e. 71 kN/m^2
15.3 **a.** 12,665 lb/ft^2 **b.** 5625 lb/ft^2
　　　c. 104 kN/m^2 **d.** 6664 lb/ft^2
　　　e. 73.6 kN/m^2
15.5 $B = 1.3$ m
15.7 54,155 lb

15.9 100,245 lb
15.11 143 kN/m^2
15.13 160.4 kN
15.15 100.26 kip
15.17 439.5 kN

Chapter 17

17.1 5.48%

17.3

Depth (ft)	N_{cor}
1.5	16
3	10
4.5	14
6	14
7.5	12

17.5

Depth (ft)	N_{cor}
5	11
10	11
15	10
20	9
25	13
30	11

17.7 50 kN/m^2
17.9 410 to 615 kN/m^2

Index